Fourth Edition

Analog and Digital Communication Systems

Martin S. Roden

Department of Electrical and Computer Engineering
California State University, Los Angeles

PRENTICE HALL
Upper Saddle River, New Jersey 07458

Library of Congress Cataloging-in-Publication Data

Roden, Martin S.
  Analog and digital communication systems / Martin S. Roden.—4th ed.
    p.   cm.
  Includes bibliographical references and index.
  ISBN 0-13-372046-2
  1. Telecommunication.   2. Digital communications.   I. Title.
TK5105.R64 1996
621.382—dc20   95-11659
                      CIP

Acquisitions editor: **TOM ROBBINS**
Production editor: **IRWIN ZUCKER**
Copy editor: **BRIAN BAKER**
Cover designer: **BRUCE KENSELAAR**
Buyer: **DONNA SULLIVAN**
Editorial assistant: **PHYLLIS MORGAN**

©1996 by Prentice-Hall, Inc.
Simon & Schuster / A Viacom Company
Upper Saddle River, New Jersey 07458

The author and publisher of this book have used their best efforts in preparing this book. These efforts include the
development, research, and testing of the theories and programs to determine their effectiveness. The author and publisher
make no warranty of any kind, expressed or implied, with regard to these programs or the documentation contained in this
book. The author and publisher shall not be liable in any event for incidental or consequential damages in connection with,
or arising out of, the furnishing, performance, or use of these programs.

Printed in the United States of America

10 9 8 7 6 5

ISBN 0-13-372046-2

Prentice-Hall International (UK) Limited, *London*
Prentice-Hall of Australia Pty. Limited, *Sydney*
Prentice-Hall Canada Inc., *Toronto*
Prentice-Hall Hispanoamericana, S.A., *Mexico*
Prentice-Hall of India Private Limited, *New Delhi*
Prentice-Hall of Japan, Inc., *Tokyo*
Simon & Schuster Asia Pte. Ltd., *Singapore*
Editora Prentice-Hall do Brasil, Ltda., *Rio de Janeiro*

Contents

Preface

The fourth edition of *Analog and Digital Communication Systems* is a greatly modified and enhanced version of the earlier editions. The digital communications portion of the text has been considerably expanded and reorganized. The material is presented in a manner that emphasizes the unifying principles governing all forms of communication, whether analog or digital. Practical design applications and computer exercises have been expanded. In particular, many of the graphs are prepared and formulas are solved using MATLAB™ or Mathcad™. In such cases, the instruction set is presented to give the student practice in using these important tools.

In addition to presenting a unified approach to analog and digital communication, this text strikes a balance between theory and practice. While the undergraduate engineering student needs a firm foundation in the theoretical aspects of the subject, it is also important that he or she be exposed to the real world of *engineering design*. This serves two significant purposes. The first is that an introduction to the real world acts as a strong motivating factor: The "now" generation needs some "touchy-feely" to motivate the spending of hours digging through mathematical formulae. The second purpose of real-world engineering is to ease the transition from academia to the profession. A graduate's first experience with real-world design should not be a great shock, but instead should be a natural transition from the classroom environment.

The book is intended as an introductory text for the study of analog and/or digital communication systems, with or without noise. Although all necessary background material has been included, prerequisite courses in linear systems analysis and in probability are helpful.

The text stresses a mathematical systems approach to all phases of the subject matter. The mathematics used throughout is as elementary as possible, but is carefully chosen so as not to contradict any more sophisticated approach that may eventually be required. An attempt is made to apply intuitive techniques prior to grinding through the mathematics. The style is informal, and the text has been thoroughly tested in the classroom with excellent success.

Chapter 1, which is for motivational purposes, outlines the environmental factors that must be considered in communicating information. The chapter also clarifies the differences between analog, sampled, and digital signals. The final section lays out a block diagram of a comprehensive communication system. This diagram could serve as a table of contents for the remainder of the text.

Chapter 2 sets forth the mathematical groundwork of signal analysis. It should be review for most students taking the course. Chapter 3 applies signal analysis results to linear systems, with an emphasis on filters. The material in Chapters 2 and 3 is covered in most linear systems courses.

Chapter 4 introduces probability and random analysis and applies these to the study of narrowband noise. The matched filter is introduced as a technique to maximize the signal-to-noise ratio. Chapter 5 deals with baseband communication. Although it concentrates on analog communication, the sampled systems form an important transition into digital communication.

Chapter 6 is a thorough treatment of amplitude modulation (AM), including applications to broadcast radio, television, and AM stereo. Chapter 7 parallels Chapter 6, but for angle instead of amplitude modulation. A section on broadcast frequency modulation (FM) and FM stereo is included. The final section of the chapter compares various angle modulation schemes.

Source encoding is the subject of Chapter 8. In addition to a thorough discussion of the analog-to-digital conversion process, and of the associated round-off errors, the chapter presents baseband forms of digital transmission. Chapter 9 focuses on channel encoding, including data compression, entropy coding, and forward error correction. Both block codes and convolutional codes are analyzed. Baseband forms of digital transmission and reception are the topic of Chapter 10, and modulated forms of transmission are examined in Chapter 11. Transmitters, receivers, error analysis, and timing considerations are treated for amplitude shift keying (ASK), frequency shift keying (FSK), and phase shift keying (PSK). Hybrid signalling techniques and modems conclude the chapter.

The final chapter, Chapter 12, summarizes important design considerations for both analog and digital communication systems. The chapter concludes with five contemporary case studies.

Problems are presented at the end of Chapters 1 through 11. Numerous solved examples are given within each chapter.

The text is ideally suited for a two-term sequence at the undergraduate level. Chapters 2 and 3 can be omitted if a course in linear systems forms a prerequisite. Chapter 4 can be omitted either if a probability course is a prerequisite or if noise analysis is not part of the course.

If the text is used for a digital communication course, the following sequence of sections and chapters is recommended: Sections 4.7, 5.2, and 5.3, and Chapter 8, Chapter 10, and Chapter 11. Channel encoding (Chapter 9) is normally not included in an introductory digital communication course.

It gives me a great deal of pleasure to acknowledge the assistance and support of many people without whom this text would not have been possible:

To the many classes of students who are responsive during lectures and helped indicate the clearest approach to each topic.

To the faculty around the world who have written to me with comments and suggestions.

To my colleagues at Bell Telephone Labs, Hughes Aircraft Ground Systems, and California State University, Los Angeles. Special thanks go to Professors Roy Barnett, Fred Daneshgaran, George Killinger, and Lili Tabrizi for their many helpful suggestions.

To Dennis J. E. Ross for guidance and assistance.

To Professor A. Papoulis, who played a key role during the formative years of my education.

I sincerely hope this text is the answer to your prayers. If it is, please let me know. If it isn't, please also communicate with me so that, together, we can improve engineering education.

Introduction
for the Student

Communication is perhaps the oldest applied area within electrical engineering. As is the case in many technical disciplines, the field of communication is experiencing a revolution several times each decade. Some important milestones of the past include the following:

The television revolution: After only five decades, television has become a way of life. Wristwatch television, wall-size television, and fully interactive cable television are all available.

The space revolution: This has been the catalyst for many innovations in long-distance communication. Satellite communication permits universal access.

The digital revolution: The overriding emphasis on digital electronics and processing has resulted in a rapid change in direction from analog communication to digital communication.

The computer revolution: Microprocessors are changing the shape of everything related to computing and control, including many phases of communication. Home and portable computers (e.g., the "notebook") with modems are popular. They permit direct data communication by the general public.

The consumer revolution: Consumers first discovered calculators and digital watches. They then expanded to sophisticated TV recording systems, CB radio, cellular phones, facsimile (FAX) systems, and video games. When digital watches were not enough, numerous other functions were added to the wrist device, such as calculators, TV, pager receivers, and pulse and temperature indicators.

The personal communication revolution: Cellular telephones began as a tool of the business person. They then expanded into the public sector, with concerns for safety. Instant worldwide analog and digital communication is now desired by a broad spectrum of the public.

There is every reason to expect the rate of revolution to continue or, indeed, accelerate. Two-way interactive cable, high-definition television (HDTV), videotext, and personal communication will greatly reduce our need for hard copies of literature and for traveling. People are juggling bank accounts, ordering groceries, sending electronic mail, and holding business-related conferences without leaving their homes. The problems of urban centers are being attacked through communication. Indeed, adequate communication can make cities unnecessary. The potential impact of these developments is mind boggling, and the possibilities involving communications are most exciting.

These facts should inspire you to consider the field of communication as a career. But even if you decide not to approach the field any more deeply than in this text, you will be a far more aware person for having experienced even the small amount contained herein. If nothing more, the questions arising in modern communication, as it relates to everything from the space program to home entertainment, will take on a new meaning for you. The devices around you will no longer appear mysterious. You will learn what "magic force" lights the stereo indicator on your receiver, the basics of how a video game works, the inner workings of a FAX machine, and the mechanism by which your voice can travel to any part of the globe in a moment of time.

Enjoy the book! Enjoy the subject! And please, after studying this text, communicate any comments that you may have (positive, negative, or neutral) to me at California State University, Los Angeles, CA 90032. Thank you!

Martin S. Roden
Los Angeles, California

1

Introduction

1.0 PREVIEW

What We Will Cover and Why You Should Care

You will not encounter your first communication system until Chapter 5 of this text. The first four chapters form the framework and define the parameters under which we operate.

Chapter 1 begins with a brief history of the exciting revolution in communication, a revolution that is accelerating at an ever-increasing rate.

We then turn our attention to an investigation of the environment under which our systems must operate. In particular, the characteristics of various communication channels are explored, and we present analytical techniques for the resulting signal distortion. It would not make much sense to start designing communication systems without knowing something about the environment in which they must operate.

The third section of the chapter defines the three types of signals with which we will operate. We begin with analog signals and make a gradual transition to digital signals through the intermediate step of discrete time (sampled) signals.

The block diagram of a communication system is presented and discussed in the final section. The various blocks in the transmitter and receiver form the road map for our excursion through this exciting subject. We present a single comprehensive system block diagram that applies to both analog and digital communication systems. This block diagram forms an abbreviated table of contents for the remainder of the text.

Necessary Background

There are no prerequisites to understanding most of this introductory chapter. The only exception is the discussion of channel distortion in Section 1.2. Understanding the equations in this section requires a basic knowledge of Fourier transforms and systems theory. Such material is reviewed in Chapter 3.

1

1.1 THE NEED TO COMMUNICATE

Among the earliest forms of communication were vocal-cord sounds generated by animals and human beings, with reception via the ear. When greater distances were required, the sense of sight was used to augment that of hearing. In the second century B.C., Greek telegraphers used torch signals to communicate. Different combinations and positions of torches were used to represent the letters of the Greek alphabet. These early torch signals represent the first example of data communication. Later, drum sounds were used to communicate over greater distances, again calling upon the sense of hearing. Increased distances were possible because the drum sounds were more easily distinguished from background noise than were human vocal-cord sounds.

In the 18th century, communication of letters was accomplished using semaphore flags. Like the torches of ancient Greece, these flags relied on the human eye to receive the signal. This reliance on the eye, of course, severely limited the transmission distances.

In 1753, Charles Morrison, a Scottish surgeon, devised an electrical transmission system using one wire (plus ground) for each letter of the alphabet. A system of pith balls and paper with letters printed on it was used at the receiver.

In 1835, Samuel F. B. Morse began experimenting with telegraphy. Two years later, in 1837, the telegraph was invented by Morse in the United States and by Sir Charles Wheatstone in Great Britain. The first public telegram was sent in 1844, and electrical communication was established as a major component of life. These early forms of communication consist of individual message components such as the letters of the alphabet. (We would later call it digital communication.) It was not until Alexander Graham Bell invented the telephone in 1876 that analog electrical communication became common.

Experimental radio broadcasts began about 1910, with Lee De Forest producing a program from the Metropolitan Opera House in New York City. Five years later, an experimental radio station opened at the University of Wisconsin in Madison. Stations WWJ in Detroit and KDKA in Pittsburgh were among the first to conduct regular broadcasts, in the year 1920.

Public television had its beginning in England in 1927. In the United States, it started three years later. During the early period, broadcasts did not follow any regular schedule. Regular scheduling did not begin until 1939, during the opening of the New York World's Fair.

Satellite communication was launched in the 1960s, with Telstar I being used to relay TV programs starting in 1962 and the first commercial communications satellites being launched in the mid-1960s.

The 1970s saw the beginning of the computer communication revolution. Data transfer is an integral part of our daily lives and has led to a merging of the disciplines of *communication and computer engineering. Computer networking* is one of the fastest growing areas of communication.

The *personal communication* revolution began in the 1980s. Before the decade of the nineties is over, the average professional will have a cellular telephone in the car, a portable telephone (no larger than the Star Trek® communicators of the original series), a paging system, a modem in the home computer for use in paying bills or accessing the daily news, and a home FAX machine. Consumers will use fully interactive compact disk technology, laptops will be networked with worldwide data services, and the global-positioning satellite (GPS) will assist in navigating cars through traffic jams.

The coming millennium is certain to bring a new set of applications and innovations as communication continues to have a significant impact on our lives.

1.2 THE ENVIRONMENT

Before we can begin designing systems to communicate information, we need to know something about the channel through which the signals must be transmitted. We start by exploring the ways in which the channel can change our signals, and then we discuss some common types of channels.

1.2.1 Distortion

Anything that a channel does to a signal other than delaying it and multiplying it by a constant is considered to be *distortion*. (See Chapter 3 for a discussion of *distortionless linear systems*.) Let us assume that the channels we will encounter are *linear* and therefore cannot change the frequencies of their input. Some *nonlinear* forms of distortion are significant at higher transmission frequencies. Indeed, the higher frequencies are affected by air turbulence, which causes a *frequency variation. Doppler radar systems* used for monitoring weather capitalize on this phenomenon.

Linear distortion can cause problems in pulse transmission systems of the type used in pulse modulation or in digital communication. This distortion is characterized by *time dispersion* (spreading), due either to multipath effects or to the characteristics of the channel. For now, we look at the effects that can be readily characterized by the system function of the channel. The channel can be characterized by a system transfer function of the form

$$H(f) = A(f)e^{-j\theta(f)} \tag{1.1}$$

The *amplitude factor* is $A(f)$, and the *phase factor* is $\theta(f)$.

Distortion arises from these two frequency-dependent quantities as follows: If $A(f)$ is not a constant, we have what is known as *amplitude distortion;* if $\theta(f)$ is not linear in f, we have *phase distortion*.

Amplitude Distortion

Let us first assume that $\theta(f)$ is linear with frequency. The transfer function is therefore of the form

$$H(f) = A(f)e^{-j2\pi f t_0} \tag{1.2}$$

where the phase proportionality constant has been denoted as t_0 because it represents the channel delay.

One way to analyze Eq. (1.2) is to expand $A(f)$ into a series—for example, a Fourier series. This can be done if $A(f)$ is bandlimited to a certain range of frequencies. In such cases, we can write

$$H(f) = \sum_{n=0}^{\infty} H_n(f) \tag{1.3}$$

where the terms in the summation are of the form

$$H_n(f) = a_n \cos\left(\frac{n\pi f}{f_m}\right) e^{-j2\pi f t_0} \tag{1.4}$$

These terms are related to the *cosine filter*, whose amplitude characteristic follows a cosine wave in the passband. This filter is shown in Fig. 1.1 for $n = 2$. The system function for this filter is

$$H(f) = \left(A + a\cos\frac{2\pi}{f_m}f\right)e^{-j2\pi f t_0}$$

$$= Ae^{-j2\pi f t_0} + \frac{a}{2}\exp\left[j2\pi f\left(\frac{1}{f_m} - t_0\right)\right] \tag{1.5}$$

$$+ \frac{a}{2}\exp\left[j2\pi f\left(-\frac{1}{f_m} - t_0\right)\right]$$

Computer Exercise:

Plot Eq. (1.4) for representative values of f_m and t_0.
Solution: We present the instruction steps both for Mathcad and for MATLAB.
Mathcad: We illustrate the instructions to type and the resulting expression on the screen:

```
fm:1                      fm:= 1
f:-2,-1.95;2              f:=-2,-1.95..2
H(f):cos(2*π*f/fm)        H(f):=cos(2.π.f/fm)
```

NOTES: We set fm to unity and step the frequency from -2 to $+2$ in steps of 0.05. Enter π by pressing CONTROL+P. You then enter the plotting mode by pressing "@". Insert "f" for the abscissa and "H(f)" for the ordinate, and the plot results.

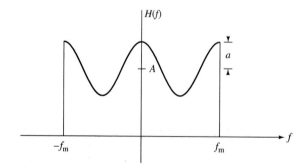

Figure 1.1   Cosine filter.

MATLAB: The following instructions are typed:

```
fm=1
f=-2:.05:2
H=cos(2*pi*f/fm)
plot(f,H)
```

The result is shown in Fig. 1.2.

If the input $r(t)$ to the cosine filter is bandlimited, the output is

$$s(t) = Ar(t - t_0) + \frac{a}{2}r\left(t + \frac{1}{f_m} - t_0\right) + \frac{a}{2}r\left(t - \frac{1}{f_m} - t_0\right) \qquad (1.6)$$

Equation (1.6) indicates that the response is in the form of an undistorted version of the input, added to two time-shifted versions *(echoes,* or a *multipath).*

Returning to the case of a general filter, we see that the output of a system with amplitude distortion is a sum of shifted inputs. Thus, with

$$H(f) = \sum_{n=0}^{\infty} a_n \cos\left(\frac{n\pi f}{f_m}\right) e^{-j2\pi f t_0} \qquad (1.7)$$

the output due to an input $r(t)$ is

$$s(t) = \sum_{n=0}^{\infty} \frac{a_n}{2}\left[r\left(t + \frac{n}{2f_m} - t_0\right) + r\left(t - \frac{n}{2f_m} - t_0\right)\right] \qquad (1.8)$$

Equation (1.8) can be computationally difficult to evaluate. This approach is therefore usually restricted to cases where the Fourier series contains relatively few significant terms.

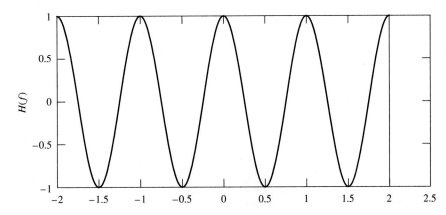

Figure 1.2   Computer plot of $H(f)$.

Example 1.1

Consider the triangular filter characteristic shown in Fig. 1.3. Assume that the phase characteristic is linear with slope $-2\pi t_0$. Find the output of this filter when the input signal is

$$r(t) = \frac{\sin 400 \pi t}{t}$$

Solution: We must first expand $H(f)$ in a Fourier series to get

$$H(f) = \frac{1}{2} + \frac{4}{\pi^2} \cos \frac{\pi f}{1000} + \frac{4}{9\pi^2} \cos \frac{2\pi f}{1000} + \frac{4}{25\pi^2} \cos \frac{5\pi f}{1000} + \cdots$$

The signal $r(t)$ is bandlimited, so that all frequencies are passed by the filter. This is true because $R(f)$ is zero at frequencies above 200 Hz and the filter cuts off at $f = 1,000$ Hz. If we retain the first three nonzero terms in the series, the output becomes

$$s(t) = \frac{1}{2} r(t - t_0) + \frac{2}{\pi^2} \left[r\left(t - \frac{1}{2000} - t_0 \right) + r\left(t + \frac{1}{2000} - t_0 \right) \right]$$

$$+ \frac{2}{9\pi^2} \left[r\left(t - \frac{3}{2000} - t_0 \right) + r\left(t + \frac{3}{2000} - t_0 \right) \right]$$

This result is sketched as Fig. 1.4 for $t_0 = 0.005$ second.

Computer Exercise

Plot the result of Example 1.1 using both Mathcad and MATLAB. The MATLAB instructions are:

```
t=-.01:.0005:.02;
to=.005;
t1=t-to;
t2=t1-.001;
t3=t1+.001;
t4=t1-.003;
t5=t1+.003;
a1=400*pi*t1;
a2=400*pi*t2;
a3=400*pi*t3;
a4=400*pi*t4;
```

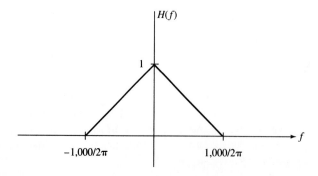

Figure 1.3   Triangular filter characteristic.

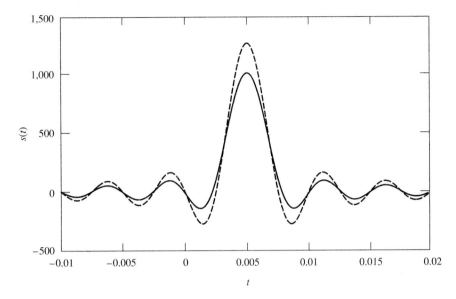

Figure 1.4   Result of Example 1.1.

```
a5=400*pi*t5;
s=sin(a1)./t1+(2/pi^2)*(sin(a2)./t2+sin(a3)./t3))
  +(2/(9*pi^2))*(sin(a4)./t4+sin(a5)./t5))
plot(t,s)
```

Note the presence of the period (.) before the division in the next-to-last statement. In MAT-LAB, it is important to realize that we deal with matricies. In this case, t is a (1×61) vector, so s has the same dimensions. If we attempt to execute an expression such as s=sin(t)/t, we get an error, since we are trying to divide one (1×61) vector by another (1×61) vector. To force the division to be scalar, we precede the division sign by a period.

In Mathcad, the equations can be entered directly. The instructions are:

```
t:=-.01,-.0099.. .02
r(t):=sin(400*π*t)/t
rd(t):=r(t-.005)
s(t):=r(t-.005)/2+(2/π^2)*(r(t-.001-005)+r(t+.001-.005))+
      (2/(9*π^2))*(r(t-.003-.005)+r(t+.003-.005))
@[enter t for abscissa, and s(t),rd(t) for ordinate]
```

Note that we have defined rd(t) as the undistorted delayed version of r(t) so that we could plot the latter. Figure 1.4 was generated using Mathcad.

The undistorted function is plotted in Figure 1.4 as a dashed line. Examining the differences between the two curves, we see that the distortion is manifest as an attenuation and a slight time delay. That is, the peaks of the output are smaller and slightly to the right of the corresponding peaks of the undistorted waveform.

Phase Distortion

Deviations of the phase away from the distortionless (linear phase) case can be characterized by variations in the slope of the phase characteristic and in the slope of a line from the origin to a point on the curve. We define *group delay* (also known as *envelope delay*) and *phase delay* as follows:

$$t_{ph}(f) = \frac{\theta(f)}{2\pi f} \qquad t_{gr}(f) = \frac{d\theta(f)}{2\pi df} \tag{1.9}$$

Figure 1.5 illustrates these definitions for a representative phase characteristic. If the channel were ideal and distortionless, the phase characteristic would be linear, and the group and phase delays would both be constant for all f. In fact, for this ideal case, both delays would be equal to the time delay t_0 from input to output.

The phase characteristic can be approximated as a piecewise linear curve. As an example, examine the phase characteristic of Figure 1.5. If we were to operate in a relatively narrow band around f_0, the phase could be approximated as the first two terms in a Taylor series expansion:

$$\theta(f) \approx \theta(f_0) + \frac{d\theta(f_0)}{df}(f - f_0)$$
$$= t_{ph}(f_0)f_0 + (f - f_0)t_{gr}(f_0) \tag{1.10}$$

Equation (1.10) applies for positive frequency, and its negative applies for negative frequency. This is so because the phase characteristic for a real system must be an odd function of frequency.

Now suppose that the amplitude factor is constant, that is, $A(f) = A$, and a wave of the form $r(t)\cos 2\pi f_0 t$ forms the input to the system. The Fourier transform of the input is found from the modulation property of the transform (see Chapter 3):

$$r(t)\cos 2\pi f_0 t \leftrightarrow \frac{1}{2}[R(f - f_0) + R(f + f_0)] \tag{1.11}$$

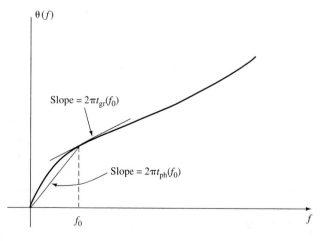

Figure 1.5 Group and phase delay.

The Fourier transform of the output, $S(f)$, is given by the product of the input transform with the system function. Thus,

$$S(f) = \frac{1}{2}R(f - f_0)A\exp[jt_{ph}(f_0)2\pi f_0]\exp[j2\pi(f - f_0)t_{gr}(f_0)] \tag{1.12}$$

Equation (1.12) has been written for positive f. For negative f, the value of the expression is the complex conjugate of its value at positive f. We now find the time function corresponding to this Fourier transform. The transform can be simplified by taking note of the following three Fourier transform relationships (we ask you to prove them as part of Problem 1.2.1):

$$r(t - t_0)\cos2\pi f_0 t \leftrightarrow \frac{1}{2}R(f - f_0)e^{-j2\pi(f-f_0)t_0}$$

$$+ \frac{1}{2}R(f + f_0)e^{-j2\pi(f+f_0)t_0}$$

$$r(t - t_1)\cos2\pi f_0(t - t_1) \leftrightarrow \frac{1}{2}[R(f - f_0) + R(f + f_0)]e^{-j2\pi ft_1} \tag{1.13}$$

$$r(t - t_0)\cos2\pi f_0(t - t_1) \leftrightarrow \frac{1}{2}R(f - f_0)e^{-j2\pi(f-f_0)(t_0-t_1)}$$

$$+ \frac{1}{2}R(f + f_0)e^{-j2\pi(f+f_0)(t_0-t_1)}e^{-j2\pi ft_1}$$

Using these relationships to simplify Eq. (1.12), we find that

$$s(t) = Ar[t - t_{gr}(f_0)]\cos2\pi f_0[t - t_{ph}(f_0)] \tag{1.14}$$

Recall that this is the output due to an input $r(t)\cos2\pi f_0 t$. This result indicates that the time-varying amplitude of the input sinusoid is delayed by an amount equal to the group delay and the oscillating portion is delayed by an amount equal to the phase delay. Both group and phase delay are evaluated at the frequency of the sinusoid. Equation (1.14) will prove significant later. Getting ahead of the game, we will work with two types of receiver: coherent and incoherent. Incoherent receivers operate only upon the amplitude of the received signal, so the group delay is critical to the operation of the receiver. On the other hand, coherent receivers use all of the information about the waveform, so phase delay is also important.

1.2.2 Typical Communication Channels

All communication systems contain a *channel,* which is the medium that connects the receiver to the transmitter. The channel may consist of copper wires, coaxial cable, fiber optic cable, waveguide, air (including the upper atmosphere in the case of satellite transmission), or a combination of these. All channels have a maximum frequency beyond which input signal components are almost entirely attenuated. This is due to the presence of distributed capacitance and inductance. As frequencies increase, the parallel capacitance approaches a short circuit and the series inductance approaches an open circuit.

Many channels also exhibit a low-frequency cutoff due to the dual of the foregoing effects. If there is a low-frequency cutoff, the channel can be modeled as a bandpass filter. If there is no low-frequency cutoff, the channel model is a lowpass filter.

Communication channels are categorized according to bandwidth. There are three generally used grades of channel: narrowband, voiceband, and wideband.

Bandwidths up to 300 Hz are in the *narrow band;* that is, they are *telegraph grade.* They can be used for slow data transmission, on the order of 600 bits per second *(bps).* Narrowband channels cannot reliably be used for unmodified voice transmissions.

Voice-grade channels have bandwidths between 300 Hz and 4 kHz. While they were originally designed for analog voice transmission, they are regularly used to transmit data at rates on the order of 10 kilobits per second *(kbps).* Some forms of compressed video can be sent on voice-grade channels. The public telephone (subscriber loop) circuits are voice-band.

Wideband channels have bandwidths greater than 4 kHz. They can be leased from a carrier (e.g., a telephone company) and can be used for high-speed data, video, or multiple voice channels.

We now give a brief overview of the variety of communication channels in use today. We then focus on telephone channels, since their use in both analog and digital communication predominates over that of the other types of channels.

Wire, Cable, and Fiber

Copper wire, coaxial cable, or optical fibers can be used in *point-to-point* communication. That is, if we know the location of the transmitter and the location of the receiver, and if the two devices can be conveniently connected to each other, a wire connection is possible. Copper wire pairs, twisted to reduce the effects of incident noise, can be used for low-frequency communication. The bandwidth of this system is dependent upon length. The attenuation (in dB/km) follows a curve similar to that shown in Fig. 1.6.

An improvement over twisted copper pairs is realized when one moves to *coaxial cable.* The bandwidth of the channel is much higher than that of twisted wire, and multiple pairs of wires can be enclosed within a single cable sheath. The sheath which surrounds the wires shields them from incident noise, so coaxial cables can be used over longer distances than can twisted pair.

Fiber optics offers advantages over metal cable, both in bandwidth and in noise immunity. Fiber optics is particularly attractive for data communication, where the bandwidth permits much higher data rates than those achievable with metallic connectors.

Air (terrestrial) communication has both advantages and disadvantages when used as a transmission channel. The most important advantage is the ability to *broadcast* signals. You do not need to know the exact location of the receiver in order to set up a communication link. Mobile communication would not be possible without that capability. Among the disadvantages are channel characteristics that are highly dependent on frequency, additive noise, limited allocation of available frequency bands, and susceptibility to intentional interference (jamming).

Attenuation (at sea level) is a function of frequency, barometric pressure, humidity, and weather conditions. A typical curve for fair-weather conditions would resemble Fig.

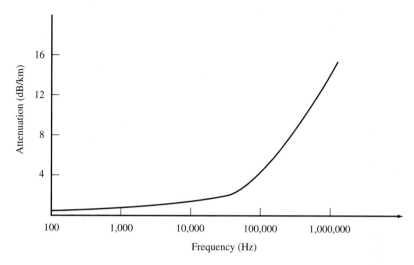

Figure 1.6   Copper wire attenuation vs. distance.

1.7(a). Visible light occupies the range of frequencies from about 40,000–75,000 GHz. Figure 1.7(b) amplifies the lower frequency portion of the attenuation curve.

Additive noise and transmission characteristics also depend upon frequency. The higher the frequency, the more the transmission takes on the characteristics of light. For example, at radio frequencies *(rf)*, in the range of 1 MHz, transmission is *not* line of sight, and reception beyond the horizon is possible. However, at *ultrahigh frequencies (uhf)*, in the range of 500 MHz and above, transmission starts acquiring some of the characteristics of light. Line of sight is needed, and humidity and obstructions degrade transmission.

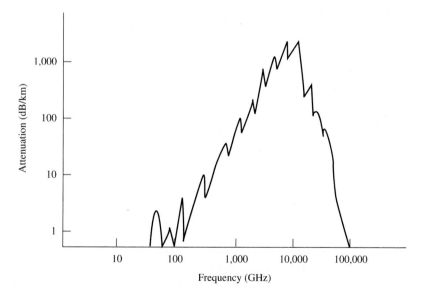

Figure 1.7(a)   Attenuation vs. frequency for air.

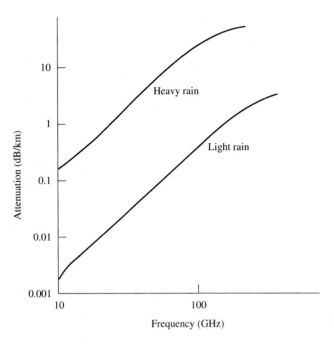

Figure 1.7(b).

At *microwave* frequencies, transmission is line of sight, and antennas must be situated in a manner to avoid obstructions.

Satellite communication provides advantages in long-distance communication. The signal is sent to the satellite via an *uplink,* and the electronics in the satellite (transponder) retransmits this signal to the *downlink.* It is just as easy for the satellite to retransmit the signal to a receiver immediately adjacent to the uplink as it is to transmit to a receiver 5,000 km away from the uplink. The only requirement is that the receiver be within the *footprint of the satellite*—that is, the area of the earth's surface covered by the satellite transmitting antenna pattern. Satellite communication has several major disadvantages. Satellites are usually located in assigned orbital slots in geosynchronous orbit, about 35,000 km (22,300 miles) above the earth's surface. This results in a round-trip travel time of approximately ½ second, making rapid interactive two-way communication difficult. The medium is relatively expensive, although signal compression techniques are bringing down the price.

All broadcast transmission systems suffer from a lack of privacy: Anyone within the reception area of the transmitting antenna can tap into the conversation. This shortcoming can be partially alleviated with scrambling and encryption.

Telephone Channels

There are two types of phone lines in use today. The *dial-up line* is routed through voice-switching offices. The switching operations can add impulse noise, which is heard as occasional clicks during a phone conversation—not terribly devastating to the conversation. But it should not be too surprising to learn that this causes serious problems in data communication. The alternative is the *leased line,* which is a permanent circuit that is not subject to the public type of switching. Since the same line is used every time, some types of distortion can be predicted and compensated for (with *equalizers*).

In the dial-up circuit, assorted problems arise due to procedures adopted for voice channels. For example, the system has bridge taps. A *bridge tap* is a jumper that is installed when a phone is removed or when extra phone jacks have been installed for possible later expansion. These are not serious causes of distortion in voice transmission, but their capacitance leads to delays that can destroy data. (Recall your transmission line theory and what happens if the termination is not properly matched.)

The phone system is specifically tailored to audio signals with an upper frequency in the vicinity of 4 kHz. When such lines are used for data, the upper cutoff is often stretched to provide data rates above 10 kbps. *Loading coils* in the line improve performance in the voiceband, but cause additional amplitude distortion above 4 kHz, thus making higher bit rates more difficult to achieve.

Long-distance phone channels contain *echo suppressors* that are voice activated. These prevent a speaker from receiving an echo due to reflections from transitions in the channel. The time delay in activating the echo suppressors can make certain types of data operation impossible. Many telephone line data sets contain a provision to disable the echo suppressors using a tone of about 2,000 Hz.

Phone lines have amplitude characteristics that are not constant with frequency, and they therefore contribute to amplitude distortion. Figure 1.8 shows a typical attenuation, or loss, curve. The loss is given in decibels *(dB)* and is relative to attenuation at about 1,000 Hz, where the minimum loss occurs.

Phase distortion also occurs in the phone line. A typical phase characteristic for about 7 kilometers of phone line is shown in Fig. 1.9.

Voice-grade channels are classified according to the maximum amount of attenuation distortion and maximum envelope delay distortion within a particular frequency range. Telephone companies can provide *conditioning* to reduce the effects of particular types of distortion. In leasing a telephone channel, a particular conditioning is specified,

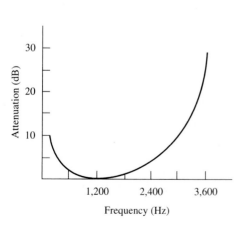

Figure 1.8   Typical telephone channel attenuation.

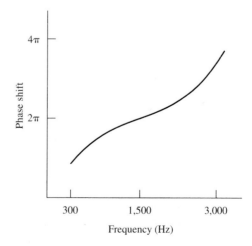

Figure 1.9   Typical telephone channel phase.

and the company guarantees a certain performance level. Naturally, the better the channel, the higher the cost will be. Table 1.1 lists typical channel conditioning characteristics for representative types of conditioning.

As an example, let us examine the C2 channel. If you were to purchase such a channel and use it within the band of frequencies between 500 Hz and 2,800 Hz, you would be guaranteed that the attenuation would not vary beyond the range of -1 dB to $+3$ dB (relative to response at 1,004 Hz). If you use the channel in the wider band between 300 Hz and 3,000 Hz, the guaranteed attenuation range increases to -2 dB to $+6$ dB. Similarly, the envelope delay variation will be less than 3,000 microseconds if you operate within the band between 500 Hz and 2,800 Hz.

In addition to the parameters given in Table 1.1, the various lines have specified losses, loss variations, maximum frequency error, and phase jitter. For example, the C1 channel specifies a loss of 16 dB $\pm$ 1 dB at 1,000 Hz. The variation in loss is limited to 4 dB over long periods of time. The frequency error is limited to 5 Hz and the phase jitter to 10°.

TABLE 1.1   TYPICAL TELEPHONE CHANNEL PARAMETERS

Conditioning	Attenuation Frequency Range	Distortion Variation	Envelope Delay Frequency Range	Distortion Variation (μs)
Basic	500–2,500	−2 to +8	800–2,600	1,750
	300–3,000	−3 to +12		
C1	1,000–2,400	−1 to +3	1,000–2,400	1,000
	300–2,700	−2 to +6	800–2,600	1,750
	300–3,000	−3 to +12		
C2	500–2,800	−1 to +3	1,000–2,600	500
	300–3,000	−2 to +6	600–2,600	1,500
			500–2,800	3,000
C4	500–3,000	−2 to +3	1,000–2,600	300
	300–3,200	−2 to +6	800–2,800	500
			600–3,000	1,500
			500–3,000	3,000

1.3 TYPES OF SIGNALS

The signals we wish to transmit either come directly from the source or result from signal-processing operations. Examples of signals that come directly from the source are the pressure wave emitted by human vocal cords and the electrical signal resulting from a power source connected to a computer keyboard. Processed signals can result from analog-to-digital converters, encoding and encryption devices, and signal-conditioning circuitry. The manner in which we communicate information is dependent on the form of the signal. There are two broad signal classifications: analog and digital. Within these are a number of more detailed subdivisions.

1.3.1 Analog Signals

An *analog signal* can be viewed as a waveform that can take on a continuum of values for any time within a range of times. Although our measuring device may be limited in resolution (e.g., it may not be possible to read an analog voltmeter more accurately than to the nearest hundredth of a volt), the actual signal can take on an infinity of possible values. For example, you might read the value of a voltage waveform at a particular time to be 13.45 volts. If the voltage is an analog signal, the actual value would be expressed as an extended decimal with an infinite number of digits to the right of the decimal point.

Just as the ordinate of the function contains an infinity of values, so does the time axis. Although we may conveniently resolve the time axis into points (e.g., every microsecond on an oscilloscope), the function has a defined value for any of the infinite number of time points between any two resolution points.

An example of an analog signal is a human speech waveform. We illustrate a representative waveform and its Fourier transform in Fig. 1.10. Note that we show only the magnitude of the Fourier transform. If the speech waveform resulted from someone whistling into a microphone, the time waveform would be a sinusoid, and the Fourier transform would be an impulse at the whistling frequency. If the person hummed into a microphone, the time waveform would be periodic with fundamental frequency equal to the frequency at which the person is humming. The Fourier transform would consist of impulses at the fundamental frequency and at its harmonics.

1.3.2 Analog Sampled Signals

Suppose that an analog time signal is defined only at discrete time points. For example, suppose you read a voltage waveform by sending values to a voltmeter every microsecond. The resulting function is known only at these discrete points in time. This results in a *discrete time function,* or a *sampled waveform.* It is distinguished from a continuous analog waveform by the manner in which we specify the function. In the case of the continuous analog waveform, we must either display the function (e.g., graphically, on an oscilloscope) or give a functional relationship between the variables. In contrast to this, the discrete signal can be thought of as a list or sequence of numbers. Thus, while an analog waveform can be expressed as a function of time, $v(t)$, the discrete waveform is a sequence of the form, v_n or $v[n]$, where n is an integer or index.

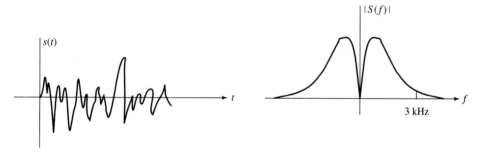

Figure 1.10   Representative speech waveform.

Discrete signals can be visualized as pulse waveforms. Figure 1.11 shows an analog time function and the resulting sampled pulse waveform. (We will refer to this sampled waveform later as *pulse amplitude modulation,* or *PAM.*)

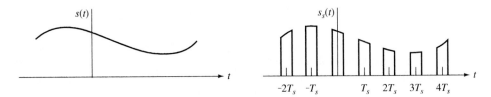

Figure 1.11   Discrete waveform derived from analog time function.

1.3.3 Digital Signals

A *digital signal* is a form of sampled or discrete waveform, but each number in the list can now take on only specific values. For example, if we were to take a sampled voltage waveform and round each value to the nearest tenth of a volt, the result would be a digital signal.

We can use a thermometer as an example of all three types of signal. If the thermometer has a dial or a tube of mercury, the output is an analog signal: We can read the temperature at any time and to any desired degree of accuracy (limited, of course, by the resolution of the reader—human or mechanical).

Suppose now that the thermometer consists of a dial, but that it is updated only once every minute. The result is an analog sampled signal.

If the transducer now takes the form of a numerical readout, the thermometer becomes digital. The readout is the result of sampling the temperature (perhaps every minute) and then displaying the sampled temperature to a predetermined resolution (perhaps the nearest tenth of a degree).

Digital signals result from many devices. For example, dialing[1] a telephone number produces 1 of 12 possible signals, depending on which button is pressed. Other examples include pressing keys on a bank automated teller machine (ATM) and using a computer keyboard. Digital signals also result from performing analog-to-digital conversion operations.

1.4 ELEMENTS OF A COMMUNICATION SYSTEM

A communication system consists of a transmitter, a channel, and a receiver, as shown in Fig. 1.12. The purpose of the *transmitter* is to change the raw information into a format that is matched to the characteristics of the channel. For example, if the information is a speech waveform and the channel is the air, the signal must be modified prior to insertion into the channel. The reason is that the basic speech waveform will not propagate efficiently through the air because its frequency range is below that of the passband of the channel.

---

[1]The word *dialing* is as obsolete as the word *clockwise.* It is a carryover from the early days of telephony when telephone sets contained a rotary dial used to enter numbers and control a *step-by-step* rotary switch.

Figure 1.12   Block diagram of communication system.

The *channel* connects the transmitter to the receiver and may take the form of one of the channels described in Section 1.2.

The *receiver* accepts the signal from the channel and processes it to permit interfacing with the final destination (e.g., the human ear or a computer monitor).

Figure 1.13 shows an expansion of the simplified block diagram of a communication system. We shall briefly describe the function of each block. These functions will be expanded upon in later chapters of the text.

The *signal source* (or *transducer*) is the starting point of our system. It may be a microphone driven by a pressure wave from a human source or a musical instrument, a measuring device that is part of a monitoring system, or a data source such as a computer keyboard or a numeric pad on an ATM. Its output is a time waveform, usually electrical.

The *source encoder* operates upon one or more signals to produce an output that is compatible with the communication channel. The device could be as simple as a lowpass filter in an analog communication system, or it could be as complex as a converter that accepts analog signals and produces a periodic train of output *symbols*. These symbols may be binary (1's and 0's), or may be members of a set with more than two elements. When channels are used to communicate signals from more than one source at the same time, the source encoder contains a *multiplexer.*

With electrical communication replacing written communication, security has become increasingly important. We must assure that only the intended receiver can understand the message and that only the authorized sender can transmit it. *Encryption* provides such security. As unauthorized receivers and transmitters become more sophisticated and computers become larger and faster, the challenges of secure communication become greater. In analog systems, security is often provided using *scrambling* systems, as in pay television, and privacy devices, as with telephones.

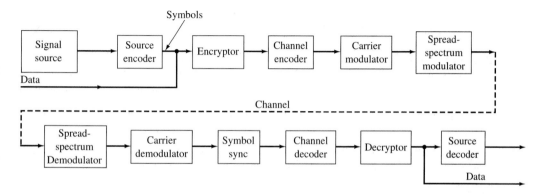

Figure 1.13   Expanded communication system block diagram.

The *channel encoder* provides a different type of communication security than that provided by the encryptor. It increases efficiency and/or decreases the effects of transmission errors. Whenever noise is introduced into a communication channel, errors occur at the receiver. These errors can take the form of changes in an analog signal, or they can make it possible for one transmitted symbol to be interpreted as a different symbol at the receiver in a digital system. We can decrease the effects of errors by providing structure to the messages in the form of redundancy. In its simplest form, this would require that the messages be repeated. We sometimes intentionally distort an analog signal to decrease the effects of frequency-sensitive noise (as, for example, in pre-emphasis/de-emphasis systems and Dolby sound systems). In digital communication, we often use *forward error correction,* in which encoding permits error correction without the necessity of the receiver asking the transmitter for additional information.

The output of the channel encoder is either a processed analog signal or a digital signal composed of symbols. For example, in a binary system, the output would be a train of 1's and 0's. We need to modify the channel encoder output signal in a manner that matches the characteristics of the channel. The *carrier modulator* produces an analog waveform that is transmitted efficiently through the system. The waveform is selected in order to provide for efficiency and also to permit multiple use of the channel by several transmitters. For analog signal sources, the carrier modulator modifies the range of signal frequencies in order to allow efficient transmission. In the case of digital signal sources, the modulator produces signal segments (bursts) corresponding to the discrete symbols at its input.

Spread spectrum is a technique for providing some immunity to frequency-selective effects such as interference and fading. A signal is spread over a wide range of frequencies so that single-tone interference affects only a small portion of the signal. Spread spectrum also has other advantages, related to simplified methods of sharing a channel among multiple users. Since an unauthorized listener may mistake a spread spectrum signal for wideband noise, the technique provides some (limited) additional security beyond that afforded by encryption.

We have been describing the blocks that form the first half of Fig. 1.13. The second half of the figure comprises the receiver, which is simply a mirror image of the transmitter. It must "undo" each operation that was performed at the transmitter. The only variation from this one-to-one correspondence is that the *carrier modulator* of the transmitter has been replaced by two blocks in the receiver: the *carrier demodulator* and the *symbol synchronizer.* The symbol synchronizer is needed only in digital systems; it thus represents a major distinction between analog and digital communication systems. Once the analog waveforms are reproduced at the receiver, it is critical that the overall signal be properly partitioned into segments corresponding to each symbol and to each message. This partitioning is the function of the synchronizer.

PROBLEMS

1.2.1 Show that the function of time corresponding to the transform of Eq. (1.12) is

$$s(t) = Ar[t - t_{\text{gr}}(f_0)]\cos 2\pi f_0[t - t_{\text{ph}}(f)_0)]$$

Hint: You may find it useful to prove the following relationships first:

$$r(t - t_0)\cos 2\pi f_0 t \leftrightarrow \frac{1}{2}[R(f - f_0)e^{-j2\pi(f-f_0)t_0} + R(f + f_0)e^{-j2\pi(f+f_0)t_0}]$$

$$r(t - t_1)\cos 2\pi f_0(t - t_1) \leftrightarrow \frac{1}{2}[R(f - f_0) + R(f + f_0)]e^{-j2\pi f t_1}$$

$$r(t - t_0)\cos 2\pi f_0(t - t_1) \leftrightarrow \frac{1}{2}[R(f - f_0)e^{-j2\pi(f-f_0)(t_0-t_1)}$$

$$+ R(f + f_0)e^{-j2\pi(f+f_0)(t_0-t_1)}]e^{-j2\pi f t_1}$$

1.2.2 Find the output of the filter of Fig. 1.3 when the input is

$$r(t) = \frac{\sin 200\,\pi t}{t} + \frac{5\sin 600\,\pi t}{t}$$

1.2.3 Find the output of a typical telephone line when the input is

(a) $$\cos 2\pi \times 500\,t + \cos 2\,\pi \times 1{,}000\,t$$

(b) A periodic triangle of frequency 1 kHz.

1.2.4 White noise forms the input to a telephone line with amplitude and phase as shown in Figs. 2.12 and 2.13. The height of the two-sided noise power spectral density is K. Find the output power.

2

Signal Analysis

2.0 PREVIEW

What We Will Cover and Why You Should Care

By this stage of your education, you have probably heard of Fourier series expansions. You have probably also learned that the earlier part of any course of study tends to be the least interesting; certain groundwork must be laid prior to getting into the interesting applications.

The purpose of this chapter is to put the study of signals into proper perspective in the much broader area of applied mathematics. Signal analysis, and indeed most of communication theory, is a mathematical science. Probability theory and transform analysis techniques form the backbone of all communication theory. Both of these disciplines fall clearly within the realm of mathematics.

It is quite possible to study communication systems without relating the results to more general mathematical concepts, but this approach is narrow minded and tends to downgrade engineers (a phenomenon to which we need not contribute). More important, the narrow-minded approach reduces a person's ability to extend existing results to new problems.

After studying the material in this chapter, you will:

- Understand the theory of applications of Fourier series
- Be able to find the Fourier transform of various functions
- Appreciate the value of the Fourier transform
- Know the important properties of the Fourier transform

Necessary Background

The only prerequisite you need to understand the material in this chapter is basic calculus.

2.1 FOURIER SERIES

A function can be represented approximately over a given interval by a linear combination of members of an orthogonal *set* of functions. If the set of functions is denoted as $g_n(t)$, this statement can be written as

$$s(t) \approx \sum_{n=-\infty}^{\infty} c_n g_n(t) \tag{2.1}$$

An *orthogonal* set of functions is a set with the property that a particular operation performed between any two distinct members of the set yields zero. You have learned that vectors are orthogonal if they are at right angles to each other. The *dot product* of any two distinct vectors is zero. This means that one vector has nothing in common with the other. The projection of one vector onto another is zero. A function can be considered an infinite-dimensional vector (think of forming a sequence by sampling the function), so the concepts from vector spaces have direct application to function spaces. There are many possible orthogonal sets of functions, just as there are many possible orthogonal sets of three-dimensional vectors (e.g., consider any rotation of the three rectangular unit vectors).

One such set of orthogonal functions is the set of harmonically related sines and cosines. That is, the functions

$$\sin 2\pi f_0 t, \sin 4\pi f_0 t, \sin 6\pi f_0 t \ldots$$

$$\cos 2\pi f_0 t, \cos 4\pi f_0 t, \cos 6\pi f_0 t \ldots$$

form an orthogonal set for any choice of f_0. These functions are orthogonal over the interval between any starting point t_0 and $t_0 + 1/f_0$. That is,

$$\int_{t_0}^{t_0 + \frac{1}{f_0}} g_n(t) g_m(t) dt = 0 \quad \textit{for all } n \neq m \tag{2.2}$$

where $g_n(t)$ is any member of the set of functions and $g_m(t)$ is any other member. This equation can be verified by a simple integration using the cosine-of-sum and cosine-of-difference trigonometric identities.

Example 2.1

Show that the set made up of the functions $\cos 2n\pi f_0 t$ and $\sin 2n\pi f_0 t$ is an orthogonal set over the interval $t_0 < t \leq t_0 + 1/f_0$ for any choice of t_0.
Solution: We must show that

$$\int_{t_0}^{t_0 + \frac{1}{f_0}} g_n(t) g_m(t) dt = 0$$

for any two distinct members of the set. Three cases must be considered:
(a) Both $g_n(t)$ and $g_m(t)$ are sine waves.

(b) Both $g_n(t)$ and $g_m(t)$ are cosine waves.

(c) One of the two is a sine wave and the other a cosine wave.

Considering each of these cases in turn, we have the following:

Case (a):

$$\int_{t_0}^{t_0+\frac{1}{f_0}} \sin 2\pi n f_0 t \sin 2\pi m f_0 t \, dt = \frac{1}{2}\int_{t_0}^{t_0+\frac{1}{f_0}} \cos(n-m)2\pi f_0 t \, dt$$

$$-\frac{1}{2}\int_{t_0}^{t_0+\frac{1}{f_0}} \cos(n+m)2\pi f_0 t \, dt$$

We have used the trigonometric identity[1]

$$\sin A \sin B = \frac{1}{2}[\cos(A-B) - \cos(A+B)]$$

For $n \neq m$, both $n-m$ and $n+m$ are nonzero integers. We note that, for the function $\cos 2\pi k f_0 t$, the interval $t_0 < t \leq t_0 + 1/f_0$ represents exactly k periods. The integral of a cosine function over any whole number of periods is zero, so we have completed this case. Note that it is important that $n \neq m$, since if $n = m$, we have

$$\frac{1}{2}\int_{t_0}^{t_0+\frac{1}{f_0}} \cos(n-m)2\pi f_0 t \, dt = \frac{1}{2}\int_{t_0}^{t_0+\frac{1}{f_0}} 1 \, dt = \frac{1}{2f_0} \neq 0$$

Case (b):

$$\int_{t_0}^{t_0+\frac{1}{f_0}} \cos 2\pi n f_0 t \cos 2\pi m f_0 t \, dt = \frac{1}{2}\int_{t_0}^{t_0+\frac{1}{f_0}}[\cos(n-m)2\pi f_0 t$$

$$+\frac{1}{2}\cos(n+m)2\pi f_0 t]dt$$

Here we have used the trigonometric identity

$$\cos A \cos B = \frac{1}{2}[\cos(A+B) + \cos(A-B)]$$

This is equal to zero by the same reasoning applied to case (a).

Case (c):

$$\int_{t_0}^{t_0+\frac{1}{f_0}} \sin 2\pi n f_0 t \cos 2\pi m f_0 t \, dt = \frac{1}{2}\int_{t_0}^{t_0+\frac{1}{f_0}} \sin(n+m)2\pi f_0 t \, dt$$

$$+\frac{1}{2}\int_{t_0}^{t_0+\frac{1}{f_0}} \sin(n-m)2\pi f_0 t \, dt$$

The applicable trigonometric identity for this case is

---

[1]We will use three different trigonometric identities in this example. However, all necessary results could be derived from the two basic identities for $\cos(A+B)$ and $\sin(A+B)$.

$$sinAcosB = \frac{1}{2}[sin(A + B) + sin(A - B)]$$

To verify case (c), we note that

$$\int_{t_0}^{t_0 + \frac{1}{f_0}} \sin k2 \, \pi f_0 t \, dt = 0$$

for all integer values of k. This is true because the integral of a sine function over any whole number of periods is zero. (There is no difference between a sine function and a cosine function other than a shift.) Each term present in case (c) is therefore equal to zero.

It follows that the given set is an orthogonal set of time functions over the interval $t_0 < t \le t_0 + 1/f_0$.

In the first sentence of this section, we used the word *approximately*. By that term, we are implying that Eq. (2.1) cannot always be made an equality. An orthogonal set of time functions is said to be a *complete set* if the approximation can in fact be made into an equality (with the word *equality* being interpreted in some special sense) through proper choice of the c_n weighting factors and for $s(t)$ being any member of a certain class of functions. The three rectangular unit vectors form a complete orthogonal set in three-dimensional space, while unit vectors in the x- and y-directions, by themselves, form an orthogonal set that is not complete.

We state without proof that the set of harmonic time functions

$$\cos 2\pi n f_0 t, \sin 2\pi n f_0 t$$

where n can take on any integer value between zero and infinity, is an orthogonal *complete* set in the space of time functions defined in the interval between t_0 and $t_0 + 1/f_0$. Therefore, a time function[2] can be expressed, in the interval between t_0 and $t_0 + 1/f_0$, by a linear combination of sines and cosines. In this case, the word *equality* is interpreted not as a pointwise equality, but in the sense that the *distance* between $s(t)$ and the series representation approaches zero as more and more terms are included in the sum. The distance is defined as

$$\int_{t_0}^{t_0 + \frac{1}{f_0}} \left| s(t) - \sum_{n-0}^{\infty} c_n g_n(t) \right|^2 dt \tag{2.3}$$

The preceding is what we will mean when we talk of equality of two time functions. This type of equality will be sufficient for all of our applications.

For convenience, we define the period of the function as

$$T = \frac{1}{f_0} \tag{2.4}$$

Any time function $s(t)$ can then be written as

---

[2] In the case of Fourier series, the class of time functions is restricted to be that class which has a finite number of discontinuities and a finite number of maxima and minima in any one period. Also, the integral of the magnitude of the function over one period must exist (i.e., be finite).

$$s(t) = a_0\cos(0) + \sum_{n=1}^{\infty}[a_n\cos2\pi nf_0t + b_n\sin2\pi nf_0t] \qquad (2.5)$$

for

$$t_0 < t < t_0 + T$$

An expansion of this type is known as a *Fourier series*. We note that the first term in Eq. (2.5) is simply a_0, since $\cos(0) = 1$. The proper choice of the constants a_n and b_n is indicated by the following relationships:

$$a_0 = \frac{1}{T}\int_{t_0}^{t_0+T} s(t)dt$$

$$a_n = \frac{2}{T}\int_{t_0}^{t_0+T} s(t)\cos2\pi nf_0t \, dt \qquad (2.6)$$

$$b_n = \frac{2}{T}\int_{t_0}^{t_0+T} s(t)\sin2\pi nf_0t \, dt$$

The expression for a_0 in Eq. (2.6) can be derived by integrating both sides of Eq. (2.5). The expressions for a_n and b_n are derived from Eq. (2.5) by multiplying both sides by the appropriate sinusoid and integrating.

Note that a_0 is the *average* of the time function $s(t)$. It is reasonable to expect this term to appear by itself in Eq. (2.5), since the average value of the sines or cosines is zero. In any equality, the time average of the left side must equal the time average of the right side.

A more compact form of the Fourier series just described is obtained if one considers the orthogonal, complete set of complex harmonic exponentials, that is, the set made up of the time functions

$$\exp(j2\pi nf_0t)$$

where n is any integer, positive or negative. This set is orthogonal over a period of $1/f_0$ sec. Recall that the complex exponential can be viewed as (actually, it is defined as) a vector of length 1 and angle $n2\pi f_0t$ in the complex two-dimensional plane. Thus,

$$\exp(j2\pi nf_0t) = \cos2\pi nf_0t + j\sin2\pi nf_0t \qquad (2.7)$$

As before, the series expansion applies in the time interval between t_0 and $t_0 + 1/f_0$. Therefore, any time function $s(t)$ can be expressed as a linear combination of these exponentials in the interval between t_0 and $t_0 + T$ where $(T = 1/f_0$, as before):

$$s(t) = \sum_{n=-\infty}^{\infty} c_n e^{j2\pi nf_0t} \qquad (2.8)$$

The c_n are given by

$$c_n = \frac{1}{T} \int_{t_0}^{t_0 + T} s(t) e^{-j2\pi n f_0 t} \, dt \tag{2.9}$$

This is easily verified by multiplying both sides of Eq. (2.8) by $e^{-j2\pi f_0 t}$ and integrating both sides.

The basic results are summed up in Eqs. (2.5) and (2.8): Any time function can be expressed as a weighted sum of sines and cosines or a weighted sum of complex exponentials in an interval. The rules for finding the weighting factors are given in Eqs. (2.6) and (2.9).

The right side of Eq. (2.5) represents a periodic function outside of the interval $t_0 < t < t_0 + T$. In fact, the period of the function is T. Therefore, if $s(t)$ happened to be periodic with period T, even though Eq. (2.5) was written to apply only within the interval $t_0 < t < t_0 + T$, *it actually applies for all time*. (Think about it!)

In other words, if $s(t)$ is periodic, and we write a Fourier series that applies over one complete period, the series is equivalent to $s(t)$ for all time.

Example 2.2

Evaluate the trigonometric Fourier series expansion of $s(t)$ as shown in Fig. 2.1. This series must apply in the interval $-\pi/2 < t < \pi/2$.

Solution: We use the trigonometric Fourier series form with $T = \pi$ and $f_0 = 1/T = 1/\pi$. The series is therefore of the form

$$s(t) = a_0 + \sum_{n=1}^{\infty} [a_n \cos 2nt + b_n \sin 2nt]$$

where

$$a_0 = \frac{1}{\pi} \int_{-\frac{\pi}{2}}^{\frac{\pi}{2}} \cos t \, dt = \frac{2}{\pi}$$

$$a_n = \frac{2}{\pi} \int_{-\frac{\pi}{2}}^{\frac{\pi}{2}} \cos t \cos 2nt \, dt = \frac{2}{\pi} \left[\frac{(-1)^{n+1}}{2n-1} + \frac{(-1)^n}{2n+1} \right]$$

$$b_n = \frac{2}{\pi} \int_{-\frac{\pi}{2}}^{\frac{\pi}{2}} \cos t \, \sin 2nt \, dt = 0$$

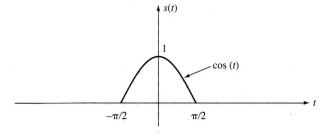

Figure 2.1 $s(t)$ for Example 2.2.

We actually did not need to evaluate the integral for b_n: Since $s(t)$ is an even function of time [i.e., $s(t) = s(-t)$], $s(t)\sin 2nt$ is an odd function, and the integral from $-T/2$ to $+T/2$ is zero. In fact, $a_n = 0$ for any odd $s(t)$. The Fourier Series is then given by

$$s(t) = \frac{2}{\pi} + \sum_{n=1}^{\infty} \frac{2}{\pi} \left[\frac{(-1)^{n+1+}}{2n-1} + \frac{(-1)^n}{2n+1} \right] \cos 2nt$$

Note that this series is also the expansion of the periodic function $s_p(t)$ shown in Fig. 2.2.

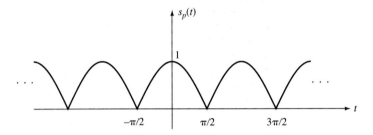

Figure 2.2 $s_p(t)$ represented by Fourier Series.

Suppose we now calculate the Fourier Series of $g(t)$ shown in Fig. 2.3, with the series required to apply in the interval $-2 < t < +2$. The result will clearly be different from that of Example 2.2. One is readily convinced of this difference once it is noted that the frequencies of the various sines and cosines will be different from those of Example 2.2. Nonetheless, for t between $-\pi/2$ and $+\pi/2$, both series expansions represent the same function of time. Both series do not, however, represent $s_p(t)$ of Fig. 2.2. The periodic function corresponding to $g(t)$, denoted as $g_p(t)$, is sketched in Fig. 2.4.

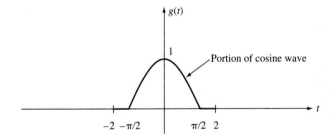

Figure 2.3 $g(t)$ similar to $s(t)$ of Example 2.2.

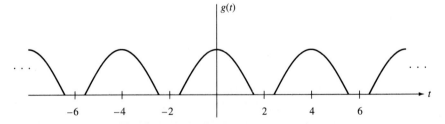

Figure 2.4   Periodic repetition of $g(t)$ of Fig. 2.3.

We conclude that the series expansion of a function in a finite interval is not unique. There are situations in which one takes advantage of this fact in order to choose the type of series that simplifies the results. (The solution of partial differential equations by separation of variables is one example.)

Example 2.3

Approximate the time function

$$s(t) = |\cos t|$$

by a constant. This constant is to be chosen so as to minimize the *error*, which is defined as the average of the square of the difference between $s(t)$ and the approximating constant. Find the "best" value of the constant.

Solution: The square error between $s(t)$ and the approximating constant C is

$$e^2(t) = [|\cos t| - C]^2$$

The average square error is found by integrating the square error over one period and dividing by the period:

$$\{e^2(t)\}_{avg} = \frac{1}{T} \int_{-\frac{T}{2}}^{\frac{T}{2}} [|\cos t| - C]^2 \, dt$$

We could evaluate this integral and differentiate with respect to C in order to minimize the error, or we can borrow a result from orthogonal vector spaces. We used orthogonal sets to approximate a general function. *Even if we do not use a sufficient number of terms to describe the function exactly, the weighting coefficients are chosen as if there were enough terms.* This is an important property of orthogonal sets. It is easier to visualize using vectors. If you think of approximating a three-dimensional vector with a sum of vectors in only two directions, you would not be able to make the summation identical to the vector. However, you would choose the weighting terms for the two given dimensions just as if there were sufficient terms (i.e., they would still be given by the projection of the vector on the appropriate axis). This is so because the missing vector is orthogonal to the vectors that are present and therefore has nothing in common with them. The same is true of the Fourier series: If terms are missing, the coefficients are chosen in the same manner as if all terms were present.

Accordingly, we must approximate $s(t)$ by the first term in its Fourier series expansion. The best value to choose for the constant C is the a_0, or constant, term in the Fourier series expansion. This particular expansion was evaluated in Example 2.2, where the value of a_0 was found to be $2/\pi$. The function $s(t)$ and its dc approximation are shown in Fig. 2.5.

In sum, if we wish to approximate $|\cos t|$ by a constant so as to minimize the mean square error, the best value of the constant to choose is $2/\pi$.

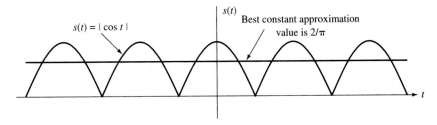

Figure 2.5   Result of Example 2.3.

2.2 COMPLEX FOURIER SPECTRUM (LINE SPECTRUM)

In finding the complex Fourier series representation of a function of time, we assign a complex weighting factor c_n to each value of n. These c_n can be plotted as a function of n. Note that this really requires two graphs, since the c_n are, in general, complex numbers. One plot can represent the magnitude of c_n and the second plot the phase. Alternatively, the real and imaginary parts could be plotted. Note further that this graph would be discrete; that is, it has nonzero value only for discrete values of the abscissa (e.g., $c_{1/2}$ has no meaning).

A more meaningful quantity than n to plot as the abscissa would be n times f_0, a quantity corresponding to the frequency of the complex exponential for which c_n is a weighting coefficient. This plot of c_n vs. nf_0 is called the *complex Fourier spectrum*.

Example 2.4

Find the complex Fourier spectrum of a full-wave rectified cosine wave. This wave is shown in Fig. 2.6 and is given by

$$s(t) = \left| \cos t \right|$$

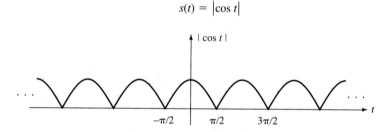

Figure 2.6　$s(t)$ for Example 2.4.

Solution: To find the complex Fourier spectrum, we must first find the exponential (complex) Fourier series expansion of the given waveform. As in Example 2.2, $f_0 = 1/\pi$. We could evaluate the c_n from Eq. (2.9) and find the Fourier series directly. However, we have already found the trigonometric Fourier series of this function in Example 2.2, namely,

$$s(t) = \frac{2}{\pi} + \sum_{n=1}^{\infty} \frac{2}{\pi}\left[\frac{(-1)^{n+1}}{2n-1} + \frac{(-1)^n}{2n+1} \right]\cos 2nt$$

We can expand the cosine function into complex exponentials by using Euler's identity. That is,

$$\cos 2nt = \frac{1}{2}\left[e^{j2nt} + e^{-j2nt} \right]$$

The exponential Fourier series is then given by

$$s(t) = \frac{2}{\pi} + \sum_{n=1}^{\infty} \frac{a_n}{2} e^{j2nt} + \sum_{n=-\infty}^{-1} \frac{a_n}{2} e^{-j2nt}$$

$$= \frac{2}{\pi} + \sum_{n=1}^{\infty} \frac{a_n}{2} e^{j2nt} + \sum_{n=1}^{\infty} \frac{a_{-n}}{2} e^{j2nt}$$

We have made a change of variables in the last summation. We see that the c_n are related to the a_n by

$$c_n = \frac{a_n}{2} \quad \text{for } n > 0$$

$$c_n = \frac{a_{-n}}{2} \quad \text{for } n < 0$$

$$c_0 = \frac{2}{\pi}$$

The resulting complex Fourier spectrum (line spectrum) is sketched in Fig. 2.7.
Note that only one plot is necessary, since in this particular example, the c_n are all real numbers.

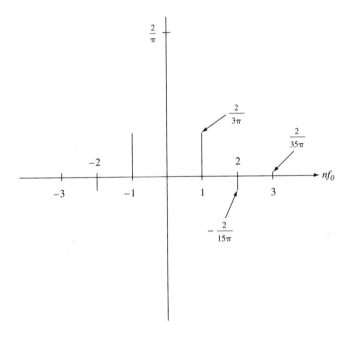

Figure 2.7   Line spectrum for Example 2.4.

2.3 FOURIER TRANSFORM

The vast majority of interesting signals extend for all time and are nonperiodic. One would certainly not go through any great effort to transmit a periodic waveform, since all of the information is contained in one period. Instead, one could either transmit the signal over a single period or transmit the values of the Fourier series coefficients in the form of a list. The question therefore arises as to whether we can write a Fourier series for a nonperiodic signal.

A nonperiodic signal can be viewed as a limiting case of a periodic signal whose period approaches infinity. Since the period approaches infinity, the fundamental frequency

f_0 approaches zero. The harmonics get closer and closer together, and in the limit, the Fourier series summation representation of $s(t)$ becomes an integral. In this manner, we could develop the Fourier integral (transform) theory.

To avoid the limiting processes required to go from Fourier series to Fourier integral, we will take an axiomatic approach. That is, we will *define* the Fourier transform and then show that the definition is extremely useful. There need be no loss in motivation by approaching the transform in this "pull out of a hat" manner, since its extreme versatility will rapidly become obvious.

What is a *transform*? Recall that a common everyday *function* is a set of rules that substitutes one number for another number. That is, $s(t)$ is a set of rules that assigns a number $s(t)$ in the *range* to any number t in the *domain*. You can think of a function as a box that spits out a number whenever you stick in a number. In a similar manner, a *transform* is a set of rules that substitutes one function for another function. It can be thought of as a box that spits out a function whenever you stick in a function.

We define one particular transform as

$$S(f) = \int_{-\infty}^{\infty} s(t)e^{-j2\pi ft}\, dt \tag{2.10}$$

Since t is a dummy variable of integration, the result of the integral evaluation (after the limits are plugged in) is not a function of t, but only a function of f. We have therefore given a rule that assigns, to every function of t (with some broad restrictions required to make the integral in Eq. (2.10) converge), a function of f.

The extremely significant Fourier transform theorem states that, given the Fourier transform of a function of time, the original time function can always be uniquely recovered. In other words, the transform is unique: Either $s(t)$ or its transform $S(f)$ uniquely characterizes a function. This is crucial! Were it not true, the transform would be useless.

An example of a useless transform (the *Roden transform*) is the following:

To every function $s(t)$, assign the function

$$R(f) = f^2 + 1.3$$

This transform defines a function of f for every function of t. The reason it has not become famous is that, among other factors, it is not unique: Given that the Roden transform of a function of time is $f^2 + 1.3$, you have not got a prayer of finding the $s(t)$ which led to that transform.

Actually, the Fourier transform theorem goes one step further than stating uniqueness: It gives the rule for recovering $s(t)$ from its Fourier transform. This rule exhibits itself as an integral and is almost of the same form as the original transform rule. That is, given $S(f)$, one can recover $s(t)$ by evaluating the integral

$$s(t) = \int_{-\infty}^{\infty} S(f)e^{j2\pi ft}\, df \tag{2.11}$$

Equation (2.11) is sometimes referred to as the *inverse transform* of S(f). It follows that this is also unique.

There are infinitely many unique transforms.[3] Why, then, has the Fourier transform achieved such widespread fame and use. Certainly, it must possess properties that make it far more useful than other transforms.

Indeed, we shall presently discover that the Fourier transform is useful in a way that is analogous to the usefulness of the common logarithm. (Remember them from high school?) In order to multiply two numbers together, we can find the logarithm of each of the numbers, add the logarithms, and then find the number corresponding to the resulting logarithm. One goes through all of this trouble in order to avoid multiplication (a frightening prospect to students). We have

$$a \quad \times \quad b \quad = \quad c$$
$$\Downarrow \quad \Downarrow \quad \Downarrow \qquad \Uparrow$$
$$\log(a) \quad + \quad \log(b) \quad = \quad \log(c)$$

An operation that often must be performed between two functions of time is *convolution*. This is enough to scare even those few who are not frightened by multiplication! In Section 2.5, we will show that if the Fourier transform of each of the two functions of time is found first, a much simpler operation can be performed upon the transforms that corresponds to convolution of the original functions. The operation that corresponds to convolution of the two functions of time is multiplication of their two transforms. (Multiplication is no longer difficult once one graduates from high school.) Thus, we will multiply the two transforms together and then find the function of time that corresponds to the resulting transform.

Notation

We will usually use the same letter for the function of time and its corresponding transform, with uppercase used for the transform. That is, if we have a function of time called $g(t)$, we shall call its transform $G(f)$. In cases where this is not possible, we find it necessary to adopt some alternative notational forms to associate the function with its transform. The script capital $\mathfrak{F}$ and $\mathfrak{F}^{-1}$ are often used to denote taking the transform and the inverse transform, respectively. Thus, if $S(f)$ is the transform of $s(t)$, we can write

$$\mathfrak{F}[s(t)] = S(f)$$

$$\mathfrak{F}^{-1}[S(f)] = s(t)$$

A double-ended arrow is also often used to relate a function of time to its transform, the two together being known as a *transform pair*. Thus, we would write

$$s(t) \leftrightarrow S(f)$$

$$S(f) \leftrightarrow s(t)$$

---

[3]As two examples, consider either time scaling or multiplication by a constant. That is, define $S_1(f) = s(2f)$ or $S_2(f) = 2s(f)$. For example, if $s(t) = \sin t$, $S_1(f) = \sin 2f$ and $S_2(f) = 2\sin f$. The extension to an infinity of possible transform rules should be obvious.

2.4 SINGULARITY FUNCTIONS

We must introduce a new kind of function before proceeding to applications of Fourier theory. The new function arises whenever we analyze periodic functions. This new entity is part of the class of functions known as *singularities*. These can be thought of as derivatives of the unit step function. We begin by finding the Fourier transform of a gating function.

Example 2.5

Evaluate the Fourier transform of

$$s(t) = \begin{cases} A & |t| < \alpha \\ 0 & \text{otherwise} \end{cases}$$

Find the transform both by performing the integration and by using a computer solution. The function $s(t)$ is illustrated in Fig. 2.8.

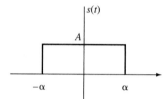

Figure 2.8 $s(t)$ for Example 2.5.

Solution: From the definition of the Fourier transform, we have

$$S(f) = \int_{-\infty}^{\infty} s(t)e^{-j2\pi ft} \, dt$$

$$= \int_{-\alpha}^{\alpha} Ae^{-j2\pi ft} \, dt = A\frac{e^{j2\pi f\alpha} - e^{-j2\pi f\alpha}}{j2\pi f}$$

$$= A\frac{\sin 2\pi f\alpha}{\pi f}$$

This transform is sketched in Fig. 2.9(a).

Note that while the Fourier transform is, in general, a complex function, the solution here turned out to be purely real. We will see the reasons for this in Section 2.6.

We now attempt to obtain the same result using computer software. Although software exists to do equations in symbolic form, we shall use Mathcad and substitute values for A and α. For purposes of illustration, we shall let $\alpha = 0.05$ and $A = 1/2\alpha$. (This should result in a Fourier transform with maximum amplitude equal to unity.)

The Mathcad instructions are as follows (the actual format for entering them depends on whether you are using Windows™ and whether you have a mouse):

$$a := .05$$
$$Fmax := \frac{2.5}{a}$$
$$A := \frac{1}{2 \cdot a}$$
$$f := 0, \frac{.01}{a} \cdot \cdot Fmax$$
$$F(f) := \int_{-a}^{a} A \cdot e^{(-2 \cdot j \cdot \pi \cdot f \cdot t)} \, dt$$

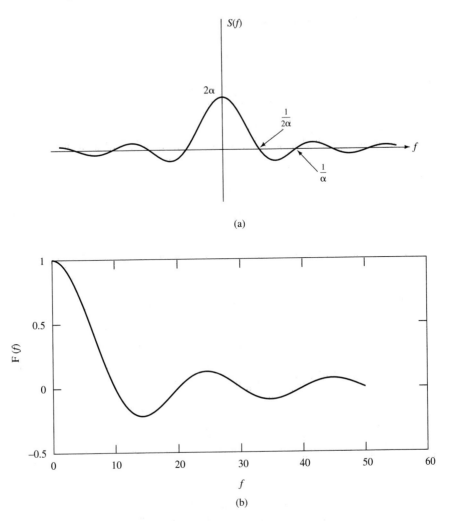

Figure 2.9   Transform of $s(t)$ for Example 2.5.

The resulting graph is shown in Fig. 2.9(b). Note that we have set Fmax so as to obtain five zeros of the function. The increment on f determines the smoothness of the curve. We have set this increment at 100 points between each pair of zeros. Of course, the smaller the increment, the longer it takes to run the simulation.

Mathcad also has a fast Fourier transform (FFT) option. There are several forms of the transform depending on whether the original data are real or complex. However, application of the FFT function requires that the number of data points be a power of 2. Since the preceding problem was so simple, and the data were given in the form of an equation, we chose to simply program the Fourier transform integral rather than use the FFT.

Functions of the type illustrated in Fig. 2.9(a) are common in communication studies. To avoid having to rewrite this type of function repeatedly, we define the *sinc* function

$$sinc(x) = \frac{sinx}{x} \tag{2.12}$$

In terms of this, the transform of Example 2.5 becomes

$$S(f) = A\frac{sin2\pi f\alpha}{\pi f} = 2A\alpha sinc\,(2\pi f\alpha)$$

Suppose we now wish to find the Fourier transform of a constant, $s(t) = A$, for all t. We could consider this to be the limit of the pulse of Fig. 2.8 as $\alpha \rightarrow \infty$. We attempt this roundabout approach because the straightforward technique fails in this case. That is, plugging $s(t) = A$ into the defining integral for the Fourier transform yields

$$S(f) = \int_{-\infty}^{\infty} Ae^{-j2\pi ft}dt \tag{2.13}$$

an integral that does not converge. From the result of Example 2.5, we see that as $\alpha \rightarrow \infty$, the Fourier transform approaches infinity at the origin, and the zero-axis crossings become infinitesimally spaced. Thus, it can only be timidly suggested that, in the limit, the height of the transform goes to infinity and the width to zero. This sounds like a pretty ridiculous function. Indeed, it is not a function at all, since it is not defined at $f = 0$. If we insist upon saying anything about the Fourier transform of a constant, we must restructure our way of thinking.

The restructuring begins by defining what is mathematically known as the *impulse*. This is a member of a class of operations known as *distributions*. We will see that when the impulse operates upon a function, the result is a number. This puts the distribution somewhere between a function (which operates upon numbers to produce numbers) and a transform (which operates upon functions to produce functions).

We use the Greek letter *delta* (δ) to denote the impulse. While we will write $\delta(t)$ as if it were a function, we avoid difficulties by *defining* the behavior of $\delta(t)$ in all possible situations.

The usual, though nonrigorous, definition of the impulse is formed by making three simple observations. Two of these, already mentioned, are

$$\delta(t) = 0, \quad t \neq 0$$

$$\delta(t) \rightarrow \infty, \quad t = 0$$

The third property is that the total area under the impulse is equal to unity:

$$\int_{-\infty}^{\infty} \delta(t)\,dt = 1 \tag{2.14}$$

Since all of the area of $\delta(t)$ is concentrated at one point, the limits on this integral can be moved toward the origin without changing the value of the integral. Thus,

$$\int_a^b \delta(t)\, dt = 1 \tag{2.15}$$

as long as $a < 0$ and $b > 0$.

One can also observe that the integral of $\delta(t)$ is $U(t)$, the unit step function. That is,

$$\int_{-\infty}^{t} \delta(\tau) d\tau = \begin{cases} 1, & t > 0 \\ 0, & t < 0 \end{cases} = U(t) \tag{2.16}$$

We mentioned that the definition comprised of the foregoing three observations was nonrigorous. This is because an elementary study of *singularity functions* shows that these properties do not uniquely define the impulse function. That is, there are other functions in addition to the impulse that satisfy the three conditions. However, the conditions can be used to indicate (not prove) a fourth condition. This condition then serves as a unique definition of the delta function and will represent the only property of the impulse we ever make use of. We shall integrate the product of an arbitrary function of time with $\delta(t)$. We obtain

$$\int_{-\infty}^{\infty} s(t)\delta(t)\, dt = \int_{-\infty}^{\infty} s(0)\delta(t)\, dt \tag{2.17}$$

Equation (2.17) says that we could replace $s(t)$ by a constant function equal to $s(0)$ without changing the value of the integral. This requires some justification.

Suppose that we have two new functions $g_1(t)$ and $g_2(t)$, and we consider the product of each of these with a third function $h(t)$. Suppose further that $h(t)$ is zero over a portion of the time axis. Then as long as $g_1(t) = g_2(t)$ at all values of t for which $h(t)$ is nonzero, it follows that $g_1(t)h(t) = g_2(t)h(t)$. At those values of t for which $h(t) = 0$, the values of $g_1(t)$ and $g_2(t)$ have no effect on the product. One possible example is illustrated in Fig. 2.10. For the functions shown, we see that

$$g_1(t)h(t) = g_2(t)h(t) \tag{2.18}$$

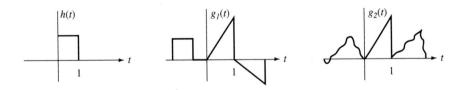

Figure 2.10   Example of $h(t)$, $g_1(t)$, and $g_2(t)$.

Returning to Eq. (2.17), we note that $\delta(t)$ is zero for all $t \neq 0$. Therefore, the product of $\delta(t)$ with any function of time depends only upon the value of that function at $t = 0$. Figure 2.11 illustrates several possible functions that have the same product with $\delta(t)$ as does $s(t)$.

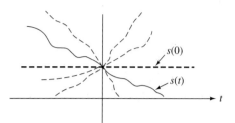

Figure 2.11   Functions having same product with $\delta(t)$.

Out of the infinity of possibilities, the constant function of time is a wise choice, since we can factor it out of the integral to get

$$\int_{-\infty}^{\infty} s(t)\delta(t)\ dt = s(0) \int_{-\infty}^{\infty} \delta(t)\ dt = s(0) \qquad (2.19)$$

This is a significant result, and we will refer to it as the *sampling property* of the impulse. Note that a great deal of information about $s(t)$ has been lost, since the result depends only upon the value of $s(t)$ at one point.

A change of variables yields a shifted impulse with the analogous sampling property:

$$\int_{-\infty}^{\infty} s(t)\delta(t - t_0)\ dt = \int_{-\infty}^{\infty} s(k + t_0)\delta(k)\ dk = s(t_0) \qquad (2.20)$$

Figure 2.12 shows $\delta(t)$ and $\delta(t - t_0)$. The upward-pointing arrow is a generally accepted technique to indicate an actual value of infinity. The number next to the arrow indicates the total area under the impulse, known as the *strength*. In sketching the impulse, the height of the arrow is made equal to the strength of the impulse.

Figure 2.12   Pictorial representation of impulse.

Equations (2.19) and (2.20) are the only things one must know about the impulse. Indeed, either of them can be treated as the definition of the impulse.

Example 2.6

Evaluate the following integrals:

$$\text{(a)} \int_{-\infty}^{\infty} \delta(t)(t^2 + 1)\, dt$$

$$\text{(b)} \int_{-1}^{2} \delta(t - 1)(t^2 + 1)\, dt$$

$$\text{(c)} \int_{3}^{5} \delta(t - 1)(t^3 + 4t + 2)\, dt$$

$$\text{(d)} \int_{-\infty}^{\infty} \delta(1 - t)(t^4 + 2)\, dt$$

Solution: (a) Straightforward application of the sampling property yields

$$\int_{-\infty}^{\infty} \delta(t)(t^2 + 1)dt = 0^2 + 1 = 1$$

(b) Since the impulse falls within the range of integration,

$$\int_{-1}^{2} \delta(t - 1)(t^2 + 1)dt = 1^2 + 1 = 2$$

(c) The impulse occurs at $t = 1$, which is outside the range of integration. Therefore,

$$\int_{3}^{5} \delta(t - 1)(t^3 + 4t + 2)dt = 0$$

(d) $\delta(1 - t)$ falls at $t = 1$, since this is the value of t that makes the argument equal to zero. Therefore,[4]

$$\int_{-\infty}^{\infty} \delta(1 - t)(t^4 + 2)dt = 1^4 + 2 = 3$$

It is now a simple matter to find the Fourier transform of the impulse:

$$\delta(t) \leftrightarrow \int_{-\infty}^{\infty} \delta(t)e^{-j2\pi ft}dt = e^{-j2\pi ft}\Big|_{t=0} = e^0 = 1 \qquad (2.21)$$

This is indeed a very nice Fourier transform for a function to have. One can guess that it will prove significant, since it is the unity, or identity, multiplicative element. That is, anything multiplied by 1 is left unchanged.

[4]Many arguments can be advanced for regarding the impulse as an even function. For one thing, it can be considered the derivative of an odd function. For another, the fact that the transform is real and even indicates that the impulse is an even time function, as will be explored in Section 2.6.

It is almost time to apply all of the theory we have been laboriously developing to a practical problem. Before doing that, we need simply evaluate several transforms that involve impulses.

Let us return to the evaluation of the transform of a constant, $s(t) = A$. We observed earlier that the defining integral

$$A \leftrightarrow \int_{-\infty}^{\infty} Ae^{-j2\pi ft}\, dt \tag{2.22}$$

does not converge. For $f \neq 0$, this integral is bounded by $A/\pi f$. For $f = 0$, the integral "blows up."

Since the integral defining the Fourier transform and that used to evaluate the inverse transform are quite similar, one might guess that the transform of a constant is an impulse. That is, since an impulse transforms to a constant, a constant should transform to an impulse. In the hope that this guess is valid, let us find the inverse transform of an impulse. We have

$$\delta(f) \leftrightarrow \int_{-\infty}^{\infty} \delta(f)e^{j2\pi ft}df = 1 \tag{2.23}$$

Our guess was correct: The inverse transform of $\delta(f)$ is a constant. Therefore, by applying a scaling factor, we have

$$A \leftrightarrow A\delta(f) \tag{2.24}$$

If we take the inverse transform of a shifted impulse, we develop the additional transform pair,

$$Ae^{j2\pi f_0 t} \leftrightarrow A\delta(f - f_0) \tag{2.25}$$

This guess-and-check technique deserves some comment. We stressed earlier that the uniqueness of the Fourier transform is extremely significant. That is, given $S(f)$, $s(t)$ can be uniquely recovered. Therefore, the guess-and-check technique is a perfectly rigorous one to use to find the Fourier transform of a function of time. If we can somehow guess at an $S(f)$ that yields $s(t)$ when $S(f)$ is plugged into the inversion integral, we have found the one and only transform of $s(t)$. As in the preceding example, this technique is very useful if the transform integral cannot be easily evaluated, whereas the inverse transform integral can.

Example 2.7

Find the Fourier transform of $s(t) = \cos 2\pi f_0 t$.
Solution: We make use of Euler's identity to express the cosine function in the form

$$\cos 2\pi f_0 t = \frac{1}{2}e^{j2\pi f_0 t} + \frac{1}{2}e^{-j2\pi f_0 t}$$

The Fourier transform of the cosine is then the sum of the transforms of the two exponentials, which we found in Eq. (2.25). Therefore,

$$\cos 2\pi f_0 t \leftrightarrow \frac{1}{2}\delta(f - f_0) + \frac{1}{2}\delta(f + f_0)$$

This transform is sketched in Fig. 2.13.

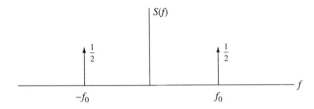

Figure 2.13   Fourier transform of cosine wave.

We can now reveal a deception we have been guilty of: Although the Fourier transform is specified by a strictly mathematical definition, and f is just an independent functional variable, we have slowly tried to brainwash you into thinking of this variable as *frequency*. Indeed, the choice of f for the symbol of the variable brings the word *frequency* to mind. We have seen several Fourier transforms that are not identically zero for negative values of f. In fact, we shall see in Section 2.6 that the transform *cannot* be zero for negative f in the case of real functions of time. Since the definition of frequency (repetition rate) has no meaning for negative values, we could never be completely correct in calling f a frequency variable.

Suppose we view the positive f-axis only. For this region, the transform of $\cos 2\pi f_0 t$ is nonzero only at the point $f = f_0$. (See Example 2.7.) Probably the only time in your earlier education that you experienced the definition of frequency is the case where the function of time is a pure sinusoid. For this function, the positive f-axis appears to have meaning when interpreted as frequency, so we shall consider ourselves justified in calling f a frequency variable.

Another transform pair that will be needed in our later work is that of a unit step function and its transform. Here, as in the case of a constant, if one simply plugs the function of time into the transform definition, the resulting integral does not converge. We could again attempt the guess-and-check technique, but due in part to the discontinuity of the step function, the technique becomes not very hopeful. The transform is relatively easy to evaluate once one realizes that

$$U(t) = \frac{1 + \operatorname{sgn}(t)}{2}$$

where sgn is the *sign* function, defined by

$$\operatorname{sgn}(t) = \begin{cases} +1, & t > 0 \\ -1, & t < 0 \end{cases}$$

$U(t)$ is illustrated in Fig. 2.14.

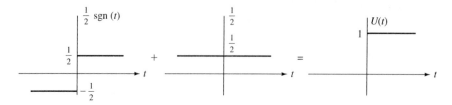

Figure 2.14   Representation of $U(t)$ in terms of sgn(t).

The transform of $\frac{1}{2}$ is $(\frac{1}{2})\delta(f)$. The transform of sgn(t) can be evaluated by expressing sgn(t) as a limit of exponentials, as shown in Fig. 2.15. We have

$$\text{sgn}(t) = \lim_{a \to 0} [e^{-a|t|}\text{sgn}(t)]$$

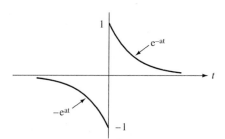

Figure 2.15   sgn(t) as a limit of exponentials.

Assuming that the order of taking the limit and taking the transform can be interchanged (generally, if the result is bounded, we can do this), we obtain

$$\mathscr{F}[\text{sgn}(t)] = \lim_{a \to 0} \mathscr{F}[e^{-a|t|}\text{sgn}(t)] \tag{2.27}$$

$$= \lim_{a \to 0} \left[\frac{1}{j2\pi f + a} + \frac{1}{j2\pi f - a} \right] = \frac{1}{j\pi f}$$

The transform of the unit step is then given by

$$U(t) \leftrightarrow \frac{1}{j2\pi f} + \frac{1}{2}\delta(f) \tag{2.28}$$

If you have been exposed to Laplace transforms, you may recall that the Laplace transform of a unit step is $1/s$. At first glance, it appears that the Fourier transform of any function that is zero for negative t should be the same as the one-sided Laplace transform, with s replaced by $j2\pi f$. However, we see that in the case of $s(t) = U(t)$, the two transforms differ by a very important factor, $(\frac{1}{2})\delta(f)$. The explanation of this apparent discrepancy requires a study of convergence in the complex s-plane.

2.5 CONVOLUTION

We are now ready to investigate the "scary" operation referred to at the end of Section 2.3. The *convolution* of two time functions $r(t)$ and $s(t)$ is defined by the integral operation

$$r(t)*s(t) = \int_{-\infty}^{\infty} r(\tau)s(t - \tau)d\tau = \int_{-\infty}^{\infty} s(\tau)r(t - \tau)d\tau \qquad (2.29)$$

The asterisk notation is conventional, $r(t)*s(t)$ and is read "$r(t)$ convolved with $s(t)$." The second integral in Eq. (2.29) results from a change of variables, and it proves that convolution is *commutative*. That is, $r(t)*s(t) = s(t)*r(t)$.

Convolution is basic to almost any linear system.

Note that the convolution of two functions of t is itself a function of t, since τ is a dummy variable of integration. The integral of Eq. (2.29) is, in general, very difficult to evaluate in closed form, as is demonstrated in the following example.

Example 2.8

Evaluate the convolution of $r(t)$ with $s(t)$, where $r(t)$ and $s(t)$ are the square pulses shown in Fig. 2.16.

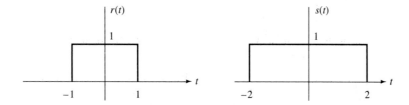

Figure 2.16   Functions for Example 2.8.

Solution: We note that the functions can be written in the form

$$r(t) = U(t + 1) - U(t - 1)$$

$$s(t) = U(t + 2) - U(t - 2)$$

where $U(t)$ is the unit step function defined by

$$U(t) = \begin{cases} 1, & t > 0 \\ 0, & t < 0 \end{cases}$$

The convolution is defined by

$$r(t)*s(t) \overset{\Delta}{=} \int_{-\infty}^{\infty} r(\tau)s(t - \tau)\, d\tau$$

We see that

$$r(\tau) = U(\tau + 1) - U(\tau - 1)$$

and

$$s(t - \tau) = U(t - \tau + 2) - U(t - \tau - 2)$$

$$r(\tau)s(t - \tau) = U(\tau + 1)U(t - \tau + 2) - U(\tau + 1)U(t - \tau - 2)$$

$$- U(\tau - 1)U(t - \tau + 2) + U(\tau - 1)U(t - \tau - 2)$$

Therefore, breaking the integral into parts, we have

$$r(t)*s(t) = \int_{-\infty}^{\infty} U(\tau + 1)U(t - \tau + 2) \, d\tau$$

$$- \int_{-\infty}^{\infty} U(\tau + 1)U(t - \tau - 2) \, d\tau$$

$$- \int_{-\infty}^{\infty} U(\tau - 1)U(t - \tau + 2) \, d\tau$$

$$+ \int_{-\infty}^{\infty} U(\tau - 1)U(t - \tau - 2) \, d\tau$$

We now note that $U(\tau + 1)$ is equal to zero for $\tau < -1$, and $U(\tau - 1)$ is zero for $\tau < 1$. Taking this into account, we can reduce the limits of integration to yield

$$r(t)*s(t) = \int_{-1}^{\infty} U(t - \tau + 2) \, dr - \int_{-1}^{\infty} U(t - \tau - 2) \, d\tau$$

$$- \int_{1}^{\infty} U(t - \tau + 2) \, d\tau + \int_{1}^{\infty} U(t - \tau - 2) \, d\tau$$

To derive this, we have replaced one of the step functions by its value, unity, in the range in which the substitution applies. We now try to evaluate each integral separately. Note that

$$U(t - \tau + 2) = 0, \quad \tau > t + 2$$

and

$$U(t - \tau - 2) = 0, \quad \tau > t - 2$$

Using these facts, we have

$$\int_{-1}^{\infty} U(t - \tau + 2) \, d\tau = \int_{-1}^{t+2} d\tau = t + 3$$

provided that $t + 2 > -1$, or equivalently, $t > -3$. Otherwise, the integral evaluates to zero. Likewise, if $t - 2 > -1$, that is, $t > 1$, then

$$\int_{-1}^{\infty} U(t - \tau - 2) \, d\tau = \int_{-1}^{t-2} d\tau = t - 1$$

If $t + 2 > 1$, that is, $t > -1$, then

$$\int_1^\infty U(t - \tau + 2)\, dr = \int_1^{t+2} d\tau = t + 1$$

If $t - 2 > 1$, that is, $t > 3$, then

$$\int_1^\infty U(t - \tau - 2)\, d\tau = \int_1^{t-2} d\tau = t - 3$$

Using these four results, we find that

$$r(t)*s(t) = (t + 3)U(t + 3) - (t - 1)U(t - 1) - (t + 1)U(t + 1) + (t - 3)U(t - 3)$$

The four terms on the right-hand side, together with their sum, are sketched in Fig. 2.17. From this modest example, we can see that, if either $r(t)$ or $s(t)$ contains step functions, the evaluation of the convolution becomes quite involved.

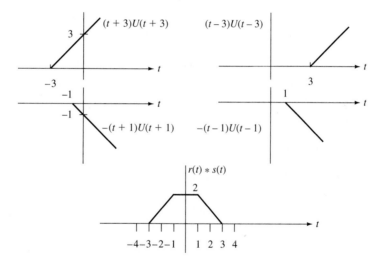

Figure 2.17   Result of Example 2.8.

2.5.1 Graphical Convolution

We claim that, for simple $r(t)$ and $s(t)$ (what we mean by *simple* should be clear at the end of this section), the result of the convolution can be obtained almost by inspection. Even in cases where $r(t)$ and $s(t)$ are quite complicated or the waveshapes are not precisely known, certain observations can be made about the convolution without actually performing the detailed integration. In many communication applications, these general observations will be sufficient, and the exact convolution will not be required.

The inspection procedure is known as *graphical convolution*. We will arrive at the technique by examining the definition of convolution. We repeat the left-hand equality of Eq. (2.29):

$$r(t)*s(t) = \int_{-\infty}^{\infty} r(\tau)s(t - \tau)d\tau$$

One of the original functions is $r(\tau)$, where the independent variable is now called τ.

The mirror image of $s(\tau)$ is represented by $s(-\tau)$, that is, $s(\tau)$ reflected around the vertical axis.

The convolution equation now tells us that for a given t, we form $s(t - \tau)$, which represents the function $s(-\tau)$ shifted to the right by t. We then take the product

$$r(\tau)s(t - \tau)$$

and integrate it (i.e., find the area under it) in order to find the value of the convolution for that particular value of t. The procedure is illustrated in Fig. 2.18 for the two functions of Fig. 2.16 from Example 2.8. The ideal way to demonstrate graphical convolution is with an animated motion picture that shows the two functions, one of which is moving across the τ-axis. The motion picture would show the two functions overlapping by varying amounts as t changes.

Unfortunately, the constraints of book publishing do not allow us to present a motion picture. Instead, we illustrate a stop-action view of the phenomenon; that is, we show a number of frames.

Figure 2.18 shows 12 separate frames of the would-be motion picture. In this particular example, it is not obvious that $s(t)$ was reflected to form the mirror image, since the original $s(t)$ was an even function of t.

It is necessary to interpolate between each pair of values of t shown in the figure. Note that it is the *area* under the product that represents the result of the convolution. Figure 2.19 plots this area as a series of points, with straight-line interpolation between pairs of points.

It should not be surprising that the result is piecewise linear; that is, the interpolation results in straight lines. This is so because the convolution becomes an integration of a constant, which results in a ramp function. With practice, only a few points are needed to plot the resulting convolution. It is comforting to compare Fig. 2.19 with Fig. 2.17 and to note that we obtained the same answer in both cases.

Example 2.9

Find the convolution of $r(t)$ with itself, where the only information given about $r(t)$ is that it is zero for $|t| > a$. That is, $r(t)$ is limited to the range between $t = -\alpha$ and $t = +\alpha$. A representative $r(t)$ is sketched in Fig. 2.20.

Solution: Not knowing $r(t)$ exactly, we certainly cannot find the resulting convolution, $r(t)*r(t)$, exactly. To see how much information we can obtain concerning this convolution, we attempt graphical convolution. Figure 2.21 is a sketch of $r(\tau)$ and $r(t - \tau)$.

We note that as t increases from zero, the two functions illustrated have less and less of the τ-axis in common. When t reaches $+2\alpha$, the two functions separate; that is, at $t = 2\alpha$, we have the situation sketched in Fig. 2.22. The two functions continue to have nothing in common for all t greater than 2α. Likewise, for negative t, one can see that the product of the two functions is zero as long as $t < -2\alpha$.

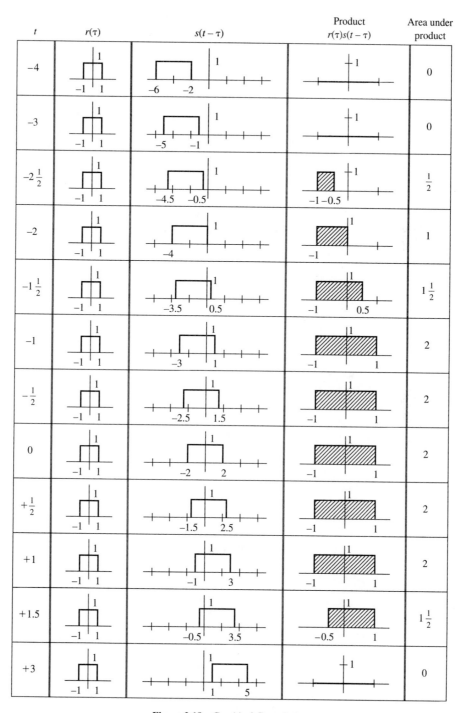

Figure 2.18   Graphical Convolution.

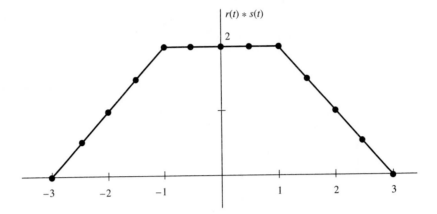

Figure 2.19   Result of Graphical Convolution.

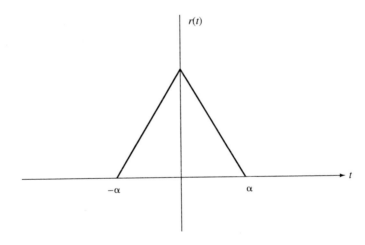

Figure 2.20   Time limited $r(t)$ for Example 2.9.

Figure 2.21 $r(\tau)$ and $r(t - \tau)$ for Example 2.9.

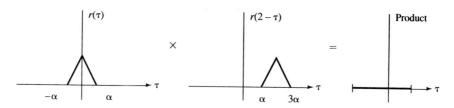

Figure 2.22   Convolution product when t = 2α.

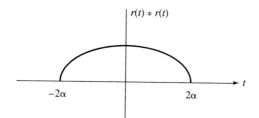

Figure 2.23   Typical Result of Example 2.9.

Although we do not know $r(t)*r(t)$ exactly, we have found the range it occupies along the t-axis. The range is zero for $|t| > 2\alpha$. A possible form of this result is sketched in Fig. 2.23. We emphasize that the sketch is not intended to indicate the exact shape of the function. (Indeed, if the original function were triangular, the convolution would be parobolic.) The results of this example will be used later in the study of modulation systems.

Example 2.10

Use graphical techniques to convolve the two functions shown in Fig. 2.24. Verify your answer using a computer approach.

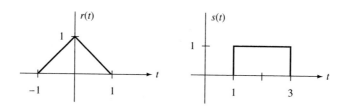

Figure 2.24   Two functions for Example 2.10.

Solution: Figure 2.25 shows the products and integrals for various values of t. The samples of the convolution, together with the interpolation between them, are shown in Fig. 2.26(a). The resulting curve is parabolic, since the function being integrated can be thought of as a ramp function.

Again, with practice, the foregoing result could be sketched with only a few key points. For example, observing the shaded region of Fig. 2.25, we see that the result increases slowly at first and then accelerates until the right edge of the square reaches the origin. The rate of increase then gets smaller again, as each incremental move to the right adds a shorter and shorter strip to the product.

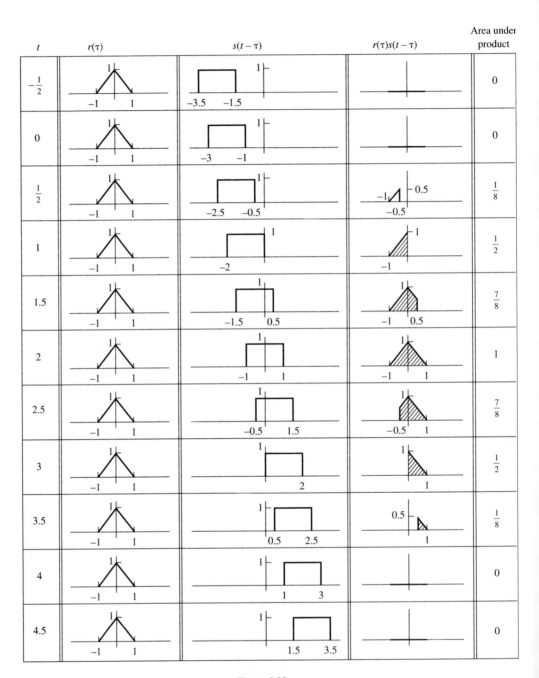

Figure 2.25

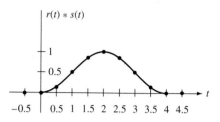

Figure 2.26(a) Graphical Convolution for Example 2.10.

We now use Mathcad to verify the graphical solution. The input code is as follows:

```
t:=-2,-1.99..2
U(t):=F(t)
r(t):-(t+1)?(U(t+1)-U(t))+(-t+1)?(U(t)-U(t-1))
s(t):U(t-1)-U(t-3)
t1:-=1,-.9..5
        1
c(t1):=er(tau)?s(t1-tau)dtau
       -1
```

Note that r(t) is the triangular pulse defined using gated ramps and s(t) is the square pulse. F(t) is the Mathcad expression for the unit step. The resulting convolution, c(t1), is shown in Figure 2.26(b). Note that it matches the result we obtained using graphical convolution.

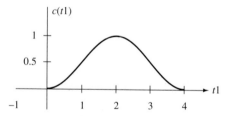

Figure 2.26(b) Result for Example 2.10.

We now investigate the operation of convolving an arbitrary time function with $\delta(t)$:

$$\delta(t)*s(t) = \int_{-\infty}^{\infty} \delta(\tau)s(t-\tau)d\tau = s(t-0) = s(t) \tag{2.30}$$

This shows that any function convolved with an impulse remains unchanged.

If we convolve $s(t)$ with the shifted impulse, $\delta(t-t_0)$, we obtain

$$\delta(t-t_0)*s(t) = \int_{-\infty}^{\infty} \delta(\tau-t_0)s(t-\tau)d\tau = s(t-t_0) \tag{2.31}$$

In sum, convolution of $s(t)$ with an impulse function does not change the functional form of $s(t)$. The only thing it may do is cause a time shift in $s(t)$ if the impulse does not occur at $t = 0$.

Now that we have a feel for the operation known as convolution, let us return to our study of the Fourier transform.

The *convolution theorem* states that the Fourier transform of a function of time that is the convolution of two functions of time is equal to the product of the two corresponding Fourier transforms. That is, if

$$r(t) \leftrightarrow R(f)$$

$$s(t) \leftrightarrow S(f)$$

then

$$r(t)*s(t) \leftrightarrow R(f)S(f) \tag{2.32}$$

The proof of the theorem is straightforward. We simply evaluate the Fourier transform of the convolution:

$$
\begin{aligned}
\mathcal{F}[r(t)*s(t)] &= \int_{-\infty}^{\infty} e^{-j2\pi ft} \left[\int_{-\infty}^{\infty} r(\tau)s(t-\tau)d\tau \right] dt \\
&= \int_{-\infty}^{\infty} r(\tau) \left[\int_{-\infty}^{\infty} e^{-j2\pi ft}s(t-\tau)dt \right] d\tau
\end{aligned}
\tag{2.33}
$$

We now make a change of variables in the inner integral by letting $t - \tau = k$. We then have

$$\mathcal{F}[r(t)*s(t)] = \int_{-\infty}^{\infty} r(\tau)e^{-j2\pi f\tau} \left[\int_{-\infty}^{\infty} s(k)e^{-j2\pi fk}dk \right] d\tau \tag{2.34}$$

The integral in the brackets is simply $S(f)$. Since $S(f)$ is not a function of τ, it can be pulled out of the outer integral. This yields the desired result,

$$\mathcal{F}[r(t)*s(t)] = S(f)\int_{-\infty}^{\infty} r(\tau)e^{-j2\pi f\tau}d\tau = S(f)R(f) \tag{2.35}$$

and the theorem is proved.

Convolution is an operation performed between two functions. These need not be functions of the independent variable t; we could just as easily have convolved two Fourier transforms together to get a third function of f:

$$H(f) = R(f)*S(f) = \int_{-\infty}^{\infty} R(k)S(f-k)dk \tag{2.36}$$

Since the integral defining the Fourier transform and that yielding the inverse transform are quite similar, one might guess that convolution of two transforms corresponds to multiplication of the two corresponding functions of time. Indeed, one can prove, in an analogous way to the previous proof, that

$$R(f)*S(f) \leftrightarrow r(t)s(t) \tag{2.37}$$

To prove this, simply calculate the inverse Fourier Transform of $R(f)*S(f)$. Equation (2.35) is called the *time convolution theorem* and Eq. (2.37) the *frequency convolution theorem*.

Example 2.11

Use the convolution theorem to evaluate the integral

$$\int_{-\infty}^{\infty} \frac{\sin 3\tau}{\tau} \frac{\sin(t-\tau)}{t-\tau} d\tau$$

Solution: We recognize that the integral represents the convolution

$$\frac{\sin 3t}{t} * \frac{\sin t}{t}$$

The transform of the integral is therefore the product of the transforms of the two functions (sin 3*t*)/*t* and (sin *t*)/*t*. These two transforms may be found in Appendix II. They and their product are sketched in Fig. 2.27.

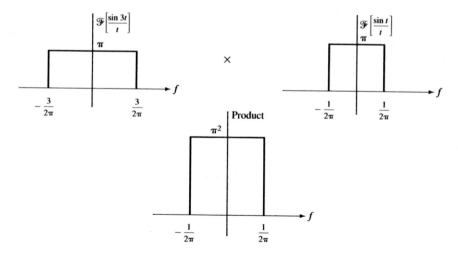

Figure 2.27   Transforms and Product for Example 2.11.

The function of time corresponding to the convolution is simply the inverse transform of the product. This is seen to be

$$\frac{\pi \sin t}{t}$$

Note that when $(\sin t)/t$ is convolved with $(\sin 3t)/t$, the only change that takes place is the addition of a scale factor π. In fact, if $(\sin t)/t$ were convolved with $(\sin 3t)/\pi t$, it would not have changed at all! This surprising result is no accident: There are entire classes of functions that remain unchanged after convolution with $(\sin t)/\pi t$. If this were not true, many of the most basic communication systems could never function.

2.5.2 Parseval's Theorem

There is little similarity between the waveshape of a function and that of its Fourier transform. However, certain relationships do exist between the energy of a function of time and the energy of its transform. Here, we use *energy* to denote the integral of the square of the function. This represents the amount of energy, in watt-seconds, dissipated in a 1Ω resistor if the time signal represents the voltage across or the current through the resistor. Such a relationship proves useful if we know the *transform* of a function of time and wish to know the energy of the function: We do not need to go through the effort of evaluating the inverse transform.

Parseval's theorem, which states this kind of relationship, is derived from the frequency convolution theorem. Starting with that theorem, we have

$$r(t)s(t) \leftrightarrow R(f)*S(f)$$

$$\mathscr{F}[r(t)s(t)] = \int_{-\infty}^{\infty} r(t)s(t)e^{-j2\pi ft}\, dt \tag{2.38}$$

$$= \int_{-\infty}^{\infty} R(k)S(f - k)\, dk$$

Since Eq.(2.38) holds for all values of f, we can let $f = 0$. For this value of f, we then obtain

$$\int_{-\infty}^{\infty} r(t)s(t)dt = \int_{-\infty}^{\infty} R(k)S(-k)\, dk \tag{2.39}$$

Equation (2.39) is one form of Parseval's formula. It can be made to relate to energy by further taking the special case of

$$s(t) = r*(t)$$

The Fourier transform of the conjugate, $\mathscr{F}[r*(t)]$, is given by the conjugate of the transform reflected around the vertical axis, that is, $R*(-f)$. You should take the time now to prove this statement.

Using the preceding result in Eq. (2.39), we find that

$$\int_{-\infty}^{\infty} \left|r^2(t)\right| dt = \int_{-\infty}^{\infty} \left|R^2(f)\right| df \tag{2.40}$$

We have used the fact that the product of a function and its complex conjugate is equal to the square of the magnitude of the function.[5] (Convince yourself that the square of the magnitude is the same as the magnitude of the square of a complex number.)

Equation (2.40) shows that the energy of a function of time is equal to the energy of its Fourier transform.

2.6 PROPERTIES OF THE FOURIER TRANSFORM

We now illustrate some of the more important properties of the Fourier transform. One can certainly go through technical life without making use of any of these properties, but to do so would involve considerable repetition and extra work. The properties allow us to derive something *once* and then to use the result for a variety of applications. They also allow us to predict the behavior of various systems.

2.6.1 Real/Imaginary–Even/Odd

The following table summarizes properties of the Fourier transform based upon observations made upon the function of time.

	Function of Time	Fourier Transform
A	Real	Real part even, imaginary part odd
B	Real and even	Real and even
C	Real and odd	Imaginary and odd
D	Imaginary	Real part odd, imaginary part even
E	Imaginary and even	Imaginary and even
F	Imaginary and odd	Real and odd

We now prove these properties. The defining integral of the Fourier transform can be expanded using Euler's identity as follows:

$$S(f) = \int_{-\infty}^{\infty} s(t)e^{-j2\pi ft}\, dt$$

$$= \int_{-\infty}^{\infty} s(t)\cos 2\pi ft\, dt - j \int_{-\infty}^{\infty} s(t)\sin 2\pi ft\, dt \qquad (2.41)$$

$$= R + jX$$

---

[5]The time signals we deal with in the real world of communication are real functions of time. However, as in basic circuit analysis, complex mathematical functions are often used to represent sinusoids. A complex number is used for the magnitude and phase angle of a sinusoid. Therefore, although complex signals do not exist in real life, they are often used in "paper" solutions of problems.

R is an even function of f, since, when f is replaced with $-f$, the function does not change. Similarly, X is an odd function of f.

If $s(t)$ is first assumed to be real, R becomes the real part of the transform and X is the imaginary part. Thus, *property A* is proved.

If, in addition to being real, $s(t)$ is even, then $X = 0$. This is true because the integrand in X is odd (the product of an even and an odd function) and integrates to zero. Hence, *property B* is proved.

If $s(t)$ is now real and odd, the same argument applies, but $R = 0$. This proves *property C*.

Now we let $s(t)$ be imaginary. X then becomes the imaginary part of the transform, and R is the real part. From this simple observation, *properties D, E, and F* are easily verified.

2.6.2 Time Shift

The Fourier transform of a shifted function of time is equal to the product of the transform of that function with a complex exponential. That is,

$$s(t - t_0) \leftrightarrow e^{-j2\pi f t_0} S(f) \tag{2.42}$$

Proof. The proof follows directly from evaluation of the transform of $s(t - t_0)$.

$$\mathcal{F}[s(t - t_0)] = \int_{-\infty}^{\infty} s(t - t_0)e^{-j2\pi f t} dt = \int_{-\infty}^{\infty} s(\tau)e^{-j2\pi f(\tau + t_0)} d\tau \tag{2.43}$$

The second integral follows from a change of variables, letting $\tau = (t - t_0)$. We now pull the part that does not depend on τ to the front of the integral and note that the remaining part is a Fourier transform of $s(t)$. Finally, we get

$$\mathcal{F}[s(t - t_0)] = e^{-j2\pi f t_0} S(f) \tag{2.44}$$

Example 2.12

Find the Fourier transform of

$$s(t) = \begin{cases} 1, & 0 < t < 2 \\ 0, & \text{otherwise} \end{cases}$$

The function $s(t)$ is sketched in Fig. 2.28.

Solution: From the definition of the Fourier transform, we have

$$S(f) = \int_0^2 e^{-j2\pi f t} dt = \frac{e^{-j2\pi f}}{j2\pi f}(e^{j2\pi f} - e^{-j2\pi f})$$

$$= e^{-j2\pi f} \frac{\sin 2\pi f}{\pi f}$$

As expected, $S(f)$ is complex, since $s(t)$ is neither even nor odd.

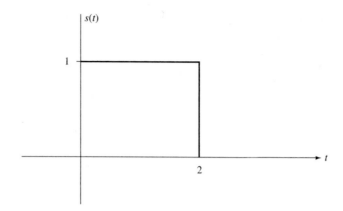

Figure 2.28 $s(t)$ for Example 2.12.

The result of Example 2.12 could have been derived in one step using the answer from Example 2.5 and the time-shift property. The $s(t)$ of Example 2.12 is the same as that of Example 2.5 (with $A = \alpha = 1$) except for a time shift of 1 sec.

2.6.3 Frequency Shift

The function of time corresponding to a shifted Fourier transform is equal to the product of the function of time corresponding to the unshifted transform and a complex exponential. That is,

$$S(f - f_0) \leftrightarrow e^{j2\pi f_0 t} s(t) \tag{2.45}$$

Proof. The proof follows directly from evaluation of the inverse transform of $S(f - f_0)$:

$$\int_{-\infty}^{\infty} S(f - f_0) e^{j2\pi ft} df = \int_{-\infty}^{\infty} S(k) e^{j2\pi t(k + f_0)} dk \tag{2.46}$$

In the second integral, we have made a change of variables, letting $k = f - f_0$. We now pull the part of the integrand that does not depend upon k in front of the integral and recognize that the remaining integral is the inverse Fourier transform of $s(t)$. This yields

$$S(f - f_0) \leftrightarrow e^{j2\pi f_0 t} s(t) \tag{2.47}$$

Example 2.13

Find the Fourier transform of

$$s(t) = \begin{cases} e^{j2\pi t}, & |t| < 1 \\ 0, & \text{otherwise} \end{cases}$$

Solution: This $s(t)$ is the same as that of Example 2.5 (with $A = \alpha = 1$), except for a multi-plying factor of $e^{j2\pi t}$. The frequency shift theorem is used to find that the transform is the original transform shifted by 1 unit of frequency. We therefore take the transform found in Example 2.5 and substitute $(f - 1)$ for f:

$$S(f) = \frac{\sin 2\,\pi(f - 1)}{\pi(f - 1)}$$

This result is illustrated in Fig. 2.29.

Figure 2.29 $S(f)$ for Example 2.13.

Note that in Example 2.13, the function of time is neither even nor odd. However, the Fourier transform turned out to be a real function of time. Such a situation arises only when the function of time is not real.

2.6.4 Linearity

Linearity is undoubtedly the most important property of the Fourier transform.

The Fourier transform of a linear combination of functions of time is a linear combination of the corresponding Fourier transforms. That is,

$$as_1(t) + bs_2(t) \leftrightarrow aS_1(f) + bS_2(f) \tag{2.48}$$

where a and b are any constants.

Proof. The proof follows directly from the definition of the Fourier transform and from the fact the integration is a linear operation:

$$\int_{-\infty}^{\infty} [as_1(t) + bs_2(t)]e^{-j2\pi ft}dt = a\int_{-\infty}^{\infty} s_1(t)e^{-j2\pi ft}dt + b\int_{-\infty}^{\infty} s_2(t)e^{-j2\pi ft}dt \tag{2.49}$$

$$= aS_1(f) + bS_2(f)$$

Example 2.14

Find the Fourier transform of

$$s(t) = \begin{cases} 1, & -1 < t < 0 \\ 2, & 0 < t < 1 \\ 1, & 1 < t < 2 \\ 0, & \text{otherwise} \end{cases}$$

This function is sketched in Fig. 2.30.

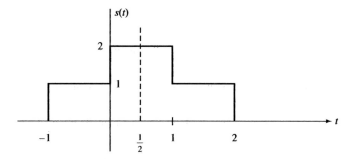

Figure 2.30 $s(t)$ for Example 2.14.

Solution: We use the linearity property and observe that $s(t)$ is the sum of the function of time in Example 2.5 with that in Example 2.12. Therefore, the transform is given by the sum of the two transforms:

$$S(f) = \frac{\sin 2\pi f}{\pi f}[1 + e^{-j2\pi f}]$$

Since the given function of time would be even if shifted to the left by 0.5 sec, we can rewrite this equation in a more descriptive form by factoring out $e^{-j\pi f}$.

$$S(f) = 2\frac{\sin 2\pi f \cos \pi f}{\pi f}e^{-j\pi f}$$

2.6.5 Modulation Theorem

The modulation theorem is very closely related to the frequency shift theorem. We treat it separately because it forms the basis of the entire study of amplitude modulation.

The result of multiplying a function of time by a pure sinusoid is to shift the original transform both up and down by the frequency of the sinusoid (and to cut the amplitude in half).

Proof. We start the proof of this theorem by assuming that $s(t)$ is given, together with its associated Fourier transform. The function $s(t)$ is then multiplied by a cosine waveform to yield

$$s(t)\cos 2\pi f_0 t$$

where the frequency of the cosine is f_0. The Fourier transform of this waveform is given by

$$\mathscr{F}[s(t)\cos 2\pi f_0 t] = \frac{1}{2} S(f - f_0) + \frac{1}{2} S(f + f_0) \tag{2.50}$$

The proof of the modulation theorem follows directly from the frequency shift theorem. We split $\cos 2\pi f_0 t$ into two exponential components and then apply the frequency shift theorem to each component:

$$s(t)\cos 2\pi f_0 t = \frac{1}{2} s(t)e^{j2\pi f_0 t} + \frac{1}{2} s(t)e^{-j2\pi f_0 t}$$
$$\leftrightarrow \frac{1}{2} S(f - f_0) + \frac{1}{2} S(f + f_0) \tag{2.51}$$

2.6.6 Scaling in Time and Frequency

We are rapidly approaching the point of diminishing returns in presenting properties of the Fourier transform. At some point, it is worth dealing with the individual properties as they arise. We shall terminate our exploration with a companion set of two properties referred to as *time and frequency scaling*. The usefulness of these properties arises when you take a function of time or a Fourier transform and either stretch or compress it along the horizontal axis. Thus, if, for example, you already know the Fourier transform of a pulse with width two units (as in Fig. 2.28), you need not do any further calculations to find the Fourier transform of a pulse of any other width.

Time Scaling

Suppose we already know that the Fourier transform of $s(t)$ is $S(f)$. We wish to find the Fourier transform of $s(at)$, where a is a real scaling factor. Thus, if, for example, $a = 2$, we are compressing the function by a factor of 2 along the t-axis, and if $a = 0.5$, we are expanding it by a factor of 2. The result can be derived directly from the definition of the Fourier transform as follows:

$$\mathscr{F}[s(at)] = \int_{-\infty}^{\infty} s(at)e^{-j2\pi ft}\, dt = \int_{-\infty}^{\infty} s(\tau)e^{-j2\pi f\tau/a}\frac{d\tau}{a} \tag{2.52}$$

The latter integral results from a change of variables, letting $s = at$. The integral is now recognized as

$$\mathscr{F}[s(at)] = \frac{1}{a} S\left(\frac{f}{a}\right) \tag{2.53}$$

The result represents a complementary operation on the frequency axis and an amplitude scaling. Thus, if, for example, $a = 2$, the time axis is compressed. In finding the transform, we expand the frequency axis by a factor of 2 and scale the amplitude of the transform by dividing by 2.

Frequency Scaling

If it is already known that the Fourier transform of $s(t)$ is $S(f)$, then the time signal that has $S(af)$ as its transform, where a is a real scaling factor, is given by

$$\mathcal{F}^{-1}[S(af)] = \frac{1}{a} s\left(\frac{t}{a}\right) \tag{2.54}$$

This is proved directly from the inverse Fourier transform integral and is left as an exercise for the student.

Think about how you might have proven the frequency-scaling result from the time-scaling property. (*Hint*: What if $a' = 1/a$ in the time scale equation?)

2.7 PERIODIC FUNCTIONS

In Example 2.7, the Fourier transform of the cosine function was found to be composed of two impulses, one occurring at the frequency of the cosine and the other at the negative of this frequency. We will now show that the Fourier transform of any periodic function of time is a discrete function of frequency. That is, the transform is nonzero only at discrete points along the f-axis. The proof follows from Fourier series expansions and the linearity of the Fourier transform.

Suppose we find the Fourier transform of a function $s(t)$ that is periodic with period T. We can express the function in terms of the complex Fourier series representation

$$s(t) = \sum_{n=-\infty}^{\infty} c_n e^{jn2\pi f_0 t} \tag{2.55}$$

where

$$f_0 = \frac{1}{T}$$

Previously, we established the transform pair

$$Ae^{j2\pi f_0 t} \leftrightarrow A\delta(f - f_0) \tag{2.56}$$

From this transform pair and the linearity property of the transform, we have

$$\mathcal{F}[s(t)] = \sum_{n=-\infty}^{\infty} c_n \mathcal{F}[e^{jn2\pi f_0 t}] \tag{2.57}$$

This transform is shown in Fig. 2.31 for a representative $s(t)$. Note that the c_n are complex numbers, so the sketch is intended for conceptual purposes only. If the function of time is real and even, the c_n will be real.

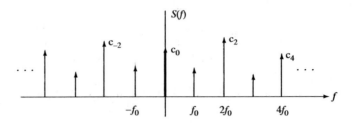

Figure 2.31   Transform of periodic $s(t)$.

The foregoing proof shows that the Fourier transform of a periodic function of time is a train of equally spaced impulses, each of whose strength is equal to the corresponding Fourier coefficient c_n.

Example 2.15

Find the Fourier transform of the periodic function made up of unit impulses, as shown in Fig. 2.32. The function is

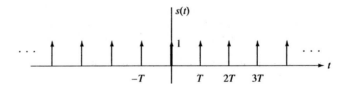

Figure 2.32   Periodic $s(t)$ for Example 2.15.

$$s(t) = \sum_{n=-\infty}^{\infty} \delta(t - nT)$$

Solution: The Fourier transform is found directly from Eq. (2.57). We have

$$S(f) = \sum_{n=-\infty}^{\infty} c_n \delta(f - nf_0)$$

where

$$f_0 = \frac{1}{T}$$

$$c_n = \frac{1}{T} \int_{-\frac{T}{2}}^{\frac{T}{2}} s(t) e^{-jn2\pi f_0 t} dt$$

Within the range of integration, the only contribution of $s(t)$ is that due to the impulse at the origin. Therefore,

$$c_n = \frac{1}{T} \int_{-\frac{T}{2}}^{\frac{T}{2}} \delta(t) e^{-jn2\pi f_0 t} dt = \frac{1}{T}$$

Finally, the Fourier transform of the pulse train is given by

$$S(f) = \frac{1}{T} \sum_{n=-\infty}^{\infty} \delta(f - nf_0)$$

where

$$f_0 = \frac{1}{T}$$

The function of Example 2.15 has an interesting Fourier series expansion. All of the coefficients are equal. Each frequency component possesses the same amplitude as every other component. This is analogous to the observation that the Fourier transform of a single impulse is a constant. The similarity leads us to examine the relationship between the Fourier transform of a periodic function and the Fourier transform of one period of the function.

Suppose that $s(t)$ represents a single period of the periodic function $s_p(t)$. Then we can express the periodic function as a sum of shifted versions of $s(t)$:

$$s_p(t) = \sum_{n=-\infty}^{\infty} s(t - nT) \tag{2.58}$$

Since convolution with an impulse simply shifts the original function, Eq. (2.58) can be rewritten as

$$s_p(t) = s(t) * \sum_{n=-\infty}^{\infty} \delta(t - nT) \tag{2.59}$$

Convolution in the time domain is equivalent to multiplication of the Fourier transforms. The Fourier transform of the train of impulses was found in Example 2.15. Transforming Eq. (2.59) then yields

$$S_p(t) = S(f) \sum_{n=-\infty}^{\infty} \frac{1}{T} \delta(f - nf_0) \tag{2.60}$$

Equation (2.60) shows that the Fourier transform of the periodic function is simply a sampled and scaled version of the transform of a single period of the waveform.

PROBLEMS

2.1.1 Evaluate the Fourier series expansion of each of the periodic functions shown in Fig. P2.1.1. Use either the complex exponential or the trigonometric form of the series.

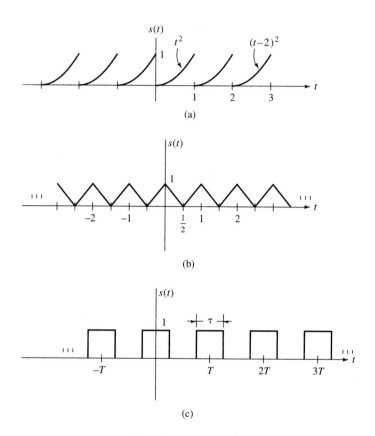

(a)

(b)

(c)

Figure P2.1.1

2.1.2 Evaluate the Fourier series expansion of the function shown in Fig. 2.3.

2.1.3 Evaluate the Fourier series representation of the periodic function

$$s(t) = 2\sin \pi t + 3\sin 2\pi t$$

2.1.4 The periodic function of Fig. P2.1.1(c) is expressed in a trigonometric Fourier series. Find the error if only three terms of the Fourier series are used. Repeat for four terms and five terms.

2.1.5 Find the Fourier series expansion of a gated sinusoid, as shown in Fig. P2.1.5.

2.1.6 The triangular waveform shown in Fig. P2.1.1(b) forms the input to a diode circuit, as shown in Fig. P2.1.6. Assume that the diode is ideal. Find the Fourier series expansion of the current $i(t)$.

2.1.7 In Fig. P2.1.7, find a Fourier series expansion of $s(t)$ that applies for $-\tau/2 < t < \tau/2$.

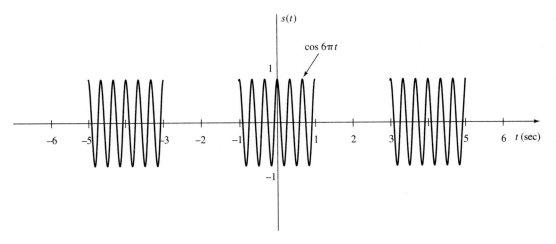

Figure P2.1.5

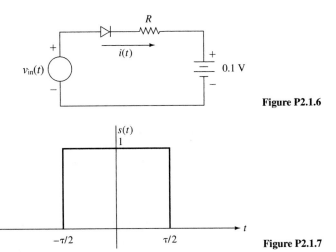

Figure P2.1.6

Figure P2.1.7

2.1.8 Find the complex Fourier series representation of $s(t) = t^2$ that applies in the interval $0 < t < 1$. How does this compare with your answer to Problem 2.1.1, Fig. P2.1.1(a)?

2.1.9 Find a trigonometric Fourier series representation of the function

$$s(t) = \cos \pi t$$

in the interval $0 < t < 2$.

2.1.10 Which of the following could *not* be the Fourier series expansion of a periodic signal?

$$s_1(t) = 2\cos t + 3\cos 3t$$

$$s_2(t) = 2\cos 0.5t + 3\cos 3.5t$$

$$s_3(t) = 2\cos 0.25t + 3\cos 0.00054t$$

$$s_4(t) = 2\cos \pi t + 3\cos \ 2t$$

$$s_5(t) = 2\cos \pi t + 3\cos \ 2\pi t$$

2.1.11 Find the error made in approximating $|\cos t|$ by a constant, $2/\pi$. Then include the first sinusoidal term in the Fourier series expansion of $|\cos t|$, and recalculate the error.

2.2.1 Find the complex Fourier spectrum of a half-wave rectified cosine wave (i.e., a waveform with every other pulse as in Fig. 2.6).

2.3.1 Evaluate the Fourier transform of each of the functions of time given in Fig. P2.3.1.

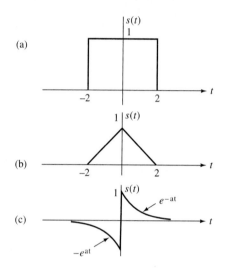

Figure P2.3.1

2.3.2 Evaluate the Fourier transform of the following functions of time:

$$s_1(t) = e^{-at}U(t)$$

$$s_2(t) = \cos 2tU(t)$$

$$s_3(t) = te^{-at}U(t)$$

2.5.1 Convolve $e^{at}U(-t)$ with $e^{-at}U(t)$. These two functions are shown in Fig. P2.5.1.

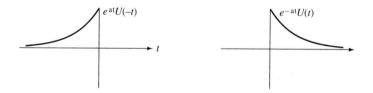

Figure P2.5.1

2.5.2 Convolve together the two functions shown in Fig. P2.5.2, using the convolution integral. Repeat using graphical techniques.

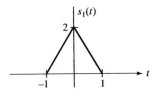 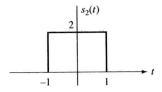

Figure P2.5.2

2.5.3 Find

$$\int_{-\infty}^{\infty} e^{-2t}U(t)\, dt$$

using Parseval's theorem. *Hint*: Use the fact that

$$e^{-2t}U(t) = \left| e^{-t}U(t) \right|^2$$

2.5.4 Evaluate the following integral using Parseval's theorem:

$$\int_{-\infty}^{\infty} \frac{\sin 2t}{t} \cos 1{,}000\, t \ \frac{\sin t}{t} \cos 2{,}000\, t\, dt$$

2.5.5 Show that if $S(f) = 0$ for $|f| > f_m$, then

$$s(t) * \frac{\sin at}{\pi t} = s(t)$$

provided that $a/2\pi > f_m$.

2.5.6 Evaluate the following integrals:

$$\int_{-\infty}^{\infty} \frac{\sin 3\tau}{\tau}\, \delta(t - \tau)\, d\tau$$

$$\int_{-\infty}^{\infty} \frac{\sin 3\,(\tau - 3)}{\tau - 3}\, \frac{\sin 5\,(t - \tau)}{t - \tau}\, d\tau$$

$$\int_{-\infty}^{\infty} \delta(t - 1)(t^3 + 4)\, dt$$

2.6.1 (a) Write the convolution of $s(t)$ with $U(t)$ in integral form. See whether you can identify this as the integral of $s(t)$.

(b) What is the transform of $s(t)*U(t)$? Solve this using the convolution theorem.

2.6.2 Any arbitrary function can be expressed as the sum of an even and odd function; that is,

$$s(t) = s_e(t) + s_0(t)$$

where

$$s_e(t) = \frac{s(t) + s(-t)}{2}$$

$$s_0(t) = \frac{s(t) - s(-t)}{2}$$

(a) Show that $s_e(t)$ is an even function and that $s_0(t)$ is an odd function.
(b) Show that $s(t) = s_e(t) + s_0(t)$.
(c) Find $s_e(t)$ and $s_0(t)$ for $s(t) = U(t)$, a unit step function.
(d) Find $s_e(t)$ and $s_0(t)$ for $s(t) = \cos20\pi t$

2.6.3 Given a function $s(t)$ that is zero for negative t, find a relationship between $s_e(t)$ and $s_0(t)$. Can this result be used to find a relationship between the real and imaginary parts of the Fourier transform of $s(t)$?

2.6.4 Evaluate the Fourier transform of $\cos5\pi t$, starting with the Fourier transform of $\cos\pi t$ and using the time-scaling property.

2.6.5 Given that the Fourier transform of $s(t)$ is $S(f)$:
(a) What is the Fourier transform of ds/dt in terms of $S(f)$?
(b) What is the Fourier transform of

$$\int_{-\infty}^{t} s(\tau)d\tau$$

in terms of $S(f)$?

2.6.6 Use the time shift property to find the Fourier transform of

$$\frac{s(t + T) - s(t)}{T}$$

From this result, find the Fourier transform of ds/dt.

2.6.7 A signal $s(t)$ is put through a gate and truncated in time as shown in Fig. P2.6.7. The gate is closed for $1 < t < 2$. Therefore,

$$s_{\text{out}}(t) = \begin{cases} s(t), & 1 < t < 2 \\ 0, & \text{otherwise} \end{cases}$$

Find the Fourier transform of $s_{\text{out}}(t)$ in terms of $S(f)$.

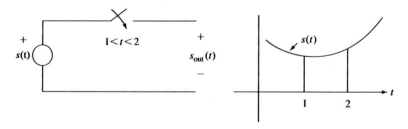

Figure P2.6.7

2.6.8 **(a)** Find the derivative with respect to time of the function shown in Fig. P2.6.8.
   (b) Find the Fourier transform of this derivative.
   (c) Find the Fourier transform of the original function from the transform of the derivative.

2.6.9 Find the Fourier transform of $s(t) = \sin 2\pi f_0 t$ from the Fourier transform of $\cos 2\pi f_0 t$ and the

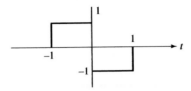

Figure P2.6.8

time shift property.

2.6.10 Find the Fourier transform of $\cos^2 2\pi f_0 t$ from the frequency convolution theorem and the transform of $\cos 2\pi f_0 t$. Check your answer by expanding $\cos^2 2\pi f_0 t$ using trigonometric identities.

2.7.1 The function $s(t)$ of Fig. P2.1.1(c) with $T = 1$ and $\tau = 0.5$ multiplies a function of time, $g(t)$, with $G(f)$ as shown in Fig. P2.7.1(a).
   (a) Sketch the Fourier transform of $g_s(t) = g(t)s(t)$.
   (b) Can $g(t)$ be recovered from $g_s(t)$?
   (c) If $G(f)$ is now as shown in Fig. P2.7.1(b), can $g(t)$ be recovered from $g_s(t)$? Explain your answer.

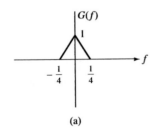

(a)

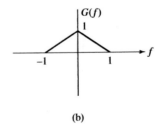

(b)

Figure P2.7.1

3

Linear Systems

3.0 PREVIEW

What We Will Cover and Why You Should Care

The previous chapter developed the basic mathematical tools required for waveform analysis. We now apply these techniques to the study of linear systems in order to determine system capabilities and features. We then will be in a position to pick those particular characteristics of linear systems which are desirable for communication applications.

This chapter begins by developing the concepts that are necessary to understand a *system function*, which is a way to describe the behavior of a particular system. We then relate the system function to the sinusoidal steady-state response of circuits. You should be familiar with this from a course on basic circuits.

The chapter then explores several specific linear systems. We concentrate on the ideal lowpass and bandpass filter, since we will be using these throughout the remainder of the text. Because ideal filters cannot be built in the real world, we then turn our attention to the approximations that must be made when practical systems are constructed.

The chapter concludes by examining the relationship between the mathematical theory and the real world. Since the Fourier transform is a mathematical concept that does not exist in real life, we examine the approximations that are possible using a laboratory spectrum analyzer.

Necessary Background

To understand the concepts presented in this chapter, you must be comfortable with the Fourier transform. You may have to review Chapter 2 from time to time.

3.1 THE SYSTEM FUNCTION

We begin by defining some common terms. A *system* is a set of rules that associates an *output* function of time with every *input* function of time. This is shown in block diagram form in Fig. 3.1.

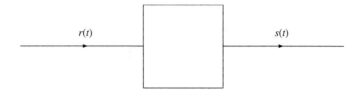

Figure 3.1   Block diagram of system.

The input, or *source*, signal is $r(t)$; $s(t)$ is the output, or *response*, signal due to the input. The actual physical structure of the system determines the exact relationship between $r(t)$ and $s(t)$.

A single-ended arrow is used as a shorthand method of relating an input to its resulting output. That is,

$$r(t) \rightarrow s(t)$$

is read, "an input $r(t)$ causes an output $s(t)$."

For example, suppose the system under study is an electric circuit. Then $r(t)$ could be an input voltage or current signal, and $s(t)$ could be a voltage or current measured anywhere in the circuit. We would not modify the block diagram representation of Fig. 3.1, even though the circuit schematic would have *two* wires for each voltage. The single lines in the figure represent signal flow.

In the special case of a two-terminal electrical network, $r(t)$ could be a sinusoidal input voltage across two terminals, and $s(t)$ could be the current flowing into one of the terminals due to the impressed voltage. In this case, the relationship between $r(t)$ and $s(t)$ is the *complex impedance* between the two terminals of the network.

Any system can be described by specifying the response $s(t)$ associated with *every* possible input $r(t)$. Obviously, this is an exhaustive process. We would certainly hope to find a much simpler way of characterizing the system.

Before introducing alternative techniques of describing systems, some additional basic definitions are needed.

A system is said to obey *superposition* if the output due to a sum of inputs is the sum of the corresponding individual outputs. That is, given that the response (output) due to an excitation (input) of $r_1(t)$ is $s_1(t)$, and that the response due to $r_2(t)$ is $s_2(t)$, then the response due to $r_1(t) + r_2(t)$ is $s_1(t) + s_2(t)$.

Restating this, we may say that a system which obeys superposition has the property that if

$$r_1(t) \rightarrow s_1(t)$$

and

$$r_2(t) \rightarrow s_2(t)$$

then

$$r_1(t) + r_2(t) \rightarrow s_1(t) + s_2(t)$$

Some thought should convince you that in order for a circuit to obey superposition, the source-free, or transient, response (the response due to the initial conditions) must be zero. (Let $r_2(t) = 0$ to prove this.) In practice, one often replaces a circuit having nonzero initial conditions with one that has zero initial conditions. Additional sources are added to simulate the contributions of the initial conditions.

A concept closely related to superposition is *linearity*. Assume again that $r_1(t) \rightarrow s_1(t)$ and $r_2(t) \rightarrow s_2(t)$. The system is said to be linear if the relationship

$$ar_1(t) + br_2(t) \rightarrow as_1(t) + bs_2(t)$$

holds for all values of the constants a and b. In the remainder of this text, we will use the words *linearity* and *superposition* interchangeably.

A system is said to be *time invariant* if the response due to an input is not dependent upon the actual time of occurrence of the input. That is, a system is time invariant if a time shift in input signal causes an equal time shift in the output waveform. In symbolic form, if

$$r(t) \rightarrow s(t)$$

then

$$r(t - t_0) \rightarrow s(t - t_0)$$

for all real t_0.

A sufficient condition for an electrical network to be time invariant is that its component values do not change with time (assuming unchanging initial conditions). That is, if all resistances, capacitances, and inductances remain constant, then the system is time invariant.

Returning to the task of characterizing a system, we shall see that for a time-invariant linear system, a very simple description is possible. That is, instead of requiring that we know the response due to *every* possible input, it will turn out that we need know only the output for one *test* input.

We showed earlier that convolution of any function with an impulse yields the original function. That is,

$$r(t) = r(t)*\delta(t) \tag{3.1}$$

$$= \int_{-\infty}^{\infty} r(\tau)\delta(t - \tau)d\tau$$

Although one must always use extra caution in working with impulses, let us assume that the integral can be considered a limiting case of a sum, so that

$$r(t) = \lim_{\Delta\tau \to 0} \sum_{n=-\infty}^{\infty} r(n\Delta\tau)\delta(t - n\Delta\tau)\Delta\tau \tag{3.2}$$

Equation (3.2) represents a weighted sum of delayed impulses. Suppose that this weighted sum forms the input to a linear time-invariant system. The output would then be a weighted sum of delayed outputs due to a single impulse.

Suppose now that we know the system's output due to a single impulse. Let us denote that output as $h(t)$, the *impulse response*. Then the output due to the input of Eq. (3.2) is given by

$$s(t) = \lim_{\Delta\tau \to 0} \sum_{n=-\infty}^{\infty} r(n\Delta\tau)h(t - n\Delta\tau)\Delta\tau \tag{3.3}$$

If we take the limit of the sum, the latter becomes an integral, and we have

$$s(t) = \int_{-\infty}^{\infty} r(\tau)h(t - \tau)d\tau = r(t)*h(t) \tag{3.4}$$

Equation (3.4) states that the output due to *any* input is found by convolving that input with the system's response to an impulse. All we need to know about the system is its impulse response. Equation (3.4) is known as the *superposition integral equation.*

The Fourier transform of the impulse is unity. Therefore, in an intuitive sense, the impulse contains all frequencies to an equal degree. This observation hints at the impulse's suitability as a test function for system behavior. On the negative side, it is not possible to produce a perfect impulse in real life. We can only approximate it with a large-amplitude, very narrow pulse.

Taking the Fourier transform of Eq. (3.4) yields

$$S(f) = R(f)H(f)$$
$$H(f) = \frac{S(f)}{R(f)} \tag{3.5}$$

The Fourier transform of the impulse response is thus the ratio of the output Fourier transform to the input Fourier transform. It is given the name *transfer function* or *system function,* and it completely characterizes a linear time-invariant system.

3.2 COMPLEX TRANSFER FUNCTION

Sinusoidal steady-state analysis defines the *complex transfer function* of a system as the ratio of the output phasor to the input phasor. A *phasor* is a complex number representing the amplitude and phase of a sinusoid. The ratio of phasors is a complex function of frequency. In the special case in which the input is a current flowing between two terminals and the output is the voltage across these terminals, the complex transfer function is the *complex impedance* between the two terminals.

As an example, consider the circuit of Fig. 3.2, where $i_1(t)$ is the input and $v(t)$ is the output. The transfer function is given by

$$H(f) = \frac{4j\pi f}{1 + 4j\pi f} \tag{3.6}$$

Alternatively, if $i_2(t)$ is the output and $i_1(t)$ is the input, the transfer function becomes

$$H(f) = \frac{1}{1 + 4j\pi f} \tag{3.7}$$

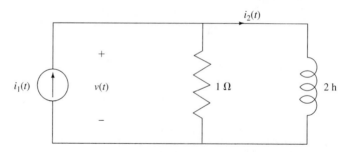

Figure 3.2   Circuit to illustrate transfer function.

We have used the same symbol, $H(f)$, for the transfer function as was used in the previous section to denote the Fourier transform of the impulse response. This is not accidental: *The two expressions are identical.* This statement can be proven by considering a system input of

$$r(t) = e^{j2\pi f_0 t} \tag{3.8}$$

This input is not physically realizable, because it is a complex function of time. Nonetheless, the system equations apply to complex inputs. We can compare the system output using sinusoidal steady-state analysis with that obtained using Fourier transform analysis. Note that the Fourier transform of the complex input function is a shifted impulse. In this manner, we can show that the two $H(f)$ functions are identical.

Example 3.1

In the circuit of Fig. 3.3, the capacitor is initially uncharged. Find $i(t)$, assuming that $v(t) = \delta(t)$.

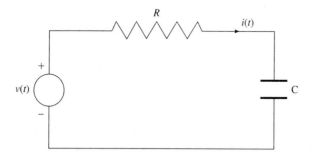

Figure 3.3   Circuit for Example 3.1.

Solution: Since the input to the circuit is an impulse, we are being asked to find the impulse response, $h(t)$.

$H(f)$ can be found using sinusoidal steady-state analysis, where the capacitor is replaced by an impedance of $1/j2\pi fC$. The transfer function is then simply the reciprocal of the circuit input impedance:

$$H(f) = \frac{1}{R + 1/j2\pi fC} = \frac{j2\pi fC}{1 + j2\pi fRC}$$

We now need to find the inverse Fourier transform of $H(f)$. There are a number of ways to do this. One way is to look in a table of Fourier transforms, such as the one in Appendix II of this text. Indeed, there are much more extensive tables available in various handbooks, and you might be fortunate enough to find an entry that applies to your problem. A second technique is to try to evaluate the inverse Fourier transform integral, either in closed form or using computer approximations (e.g., *Mathcad*). Yet another way is to use the FFT. To do so, however, you must become aware of its properties and the relationship between frequency sampling and time sampling.

We shall use tables to solve this particular problem. We rewrite the equations for $H(f)$ borrowing the expansion technique commonly used in Laplace transform analysis:

$$H(f) = \frac{1}{R} - \frac{1/R}{1 + j2\pi fRC}$$

$$h(t) = i(t) = \frac{1}{R}\delta(t) - \frac{1}{R^2 C}e^{-t/RC}U(t)$$

This waveform is illustrated in Fig. 3.4.

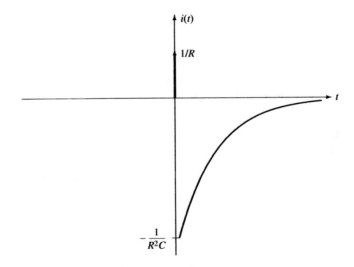

Figure 3.4 $i(t)$ for Example 3.1.

Note that the impulse in $i(t)$ has appeared without additional analysis effort. This is noteworthy, since classical circuit analysis techniques handle impulses with a great deal of difficulty.

The result of Example 3.1 is valid only for zero initial charge on the capacitor. Otherwise, superposition is violated (prove it!), and the output is not the convolution of $h(t)$ with the input.

This apparent shortcoming of Fourier transform analysis of systems with nonzero initial conditions is circumvented by treating initial conditions as sources. The consideration of initial conditions is not critical to most communication systems, so we will usually assume zero initial conditions.

3.3 FILTERS

In ordinary language, the word *filter* refers to the removal of the undesired parts of something. In linear system theory, it was probably originally applied to systems that eliminate undesired frequency components from a time waveform. The term has evolved to include systems that simply weight the various frequency components of a signal.

Many of the communication systems we discuss contain *ideal distortionless filters*. We therefore begin our study by defining *distortion*.

A *distorted* time signal is a time signal whose basic shape has been altered. $r(t)$ can be multiplied by a constant and shifted in time without changing the basic shape of the waveform.

In mathematical terms, we consider $Ar(t - t_0)$, where A and t_0 are any real constants, to be an undistorted version of $r(t)$. Of course, A cannot equal zero. The Fourier transform of $Ar(t - t_0)$ is found from the time shift property:

$$Ar(t - t_0) \leftrightarrow Ae^{-j2\pi f t_0}R(f) \tag{3.9}$$

We can consider this the output of a linear system with input $r(t)$ and system function

$$H(f) = Ae^{-j2\pi f t_0} \tag{3.10}$$

This is illustrated in Fig. 3.5.

Since $H(f)$ is complex, we have plotted its magnitude and phase. The real and imaginary parts would also have sufficed, but would not have been as instructive.

Let us view Fig. 3.5 intuitively. It seems reasonable that the magnitude function turned out to be a constant. This indicates that all frequencies of $r(t)$ are multiplied by the same factor. But why did the phase turn out to be a linear function of frequency? Why aren't the various values of phase all shifted by the same amount? The answer is clear from a simple example. Suppose we wish to shift a 1-Hz sinusoid $r(t) = \cos 2\pi t$ by 1 second. This represents a phase shift of 2π radians, or $360°$. If we now wish to shift a signal of twice the frequency, $r(t) = \cos 4\pi t$, by the same 1 second, we would have to shift the phase by 4π radians, or $720°$. If we shifted the second signal by only $360°$, it would only be delayed by 0.5 second instead of the required 1 second.

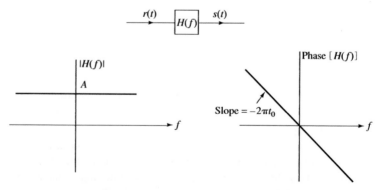

Figure 3.5   Characteristics of distortionless system.

Now consider a general signal composed of many frequency components. If we delay all components by the same angular phase, we would not be delaying them by the same amount of time, and the signal would be severely distorted. In order to delay by the same amount of time, the phase shift must be proportional to frequency.

3.3.1 Ideal Lowpass Filter

An *ideal lowpass filter* is a linear system that acts like an ideal distortionless system, provided that the input signal contains no frequency components above the *cutoff* frequency of the filter. Frequency components above this cutoff are completely blocked from appearing at the output. The cutoff frequency is the maximum frequency passed by the filter, and we denote it as f_m. The system function is then given by

$$H(f) = \begin{cases} Ae^{-j2\pi f t_0}, & |f| < f_m \\ 0, & |f| > f_m \end{cases} \tag{3.11}$$

The transfer function of the ideal lowpass filter is shown in Fig. 3.6. Note that since $h(t)$ is real, the magnitude of $H(f)$ is even and the phase is odd.

The impulse response of the ideal lowpass filter is found by computing the inverse Fourier transform of $H(f)$:

$$h(t) = \int_{-f_m}^{f_m} Ae^{-j2\pi f t_0} e^{j2\pi f t} \, df \tag{3.12}$$

$$= \frac{A \sin 2\pi f_m(t - t_0)}{\pi(t - t_0)}$$

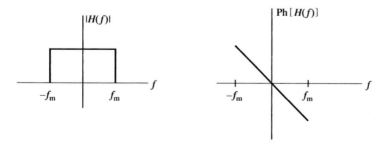

Figure 3.6   Ideal lowpass filter characteristic.

This impulse response is shown in Fig. 3.7. The amount of delay, t_0, is proportional to the slope of the phase characteristic. The cutoff frequency is proportional to the peak of $h(t)$ and inversely proportional to the spacing between zero-axis crossings of the function. That is, as f_m increases, the peak of $h(t)$ increases, and the width of the shaped pulse decreases—the response gets taller and skinnier.

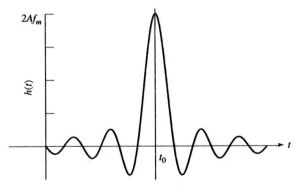

Figure 3.7   Impulse response of ideal lowpass filter.

3.3.2 Ideal Bandpass Filter

Rather than pass frequencies between zero and f_m, as in the case of the lowpass filter, the ideal bandpass filter passes frequencies between two nonzero frequencies, f_L and f_H. The filter acts like an ideal distortionless system, provided that the input signal contains no frequency components outside of the filter *passband*. The system function of the ideal bandpass filter is

$$H(f) = \begin{cases} Ae^{-j2\pi f t_0}, & f_L < |f| < f_H \\ \\ 0, & \text{otherwise} \end{cases} \tag{3.13}$$

This function is illustrated in Fig. 3.8.

The impulse response of the bandpass filter can be found by evaluating the inverse Fourier transform of $H(f)$. Alternatively, we can save a lot of work by deriving the bandpass filter impulse response from the lowpass filter impulse response and the frequency-shifting theorem. If we denote the lowpass system function as $H_{lp}(f)$, the bandpass function can be expressed as

$$H(f) = H_{lp}\left(f - \frac{f_L + f_H}{2}\right) + H_{lp}\left(f + \frac{f_L + f_H}{2}\right) \tag{3.14}$$

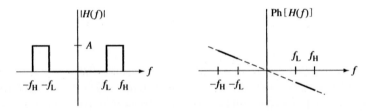

Figure 3.8   Ideal bandpass filter characteristic.

Figure 3.9 shows the relationship between $H(f)$ and $H_{lp}(f)$.

We have illustrated the system functions as if they were real functions of frequency. That is, for purposes of the derivation, we are assuming that $t_0 = 0$. This approach is justified by the time invariance of the system. When we are finished, we can simply insert a time shift and the associated phase factor. Alternatively, you can view Fig. 3.9 as a plot of the magnitudes of the functions and carry the exponential phase term through every step of the derivation.

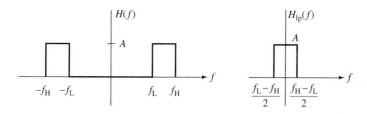

Figure 3.9   Bandpass and lowpass characteristics.

If we define the midpoint of the passband (the average of f_L and f_H) as

$$f_{av} = \frac{f_L + f_H}{2} \tag{3.15}$$

then the impulse response is

$$h(t) = h_{lp}(t)e^{j2\pi f_{av}t} + h_{lp}(t)e^{-j2\pi f_{av}t}$$

$$= 2h_{lp}(t)\cos 2\pi f_{av}t = 2h_{lp}(t)\cos\left[\pi(f_L + f_H)t\right] \tag{3.16}$$

From Eq. (3.12),

$$h_{lp}(t) = \frac{A\sin \pi(f_H - f_L)t}{\pi t} \tag{3.17}$$

Combining Eqs. (3.16) and (3.17), and reinserting the time shift (phase factor), we find the impulse response of the ideal bandpass filter:

$$h(t) = \frac{2A \sin\left[\pi(f_H - f_L)(t - t_0)\right]\cos\left[\pi(f_L + f_H)(t - t_0)\right]}{\pi(t - t_0)} \tag{3.18}$$

The impulse response is illustrated in Fig. 3.10. The outline of this waveform resembles the impulse response of the lowpass filter. Note that as the two limiting frequencies become large compared to the difference between them, the impulse response starts resembling a shaded-in version of the lowpass impulse response and its mirror image. This happens when the center frequency of the bandpass filter becomes large compared to the width of its passband. This observation will prove significant in our later studies of amplitude modulation.

Figure 3.10   Impulse response of ideal bandpass filter.

3.4 CAUSALITY

Things are going much too well. We must now darken the picture by showing that it is impossible to build ideal filters. We do this by introducing the idea of *causality*. This refers to the cause-and-effect relationship. The effect, or response, due to a cause, or input, cannot anticipate the input. That is, a causal system's output at any particular time depends only upon the input prior to that time, and not upon any future values of the input. There are no crystal balls in the real technical world.[1] For a linear system to be causal, it is necessary and sufficient that the impulse response $h(t)$ be zero for $t < 0$. It follows that the response due to a general input $r(t)$ depends only on past values of $r(t)$. To verify this, we write the expression for the output of a linear system in terms of its input and impulse response:

$$s(t) = \int_{-\infty}^{\infty} h(\tau)r(t - \tau)d\tau \tag{3.19}$$

If $h(t) = 0$ for $t < 0$, Eq. (3.19) becomes

$$s(t) = \int_{0}^{\infty} h(\tau)r(t - \tau)d\tau = \int_{-\infty}^{t} r(k)h(t - k)dk \tag{3.20}$$

The second integral in this equation results from a change of variables, where we let $k = t - \tau$. Equation (3.20) clearly shows that the output depends only on *past* values of the input. This proves sufficiency. To prove necessity, we note that $\delta(t) = 0$ for all $t < 0$. In a causal system, the inputs $r(t) = 0$ and $r(t) = \delta(t)$ must yield the same outputs, at least until time $t = 0$. That is, if the system cannot "anticipate" future values of the input, it has no way of "telling" the difference between zero and $\delta(t)$ prior to time $t = 0$. But in a linear system, an input that is identically equal to zero yields an output that is also identically equal to zero. Therefore, $h(t)$ must equal zero for $t < 0$, and the necessity part of the statement is proven.

The study of causality is important because, in general, causal systems are physically realizable and noncausal systems are not (much to the dismay of astrologers, fortune-tellers, and lottery players).

---

[1]There is a class of systems known as *prediction filters*. In fact, we use these later in the text in discussing both source encoders for digital systems and data compression of speech signals. Although we use the word *prediction*, there is nothing mysterious or noncausal about such systems.

The foregoing criterion is easy to apply if $h(t)$ is explicitly known. That is, one need simply examine $h(t)$ to see if it is zero for negative t. In many cases, however, it will be $H(f)$ that is known, and the Fourier inversion to find $h(t)$ may be difficult to perform. It would therefore be helpful if the constraint on $h(t)$ could be translated into a constraint on $H(f)$. We could then determine whether a system could be built without having to invert $H(f)$.

The *Paley-Wiener criterion* states that if

$$\int_{-\infty}^{\infty} \frac{\left| \ln H(f) \right|}{1 + (2\pi f)^2} \, df < \infty \tag{3.21a}$$

and

$$\int_{-\infty}^{\infty} \left| H(f) \right|^2 df < \infty \tag{3.21b}$$

then, for an appropriate choice of phase function for $H(f)$, $h(t) = 0$ for $t < 0$.

Note that Eqs. (3.21) do not take the phase of $H(f)$ into account. The actual form of $h(t)$ certainly does depend upon the phase of $H(f)$. Indeed, if a particular $H(f)$ corresponds to a causal system, we can make that system noncausal by shifting the original impulse response to the left on the time axis. This shift corresponds to a linear change in the phase of $H(f)$, which does not affect Eqs. (3.21). Thus, the Paley-Wiener criterion really tells us whether it is *possible* for $H(f)$ to be the transform of a causal time function (i.e., it is a necessary, but not sufficient, condition). Assuming that Eqs. (3.21) were satisfied, the phase of $H(f)$ would still have to be examined before final determination about the causality of the system could be made. Because of these considerations, we will use the Paley-Wiener condition to make only one simple, but significant, observation.

Since the logarithm of zero approaches minus infinity, if $H(f) = 0$ for any nonzero interval along the f-axis, then the first integral in Eq. (3.21a) does not converge. Therefore, the system cannot be causal. Alas, all of our ideal lowpass and bandpass filters are noncausal. Of course, we already knew this from observing the impulse responses $h(t)$. For this reason, practical filters must be, at best, approximations to their ideal counterparts.

The previous observation about $H(f)$ is a special case of a much more general theorem. Suppose $h(t)$ is such that

$$\int_{-\infty}^{\infty} \left| h(t) \right|^2 dr < \infty \tag{3.22}$$

Then if $h(t)$ is identically zero for any finite range along the t-axis, its Fourier transform cannot be zero for any finite range along the f-axis, and vice versa. This is sometimes stated as "a function that is time limited cannot be bandlimited."

The foregoing statement leads to an important observation: Any function of time that does not exist for all time (i.e., that is zero for any time interval) cannot be bandlimited.

In the next section, we discuss some of the most common approximations that are used in place of ideal filters. We will show that, provided that enough electrical elements are available, the ideal characteristics can be approached arbitrarily closely.

3.5 PRACTICAL FILTERS

We now present circuits that approximate the ideal lowpass and bandpass filters. Through-out this section, we assume that the closer $H(f)$ approaches the system function of an ideal filter, the more the filter will behave in an ideal manner in our applications. This fact is not at all obvious. A small change in $H(f)$ can lead to relatively large changes in $h(t)$. One can examine the consequences and categorize the effects of deviation from the constant-amplitude characteristic or from the linear phase characteristic of the ideal system function. (See the discussion of distortion in Section 1.2.1.)

We begin by analyzing the lowpass filter.

3.5.1 Lowpass Filter

The simplest passive approximation to a lowpass filter is the single energy-storage device circuit. An example is the RC circuit of Fig. 3.11. If the output is taken across the capacitor, this circuit approximates a lowpass filter. The reason is that, as the frequency increases, the capacitor behaves as a short circuit. The transfer function is

$$H(f) = \frac{1/j2\pi fC}{R + 1/j2\pi fC} = \frac{1}{1 + j2\pi fRC} \tag{3.23}$$

The magnitude and phase are

$$|H(f)| = \frac{1}{\sqrt{1 + (2\pi fRC)^2}} \tag{3.24}$$

$$\theta(f) = -\tan^{-1}(2\pi fRC)$$

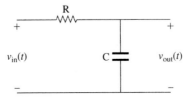

$v_{in}(t)$ C $v_{out}(t)$

Figure 3.11 *RC* circuit lowpass filter.

If we set RC to $1/2\pi$, the magnitude of the transfer function drops to $1/\sqrt{2}$ at a frequency of 1 Hz. This is the 3-dB cutoff frequency of the filter (20 $\log(1/\sqrt{2})$, which is approximately -3 dB).[2]

Figure 3.12 shows the magnitude and phase of the *RC* circuit transfer function. In Fig. 3.12(a), we use a logarithmic frequency axis, while in Fig. 3.12(b) we use a linear frequency axis. Superimposed on each set of curves is the equivalent gain curve for an ideal lowpass filter with a cutoff frequency of 1 Hz. In particular, if we view the linear frequency plot, we see that there is a dramatic difference between the *RC* approximation and the ideal lowpass filter characteristic. (Keep in mind that in a logarithmic gain curve, a drop of 20 dB is a decrease by a factor of 10.)

[2]The *decibel (dB)* is 20 times the logarithm of the amplitude ratio.

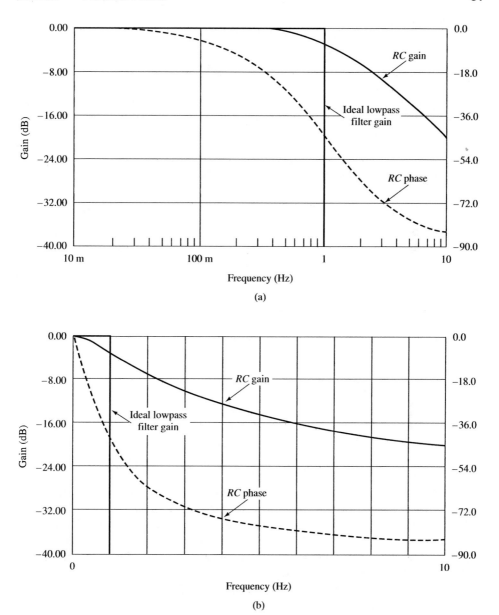

Figure 3.12   Characteristic of *RC* circuit.

These curves, and the following response waveforms, were developed using MI-CRO-CAP IV, a computer simulation program. Similar curves would result using other SPICE-based computer simulation programs or if we skip the simulation and plot the functions in Eq. (3.23) using *Mathcad* or MATLAB.

We could continue to analyze the *RC* filter distortion using the techniques derived earlier in this chapter. We choose instead to contrast the output of the *RC* circuit with that of an ideal lowpass filter for several representative inputs.

Let us first view the impulse response of the two systems. The impulse response of the ideal lowpass filter is

$$h(t) = \frac{\sin 2\pi(t - t_0)}{\pi(t - t_0)} \tag{3.25}$$

The impulse response of the *RC* circuit, with $RC = 2\pi$, is

$$h(t) = e^{-2\pi t} \tag{3.26}$$

These two impulse responses are shown in Fig. 3.13, where we have arbitrarily chosen the delay of the ideal filter to be 10 seconds so that the distinct plots can be easily seen.

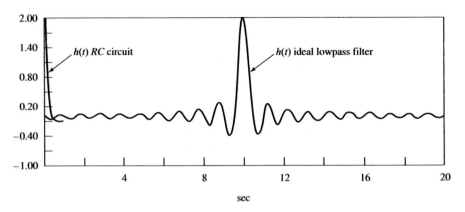

Figure 3.13   Comparison of impulse responses.

Let us now consider a square wave input to the two filters. We have simulated a square wave of fundamental frequency $\frac{1}{4}$ Hz by using the first five nonzero terms in a Fourier series expansion (up to the harmonic at a frequency of $\frac{9}{4}$ Hz).

The ideal lowpass filter with cutoff of 1 Hz passes only the first two nonzero terms (i.e., frequencies of $\frac{1}{4}$ Hz and $\frac{3}{4}$ Hz). By contrast, the *RC* filter with 3-dB frequency at 1 Hz significantly distorts these components. Figure 3.14(a) shows the input waveform, and Fig. 3.14(b) shows both the ideal lowpass filter and the *RC* output function. Not only does the *RC* filter distort the waveform in the passband, but it also admits significant energy from the signal beyond the cutoff frequency.

In the preceding examples, we have seen several types of distortion. We shall revisit distortion later, in the context of amplitude modulation, where we shall use the concepts of group and phase delay to gain an intuitive feel for the effects of the channel on a transmitted waveform. At this point, we would probably agree that an *RC* network is not a very good lowpass filter, except in limited applications. This leads us to explore more complex forms of practical filters.

There are several types of approximations to the ideal lowpass filter, each exhibiting unique characteristics. *Butterworth filters* produce no *ripple* in the passband and attenuate unwanted frequencies outside of this band. They are known as *maximally flat* filters, since they are designed to force the maximum number of derivatives of $H(f)$ (at $f = 0$) to be zero. *Chebyshev filters* attenuate unwanted frequencies more effectively than Butterworth

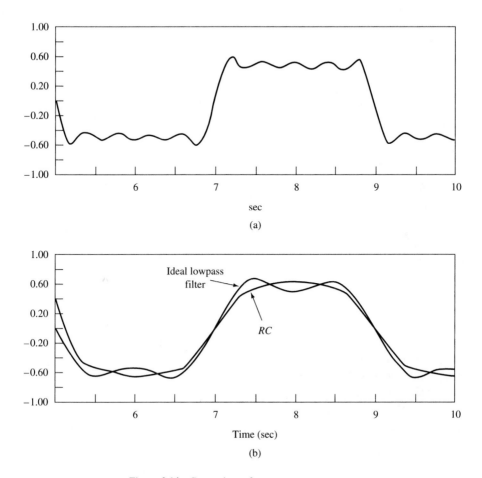

Figure 3.14 Comparison of square wave responses.

filters, but exhibit ripple in the passband. Other important classical filters include the ellip-tic, parabolic, Bessel, Papoulis, and Gaussian filters.

We limit the current discussion to Butterworth filters; for more details on the broad topic of filter design, the reader may consult various works listed in Appendix I. The am-plitude characteristic of the ideal lowpass filter can be approximated by the function

$$|H_n(f)| = \frac{1}{\sqrt{1 + (2\pi f)^{2n}}} \tag{3.27}$$

This function is sketched, for several values of n, in Fig. 3.15. We have illustrated only the positive half of the f-axis, since the function is even. We have chosen $f_m = 1/2\pi$ (1 radian/sec) for the illustration, but a simple scaling process can be used to design a filter for any cutoff frequency. Note that as n gets larger, the amplitude characteristic ap-proaches that of the ideal lowpass filter.

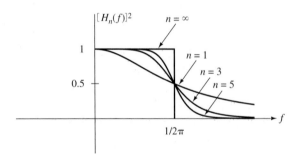

Figure 3.15 Butterworth gain functions.

We now examine the phase of $H(f)$. The result of Problem 2.6.3 indicates that, for a causal filter, the phase of the transfer function can be determined from the amplitude. Knowing one, you can derive the other. Thus, for a causal filter, we are not free to choose the amplitude and phase of $H(f)$ independently. Equation (3.27), then, contains all of the information needed to specify the filter.

If $h(t)$ is real (as it must be for a real system), the real part of $H(f)$ is even, while the imaginary part is odd. Therefore,

$$H(f) = H^*(-f) \tag{3.28a}$$

and

$$|H(f)|^2 = H(f)H^*(f) \tag{3.28b}$$

This observation, coupled with Eq. (3.27), is sufficient to design Butterworth filters.

Example 3.2

Design a third-order ($n = 3$) Butterworth filter with $f_m = 1/2\pi$.
Solution   From Eq. (3.27), we have

$$|H(f)|^2 = \frac{1}{1 + (2\pi f)^6}$$

Suppose we change this to the Laplace transform by letting $s = j2\pi f$. We will then be in a position to make observations relative to the poles and zeros of the function. We have

$$|H(s)|^2 = H(s)H(-s) = \frac{1}{1 - s^6}$$

The poles of $|H(s)|^2$ are the six roots of unity, as sketched in Fig. 3.16. They are equally spaced around the unit circle. Three of the poles are associated with $H(s)$ and the other three with $H(-s)$. Since the filter is causal, we associate the three poles in the left half-plane with $H(s)$. $H(s)$ is then found from its poles to be

$$H(s) = \frac{1}{(s - p_1)(s - p_2)(s - p_3)} = \frac{1}{s^3 + 2s^2 + 2s + 1}$$

Given $H(s)$, there are well-known techniques for synthesizing a circuit. If $v(t)$ is the response and $i(t)$ the source, the preceding system function corresponds to the circuit of Fig. 3.17(a). The component values are not realistic, since the cutoff frequency is very low. Figure 3.17(b) shows the computer-simulated frequency characteristic of this circuit.

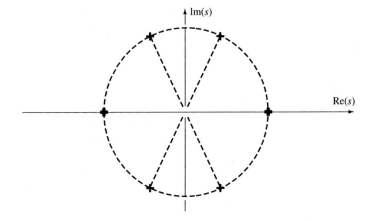

Figure 3.16   Six roots of unity.

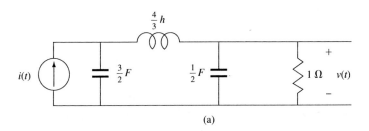

(a)

Third-order Butterworth

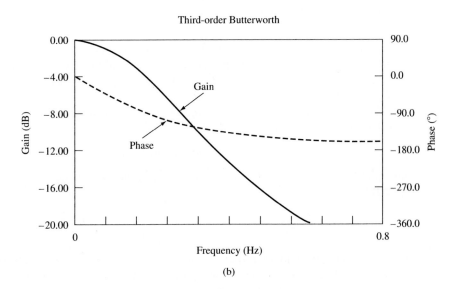

(b)

Figure 3.17   Third-order Butterworth filter.

Higher order filters would be implemented using ladder networks with additional elements. That is, additional series inductor, parallel capacitor combinations would appear.

We revisit Butterworth filters in the next section, when we examine active filters.

3.5.2 Bandpass Filter

The simplest passive approximation to a bandpass filter is the double energy-storage device circuit. An example of the *RLC* circuit is shown in Fig. 3.18.

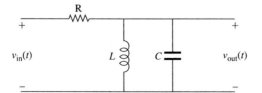

Figure 3.18 *RLC* bandpass circuit.

If the output is taken across the parallel *LC* combination, this circuit approximates a bandpass filter. The reason is that, as the frequency approaches zero, the inductor behaves as a short circuit, and as the frequency approaches infinity, the capacitor behaves as a short circuit. The circuit response therefore approaches zero at both extremes and peaks somewhere between the extremes. The transfer function is

$$H(f) = \frac{j2\pi f L}{R - (2\pi f)^2 RLC + j2\pi f L} \tag{3.29}$$

The magnitude of this is

$$|H(f)| = \frac{1}{\sqrt{R^2[1/2\pi f L) - 2\pi f C]^2 + 1}} \tag{3.30}$$

The magnitude peaks at $2\pi f = \sqrt{1/LC}$. This point is known as the *resonant frequency* of the filter. The ratio of the complex impedance to R is related to the Q of the circuit. Figure 3.19 shows a computer simulation of the circuit characteristics, where we have selected $R = L = C = 1$.

The impulse response of the *RLC* circuit is given by the inverse Fourier transform of $H(f)$. Therefore, for $R = L = C = 1$,

$$h(t) = 1.15e^{-t/2}\sin(1.15t) \tag{3.31}$$

This should be compared to the impulse response of the ideal bandpass filter derived in Eq. (3.18):

$$h(t) = \frac{2A \sin[\pi(f_H - f_L)(t - t_0)] \cos[\pi(f_L + f_H)(t - t_0)]}{\pi(t - t_0)}$$

Figure 3.20 shows the impulse response of the *RLC* circuit and the impulse response of the ideal bandpass filter, where we have chosen $f_H = 0.1$ Hz and $f_L = 0.25$ Hz, the 3-dB points of the *RLC* circuit response. Note that the Q of this filter is extremely low, since the ratio of the bandwidth to the center frequency is close to unity.

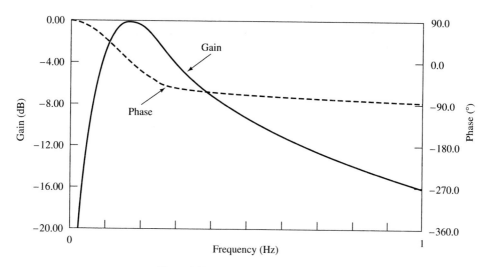

Figure 3.19   Characteristics of *RLC* filter.

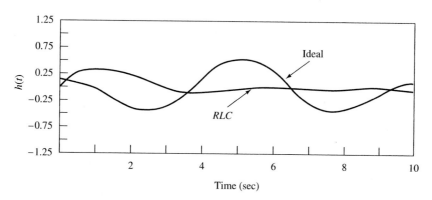

Figure 3.20   Comparison of impulse responses.

As in the case of the lowpass filter, improvements are possible by including additional components. We shall reserve our presentation of the bandpass Butterworth filter to the next section, where we show the active version of this filter.

3.6 ACTIVE FILTERS

In Section 3.5, we examined some simple realizations of filters using inductors, capacitors, and resistors. Such filters are called *passive*, since all component parts either absorb or store energy.

A filter is called *active* if it contains devices that deliver energy to the rest of the circuit. Active filters do not absorb part of the desired signal energy, as do passive filters. They are versatile and simple to design, and arbitrary causal transfer functions can be realized. For some applications, such as audio filtering, the passive filter requires an impractically large number of inductors and capacitors.

The basic building block of active filters is the *operational amplifier* (*op-amp*). The op-amp has characteristics approaching those of an ideal infinite-gain amplifier: infinite input resistance, zero output resistance, and infinite voltage gain. Practical op-amps suffer from limited bandwidths.

The basic configuration of a single op-amp is shown in Fig. 3.21. We have omitted the power-supply connections. We indicate input and feedback impedances by Z_{in} and Z_F, respectively. These could be single components (e.g., resistors, capacitors, or inductors) or combinations of components.

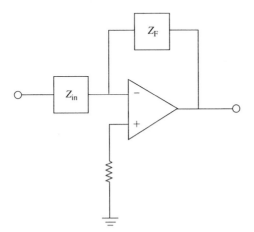

Figure 3.21 Feedback configuration of op-amp.

Active-filter analysis and design is an essential part of a study of electronics. We shall not take the time here to repeat that material. However, for purposes of illustration, we present one representative lowpass and one representative bandpass filter. In each case, we include the (computer-simulated) characteristics. The lowpass filter is shown in Fig. 3.22 and the bandpass filter in Fig. 3.23.

3.7 TIME-BANDWIDTH PRODUCT

In designing a communication system, an important consideration is the *bandwidth* of the system. The bandwidth is the range of frequencies the system is capable of handling.

The bandwidth is related to the Fourier transform of a function of time. It is not directly definable in terms of the function, unless we use intuitive statements about how quickly the function changes value.

Physical quantities of importance in communication system design include the minimum width of a time pulse and the minimum time in which the output of a system can jump from one level to another. We will show that both of these physical quantities are related to the bandwidth. We start with a specific example and then generalize the result.

The impulse response of the ideal lowpass filter is

$$h(t) = \frac{\sin 2\pi f_m(t - t_0)}{\pi(t - t_0)} \tag{3.32}$$

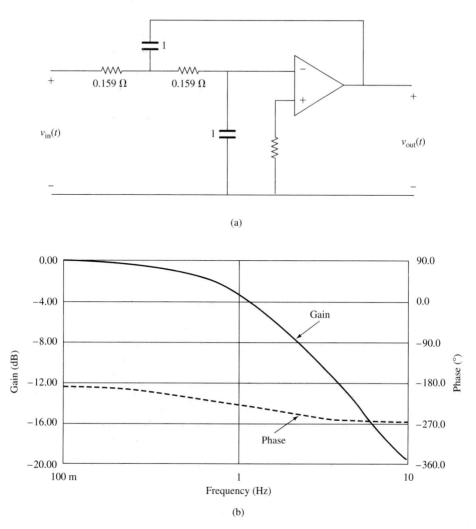

(a)

(b)

Figure 3.22   Active lowpass filter.

This $h(t)$ and the corresponding $H(f)$ are shown in Fig. 3.24. We use this transform pair to make two observations. First, the width of the largest lobe of $h(t)$ is $1/f_m$. This width is inversely proportional to the bandwidth of the signal. In fact, since the resulting bandwidth (difference between lowest and highest frequency) is f_m, the product of the pulse width and the bandwidth is unity.

The second observation regarding the lowpass filter requires that we find the step response of the filter. Since a step is the time integral of an impulse, and the lowpass filter is a linear system, the step response is the time integral of the impulse response. The step response, $a(t)$, is shown in Fig. 3.25. We now show that the *rise time* of this response is inversely proportional to the bandwidth of the filter. First we must define rise time. There are several common definitions, each of which attempts to mathematically define the length of time it takes the output to respond to a change, or jump, in the input. In practice, it is diffi-

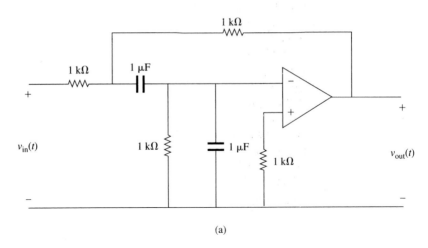

(a)

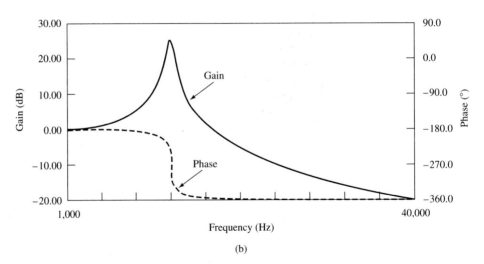

(b)

Figure 3.23   Active bandpass filter.

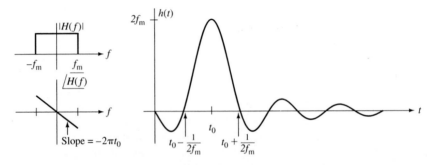

Figure 3.24   Characteristics of ideal lowpass filter.

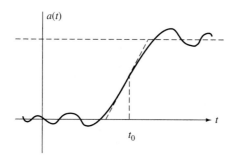

Figure 3.25   Step response of lowpass filter.

cult to define the exact time at which the output has finished responding to the input jump. We present one particular definition that is well suited to our application.

The rise time is defined as the time required for a signal to go from the initial to the final value along a ramp with constant slope equal to the maximum slope of the function. The latter is shown as a dashed line in Fig. 3.25. The maximum slope of $a(t)$ is the maximum value of the derivative $h(t)$. This maximum is given by $2f_m$. The rise time of the step response is then simply

$$t_r = \frac{1}{2f_m} \tag{3.33}$$

Since the bandwidth of the filter is f_m, the rise time and bandwidth are inversely related, and their product is equal to 0.5.

Although we have illustrated only the inverse relationship between rise time and bandwidth (or pulse width and bandwidth) for the ideal lowpass filter, the observation applies in general. That is, rise time is inversely proportional to bandwidth for any system. The product will not necessarily be 0.5, but will be a constant.

We now verify the relationship between rise time and bandwidth for a particular definition of bandwidth and pulse width. Suppose that a function of time and its Fourier transform are as shown in Fig. 3.26. We emphasize that the actual shape is not intended to be that shown in the figure. Indeed, if the pulse is time limited, the transform cannot go identically to zero over any range of frequencies. It is also unrealistic to think that the functions are monotonic. We present the pictures only to help understand the definitions of pulse width and bandwidth.

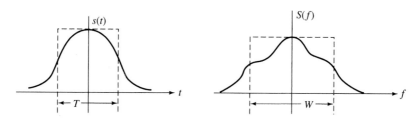

Figure 3.26   Definition of pulse width and bandwidth.

We define the pulse width T as the width of a rectangle whose height matches $s(0)$, and area is the same as that under the time pulse. This is illustrated as a dashed line in the figure. Note that the definition is not meaningful unless $s(0)$ is the maximum of the waveform.

Equivalently, we define the bandwidth BW using a pulse in the frequency domain, as illustrated in the figure. We then have

$$T = \frac{\int_{-\infty}^{\infty} s(t)dt}{s(0)}$$

$$BW = \frac{\int_{-\infty}^{\infty} S(f)df}{S(0)} \tag{3.34}$$

The product of these two is

$$T \cdot BW = \frac{\int_{-\infty}^{\infty} s(t)dt \int_{-\infty}^{\infty} S(f)df}{s(0)S(0)} \tag{3.35}$$

We now use the Fourier transform integral to find

$$S(0) = \int_{-\infty}^{\infty} s(t)e^{-j2\pi ft}dt \bigg|_{f=0} = \int_{-\infty}^{\infty} s(t)dt \tag{3.36}$$

The inverse transform integral is used to find

$$s(0) = \int_{-\infty}^{\infty} S(f)e^{-j2\pi ft}df \bigg|_{t=0} = \int_{-\infty}^{\infty} S(f)df \tag{3.37}$$

Substituting Eqs. (3.36) and (3.37) into Eq. (3.35), we find that

$$T \cdot BW = 1 \tag{3.38}$$

The product of the pulse width with the bandwidth is unity; hence, the two parameters are inversely related.

Clearly, the faster we desire a signal to change from one level to another, the more space on the frequency axis we must allow. This proves significant in digital communication, where the bit transmission rate is limited by the bandwidth of the channel.

3.8 SPECTRAL ANALYSIS

The Fourier transform does not exist in real life. It is a mathematical tool that aids in the analysis of systems. The FFT is one technique of approximating the Fourier transform of a continuous function of time, but it is a computational technique in which the Fourier transform is being evaluated by a mathematical algorithm.

There are severe limitations when one attempts to find the continuous Fourier transform of a time signal using a real analog system. First, since any real system must be causal, the best one can hope to do is find the Fourier transform based on past input values; there is no way the limits of integration can extend over all time—past and future.

Suppose that $r(t)$ forms in the input to an ideal bandpass filter with $H(f)$ as shown in Fig. 3.27. The output transform is given by

$$S(f) = \begin{cases} R(f), & f_L < |f| < f_H \\ 0, & \text{otherwise} \end{cases} \tag{3.39}$$

The function $s(t)$ is given by the inverse transform of $S(f)$:

$$s(t) = \int_{f_L}^{f_H} R(f)e^{j2\pi ft}df + \int_{-f_H}^{-f_L} R(f)e^{j2\pi ft}df \tag{3.40}$$

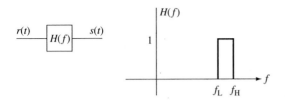

Figure 3.27   Bandpass filter.

If f_H is very close to f_L (i.e., if the filter is *narrowband*), we can assume that the integrand is approximately constant over the entire range of integration. Therefore, if f_{av} is the center of the filter passband, we have

$$s(t) \approx (f_H - f_L)[R(f_{av})e^{j2\pi f_{av}t} + R(-f_{av})e^{-j2\pi f_{av}t} \tag{3.41}$$

Now, since $R(-f_{av}) = R^*(f_{av})$, we have

$$s(t) = (f_H - f_L)|R(f_{av})| \cos[2\pi f_{av}t + \underline{/R(f_{av})}] \tag{3.42}$$

Thus, the magnitude of the output is proportional to the magnitude of the input transform evaluated at f_{av}, and the phase is shifted by the phase of $R(f_{av})$.

In many practical spectrum analyzers, the bandpass filter is swept across a range of frequencies, and the magnitude of the output varies approximately with $|R(f)|$.

There are three primary sources of error. First, while the bandpass filter is narrow, its bandwidth is not zero. This affects the *resolution* of the output. Second, the filter is causal and nonideal, which introduces error. Finally, when the center frequency of the filter varies with time (i.e., when it is *swept*), the output does not necessarily reach its steady-state value. The filter rise time is inversely proportional to its bandwidth. Therefore, the narrower the filter, the slower must be the sweep rate.

Example 3.3

Design a spectrum analyzer that can display the magnitude of the Fourier transform of the following function:

$$s(t) = 5 \cos(2\pi \times 1{,}000t) + 3 \cos(2\pi \times 1{,}100t)$$

Solution: Assuming that we can build a narrowband bandpass filter and sweep it across a range of frequencies (we will see a much better way to do this in Chapter 6), the design of the spectrum analyzer consists of choosing the frequency range and bandwidth of the filter and also choosing the rate at which the filter sweeps across the range of frequency.

Although we do not know $s(t)$ in advance (if we did, why bother with a spectrum analyzer?), we must assume that we know something about the range of frequencies $s(t)$ occupies and also about the required resolution. In fact, if we wish the approximate transform to look anything like the theoretical transform (two impulses), the bandwidth of the filter would have to be much smaller than the 100-Hz spacing between frequency components. If it were not small enough, the output would consist of components of each of the two frequencies, and we would not be capable of resolving these frequency components. Suppose we choose a filter bandwidth of 10 Hz and a center frequency sweep range of 900 Hz to 1,200 Hz.

We would probably display the result on a monitor. We do this by controlling the vertical displacement with the filter output magnitude and using a ramp generator (time scale) for the horizontal axis. The ramp must be synchronized with the filter sweep rate. The x-axis would then display frequency, while the y-axis would display the magnitude of the transform.

In order to "paint" a picture on the screen, we must either continuously repeat the sweep or add persistence (memory) to the trace. Either way, we wish to sweep across the range of frequencies in the minimum amount of time.

If the filter bandwidth is 10 Hz, the rise time is on the order of 1/10 sec. We therefore would not want to leave a particular range of frequencies in less than this period of time. Otherwise, the display would show the transient response of the filter rather than the Fourier transform magnitude function. The sweep rate should therefore be at least an order of magnitude less than 100 Hz/sec, or 10 Hz/sec. At this rate, it would take 30 seconds to sweep the entire range, so for the desired level of resolution, we would have to use a high-persistence CRT. Figure 3.28 illustrates the resulting spectrum analyzer output.

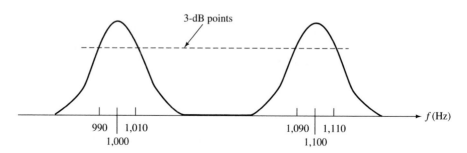

Figure 3.28   Spectrum analyzer output for Example 3.3.

PROBLEMS

3.1.1 You are given a system with input $r(t)$ and output $s(t)$. You are told that when $r(t) = 0$, $s(t)$ is not equal to zero. Show that this system cannot obey superposition.

3.1.2 A filter has the sinusoidal amplitude response shown in Fig. P3.1.2. The phase response is linear with slope $-2\pi t_0$.
   (a) Find the system response due to an input $\cos 2\pi t$.
   (b) Find the system response due to an input $(\sin 2\pi t)/t$.
   (c) Find the system response due to an input $(\sin 20\pi t)/t$.

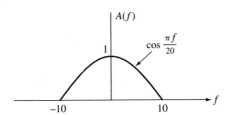

Figure P3.1.2

3.1.3 Repeat Problem 3.1.2 for the amplitude response shown in Fig. P3.1.3.

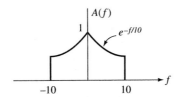

Figure P3.1.3

3.1.4 The transfer function of a cosine filter with zero delay is

$$H(f) = A + a\cos\frac{n\pi f}{f_m}$$

(a) Find the impulse response $h(t)$.
(b) Find the filter output due to an input

$$r(t) = \frac{\sin \pi t}{t}\cos 1{,}000\,\pi t$$

3.1.5 A system is shown in Fig. P3.1.5. Assume that $\theta(f) = -2\pi f t_0$. Expand $A(f)$ in a series in order to find the response due to a unit-amplitude, unit-width square pulse.

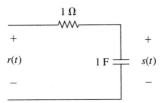

Figure P3.1.5

3.2.1 For the circuit shown in Fig. P3.2.1:
   (a) Find $H(f)$.
   (b) Plot $|H(f)|$ as a function of frequency.
   (c) What function is this circuit performing?
3.2.2 For the circuit shown in Fig. P3.2.2:
   (a) Find $H(f)$.
   (b) Plot $|H(f)|$ as a function of frequency.
   (c) What function is this circuit performing?

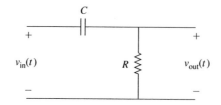

Figure P3.2.1

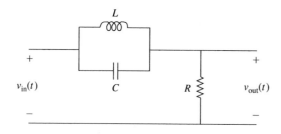

Figure P3.2.2

3.2.3 You are given the system shown in Fig. P3.2.3. The output of the system is $i(t)$. Find the phase distortion when the input is given by

$$r(t) = \frac{\sin t}{t} \cos 200\, t$$

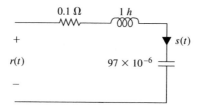

Figure P3.2.3

3.3.1 Consider the ideal lowpass filter with system function as shown in Fig. P3.3.1. Show that the response of this filter to an input $(\pi/K)\delta(t)$ is the same as that to $\sin(Kt)/Kt$.

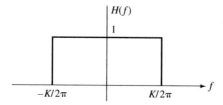

Figure P3.3.1

3.3.2 You are given the ideal lowpass filter with input as shown in Fig. P3.3.2. An error function is defined as the difference between the input and the output; that is,

$$e(t) = r(t) - s(t)$$

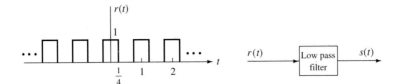

Figure P3.3.2

Find the error function if the filter cutoff frequency is
(a) $f_m = 2.5$ Hz
(b) $f_m = 3.5$ Hz
(c) $f_m = 4.5$ Hz

3.3.3 You are given the ideal lowpass filter with input as shown in Fig. P3.3.3. The mean square error is defined by

$$\text{mse} = \frac{1}{T}\int_0^T [s(t) - r(t)]^2\, dt$$

(a) Show that $s(t)$ is the average value of $r(t)$.
(b) Find the mean square error.

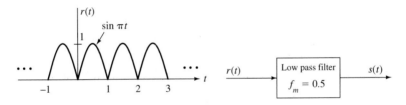

Figure P3.3.3

3.3.4 Find the impulse response for an ideal bandpass filter. Show that the filter bandwidth is inversely proportional to the width of the impulse response.

3.3.5 The periodic signal of Fig. P3.3.5 forms the input to an ideal bandpass filter with amplitude and phase as shown. Find the function of time at the output of the filter.

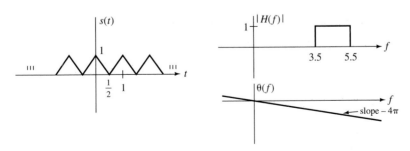

Figure P3.3.5

3.4.1 You wish to construct an ideal lowpass filter with

$$h_0(t) = \frac{\sin\,(t - 10)}{t - 10}$$

Because the system you build is causal, you actually have

$$h(t) = \begin{cases} h_0(t), & t > 0 \\ 0, & t < 0 \end{cases}$$

(a) Find $H(f)$, and compare it to the system function of the ideal filter.

(b) Find the output when the input is

$$r(t) = \frac{\sin t}{t}$$

(c) Find the error (difference between output and input) for the input of part (b).

3.5.1 Compare the step response of an RC circuit (output taken across the capacitor) with that of an ideal lowpass filter. Find the value of f_m for the ideal filter (in terms of R and C), which minimizes the integrated square error between the two step responses.

3.5.2 Design a Butterworth lowpass filter with a 3-dB cutoff at 500 Hz. The roll-off of the filter must be such that the amplitude response is attenuated by at least 50 dB at a frequency of 3 kHz.

3.8.1 Design a spectrum analyzer that can display the magnitude of the Fourier transform of the function

$$s(t) = 5\cos 2\pi \times 10^6 t + 3\cos 2\pi \times 10^7 t$$

4

Probability and Random Analysis

4.0 PREVIEW

What We Will Cover and Why You Should Care

Studying communication systems without taking noise into consideration is analogous to learning how to drive a car by practicing in a huge abandoned parking lot. While it is a valuable first step, it is not very realistic. Were it not for noise, the communication of signals would be trivial indeed. You could simply build a circuit composed of a battery and a microphone and hang the connecting wire outside the window. The intended receiver (perhaps on the other side of the globe) would simply hang a wire outside his or her window, receive a very weak signal, and use lots and lots of amplification (e.g., op-amps without feedback).

Of course, this example is unrealistic. In real life, the person at the receiver would receive a complete mess made up of an infinitesimal portion of the desired signal mixed with the signals of all other people wishing to communicate at the same time. Also mixed with the signal would be the radiated waveforms caused by automotive ignitions, people dialing telephones, lights being turned on and off, electrical storms, cats rubbing their fur on rugs, sunspots, and a virtual infinity of other spurious signals. The real challenge of communication is separating the desired signal from the undesired junk.

Anything other than the desired signal is called *noise*. Noise includes the effects just described, but it also includes some things you normally would not think of as noise. For example, a radar system attempting to track a particular object considers the signals returned from other objects to be noise. Similarly, if you tune in one channel on a television set and receive background images of adjacent channels, those background images are noise.

The most common types of noise emanate from a combination of numerous sources. In fact, the sources are so voluminous that one cannot hope to describe the resulting noise in a deterministic way (e.g., with a formula). We must therefore resort to discussing noise in terms of *averages*. The study of averages requires a knowledge of basic probability, thereby justifying this chapter.

After studying this chapter, you will:

- understand the basics of probability theory
- be able to solve problems involving random quantities

- have the tools necessary to evaluate the performance of communication systems in the presence of noise
- understand the matched filter, which is a building block in digital receivers.

Necessary Background

To understand basic probability, you need to know only elementary calculus. To understand random processes (which are discussed later in the chapter), you also need to know basic system theory.

4.1 BASIC ELEMENTS OF PROBABILITY THEORY

Probability theory can be approached either using theoretical mathematics or through empirical reasoning. The *mathematical approach* embeds probability theory within a study of *abstract set theory.* In contrast, the *empirical approach* satisfies one's intuitions. In our basic study of communication, we will find the empirical approach to be sufficient, although advanced study and references to current literature require extending these concepts using principles of set theory.

Before we define probability, we must extend our vocabulary by defining some other important terms:

An *experiment* is a set of rules governing an operation that is performed.
An *outcome* is the result realized after performing an experiment one time.
An *event* is a combination of outcomes.

Consider the experiment defined by flipping a single die (half of a pair of dice) and observing which of the six faces is at the top when the die comes to rest. (Notice how precise we are being: If you simply say "flipping a die," you could mean that you observe the *time* at which it hits the floor.) There are six possible outcomes, namely, any one of the six surfaces of the die facing upward after the performance of the experiment.

There are many possible events (64, to be precise). One event would be that of "an even number of dots showing." This event is a combination of the three outcomes of two dots, four dots, and six dots showing. Another event is "one dot showing." This event is known as an *elementary event,* since it is the same as one of the outcomes. Of the 64 possible events, six represent elementary events. You should be able to list the 64 events. Try it! If you come up with only 62 or 63, you are probably missing the combination of all outcomes and/or the combination of no outcomes.

4.1.1 Probability

We now define what is meant by the probability of an event. Suppose that an experiment is performed N times, where N is very large. Furthermore, suppose that in n of these N experiments, the outcome belongs to an event A (e.g., consider flipping a die 1,000 times, and in 495 of these flips the outcome is "even"; then $N = 1,000$ and $n = 495$). If N is large enough, the probability of event A is given by the ratio n/N. That is, the probability is the fraction of times that the event occurs. Formally, we define the probability of event A as

$$\Pr\{A\} = \lim_{N \to \infty} \frac{n_A}{N} \tag{4.1}$$

In Eq. (4.1), n_A is the number of times that the event A occurs in N performances of the experiment. This definition is intuitively satisfying. For example, if a coin were flipped many times, the ratio of the number of heads to the total number of flips would approach $\frac{1}{2}$. We therefore define the probability of a head to be $\frac{1}{2}$. This simple example shows why N must approach infinity. Suppose, for example, you flipped a coin three times and in two of these flips, the outcome were heads. You would certainly not be correct in assuming the probability of heads to be $\frac{2}{3}$!

Suppose that we now consider two different events, A and B, with probabilities

$$\Pr\{A\} = \lim_{N \to \infty} \frac{n_A}{N} \quad \text{and} \quad \Pr\{B\} = \lim_{N \to \infty} \frac{n_B}{N} \tag{4.2}$$

If A and B could not possibly occur at the same time, we call them *disjoint*. For example, the events "an even number of dots" and "two dots" are not disjoint in the die-throwing example, while the events "an even number of dots" and "an odd number of dots" are disjoint.

The probability of event A *or* event B is the number of times A or B occurs divided by N. If A and B are disjoint, this is

$$\Pr\{A \text{ or } B\} = \lim_{N \to \infty} \frac{n_A + n_B}{N} = \Pr\{A\} + \Pr\{B\} \tag{4.3}$$

Equation (4.3) expresses the additivity concept: If two events are disjoint, the probability of their sum is the sum of their probabilities.

Since each of the outcomes (elementary events) is disjoint from every other outcome, and each event is a sum of outcomes, it would be sufficient to assign probabilities only to the elementary events. We could derive the probability of any event from these given probabilities. For example, in the die-flipping experiment, the probability of an even outcome is the sum of the probabilities of "2 dots," "4 dots," and "6 dots."

Example 4.1

Consider the experiment of flipping a coin twice and observing which side is facing up when the coin comes to rest. List the outcomes, the events, and their respective probabilities.
Solution: The outcomes of this experiment are (letting H denote heads and T tails)

$$HH, HT, TH, \text{ and } TT$$

We shall assume that somebody has used intuitive reasoning or has performed this experiment enough times to establish that the probability of each of the four outcomes is $\frac{1}{4}$. There are 16 events, of combinations of these outcomes:

$$\{HH\}, \{HT\}, \{TH\}, \{TT\}$$
$$\{HH,HT\}, \{HH,TH\}, \{HH,TT\}, \{HT,TH\}, \{HT,TT\}, \{TH,TT\}$$
$$\{HH,HT,TH\}, \{HH,HT,TT\}, \{HH,TH,TT\}, \{HT,TH,TT\}$$
$$\{HH,HT,TH,TT\}, \text{ and } \{\phi\}$$

Note that the comma within the braces is read "or." Thus, the events {*HH,HT*} and {*HT,HH*} are identical, and we list this event only once. For completeness, we have included the zero event, denoted {ϕ}. This is the event made up of none of the outcomes and is called the *null* event. We also include the event consisting of all of the outcomes, the so-called *certain* event.

Using the additivity rule, the probability of each of these events is the sum of the probabilities of the outcomes comprising each event. Therefore,

$\Pr\{HH\} = \Pr\{HT\} = \Pr\{TH\} = \Pr\{TT\} = \frac{1}{4}$

$\Pr\{HH,HT\} = \Pr\{HH,TH\} = \Pr\{HH,TT\} = \Pr\{HT,TH\} = \Pr\{HT,TT\} = \Pr\{TH,TT\} = \frac{1}{2}$

$\Pr\{HH,HT,TH\} = \Pr\{HH,HT,TT\} = \Pr\{HH,TH,TT\} = \Pr\{HT,TH,TT\} = \frac{3}{4}$

$\Pr\{HH,HT,TH,TT\} = 1$

$\Pr\{ϕ\} = 0$

The next-to-last probability indicates that the event made up of all four outcomes is the *certain* event. It has probability 1 of occurring, since each time the experiment is performed, the outcome must belong to this event. Similarly, the null event (the last probability) has probability zero of occurring, since each time the experiment is performed, the outcome does not belong to the zero event.

4.1.2 Conditional Probabilities

We would like to be able to tell whether one random quantity has any effect on another. For instance, in the die experiment, if we knew the time at which the die hit the floor, would it tell us anything about which face was showing? In a more practical case, if we knew the frequency of a random noise signal, would this tell us anything about its amplitude? These questions lead naturally into a discussion of *conditional probabilities*.

Let us examine two events, *A* and *B*. The probability of event *A given that event B has occurred* is defined by

$$\Pr\{A/B\} = \frac{\Pr\{A \text{ AND } B\}}{\Pr\{B\}} \tag{4.4}$$

For example, if *A* represented two dots appearing in the die experiment and *B* represented an even number of dots, the probability of *A* given *B* would be the probability of two dots appearing, assuming that we know the outcome is either two, four, or six dots appearing. Thus, the conditional statement has reduced the scope of possible outcomes from six to three. We would intuitively expect the answer to be $\frac{1}{3}$. Now, from Eq. (4.4), the probability of "*A* AND *B*" is the probability of getting two *AND* an even number of dots simultaneously. (In set theory, this is known as the *intersection*.) It is simply the probability of two dots appearing, or $\frac{1}{6}$. The probability of *B* is the probability of two, four, or six dots appearing, which is $\frac{1}{2}$. The ratio is $\frac{1}{3}$, as expected.

Similarly, we could have defined event *A* as "an even number of dots" and event *B* as "an odd number of dots." The event "*A* AND *B*" would then be the zero event, and $\Pr\{A/B\}$ would be zero. This is reasonable because the probability of an even outcome assuming that an odd outcome occurred is clearly zero.

Two events, A and B, are said to be *independent* if

$$\Pr\{A/B\} = \Pr\{A\} \qquad (4.5)$$

Thus, if A and B are independent, the probability of A given that B occurred is simply the probability of A. Knowing that B has occurred tells nothing about A. Plugging Eq. (4.5) into Eq. (4.4) shows that if A and B are independent, then

$$\Pr\{A \text{ and } B\} = \Pr\{A\}\Pr\{B\} \qquad (4.6)$$

You have probably used this fact before in simple experiments. For example, we assumed that the probability of flipping a coin and having it land with heads facing up was $\frac{1}{2}$. Hence, the probability of flipping the coin twice and getting two heads is $\frac{1}{2} \times \frac{1}{2} = \frac{1}{4}$. This is true because the events are independent of each other.

Example 4.2

A coin is flipped twice. The following four different events are defined:

A is the event of getting a head on the first flip.
B is the event of getting a tail on the second flip.
C is the event of getting a match between the two flips.
D is the elementary event of getting a head on both flips.

(**a**) Find $\Pr\{A\}$, $\Pr\{B\}$, $\Pr\{C\}$, $\Pr\{D\}$, $\Pr\{A/B\}$, and $\Pr\{C/D\}$.
(**b**) Are A and B independent? Are C and D independent?
Solution: (**a**) The events are defined by the following combination of outcomes:

$$A = \{HH, HT\}$$
$$B = \{HT, TT\}$$
$$C = \{HH, TT\}$$
$$D = \{HH\}$$

Therefore,

$$\Pr\{A\} = \Pr\{B\} = \Pr\{C\} = \tfrac{1}{2}$$
$$\Pr\{D\} = \tfrac{1}{4}$$

(**b**) To find $\Pr\{A/B\}$ and $\Pr\{C/D\}$, we use Eq. (4.4):

$$\Pr\{A/B\} = \frac{\Pr\{A \text{ AND } B\}}{\Pr\{B\}}$$
$$\Pr\{C/D\} = \frac{\Pr\{C \text{ AND } C\}}{\Pr\{D\}}$$

The event $\{A \text{ AND } B\}$ is $\{HT\}$. The event $\{C \text{ AND } D\}$ is $\{HH\}$. Therefore,

$$\Pr\{A/B\} = \frac{\frac{1}{4}}{\frac{1}{2}} = 0.5$$

$$\Pr\{C/D\} = \frac{\frac{1}{4}}{\frac{1}{4}} = 1$$

Since $\Pr\{A/B\} = \Pr\{A\}$, the event of a head on the first flip is independent of that of a tail on the second flip. Since $\Pr\{C/D\} \neq \Pr\{C\}$, the event of a match and that of two heads are not independent.

4.1.3 Random Variables

We would like to perform several forms of analysis on probabilities. It is not too satisfying to work with symbols such as "heads," "tails," and "two dots." It would be preferable to work with numbers. We therefore associate a real number with each possible outcome of an experiment. For example, in the single-flip-of-the-coin experiment, we could associate the number 0 with "tails" and 1 with "heads." We could just as well (although we won't) associate π with "heads" and 207 with "tails."

The mapping (function) that assigns a number to each outcome is called a *random variable*.

Once a random variable is assigned, we can perform many forms of analysis. We can, for example, plot the various outcome probabilities as a function of the random variable. An extension of that type of plot is the *distribution function $F(x)$*. If the random variable is denoted by[1] X, then the distribution function $F(x_0)$ is defined by

$$F(x_0) = \Pr\{X \le x_0\} \tag{4.7}$$

We note that $\{X \le x_0\}$ defines an event, or combination of outcomes.

Example 4.3

Assign two different random variables to the "one-flip-of-the-die" experiment, and plot the two resulting distribution functions.

Solution: The first assignment we will choose is the one that is naturally suggested by this particular experiment. That is, we assign the number 1 to the outcome described by the face with one dot facing up, we assign the number 2 to "two dots," 3 to "three dots," and so on. We therefore see that the event $\{X \le X_0\}$ includes the one-dot outcome if x_0 is between 1 and 2. If x_0 is between 2 and 3, the event includes the one-dot and two-dot outcomes. Thus, the distribution function is as shown in Fig. 4.1(a).

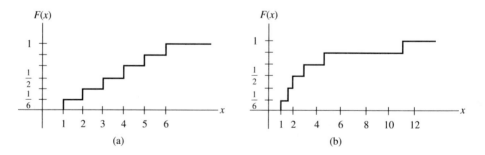

(a)   (b)

Figure 4.1   Distribution function for Example 4.3.

[1]We shall use uppercase letters for random variables and lowercase letters for the values they can take on. Thus, $X = x_0$ means that the random variable X is equal to the number x_0.

Let us now choose a different assignment of the random variable, one representing a less natural choice:

Outcome	Random Variable
One dot	1
Two dots	π
Three dots	2
Four dots	$\sqrt{2}$
Five dots	11
Six dots	5

We have chosen strange numbers to illustrate that the mapping is arbitrary. The resulting distribution function is plotted as Fig. 4.1(b). As an example, let us verify one point on the distribution function, the point for $x = 3$. The event $\{X \le 3\}$ is the event made up of the following three outcomes: one dot, three dots, and four dots. This is true because the value of the random variable assigned to each of these three outcomes is less than 3.

A distribution function can never decrease with increasing argument. The reason is that an increase in argument can only add outcomes to the event, and the probabilities of these added outcomes cannot be negative. We also easily verify that

$$F(-\infty) = 0 \quad \text{and} \quad F(+\infty) = 1 \tag{4.8}$$

4.1.4 Probability Density Function

The *probability density function* is defined as the derivative of the distribution function. Using the symbol $p_X(x)$ for the density, we have

$$p_X(x) = \frac{dF(x)}{dx} \tag{4.9}$$

Since $p(x)$ is the derivative of $F(x)$, $F(x)$ is the integral of $p(x)$:

$$F(x_0) = \int_{-\infty}^{x_0} p_X(x)\, dx \tag{4.10}$$

The random variable can be used to define any event. For example, $\{x_1 < X \le x_2\}$ defines an event. Since the events $\{X \le x_1\}$ and $\{x_1 < X \le x_2\}$ are disjoint, the additivity principle can be used to prove that

$$\Pr\{X \le x_1\} + \Pr\{x_1 < X \le x_2\} = \Pr\{X \le x_2\}$$

or $\tag{4.11}$

$$\Pr\{x_1 < X \le x_2\} = \Pr\{X \le x_2\} - \Pr\{X \le x_1\}$$

Combining Eqs. (4.10) and (4.11), we have the important result,

$$\Pr\{x_1 < X \le x_2\} = \int_{-\infty}^{x_2} P_X(x)dx - \int_{-\infty}^{x_1} P_X(x)dx \qquad (4.12)$$

$$= \int_{x_1}^{x_2} P_X(x)dx$$

We now see why $p_X(x)$ is called a density function: The probability that X is between any two limits is given by the area under the density function between these two limits.

Since the distribution function can never decrease with increasing argument, its slope, the density function, can never be negative.[2] Also, since the distribution function approaches unity as its argument approaches infinity, the integral of the density function over infinite limits must be unity.

The examples given previously (die and coin) result in density functions that contain impulses. The random variables associated with such experiments are known as *discrete random variables.* Another class of experiments gives rise to random variables with continuous density functions. This is logically called the class of *continuous random variables.* We present several frequently occurring continuous random variable density functions in Section 4.2. For now, we examine the simplest of these functions, the *uniform density function.* This function is shown in Fig. 4.2(a), where a and b are specified parameters. The height of the density must be such that the total area is unity.

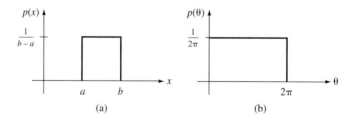

(a)   (b)

Figure 4.2   Uniform density function.

Let us look at one practical experiment that results in a uniformly distributed random variable. Suppose you were asked to turn on a sinusoidal generator. The output of the generator would be of the form

$$v(t) = A \cos(2\pi f_0 t + \theta) \qquad (4.13)$$

Since the absolute time at which you turn on the generator is random, it would be reasonable to expect that θ is uniformly distributed between 0 and 2π. (This is true provided that f_0 is much larger than the reciprocal of your reaction time.) It would thus follow the density function shown in Fig. 4.2(b).

---

[2]We could infer the same conclusion from Eq. (4.12). If the probability density function were negative over any range of values, we could integrate the curve over that range to get a negative result. This would imply that the probability of the variable being in that range is negative, but that is impossible.

Example 4.4

A random variable is uniformly distributed between 1 and 3. Find the probability that the variable is in the range between 1.5 and 2.

Solution: The density function is as shown in Fig. 4.2(a), where a is 1 and b is 3. In order for this to integrate to unity, the height of the density must be $\frac{1}{2}$. The probability that the variable is between 1.5 and 2 is simply the integral under the curve between these limits. Clearly, this is equal to $\frac{1}{4}$.

4.1.5 Expected Values

Expected values, or averages, are important in communication. The average of the square of a voltage is closely related to the power associated with that voltage. The power of a noise voltage is an important measure of the level of disturbance caused by the voltage.

The expected values that come up often enough to be given names are the *mean*, *variance*, and *moments* of a random variable. We define these terms in this subsection.

Picture yourself as a professor who has just given an examination. How would you average the resulting grades? You would probably add them all together and divide by the number of grades. If an experiment is performed many times, the average of the random variable that results would be found in the same way.

An alternative way to find the sum of grades is to take 100 multiplied by the number of students who got 100 as a grade, add this to 99 times the number of students who got 99, and continue this process for all possible grades. Then divide the sum by the total number of grades. Let us formalize this approach.

Let x_i, $i = 1, 2, \ldots, M$, represent the possible values of the random variable, and let n_i represent the number of times the outcome associated with x_i occurs. Then the average of the random variable after N performances of the experiment is

$$X_{avg} = \frac{1}{N} \sum_i n_i x_i = \sum_i \frac{n_i}{N} x_i \qquad (4.14)$$

Since x_i ranges over all possible values of the random variable,

$$\sum_i n_i = N \qquad (4.15)$$

As N approaches infinity, n_i/N becomes the probability, $\Pr\{x_i\}$. Therefore,

$$X_{avg} = \sum_i x_i \Pr\{x_i\} \qquad (4.16)$$

This average value is known as the *mean, expected value,* or *first moment* of X and is given the symbol $E\{x\}$, X_{avg}, m_x, or $\bar{x}$. The words "expected value" should not be taken too literally, since they do not always lend themselves to an intuitive definition. As an example, suppose we assign 1 to heads and 0 to tails in the coin flip experiment. Then the expected value of the random variable is $\frac{1}{2}$. However, no matter how many times you perform the experiment, you will never obtain an outcome with an associated random variable of $\frac{1}{2}$.

Now suppose that we wish to find the average value of a continuous random variable. We can use Eq. (4.16) if we first round off the continuous variable to the nearest multiple of Δx. Thus, if X is between $(k - \frac{1}{2})\Delta x$ and $(k + \frac{1}{2})\,\Delta x$, we round it off to $k\Delta x$. The probability of X being in this range is given by the integral of the probability density function:

$$\Pr\left\{\left(k - \frac{1}{2}\right)\Delta x < x \leq \left(k + \frac{1}{2}\right)\Delta x\right\} = \int_{(k - \frac{1}{2})\Delta x}^{(k + \frac{1}{2})\Delta x} p_X(x)\, dx \qquad (4.17)$$

If Δx is small, this can be approximated by $p_X(k\Delta x)\Delta x$. Therefore, Eq. (4.16) can be rewritten as

$$X_{avg} = \sum_{k=-\infty}^{\infty} k\Delta x p_X(k\Delta x)\Delta x \qquad (4.18)$$

As Δx approaches zero, this becomes

$$X_{avg} = m_x = \int_{-\infty}^{\infty} x p_X(x) dx \qquad (4.19)$$

Equation (4.19) is very important. It tells us that to find the average value of x, we simply weight x by the density function and integrate the product.

The same approach can be used to find the average of any function of a random variable. For example, suppose again that you are a professor who gave an exam, but instead of entering the raw percentage score in your grade book, you enter some function of this score, such as e^x, where x is the raw score. If you now wish to average the entries in the book, you would follow the reasoning used earlier [Eqs. (4.14) through (4.19)], with the result that $x p_X(x)$ in Eq. (4.19) gets replaced by $e^x p_X(x)$.

In general, if $y = g(x)$, the expected value of y is given by

$$Y_{avg} = [g(x)]_{avg} = \int_{-\infty}^{\infty} g(x) p_X(x) dx \qquad (4.20)$$

Equation (4.20) is extremely significant and useful. It tells us that in order to find the expected value of a function of x, we simply integrate that function weighted by the *density of* X. It is not necessary to find the density of the new random variable first.

We often seek the expected value of the random variable raised to a power. This is given the name *moment*. Thus, the expected value of x^n is known as the nth *moment* of the random variable X.

If we first shift the random variable by its mean and then take a moment of the resulting shifted variable, the *central moment* results. Thus, the nth central moment is given by the expected value of $(x - m_x)^n$.

The *second central moment* is extremely important, because it is related to power. It is given the name *variance* and the symbol σ^2. Thus, the variance is

$$\sigma^2 = E\{(x - m_x)^2\} = \int_{-\infty}^{\infty} (x - m_x)^2 p_X(x)\, dx \qquad (4.21)$$

The variance is a measure of how far we can expect the variable to deviate from its mean value. As the variance gets larger, the density function tends to "spread out." The square root of the variance, σ, is known as the *standard deviation*.

Example 4.5

Suppose X is uniformly distributed as shown in Fig. 4.3. Find $E\{x\}$, $E\{x^2\}$, $E\{\cos x\}$ and $E\{(x - m_x)^2\}$.

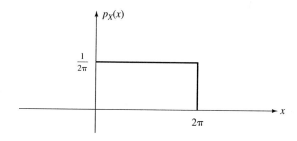

Figure 4.3   Density of x for Example 4.5.

Solution: We apply Eq. (4.20) to find

$$E\{x\} = \int_{-\infty}^{\infty} x p_X(x)dx = \frac{1}{2\pi} \int_{0}^{2\pi} x\, dx = \pi$$

$$E\{x^2\} = \int_{-\infty}^{\infty} x^2 p_X(x)dx = \frac{1}{2\pi} \int_{0}^{2\pi} x^2\, dx = \frac{4}{3}\pi^2$$

$$E\{\cos x\} = \int_{-\infty}^{\infty} \cos x p_X(x)dx = \frac{1}{2\pi} \int_{0}^{2\pi} \cos x\, dx = 0$$

$$E\{(x - \pi)^2\} = \int_{-\infty}^{\infty} (x - \pi)^2 p_X(x)dx = \frac{1}{2\pi} \int_{0}^{2\pi} (x - \pi)^2 dx = \frac{\pi^2}{3}$$

4.1.6 Functions of a Random Variable

"Everybody talks about the weather, but nobody does anything about it." As communication engineers, we ourselves would be open to the same type of criticism if all we ever did was make statements such as "There is a 42-percent probability that the noise will be annoying." A significant part of communication engineering involves changing noise from one form to another in the hope that the new form will be less annoying than the old. We must therefore study the effects of processing on random phenomena.

Consider a function of a random variable, $y = g(x)$, where X is a random variable with known density function. A representative function is shown in Fig. 4.4(a). Since X is random, Y is also random. We are interested in finding the density function of Y.

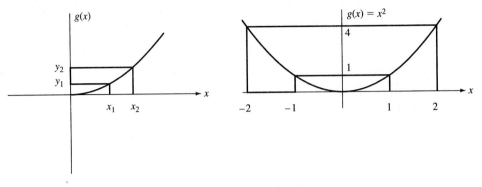

Figure 4.4   Representative $g(x)$.

The event $\{x_1 < X \le x_2\}$ corresponds to the event $\{y_1 < Y \le y_2\}$,[3] where

$$y_1 = g(x_1) \quad \text{and} \quad y_2 = g(x_2)$$

The two events are identical because they include the same outcomes. We are assuming for the moment that $g(x)$ is a single-valued function. Since the events are identical, their probabilities must also be equal. That is,

$$\Pr\{x_1 < X \le x_2\} = \Pr\{y_1 < Y \le y_2\} \tag{4.22}$$

and in terms of the densities,

$$\int_{x_1}^{x_2} p_X(x)dx = \int_{y_1}^{y_2} p_Y(y)dy \tag{4.23}$$

If we now let x_2 get very close to x_1, then in the limit, Eq. (4.23) becomes

$$p_X(x_1)dx = p_Y(y_1)dy \tag{4.24}$$

and lastly,

$$p_Y(y_1) = \frac{p_X(x_1)}{dy/dx} \tag{4.25}$$

If $y_1 > y_2$, the slope of the curve is negative, and we would find (you should prove this result) that

$$p_Y(y_1) = -\frac{p_X(x_1)}{dy/dx} \tag{4.26}$$

We can account for both of these cases by writing

$$p_Y(y_1) = \frac{p_X(x_1)}{|dy/dx|} \tag{4.27}$$

---

[3]We are assuming that y1 is less than y2 if x1 is less than x2. That is, $g(x)$ is monotonically increasing and has a positive derivative. If this is not the case, the inequalities would have to be reversed.

Finally, writing $x_1 = g^{-1}(y_1)$, and realizing that y_1 can be a variable (i.e., replace it with y), we have

$$p_Y(y) = \frac{p_X[g^{-1}(y)]}{|dy/dx|} \tag{4.28}$$

If the function $g(x)$ is not monotonic, the event $\{y_1 < Y < y_2\}$ can correspond to more than one interval of the variable X. For example, if $g(x) = x^2$, then the event $\{1 < Y \le 4\}$ is the same as the event $\{1 < X \le 2\}$ or $\{-2 < X \le -1\}$. This is shown in Fig. 4.4(b). Therefore,

$$\int_1^2 p_X(x)dx + \int_{-2}^{-1} p_X(x)dx = \int_1^4 p_Y(y)dy \tag{4.29}$$

In terms of the density functions, this means that $g^{-1}(y)$ has two values. Denoting these values as x_a and x_b, we have

$$p_Y(y) = \frac{p_X(x)}{|dy/dx|}\bigg|_{x=x_a} + \frac{p_X(x)}{|dy/dx|}\bigg|_{x=x_b} \tag{4.30}$$

Example 4.6

A random voltage v is put through a full-wave rectifier. The input voltage is uniformly distributed between -2 volts and $+2$ volts. Find the density of the output of the full-wave rectifier. **Solution:** Calling the output y, we have $y = g(v)$, where $g(v)$ and the density of V are sketched in Fig. 4.5. Note that we have let the random variable be equal to the value of voltage.

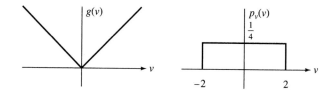

Figure 4.5 $g(v)$ and $p(v)$ for Example 4.6.

At every value of V, $|dg/dv| = 1$. For $y > 0$, $g^{-1}(y) = \pm y$. For $y < 0$, $g^{-1}(y)$ is undefined. That is, there are no values of v for which $g(v)$ is negative. Equation (4.30) is then used to find

$$p_Y(y) = p_V(y) + p_V(-y) \quad y > 0$$

$$p_Y(y) = 0 \quad y < 0$$

This result is shown in Fig. 4.6.

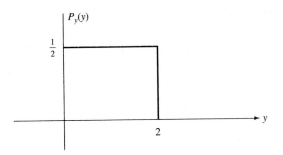

Figure 4.6 Density of output for Example 4.6.

4.2 FREQUENTLY ENCOUNTERED DENSITY FUNCTIONS

We introduced the basic concepts of probability in Section 4.1. The uniform density function was presented. While some experiments in communication lead to this density, the majority of random variables we encounter follow densities other than uniform. The current section explores several of the most frequently encountered densities.

4.2.1 Gaussian Random Variables

The most common density confronted in the real world is called the *Gaussian (or normal) density function*. The reason it is so common is attributed to the *central limit theorem*, a theorem we shall discuss in a few moments. The Gaussian density function is defined by the equation

$$p_X(x) = \frac{1}{\sqrt{2\pi}\sigma} \exp\left[\frac{-(x-m)^2}{2\sigma^2}\right] \tag{4.31}$$

where m and σ are given constants. The function is sketched in Fig. 4.7.

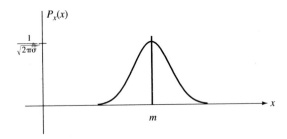

Figure 4.7 Gaussian density function.

The parameter m dictates the center position, or symmetry point, of the density. Evaluating the integral of $xp_X(x)$ would show that m is the mean value of the variable. The other parameter, σ, indicates the spread of the density. Evaluating the integral of $(x-m)^2 p_X(x)$ would show that σ^2 is the variance of the variable, so σ is the standard deviation. As σ increases, the bell-shaped curve gets wider and the peak decreases. Alternatively, as σ decreases, the density sharpens into a narrow pulse with a higher peak. (The area must always be unity.)

To evaluate probabilities that Gaussian variables are within certain ranges, we find it necessary to integrate the density. However, Eq. (4.31) cannot be integrated in closed form, although software such as *Mathcad* can easily be used. The Gaussian density is sufficiently important that this integral has been computed and tabulated under the names *error function* (*erf*) and Q-*function*. The error function is defined as

$$\text{erf}(x) = \frac{2}{\sqrt{\pi}} \int_0^x e^{-u^2}\, du \tag{4.32}$$

It can be shown that $\text{erf}(\infty) = 1$. Therefore,

$$\frac{2}{\sqrt{\pi}} \int_x^\infty e^{-u^2} du = \frac{2}{\sqrt{\pi}} \int_0^\infty e^{-u^2} du - \frac{2}{\sqrt{\pi}} \int_0^x e^{-u^2} du \tag{4.33}$$
$$= \text{erf}(\infty) - \text{erf}(x) = 1 - \text{erf}(x)$$

For convenience, this last expression is tabulated under the name *complementary error function* (*erfc*). Thus, the relationship between the error function and the complementary error function is

$$\text{erfc}(x) = 1 - \text{erf}(x) \tag{4.34}$$

Both the error function and the complementary error function are tabulated in Appendix III. The area under a Gaussian density with any values of m and σ can be expressed in terms of error functions. For example, the probability that X is between x_1 and x_2 is

$$\Pr\{x_1 < X \le x_2\} = \frac{1}{\sqrt{2\pi}\sigma} \int_{x_1}^{x_2} \exp\left[\frac{-(x-m)^2}{2\sigma^2}\right] dx \tag{4.35}$$

Making the change of variables,

$$u = \frac{x-m}{\sqrt{2}\sigma} \tag{4.36}$$

we get

$$\Pr\{x_1 < X \le x_2\} = \frac{1}{\sqrt{\pi}} \int_{\frac{x_1-m}{\sqrt{2}\sigma}}^{\frac{x_2-m}{\sqrt{2}\sigma}} e^{-u^2} du \tag{4.37}$$
$$= \frac{1}{2}\text{erf}\left(\frac{x_2-m}{\sqrt{2}\sigma}\right) - \frac{1}{2}\text{erf}\left(\frac{x_1-m}{\sqrt{2}\sigma}\right)$$

We have assumed that both x_1 and x_2 are greater than m, since the error function is not defined for negative arguments. Example 4.7 will deal with a situation where this assumption is not valid.

A companion to the error function is the Q-function. It is sometimes called the complementary error function, or *co-error function*. However, the Q-function differs from the complementary error function of Eq. (4.34) by a constant multiplier and a scaling factor. The Q-function is defined as

$$Q(x) = \int_x^\infty \frac{1}{\sqrt{2\pi}} e^{-u^2/2} \, du \tag{4.38}$$

The integrand is a unit-variance, zero-mean Gaussian density function. Note that

$$Q(-\infty) = 1$$

$$Q(-x) = 1 - Q(x)$$

We can relate the Q-function to the error function by making a change of variables:

$$Q(x) = \frac{1}{\sqrt{2\pi}} \int_x^\infty e^{-u^2/2} \, du = \frac{1}{\sqrt{\pi}} \int_{x/\sqrt{2}}^\infty e^{-v^2} \, dv$$

$$= \frac{1}{2} \operatorname{erfc}\left(\frac{x}{\sqrt{2}}\right) \tag{4.39}$$

The Q-function is tabulated in Appendix IV.

Both the Q-function and the error function contain the information necessary to evaluate integrals of Gaussian density functions. Some feel that the Q-function is more satisfying and easier to work with, since the integrand is a normalized Gaussian density.

Using the Q-function, let us now find the probability that a random variable is between two limits:

$$\Pr\{x_1 < X \le x_2\} = \frac{1}{\sqrt{2\pi}\sigma} \int_{x_1}^{x_2} \exp\left[\frac{(x - m)^2}{2\sigma^2}\right] dx \tag{4.40}$$

With the Q-function, the required change of variables yields

$$u = \frac{x - m}{\sigma}$$

$$\Pr\{x_1 < X \le x_2\} = \frac{1}{\sqrt{2\pi}} \int_{\frac{x_1 - m}{\sigma}}^{\frac{x_2 - m}{\sigma}} e^{-u^2/2} \, du$$

$$= Q\left(\frac{x_1 - m}{\sigma}\right) - Q\left(\frac{x_2 - m}{\sigma}\right) \tag{4.41}$$

Although the error function has traditionally been much more common than the Q-function, there are indications that the Q-function will predominate in the future. It makes absolutely no difference which one you use to solve a problem. (You'll get the same answer either way.) The only question you should have is which type of table is more readily available. Of course, if you use the wrong table, you will get the wrong answer.

Now that we are familiar with the Gaussian density, let us return to the discussion of why it occurs so frequently in the real world. It results whenever a large number of factors contribute to an end result, as in the case of static in broadcast radio. Two conditions must be satisfied before the sum of many random variables starts to appear Gaussian. The first relates to the individual variances and to their infinite sum: The sum must approach infinity as the number of variables added together approaches infinity. The second condition is satisfied if the component densities go to zero outside some range. (This is a sufficient, but not a necessary, condition.) Since all quantities we deal with in the real world have bounded ranges, they satisfy the second condition.

Although we do not prove the central limit theorem here, one simple example is often given to indicate the reasonableness of the theorem. Suppose that we add together independent uniform random variables, where each is distributed between -1 and $+1$, as shown in Fig. 4.8(a). This is an example of *independent identically distributed* (*iid*) variables. When the first two variables are added, it can be shown that the resulting density is the convolution of the two original densities, as shown in Fig. 4.8(b). This doesn't yet look very much like the Gaussian curve, but we have summed only two variables.

If a third variable is added, the ramp of Fig. 4.8(b) is convolved once more with the uniform density to get the parabolic curve of Fig. 4.8(c). At this point, the curve bears a re-

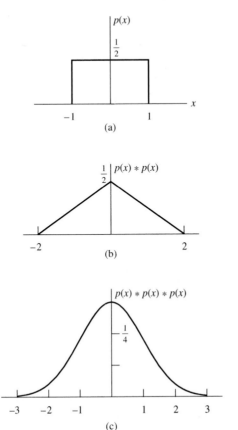

Figure 4.8   The central limit theorem.

semblance to the bell-shaped Gaussian density. As more and more variables are added, the agreement becomes closer and closer. In fact, the series converges quite rapidly to a Gaussian density, even if the densities we start with are not uniform.

Example 4.7

A binary communication system is a communication system that sends only one of two possible messages. A simple example of a binary system is one in which either *zero volts* or *one* volt is sent. Consider such a system in which the transmitted voltage is corrupted by additive atmospheric noise. If the receiver receives anything above $\frac{1}{2}$ volt (i.e., the midpoint), it assumes that a *one* was sent. If it receives anything below $\frac{1}{2}$ volt, it assumes that a *zero* was sent. Measurements show that if one volt is transmitted, the received signal level is random and has a Gaussian density with $m = 1$ and $\sigma = \frac{1}{2}$. Find the probability that a transmitted *one* will be interpreted as a *zero* at the receiver (i.e., find the probability of a *bit error*).

Solution: The received signal level has a Gaussian density with $m = 1$ and $\sigma^2 = (\frac{1}{2})^2$. Thus, if we designate the random variable as V, we have

$$P_V(v) = \sqrt{\frac{2}{\pi}} \, \exp\left[\frac{-(v-1)^2}{2(0.5)^2}\right]$$

Since any value received below a level of 0.5 is called 0, the probability that a transmitted 1 will be interpreted as a 0 at the receiver is simply the probability that the random variable V is less than 0.5. This is given by the integral

$$\int_{-\infty}^{0.5} P_V(v)dv = \sqrt{\frac{2}{\pi}} \int_{-\infty}^{0.5} \exp\left[-2(v-1)^2\right]dv$$

To reduce this equation to a form that can be found in a table of error functions, we make the change of variable

$$u = \sqrt{2}(v - 1)$$

to get

$$\Pr(\text{error}) = \frac{1}{\sqrt{\pi}} \int_{-\infty}^{-\frac{\sqrt{2}}{2}} e^{-u^2}du$$

This is not yet in the form of an error function. However, since $\exp(-u^2)$ is an even function, we can take the mirror image of the integral limits without changing the value of the integral:

$$\Pr(\text{error}) = \frac{1}{\sqrt{\pi}} \int_{\frac{\sqrt{2}}{2}}^{\infty} e^{-u^2}du$$

This is now seen to be related to the complementary error function:

$$\Pr(\text{error}) = \frac{1}{2} \, \text{erfc}\left(\frac{\sqrt{2}}{2}\right) = 0.16$$

Thus, on the average, one would expect 16 out of every 100 transmitted 1's to be misinterpreted as 0's at the receiver. This is an extremely poor level of performance.

4.2.2 Rayleigh Density Function

The *Rayleigh density function* is defined as

$$P_X(x) = \begin{cases} \dfrac{x}{K^2} \exp\left(\dfrac{-x^2}{2K^2}\right), & x > 0 \\[2ex] 0, & x < 0 \end{cases} \tag{4.42}$$

where K is a given constant. Figure 4.9 shows the density function for two different values of K.

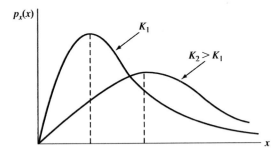

Figure 4.9   Rayleigh density function.

The Rayleigh density function is related to the Gaussian density function. In fact, the square root of the sum of the squares of two zero-mean Gaussian-distributed random variables is itself Rayleigh. If we transform from rectangular to polar coordinates, the radius is

$$r = \sqrt{x^2 + y^2}$$

If x and y are both Gaussian, in most cases (the restriction is one of *independence*) r will be Rayleigh. As an example, suppose that you were throwing darts at a target on a dart board and that the horizontal and vertical components of your error were Gaussian distributed with zero mean (a fair assumption if you are not biased by wind, gravity, or a muscle twitch). The *distance* from the center of the target to the position of the dart would then be Rayleigh distributed.

The *chi-square* distribution is closely related to the Rayleigh. If we were to consider r^2 as the variable instead of r, we would find that variable to be chi-square distributed. That is, a chi-square distribution results from summing the squares of Gaussian variables. If we sum two such variables, the result is chi square with *two degrees of freedom*. In general, if

$$z = x_1^2 + x_2^2 + x_3^2 + \ldots + x_n^2$$

and the x_i are Gaussian with $m = 0$, z will be chi square with n degrees of freedom. If the Gaussian variables all have $\sigma = 1$, the actual form of the density is given by

$$p(z) = \begin{cases} \dfrac{(z)^{n/2-1}}{2^{n/2}(n/2 - 1)!} e^{-z/2}, & z > 0 \\[2ex] 0, & z < 0 \end{cases} \tag{4.43}$$

The "!" in Eq. (4.43) indicates the *factorial* operation.

4.2.3 Exponential Random Variables

Occasionally, we deal with random variables that are exponentially distributed. This occurs in problems where we view a pattern of waiting times. (Such a pattern is important in communication network traffic studies.) It also shows up in examining the life of some systems, where we are interested in the *mean time between failures* (*MTBF*).

The exponential density is defined as

$$p_X(x) = \begin{cases} \dfrac{1}{m} e^{-x/m}, & x > 0 \\ \\ 0, & x < 0 \end{cases}$$

(4.44)

This density is shown in Fig. 4.10. The parameter m is the mean value of the variable, as can be verified by a simple integration. (Convince yourself that you can do this.)

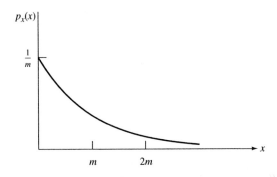

Figure 4.10   Exponential density function.

4.2.4 Ricean Density Function

In Section 4.2.2, we stated that the Rayleigh density function results when we take the square root of the sum of the squares of two zero-mean Gaussian densities. When we analyze digital communication systems in the absence of a signal (i.e., when noise alone is present), we often are dealing with variables that are Rayleigh distributed.

We also encounter situations where a transmitted signal is embedded in noise (which adds in the channel). In these cases, our receivers will sometimes effectively take the square root of the sum of the squares of two quantities. One of these will be zero-mean Guassian distributed, but the other results from an addition of signal and noise. The quantity we then observe is of the form

$$z = \sqrt{[s + x]^2 + y^2}$$

(4.45)

In this equation, x and y are zero-mean Gaussian random functions, and s is the signal. If s is zero, z follows a Rayleigh density function. If s is not zero, z follows a more complex density known as a *Ricean density*. This density is given by the equation

$$p(z) = \frac{z}{\sigma^2} \exp\left[-\frac{1}{2\sigma^2} (z^2 + s^2) \right] I_0\left(\frac{sz}{\sigma^2}\right)$$

(4.46)

In Eq. (4.46), s is the value of the signal (this will normally be a specific time sample $s(T)$), and I_0 is a *modified Bessel function of zero order*. We discuss Bessel functions in Chapter 6 (when we deal with FM). For now, you can think of them as functions you look up in a table. You simply plug in s, z, and σ^2 to find the argument of the Bessel function, and then you look in a table of Bessel functions. Note that when $s = 0$, Eq. (4.46) reduces to the Rayleigh density $[I_0(0) = 0]$.

4.3 RANDOM PROCESSES

Some of the noise encountered in a communication system is *deterministic*, as in the case of some types of jamming signals in a radar system. In these cases, the noise can be described completely by a formula, graph, or equivalent technique. Other types of noise are composed of components from so many sources that we find it more convenient to analyze them as *random processes*. Each time the experiment is performed, we observe a *function of time* as the outcome. This contrasts with the one-dimensional outcomes studied earlier in the chapter.

We usually assume that additive noise is a random process that is *Gaussian*. This means that if the process is sampled at any point in time, the probability density of the sample is Gaussian. Such an assumption proves important, because if we start with a Gaussian random process and put it through any linear system, the system output will also be a Gaussian process. This fact will be used many times in examining the performance of various digital communication systems.

In addition to knowing that the noise process is Gaussian, it will be necessary to know something about the relationship between various time samples. We do this by examining the *autocorrelation function* and its Fourier transform, known as the *power spectral density*.

Until this point, we have considered single random variables. All averages and related parameters were simply numbers. Now we shall add another dimension to the study: time. Instead of talking about numbers only, we will now be able to characterize random functions. The advantages of such a capability should be clear. Our approach to random function analysis begins with the consideration of discrete-time functions, since these will prove to be a simple extension of random variables.

Imagine a single die being flipped 1,000 times. Let X_i be the random variable assigned to the outcome of the ith flip. Now list the 1,000 values of the random variables

$$x_1, x_2, x_3, ..., x_{999}, x_{1,000}$$

For example, if the random variable is assigned to be equal to the number of dots on the top face after flipping the die, a typical list might resemble the following:

$$4,6,3,5,1,4,2,5,3,1,4,5,....$$

Suppose that all possible sequences of random variables are now listed. We would then have a collection of $6^{1,000}$ entries, each one resembling the sequence shown. This collection is known as the *ensemble* of possible outcomes. Together with associated statistical properties, the ensemble forms a *random process*. In this particular example, the process is a discrete-valued and discrete-time process.

If we were to view one digit—say, the third entry—a random variable would result. In this example, the random variable would represent that assigned to the outcome of the third flip of the die.

We can completely describe the preceding random process by specifying the probability density function

$$p(x_1, x_2, x_3, ..., x_{999}, x_{1,000})$$

This 1,000-dimensional probability density function is used in the same way as a one-dimensional probability density function. Its integral must be unity:

$$\int_{-\infty}^{\infty}\int_{-\infty}^{\infty}...\int_{-\infty}^{\infty} p(x_1,x_2,x_3,..., x_{999}, x_{1,000})dx_1 dx_2...dx_{1,000} = 1 \qquad (4.47)$$

The probability that the variables fall within any specified 1,000-dimensional volume is the integral of the probability density function over that volume. The expected value of any single variable can be found by integrating the product of that variable and the probability density function. Thus, the expected value of x_1 is

$$\int_{-\infty}^{\infty}\int_{-\infty}^{\infty}...\int_{-\infty}^{\infty} x_1 p(x_1,x_2,x_3,..., x_{999}, x_{1,000})dx_1 dx_2...dx_{1,000} \qquad (4.48)$$

This is known as a *first-order average*.

In a similar manner, the expected value of any multidimensional function of the variables is found by integrating the product of that function and the density function. Thus, the expected value of the product $x_1 x_2$ is

$$\int_{-\infty}^{\infty}\int_{-\infty}^{\infty}...\int_{-\infty}^{\infty} x_1 x_2\, p(x_1,x_2,x_3,..., x_{999}, x_{1,000})dx_1 dx_2...dx_{1,000} \qquad (4.49)$$

This is a *second-order average*.

We can continue this process for higher order averages. In many instances, however, it proves sufficient to specify only the first- and second-order averages (moments). That is, one would specify

$$E\{x_i\} \quad \text{and} \quad E\{x_i x_j\} \qquad \text{for all } i \text{ and } j$$

Since we are interested in continuous functions of time, we now extend our example to the case of an infinite number of random variables in each list and an infinite number of lists in the ensemble. It may be helpful to refer back to the simple case (i.e., 1,000 flips of the die) from time to time.

The most general form of the random process results if our simple experiment yields an infinite range of values as the outcome and if the period between performances of the experiment approaches zero.

Another way to arrive at a random process is to perform a discrete experiment, but assign a *function of time* instead of a number to each outcome. When the samples of the process are functions of time, we call this a *stochastic process*. As an example, consider an experiment defined by picking a 6-volt dc generator from an infinite inventory in a warehouse. The voltage of the selected (call it the *i*th) generator, $v_i(t)$, is then displayed on an oscilloscope. The waveform $v_i(t)$ is a *sample function* of the process. It will not be a perfect constant of 6 volts, due to imperfections in the generator construction and also due to rf pickup when the wires are acting as an antenna. Since we assume an unlimited inventory of generators, there is an infinite number of possible sample functions of the process. This infinite group of samples forms the ensemble. Each time we choose a generator and measure its voltage, a sample function from the infinite ensemble results.

If we were to sample the voltage at a specific time $t = t_0$, the sample $v(t_0)$ would be a random variable. Since $v(t)$ is assumed to be continuous with time, there is an infinity of random variables associated with the process.

Having introduced the concept with the generator example, let us now speak in general terms. Let $x(t)$ represent a stochastic process. $x(t)$ is then an infinite ensemble of all possible sample functions. For every specific value of time $t = t_0$, $x(t_0)$ is a random variable.

Suppose we now examine the first- and second-order averages of the process. Note that instead of having a discrete list of numbers, as in the die example, we have continuous functions of time. The first moment is then a function of time. This mean value is given by

$$m(t) = E\{x(t)\} \tag{4.50}$$

The second moments are found by averaging the product of two different time samples. We use the symbol R_x for this moment and call it the *autocorrelation*. The autocorrelation is then given by

$$R_x(t_1,t_2) = E\{x(t_1)x(t_2)\} \tag{4.51}$$

Both the mean and autocorrelation must be thought of as averages taken over the entire ensemble of functions of time. To find $m(t_0)$, we must average all samples across the ensemble at time t_0. To find the autocorrelation, we must average the product of $x(t_1)$ and $x(t_2)$ across the ensemble. This is generally difficult, and the ensemble averages are virtually impossible to ascertain experimentally.

In practice, we could find the mean by measuring the voltage of a great number of generators at time t_0 and average the resulting numbers. For the example of dc generators, we would expect this average to be independent of t_0. Indeed, most processes we consider have mean values that are independent of time.

A process with *overall statistics* that are independent of time is called a *stationary* (or *strict-sense stationary*) process. If only the mean and second moment are independent of time, the process is *wide-sense stationary*. Since we are interested primarily in power, and power depends upon the second moment, wide-sense stationarity will be sufficient for our analyses.

If a process $x(t)$ is stationary, then the shifted process $x(t - T)$ has the same statistics, independently of the value of T. Clearly, for a stationary process, $m(t_0)$ cannot depend upon t_0.

Viewing the autocorrelation of a stationary process, we have

$$R_x(t_1,t_2) = E\{x(t_1)x(t_2)\} = E\{x(t_1 - T)x(t_2 - T)\} \tag{4.52}$$

In the last equality, we have shifted the process by an amount T. If we now let $T = t_1$, we find that

$$R_x(t_1,t_2) = E\{x(0)x(t_2 - t_1)\} \tag{4.53}$$

Equation (4.53) indicates that the autocorrelation of a stationary (the wide sense is suffi-cient) process depends only on the time spacing t_2-t_1 between the two samples. That is, the left-hand time point can be placed anywhere, and as long as the right-hand point is separated from this by t_2-t_1, the autocorrelation remains unchanged. Since the indepen-dent variable of the autocorrelation is effectively one dimensional instead of two dimen-sional, we use the argument τ and refer to the autocorrelation of a stationary process as $R_x(\tau)$. Thus,

$$R_x(\tau) = E\{x(t)x(t - \tau)\} = E\{x(t)x(t + \tau)\} \tag{4.54}$$

The last equality results from adding τ to each of the arguments. This shows that autocor-relation is an even function.

If t_2 and t_1 are widely separated such that $x(t_1)$ and $x(t_2)$ are independent, the autocor-relation reduces to

$$R_x(\tau) = E\{x(t)x(t - \tau)\} = E\{x(t)\}\,E\{x(t - \tau)\} = m^2 \tag{4.55}$$

The average value of a random function of time is the dc value, and most communi-cation channels will not pass dc. (They contain bandpass filters.) Therefore, most of the processes we consider have mean values equal to zero. In that case, the value of τ at which $R_x(\tau)$ goes to zero represents the time over which the process is correlated. If two samples are separated by this length of time, one sample has no effect upon the other.

Example 4.8

Suppose you are given a stochastic process $x(t)$ with mean value m and autocorrelation $R_x(\tau)$. This tells you immediately that the process is at least wide-sense stationary. If it were not, the mean and autocorrelation could not have been given in the form they were. Find the mean and autocorrelation of the process

$$y(t) = x(t) - x(t - T)$$

Solution: To solve problems of this type, we need only recall the definition of the mean and autocorrelation and the fact that taking expected values is a linear operation. We then have

$$m_y = E\{y(t)\} = E\{x(t) - x(t - T)\}$$
$$= E\{x(t)\} - E\{x(t-T)\}$$
$$= m_x - m_x = 0$$
$$R_y(\tau) = E\{y(t)y(t + \tau)\}$$
$$= E\{[x(t) - x(t - T)][x(t + \tau) - x(t + \tau - T)]\}$$
$$= E\{x(t)x(t + \tau)\} - E\{x(t)x(t + \tau - T)\}$$
$$\quad - E\{x(t - T)x(t + \tau)\} + E\{x(t - T)x(t + \tau - T)\}$$
$$= R_x(\tau) - R_x(\tau - T) - R_x(\tau + T) + R_x(\tau)$$
$$= 2R_x(\tau) - R_x(\tau - T) - R_x(\tau+T)$$

We note that the process $y(t)$ is wide-sense stationary. If this were not the case, both m_y and R_y would be functions of t.

Example 4.9

Consider the experiment of starting a sinusoidal generator of deterministic frequency f_o and amplitude A. The exact starting time is random. Thus,

$$x(t) = A \sin (2\pi f_0 t + \theta)$$

where the phase θ is a random variable with uniform density, as shown in Fig. 4.11. Find the autocorrelation of the random process.

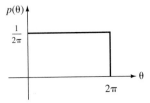

Figure 4.11   Density of phase for Example 4.9.

Solution: The autocorrelation is found directly from the definition:

$$R_x(\tau) = E\{x(t)x(t + \tau)\}$$

$$= E\{A^2 \sin (2\pi f_0 t + \theta) \sin [2\pi f_0(t + \tau) + \theta]\}$$

We can use trigonometric identities to rewrite this as

$$R_x(\tau) = E\left\{\frac{A^2}{2}(\cos 2\pi f_0 \tau - \cos [2\pi f_0(2t + \tau) + \theta])\right\}$$

The term $(A^2/2)(\cos 2\pi f_0 \tau)$ is not random, so its expected value is itself. The expected value of $\cos[2\pi f_0 (2t + \tau) + \theta]$ is found by integrating its product with the density of θ. For the entire expression, we obtain

$$R_x(\tau) = \frac{1}{2}A^2 \cos 2\pi f_0 \tau - \frac{A^2}{4\pi} \int_0^{2\pi} \cos [2\pi f_0(2t + \tau) + \theta] \, d\theta$$

The second term on the right represents the integral of a cosine function over two entire periods. This integral is equal to zero. Thus,

$$R_x(\tau) = \frac{1}{2}A^2 \cos 2\pi f_0 \tau$$

Time Averages

Suppose you were asked to find the average value of the voltage in the dc generator example. You would have to measure the voltage of many generators (at any given time) and then compute the average. Once you were told that the process is stationary, you would probably be tempted to take one generator and average its voltage over a large time interval. Using either technique, you would expect to get 6 volts as the result. That is, you would reason that

$$m_v = E\{v(t)\} = \lim_{T \to \infty} \frac{1}{T} \int_{-\frac{T}{2}}^{\frac{T}{2}} v_i(t) \, dt \tag{4.56}$$

Here, $v_i(t)$ is one sample function of the ensemble. This approach does not always result in the correct answer. Suppose, for example, that one of the generators was burned out and you happened to choose that particular generator. You would erroneously think that $m_v = 0$.

Most of the processes we encounter have the property that any sample function contains all of the essential information about the process. Such processes are known as *ergodic*. The generator process is ergodic as long as none of the generators is "exceptional."

A process that is ergodic must also be stationary. This is so because, once we agree that all averages can be found from a single time sample, the averages can no longer be a function of the time at which they are computed. Alternatively, a stationary process need not be ergodic. (Consider the aforementioned burned-out generator.)

The autocorrelation of an ergodic process is given by

$$R_x(\tau) = E\{x(t)x(t + \tau)\} = \lim_{T \to \infty} \frac{1}{T} \int_{-\frac{T}{2}}^{\frac{T}{2}} x(t)x(t + \tau) \, dt \tag{4.57}$$

The autocorrelation is a function of time. We define $G(f)$ as the Fourier transform of the autocorrelation. $G(f)$ is called the *power spectral density* for reasons that will become obvious in a moment. We have

$$G_x(f) = \mathcal{F}[R_x(t)] = \int_{-\infty}^{\infty} R_x(t)e^{-j2\pi t} dt \tag{4.58}$$

The autocorrelation is then the inverse transform of the power spectral density:

$$R_x(t) = \mathcal{F}^{-1}[G_x(f)] = \int_{-\infty}^{\infty} G_x(f) e^{j2\pi ft} df = 2 \int_{0}^{\infty} G_x(f) e^{j2\pi ft} df \tag{4.59}$$

In the last equality, we have doubled the positive half-range of the integral. This results from the fact that the power spectral density must be real and even, since the autocorrelation is even.

We can now relate the average power to the power spectral density:

$$P_{av} = E\{x^2(t)\} = R_x(0) = 2 \int_{0}^{\infty} G_x(f) \, df \tag{4.60}$$

Equation (4.60) is a very important result. It says that to find the power of a random function of time, we integrate the power spectral density over all positive values of f (the frequency variable) and then double the result.

The other important result we need is the effect that a filter has on the power of a random signal. If a stochastic process forms the input to a filter, as shown in Fig. 4.12, the out-

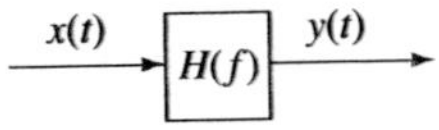

Figure 4.12   Stochastic process as input to filter.

put is also a stochastic process. That is, each sample function of the input process yields a sample function of the output process. We wish to find the statistics of the output process.

We begin with the mean value,

$$E\{y(t)\} = E\left\{\int_{-\infty}^{\infty} h(\tau)x(t - \tau)d\tau\right\} \tag{4.61}$$

The average of a sum is the sum of the averages. Therefore, with some broad restrictions (a finite mean value and a stable system), we can interchange the order of taking the expected value and integrating. If we assume that $x(t)$ is stationary, we have

$$E\{y(t)\} = \int_{-\infty}^{\infty} E\{h(\tau)x(t - \tau)\}\, d\tau = \int_{-\infty}^{\infty} h(\tau)E\{x(t - \tau)\}\, d\tau \tag{4.62}$$

$$= m_x \int_{-\infty}^{\infty} h(\tau)\, d\tau = m_x H(0)$$

Most of the random processes we encounter in this text have zero average value. If the input to the filter has zero average value, the output mean is also zero.

We now evaluate the autocorrelation of the output process. We assume that the input process is stationary. Then

$$R_y(t_1, t_2) = E\{y(t_1)y(t_2)\}$$

$$= E\left\{\int_{-\infty}^{\infty} h(\tau)x(t_1 - \tau)d\tau \int_{-\infty}^{\infty} h(\tau)x(t_2 - \tau)d\tau\right\} \tag{4.63}$$

Once again, we interchange the order of taking the expected value and integrating. We combine the two integrals (using two different symbols for the dummy variable of integration) to get

$$R_y(t_1, t_2) = \int_{-\infty}^{\infty} h(\tau_1)d\tau_1 \int_{-\infty}^{\infty} E\{x(t_1 - \tau_1)x(t_2 - \tau_2)\}h(\tau_2)d\tau_2$$

$$= \int_{-\infty}^{\infty} h(\tau_1)d\tau_1 \int_{-\infty}^{\infty} R_x(t_1 - t_2 + \tau_2 - \tau_1)h(\tau_2)d\tau_2 \tag{4.64}$$

Note that the result depends, not on the values of t_1 and t_2, but only on their difference. Therefore, the output process is wide-sense stationary. The autocorrelation is a function of only one variable, the spacing between the two time points. We have used the notation τ for this spacing. Doing the same here, we find that the result becomes

$$R_y(\tau) = \int_{-\infty}^{\infty}\int_{-\infty}^{\infty} R_x(\tau - \tau_1 + \tau_2)h(\tau_1)h(\tau_2)d\tau_1 d\tau_2 \tag{4.65}$$

Equation (4.65) shows that the output autocorrelation is the result of convolving the input autocorrelation, first with $h(t)$ and then with $h(-t)$. Therefore,

$$R_y(t) = R_x(t)*h(t)*h(-t) \tag{4.66}$$

Taking the Fourier transform of this equation, and recognizing that the Fourier transform of $h(-t)$ is the complex conjugate of the transform of $h(t)$, we have

$$G_y(f) = G_x(f)H(f)H^*(f) = G_x(f)|H(f)|^2 \tag{4.67}$$

Equation (4.67) has the intuitive interpretation. $G(f)$ is the power spectral density. It is not surprising that the output power spectral density is weighted by the square of the magnitude of the transfer function, since this is what would happen to the power of a single sinusoid that goes through the filter.

Example 4.10

A signal that is received is made up of two components: signal and noise. That is,

$$r(t) = s(t) + n(t)$$

The signal can be considered a sample of a random process because random amplitude fluctuations are introduced by turbulence in the air. You are told that the autocorrelation of the signal process is

$$R_s(\tau) = 2e^{-|\tau|}$$

The noise is a sample function of a random process with autocorrelation

$$R_n(\tau) = e^{-2|\tau|}$$

Both processes have zero mean value, and they are independent of each other.
Find the autocorrelation and total power of $r(t)$.

Solution: From the definition of autocorrelation, we have

$$R_r(\tau) = E\{r(t)r(t+\tau)\}$$
$$= E\{[s(t) + n(t)][s(t + \tau)n(t + \tau)]\}$$
$$= E\{s(t)s(t + \tau)\} + E\{s(t)n(t + \tau)\}$$
$$= E\{n(t)s(t + \tau)\} + E\{n(t)n(t + \tau)\}$$

Since the signal and noise are independent,

$$E\{s(t + \tau)n(t)\} = E\{s(t + \tau)\}E\{n(t)\} = 0$$

and

$$E\{s(t)n(t + \tau)\} = E\{s(t)\}E\{n(t + \tau)\} = 0$$

Finally, the autocorrelation is

$$R_r(\tau) = R_s(\tau) + R_n(\tau) = 2e^{-|\tau|} + e^{-2|\tau|}$$

The total power of $r(t)$ is $R_r(0)$, or 3 watts.

Example 4.11 (the random telegraph signal)

Evaluate the autocorrelation of the *random telegraph waveform*, as shown in Fig. 4.13. This is a binary waveform that can take on one of two values, $+A$ or $-A$. The probabilities of each

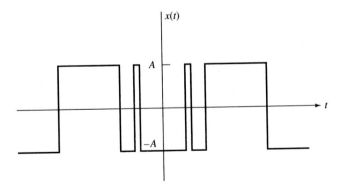

Figure 4.13   Random telegraph waveform.

of these values are equal (i.e., $\frac{1}{2}$). Assume that transitions occur randomly and that there is an average of λ transitions per second. The probability that n transitions occur in a positive time interval τ is given by a *Poisson distribution*,

$$\Pr(n,\tau) = \frac{(\lambda\tau)^n}{n!}\,e^{-\lambda\tau}$$

Solution: We first find the autocorrelation of the process:

$$R_x(\tau) = E\{x(t)x(t+\tau)\}$$

The product inside the braces is either $+A^2$ or $-A^2$. The plus sign obtains if there is an even number of transitions in the interval, and the minus sign obtains if there is an odd number. For a given value of τ, the probability of an even number of transitions is found by summing the Poisson probability distribution over all even values of n. Similarly, the probability of an odd number of transitions is the sum of the Poisson distribution over all odd values of n. Therefore,

$$PR(\text{even}) = e^{-\lambda\tau} \sum_{\substack{n=0 \\ n=\text{even}}}^{\infty} \frac{(\lambda\tau)^n}{n!}$$

$$PR(\text{odd}) = e^{-\lambda\tau} \sum_{\substack{n=1 \\ n=\text{odd}}}^{\infty} \frac{(\lambda\tau)^n}{n!}$$

The autocorrelation is then

$$R_x(\tau) = A^2\,\Pr(\text{even}) - A^2\,\Pr(\text{odd})$$

$$= A^2 e^{-\lambda\tau} \sum_{n=0}^{\infty} (-1)^n \frac{(\lambda\tau)^n}{n!}$$

$$= A^2 e^{-\lambda\tau} e^{-\lambda\tau} = A^2 e^{-2\lambda\tau}$$

The result applies for positive time invervals. We know that the autocorrelation must be an even function, so we can write the autocorrelation as

$$R_x(\tau) = A^2 e^{-2\lambda|\tau|}$$

This result is shown in Fig. 4.14.

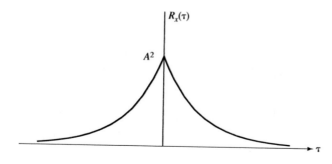

Figure 4.14 Autocorrelation of random telegraph wave.

4.4 WHITE NOISE

Suppose that $x(t)$ is a stochastic process with a constant power spectral density, as shown in Fig. 4.15. This process contains all frequencies "to an equal degree." Since white light is composed of all frequencies (colors), the process is known as *white noise*.

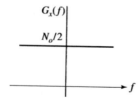

Figure 4.15 Power spectral density of white noise.

Suppose now that white noise forms the input to an ideal bandpass filter with a passband extending from a low-frequency cutoff of f_L to a high-frequency cutoff of f_H. Then the power of the output for the filter is (refer to Eq. (4.67))

$$P_{\text{out}} = 2 \int_0^\infty G_x(f)|H(f)|^2 \, df = 2 \int_{f_L}^{f_H} G_x(f) \, df = 2 \int_{f_L}^{f_H} \frac{N_0}{2} \, df = N_0(f_H - f_L) \qquad (4.68)$$

The output of the bandpass filter consists of all components of the input lying within the passband of the filter. The output power can therefore be considered to be that portion of the input power in the frequency range between f_L and f_H. We see from Eq. (4.68) that this is proportional to the bandwidth, with the proportionality factor being N_0. Therefore, N_0 is the *power per Hz* of the noise waveform. The total power in a band of frequencies is the product of N_0 with the bandwidth.[4]

[4]The power spectral density of Fig. 4.15 is known as the *two-sided power spectral density*. Since the power spectral density is real and even, and the equation for power contains a factor of two, we sometimes define a *one-sided power spectral density* with a value twice that of the two-sided density. The one-sided density of white noise therefore has a height of N_0 instead of $N_0/2$, and to find the power in a band of frequencies, we simply integrate (without the factor of two).

The autocorrelation of white noise is the inverse Fourier transform of the power spectral density. Therefore,

$$R_x(\tau) = \frac{N_0}{2} \delta(\tau) \tag{4.69}$$

The average power of a stochastic process is $R_x(0)$, which, for this case, is infinity. Therefore, white noise cannot exist in real life. (Thankfully, signals with infinite power do not exist; were this not the case, you probably would not be here to read this text.) However, many types of noise that are encountered can be assumed to be approximately white.

Example 4.12

White noise forms the input to the RC circuit of Fig. 4.16. Find the autocorrelation and power spectral density at the output of the filter.

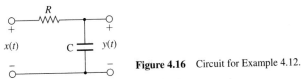

Figure 4.16   Circuit for Example 4.12.

Solution: The output power spectral density is the input density multiplied by the square of the magnitude of the transfer function:

$$G_y(f) = G_x(f)|H(f)|^2 = \frac{N_0/2}{1 + (2\pi f)^2 C^2 R^2}$$

The output autocorrelation is the inverse Fourier transform of $G_y(f)$. Therefore,

$$R_y(\tau) = \frac{N_0/2}{2RC} \exp\left(\frac{-|\tau|}{RC}\right)$$

Example 4.13

Repeat Example 4.12 for an ideal lowpass filter with cutoff frequency f_m, as shown in Fig. 4.17.

Solution: Once again, the output power spectral density is found from the input density by

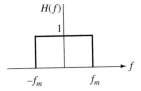

Figure 4.17   Ideal lowpass filter for Example 4.13.

multiplying it by the square of the magnitude of the transfer function:

$$G_y(f) = G_x(f)|H(f)|^2$$

$$= \begin{cases} N_0/2, & |f| < f_m \\ 0, & \text{otherwise} \end{cases}$$

The autocorrelation is the inverse Fourier transform of $G_y(f)$. Thus,

$$R_y(\tau) = \frac{N_0}{2} \frac{\sin 2\pi f_m \tau}{\pi \tau}$$

The output power spectral density and autocorrelation are shown in Fig. 4.18.

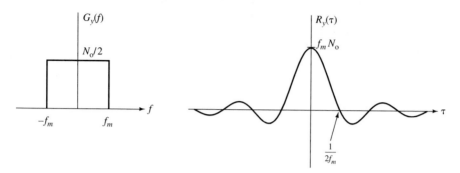

Figure 4.18   Autocorrelation and power spectrum for Example 4.13.

Suppose that in Example 4.13 the input process had a power spectral density as shown in Fig. 4.19, where $f_1 > f_m$. The output process would then be identical to that found in the example. Therefore, if a system exhibits an upper cutoff frequency, and the input noise has a flat spectrum up to this cutoff frequency, we can consider the input noise to be white. This will simplify the analysis.

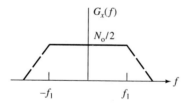

Figure 4.19   Nonwhite power spectral density.

Since the autocorrelation of white noise is zero for $t \neq 0$, two different time samples of white noise are uncorrelated even if they are taken very close together. Thus, knowing the sample value of white noise at one instant of time tells us absolutely nothing about its value an instant later.[5] From a practical standpoint, this is an unfortunate situation. It would appear to make the elimination of noise more difficult.

Up to this point, we have said nothing about the actual probability distributions of the process. We have talked only about the first and second moment. There is an infinity of random processes with the same first and second moments. (The analogy to mechanics is that, given the center of gravity and moment of inertia of an object, the exact shape can be any of an infinity of possibilities.) Each random variable $x(t_0)$ has a certain probability density. By considering only the mean and second moment, we are not telling the whole story.

---

[5]The zero correlation is a necessary, but not sufficient, condition for independence. However, for Gaussian processes, two uncorrelated samples are always independent.

Because of the central limit theorem, most processes we encounter are Gaussian. Once we know that a random variable is Gaussian, the density function is completely specified by its mean and second moment.

We now examine several types of noise encountered in communication systems and determine whether the white noise model is appropriate for these noise sources.

Thermal Noise

Thermal noise is produced by the random motion of electrons in a medium. The intensity of this motion increases with increasing temperature and is zero only at a temperature of absolute zero.

If the voltage across a resistor is examined using a sensitive oscilloscope, a random pattern will be displayed on the screen. The power spectral density of this random process is of the form

$$G(f) = \frac{A|f|}{e^{B|f|} - 1} \tag{4.70}$$

where A and B are constants that depend on temperature and other physical constants. Figure 4.20 shows the curve of Eq. (4.70). For frequencies below the knee of the curve, $G(f)$ is almost constant. If we operate in this frequency range, we can consider thermal noise to be white noise. Thermal noise appears to be approximately white up to extremely high fre-

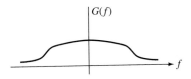

Figure 4.20   Power spectral density of thermal noise.

quencies, on the order of 10^{13} Hz. For frequencies within this range, the mean square value of the voltage across the resistor $[R(0)]$ has been shown to be

$$\overline{v^2} = R(0) = 4kTRB \tag{4.71}$$

where k is Boltzmann's constant (1.38×10^{-23} J/°K), T is the temperature in degrees Kelvin, R is the resistance value, and B is the observation bandwidth. This means that the height of the spectral density over the constant region is $2kTR$.

Of more practical concern is the actual power generated by a resistor. That is, if a resistor is connected to additional circuitry, how much noise power is generated in that additional circuitry? We know from basic circuit theory that this depends on the impedance of the external circuit. Specifically, the power transferred is a maximum when the load impedance matches the generator impedance. This yields the *maximum available power*, which (using a voltage divider relationship) is

$$N = \frac{v^2}{4R} = kTB$$

with a corresponding power spectral density of

$$G_n(f) = \frac{kT}{2} \tag{4.72}$$

Equation (4.72) yields the power spectral density of the available noise power from a resistor.

If we have a system with a number of noise-generating devices within it, we often refer to the system *noise temperature*, T_e, in degrees Kelvin. This is the temperature of a single noise source that would produce the same total noise power at the output.

If the input to the system contains noise, the system then adds its own noise to produce a larger output noise. The system *noise figure* is the ratio of noise power at the output to that at the input. It is usually expressed in dB. For example, a noise figure of 3 dB indicates that the system is adding an amount of noise equal to that which appears at the input, so the output noise power is twice that of the input.

Other Forms of Noise

Shot noise (or *quantum noise*) occurs because, although we think of current as being continuous, it is actually a discrete phenomenon. In fact, current occurs in discrete pulses each time an electron moves across an observation point. A plot of current as a function of time would resemble that of Fig. 4.21. Shot noise is the variation of current around the average value. As in the case of thermal noise, the power spectral density of shot noise is approximately flat within the range of frequencies of interest to us.

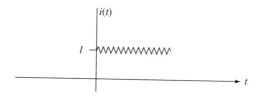

Figure 4.21 Shot noise.

Flicker noise (or *1/f noise*) occurs in electronic devices. It arises out of surface imperfections resulting from the fabrication process. Its power spectral density decreases inversely with increasing frequency. Flicker noise is most important at low frequencies (below about 100 Hz). Thus, although it cannot be approximated as white noise, it usually is negligible at the frequencies at which communication systems operate.

4.5 NARROWBAND NOISE

Most communication systems with which we deal contain bandpass filters. Therefore, white noise appearing at the input to the system will be shaped into bandlimited noise by the filtering operation. If the bandwidth of the noise is relatively small compared to the center frequency, we refer to this as *narrowband noise*. We have no problem deriving the power spectral density and autocorrelation of this noise, and these quantities are sufficient to analyze the effect of linear systems. However, we will often be dealing with multipliers, and the frequency analysis approach is not sufficient, since nonlinear operations are present. In such cases, it proves useful to have a trigonometric expansion for the noise signals. The form of this expansion is

$$n(t) = x(t)\cos 2\pi f_0 t - y(t)\sin 2\pi f_0 t \tag{4.73}$$

In equation (4.73), $n(t)$ is the noise waveform and f_0 is a frequency (often the center frequency) within the band occupied by the noise. Since the sine and cosine vary by 90 degrees, $x(t)$ and $y(t)$ are known as the *quadrature components* of the noise.

Equation (4.73) can be derived by starting with exponential notation. We have

$$n(t) = \text{Re}\{r(t)e^{j2\pi f_0 t}\} \tag{4.74}$$

where $r(t)$ is a complex function with a low-frequency bandlimited Fourier transform, Re is the real part of the expression in brackets that follows it, and the exponential function has the effect of shifting the frequencies of $r(t)$ by f_0. Expanding the exponential by means of Euler's identity and letting $x(t)$ be the real part of $r(t)$ and $y(t)$ be the imaginary part, we have

$$\begin{aligned} n(t) &= \text{Re}\{[x(t) + jy(t)] (\cos 2\pi f_0 t + j\sin 2\pi f_0 t)\} \\ &= x(t)\cos 2\pi f_0 t - y(t)\sin 2\pi f_0 t \end{aligned} \tag{4.75}$$

This is the same as Eq. (4.73).

Solving Eq. (4.73) explicitly for $x(t)$ and $y(t)$ is not simple. One way to do so is by using Hilbert transforms.

Hilbert Transform

The *Hilbert transform* of a function of time is obtained by shifting all frequency components by $-90°$. The Hilbert transform operation can therefore be represented by a linear system, with $H(f)$ as shown in Fig. 4.22.

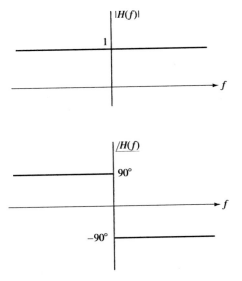

Figure 4.22   Hilbert transform operation.

Note that the phase function of a real system must be odd. The system function is then given by

$$H(f) = -j \, \text{sgn} \, (f) \tag{4.76}$$

The impulse response of this system is the inverse transform of $H(f)$. This is given by

$$h(t) = \frac{1}{\pi t} \tag{4.77}$$

The Hilbert transform of $s(t)$ is then given by the convolution of $s(t)$ with $h(t)$. Let us denote the transform by $\hat{s}(t)$. Then

$$\hat{s}(t) = -\frac{1}{\pi} \int_{-\infty}^{\infty} \frac{s(\tau)}{t - \tau} d\tau \tag{4.78}$$

If we take the Hilbert transform of a Hilbert transform, the effect in the frequency domain is to multiply the transform of the signal by $H^2(f)$. But $H^2(f) = -1$, so we return to the original signal which is a change of sign. This indicates that the inverse Hilbert transform equation is the same as the transform relationship, except with a minus sign. Therefore,

$$s(t) = -\frac{1}{\pi} \int_{-\infty}^{\infty} \frac{\hat{s}(\tau)}{t - \tau} d\tau \tag{4.79}$$

Example 4.14

Find the Hilbert transform of the following time signals:

(a) $s(t) = \cos(2\pi f_0 t + \theta)$
(b)
$$s(t) = \frac{\sin 2\pi t}{t} \cos 200 \, \pi t$$

(c)
$$s(t) = \frac{\sin 2\pi t}{t} \sin 200 \, \pi t$$

Solution: Although the Hilbert transform is defined by a convolution operation, it is almost always easier to avoid time convolution by working with Fourier transforms.
(a) The Fourier transform of $s(t)$ is

$$S(f) = \frac{1}{2}[\delta(f - f_0) + \delta(f + f_0)] \, e^{-j\theta f/f_0}$$

Note that the phase shift of θ radians is equivalent to a time shift of $\theta/2\pi f_0$ seconds. We now multiply this by $-j\text{sgn}(f)$ to get

$$\hat{S}(f) = \frac{1}{2} \, [-j\delta(f - f_0) + j\delta(f + f_0)] e^{-j\theta f/f_0}$$

The quantity in square brackets is the Fourier transform of a sine wave. Therefore,

$$\hat{s}(t) = \sin \, (2 \, \pi \, f_0 t + \theta)$$

This result is not surprising, since the Hilbert transform is a 90-degree phase-shifting operation.

(b) Let

$$x(t) = \frac{\sin 2\pi t}{t}$$

The Fourier transform of $s(t)$ is then

$$S(f) = \frac{1}{2}X(f - 100) + \frac{1}{2}X(f + 100)$$

Since $X(f)$ is bandlimited to $f = \pm 1$, the first term in $S(f)$ occupies frequencies between 99 and 101 Hz, while the second term occupies frequencies between -101 and -99 Hz. When $S(f)$ is multiplied by $-j\,\text{sgn}(f)$, we find that

$$\hat{S}(f) = -\frac{1}{2}jX(f - 100) + \frac{1}{2}jX(f + 100)$$

The inverse transform yields

$$\hat{s}(t) = x(t)\sin 200\,\pi\,t = \frac{\sin 2\pi t}{t}\,\sin 200\,\pi\,t$$

(c) We use the fact that the Hilbert transform of a Hilbert transform is the negative of the original function. Therefore, by inspection, we have

$$\hat{\hat{s}}(t) = -x(t)\cos 200\pi t = -\frac{\sin 2\pi t}{t}\,\cos 200\,\pi t$$

We are now ready to return to the solution of Eq. (4.73). If $x(t)$ and $y(t)$ are assumed to be bandlimited to frequencies below f_0, we can take the Hilbert transform of both sides of that equation to get

$$\hat{n}(t) = x(t)\sin 2\pi f_0 t + y(t)\cos 2\pi f_0 t \tag{4.80}$$

If Eq. (4.73) is multiplied by $\cos 2\pi f_0 t$ and Eq. (4.80) is multiplied by $\sin 2\pi f_0 t$, then when the two expressions are added together, $y(t)$ is eliminated, yielding

$$n(t)\cos 2\pi f_0 t + \hat{n}(t)\sin 2\pi f_0 t = x(t)[\cos{}^2 2\pi f_0 t + \sin{}^2 2\pi f_0 t] = x(t) \tag{4.81}$$

Similarly, we can reverse the multiplications to obtain

$$y(t) = \hat{n}(t)\cos 2\pi f_0 t - n(t)\sin 2\pi f_0 t \tag{4.82}$$

Example 4.15

Show that the system of Fig. 4.23 yields the quadrature components at the output when the input is narrowband noise.

Solution: The inputs to the lowpass filters are

$$n_1(t) = 2x(t)\cos{}^2 2\pi f_0 t - 2y(t)\sin 2\pi f_0 t\cos 2\pi f_0 t$$

$$= x(t) + x(t)\cos 4\pi f_0 t - y(t)\sin 4\pi f_0 t$$

and

$$n_2(t) = -2x(t)\cos 2\pi f_0 t\sin 2\pi f_0 t + 2y(t)\sin^2 2\pi f_0 t$$

$$= -x(t)\sin 4\pi f_0 t + y(t) - y(t)\cos 4\pi f_0 t$$

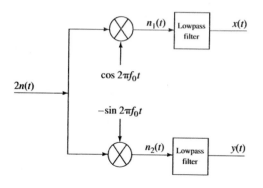

Figure 4.23 System to generate quadrature components.

The modulation theorem indicates that the Fourier transform of $x(t)\cos 4\pi f_0 t$ occupies a range around a frequency of $2f_0$. This is also true of the other terms with $4\pi f_0$ in the argument of the sinusoid. The lowpass filter is designed to pass the frequencies of $x(t)$ and $y(t)$, so it will reject these high-frequency terms. The outputs are therefore as shown on the diagram.

The autocorrelation of $x(t)$ and $y(t)$ can now be derived from Eqs. (4.81) and (4.82):

$$R_x(\tau) = R_y(\tau) = R_n(\tau)\cos 2\pi f_0 \tau + \left(R_n(\tau) * \frac{1}{\pi\tau}\right)\sin 2\pi f_0 \tau \qquad (4.83)$$

Finally, we apply the modulation theorem to Eq. (4.83) to get

$$G_x(f) = G_y(f) = G_n(f - f_0) + G_n(f + f_0)$$
$$\text{for } f_0 - f_m < |f| < f_0 + f_m \qquad (4.84)$$

Equation (4.84) is the key result that will enable us to calculate the effects of noise on AM and FM communication systems.

Example 4.16

Express the three narrowband noise processes of Fig. 4.24 in quadrature form, using f_0 as the center frequency.

Solution: We use Eq. (4.84) to immediately sketch the power spectral densities of $x(t)$ and $y(t)$. These are shown in Fig. 4.25. The noise is then expressed as

$$n(t) = x(t)\cos 2\pi f_0 t - y(t)\sin 2\pi f_0 t$$

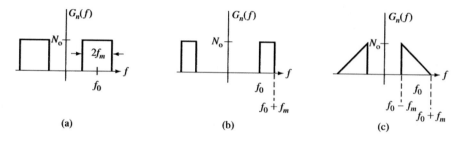

Figure 4.24   Noise processes for Example 4.16.

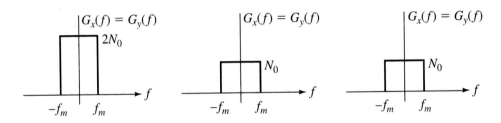

Figure 4.25   Power spectral density of quadrature components for Example 4.16.

4.6 SIGNAL-TO-NOISE RATIO

In many communication studies, the probabilistic parameter of interest is the *average power*. By itself, this quantity would not tell very much, since we could always modify the average power of a signal by putting it through an amplifier or an attenuator. A problem arises, however, because the received waveform usually consists of a desired signal plus noise. If we amplify or attenuate the total received waveform, we do the same thing to the noise as we do to the signal. The parameter of interest then is not the signal power, but the ratio of that power to the power of the unwanted noise. This is the *signal-to-noise ratio,* abbreviated as S/N or SNR.

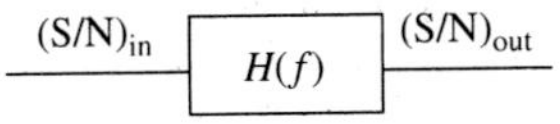

$$(S/N)_{in} \quad \boxed{H(f)} \quad (S/N)_{out}$$

Figure 4.26   Signal-to-noise ratio.

Figure 4.26 shows a block diagram of a system with the input S/N and output S/N indicated. The ratio of these two SNRs gives some measure of the effectiveness of the system. We designate the ratio as ΔSNR, the *signal-to-noise improvement* of the system:

$$\Delta SNR = \frac{(S/N)_{out}}{(S/N)_{in}} \tag{4.85}$$

ΔSNR is often expressed in *decibels*, or dB, as

$$\Delta SNR_{dB} = 10 \log_{10}(\Delta SNR) \tag{4.86}$$

Example 4.17

A signal is given by

$$r(t) = s(t) + n(t)$$

where

$$s(t) = 5 \cos 2\pi \times 1{,}000t + 10 \cos 2\pi \times 1{,}100t$$

The noise $n(t)$ is white with power $N_0 = 0.05$ watt/Hz.

The total received signal is put through a bandpass filter with passband between 990 and 1,110 Hz. Find the SNR at the filter output.

Solution: We can assume that, in the steady state, the entire signal $s(t)$ appears at the output of the filter. Thus, the output power is $25/2 + 100/2 = 62.5$ watts. The noise power at the filter output is found from

$$P_n = 2 \int_{990}^{1,110} G_n(f)\, df = 6 \text{ watts}$$

The SNR is then

$$\text{S/N} = 62.5/6 = 10.4 = 10.17 \text{ dB}$$

Example 4.18

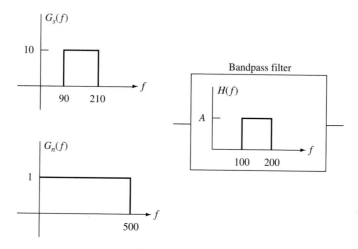

Figure 4.27   Bandpass filter and signals for Example 4.18.

The input to the bandpass filter shown in Fig. 4.27 is the sum of a signal and noise. The power spectral density of the signal and of the noise are as shown. Find the S/N improvement of the filter.

Solution: The input powers are found by integrating the corresponding densities:

Signal power in = 2,400 watts
Noise power in = 1,000 watts
S/N in = 2.4 or 4.8 dB

The output power spectral densities, which are illustrated in Fig. 4.28, result from multiplying the input densities by the square of the magnitude of the filter transfer function. Integration of these densities results in the following output powers:

Signal power out = $2,000A^2$ watts
Noise power out = $200A^2$ watts
S/N out = 10 or 10 dB

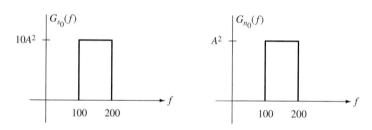

Figure 4.28   Output power spectral densities for Example 4.18.

The S/N improvement is

$$\Delta SNR = 10/2.4 = 4.17 \text{ or } 6.2 \text{ dB}$$

Note that we could have found the S/N improvement by subtracting the input S/N from the output S/N when both are expressed in dB. Note further that the answer is independent of A. This is true because both the noise and the signal are multiplied by A, so the ratio is unaffected.

4.7 MATCHED FILTER

There are a variety of operations we may wish to perform on a received signal. In some cases, we wish to filter the signal in order to remove as much noise as possible and therefore be left with a function of time that resembles the desired signal as closely as possible. In other situations, we may wish to maximize the output SNR without regard to preserving the shape of the signal waveform. That is, we may use a filter that significantly alters the shape of both the signal and the noise in a way that increases the SNR. Such a filter distorts the signal.

Analog receivers typically try to reconstruct the waveform as closely as possible, while digital receivers attempt to "pull" the signal out of background noise without regard to distortion.

In order to motivate this study, let us get way ahead of the game. Suppose you wish to send a list of 1's and 0's by speaking the words *one* and *none* into a microphone. Then, to send 1010, you would speak *one-none-one-none*. Noise adds to the transmitted signal, so let us assume that the receiver has difficulty in distinguishing between the two different words sent. Now suppose that you filter the received signal plus noise in a manner that blocks a good portion of the noise. But in the process, you change the transmitted *one* signal into a waveform that, when placed into a speaker, generates the word *start*. The same filter changes *none* to *halt*. Therefore, instead of hearing a highly noise-corrupted sequence consisting of repetitions of the words *one* and *none*, you hear a relatively uncorrupted signal consisting of *start-halt-start-halt*. You could still recover the original binary sequence in spite of what would be considered severe distortion of the signal waveform.

The *matched filter* is a linear system that maximizes the output SNR. We designate the input to the filter as $s(t) + n(t)$, and the resulting output is $s_o(t) + n_o(t)$. This is shown in Fig. 4.29.

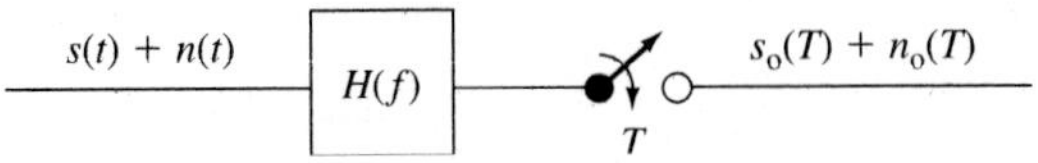

Figure 4.29 Matched filter.

Since the system is assumed to be linear, $s_o(t)$ is the output due to an input of $s(t)$, and $n_o(t)$ is the output due to $n(t)$. The filter is designed to maximize the ratio $s^2_o(T)/n^2_o(T)$. Because the denominator of this expression is random, we use the average value. The output SNR is then

$$\rho = \frac{s_o^2(T)}{n_o^2(T)} = \frac{\left| \int_{-\infty}^{\infty} S(f)H(f)e^{j2\pi fT}\, df \right|^2}{\int_{-\infty}^{\infty} |H(f)|^2 G_n(f)\, df} \tag{4.87}$$

The numerator of Eq. (4.87) is the square of the inverse Fourier transform of the product of the input transform with the system function. Thus, it is the square of the deterministic time sample $s_o(T)$. The $G_n(f)$ in the denominator is the power spectral density of the input noise. Thus, the denominator integrand is the power spectral density of the output noise. Note that we are integrating from $-\infty$ to $+\infty$ instead of doubling the integral from zero to ∞. We do this for reasons that will soon become clear.

We wish to choose $H(f)$ so as to maximize Eq. (4.87). This is a difficult maximization problem. (You cannot simply take the derivative and set it to zero, since we are trying to find a function rather than a value.) The choice is simplified if we apply *Schwartz's inequality* to the numerator. In doing so, we will be able to solve for $H(f)$ almost by inspection.

Schwartz's inequality states that for all functions $f(x)$ and $g(x)$,

$$\left| \int f(x)g(x)\, dx \right|^2 \leq \int |f^2(x)|\, dx \int |g^2(x)|\, dx \tag{4.88}$$

We will derive Schwartz's inequality for the special case of real functions by starting with the observation that

$$\int [f(x) - Tg(x)]^2\, dx \geq 0$$

for all real $f(x)$, $g(x)$, and T (i.e., we are integrating a non-negative function). Expanding this expression, we obtain

$$T^2 \int g^2(x)dx - 2T \int f(x)g(x)\, dx + \int f^2(x)\, dx \geq 0 \tag{4.89}$$

The left side of Eq. (4.89) is quadratic in T. Since the value can never be negative, the quadratic cannot have distinct real roots. Therefore, the discriminant cannot be positive. Thus,

$$4\left[\int f(x)g(x)\, dx \right]^2 - 4\int f^2(x)\, dx \int g^2(x)\, dx \leq 0 \tag{4.90}$$

and the inequality is established. Proving the inequality for complex functions is more difficult, so we ask you to accept the result in Eq. (4.88).

We now apply Schwartz's inequality to Eq. (4.87). Hindsight is a wonderful thing: Had you already solved this equation, you would know that we wish to cancel terms from the numerator and denominator. To begin this process, we rewrite the numerator of Eq. (4.87) as

$$\left| \int_{-\infty}^{\infty} S(f)H(f)e^{j2\pi fT}\,df \right|^2 = \left| \int_{-\infty}^{\infty} \frac{S(f)}{\sqrt{G_n(f)}} H(f)\sqrt{G_n(f)}e^{j2\pi fT}\,df \right|^2 \qquad (4.91)$$

The square root operation is unambiguous, since $G_n(f)$ can never be negative. Now applying Schwartz's inequality, we get

$$\left| \int_{-\infty}^{\infty} S(f)H(f)e^{j2\pi fT}\,df \right|^2 \le \int_{-\infty}^{\infty} |H(f)|^2 G_n(f)df \int_{-\infty}^{\infty} \frac{|S(f)|^2}{G_n(f)}df \qquad (4.92)$$

Combining this with Eq. (4.87), we have

$$\rho \le \frac{\displaystyle\int_{-\infty}^{\infty} |H(f)|^2 G_n(f)df \int_{-\infty}^{\infty} |S(f)|^2 / G_n(f)df}{\displaystyle\int_{-\infty}^{\infty} |H(f)|^2 G_n(f)df} \qquad (4.93)$$

$$= \int_{-\infty}^{\infty} \frac{|S(f)|^2}{G_n(f)}\,df$$

Equation (4.93) fixes an upper bound on the SNR at the output of the filter. If we can somehow guess at an $H(f)$ that yields this maximum, we need look no further.

Before attempting the guess, let us recap the approach we are taking. We wish to choose $H(f)$ so as to maximize Eq. (4.87). This is a difficult mathematical problem to solve. Instead of attempting a direct solution, we have placed an upper bound upon the equation. If the SNR cannot exceed that bound, and we somehow find an $H(f)$ that achieves the bound, we have solved the original problem. (In general, there is no guarantee that we can even achieve a bound of this type; however, in this case, we can.)

We have now reduced the problem to finding an $H(f)$ that reduces Eq. (4.87) to Eq. (4.93). We are asking you to be creative, and there is no road map for doing so. You need to stare at the two expressions, hoping for an inspiration. [Indeed, except by hindsight, we have no assurance that an $H(f)$ exists that will achieve the bound of Eq. (4.93).]

The answer is (if you figured this out, your insight is excellent)

$$H(f) = e^{-j2\pi fT}\,\frac{S^*(f)}{G_n(f)} \qquad (4.94)$$

$S^*(f)$ is the complex conjugate of $S(f)$. If you were not able to see this answer, you might wish to go back and assume that the noise is white (as we shall in a moment). That is, assume that $G_n(f)$ is a constant. This makes the creative inspiration easier to achieve.

Since $H(f)$ appears as a square in both the numerator and denominator of Eq. (4.87), any scaling factor can be applied to $H(f)$ without affecting the SNR. That is, the $H(f)$ of Eq. (4.94) can be multiplied by any constant. We shall therefore rewrite this equation, inserting C for an arbitrary constant:

$$H(f) = Ce^{-j2\pi fT}\frac{S^*(f)}{G_n(f)} \tag{4.95}$$

Alas, simple amplification does not improve the SNR, since both the signal and noise are multiplied by the same amount.

Now let us assume that the input noise is white, so that $G_n(f) = N_0/2$. The matched filter of Eq. (4.95) then becomes

$$H(f) = \frac{2C}{N_0}e^{-j2\pi fT}S^*(f) \text{ "=" } Ce^{-j2\pi fT}S^*(f) \tag{4.96}$$

Note that because C is an arbitrary constant, it would create unnecessary bookkeeping to write $2C/N_0$ in Eq. (4.96). That is why we have replaced $2C/N_0$ with C. (We put the equals sign in quotes so that you don't draw the conclusion that C must be zero.)

We can find the SNR at the output of the matched filter (with white noise at the input) directly from Eq. (4.93):

$$\rho = \frac{2}{N_0}\int_{-\infty}^{\infty}|S(f)|^2\,df = \frac{2}{N_0}\int_{-\infty}^{\infty}s^2(t)\,dt \tag{4.97}$$

The last equality in Eq. (4.97) results from Parseval's theorem.

The inverse Fourier transform of Eq. (4.96) yields the impulse response of the matched filter:

$$h(t) = Cs(T - t) \tag{4.98}$$

This is found by noting that the inverse transform of $S^*(f)$ is $s(-t)$ and the exponential leads to a time shift. At this point, there is no assurance that this filter is physically realizable (i.e., causal).

Example 4.19

Find the impulse response of the matched filter for the two functions of time shown in Fig. 4.30.

Solution: $h(t)$ is derived directly from Eq. (4.98). The result is shown in Fig. 4.31.

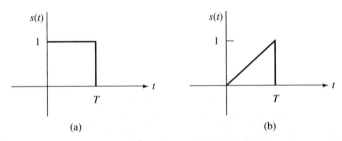

(a)   (b)

Figure 4.30   Functions of time for Example 4.19.

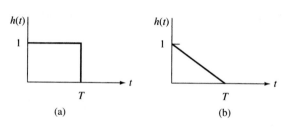

Figure 4.31   Matched filter for Example 4.19.

(a)   (b)

The actual function of time at the output of the matched filter can be found by convolving the input function with the impulse response. Therefore,

$$s_o(t) \; + \; n_o(t) \; = \; [s(t) \; + \; n(t)] * h(t)$$

and at time $t = T$, we have

$$s_o(T) \; + \; n_o(T) \; = \; \int_0^T [s(\tau) \; + \; n(\tau)]s(\tau) \, d\tau$$

Thus, the matched filter is equivalent to the system of Fig. 4.32.

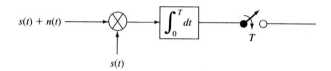

Figure 4.32   Correlator.

The operation being performed by this system is called *correlation* (i.e., multiply two functions of time together and integrate the product). For that reason, the matched filter is often referred to as a *correlator*. In a generalized sense, the filter is finding the projection of the input signal in the direction of $s(t)$. Since the system is aligned in that direction, the output SNR is maximized.

Example 4.20

Find the output SNR of a matched filter, where the signal is

$$s(t) = A, \quad \text{for} \quad 0 < t < T$$

The noise is white with power spectral density $N_0/2$.
Solution: The matched filter achieves the SNR of Eq. (4.97). Therefore,

$$\text{SNR} \; = \; \frac{2}{N_0} \int_{-\infty}^{\infty} s^2(t) \, dt \; = \; \frac{2A^2T}{N_0}$$

PROBLEMS

4.1.1 Three coins are tossed at the same time. List all possible outcomes of this experiment. List five representative events. Find the probabilities of the following events:
{all tails}; {one head only}; {three matches}

4.1.2 Find the probability that, if a perfectly balanced die is rolled, the number of spots on the face turned up is greater than or equal to 2.

4.1.3 An urn contains 4 white balls and 7 black balls. An experiment is performed in which 3 balls are drawn out in succession without replacing any. List all possible outcomes and assign probabilities to each.

4.1.4 In Problem 4.1.3, you are told that the first two draws are white balls. Find the probability that the third is also a white ball.

4.1.5 An urn contains 4 red balls, 7 green balls, and 5 white balls. Another urn contains 5 red balls, 9 green balls, and 2 white balls. One ball is drawn from each urn. What is the probability that both balls will be of the same color?

4.1.6 Three people, A, B, and C, live in the same neighborhood and use the same bus line to go to work. Each of the three has a probability of $\frac{1}{4}$ of making the 6:10 bus, a probability of $\frac{1}{2}$ of making the 6:15 bus, and a probability of $\frac{1}{4}$ of making the 6:20 bus. Assuming independence, what is the probability that they all take the same bus?

4.1.7 The probability density function of a certain voltage is given by

$$p_v(v) \ = \ ve^{-v}U(v)$$

where $U(v)$ is the unit step function.
(a) Sketch this probability density function.
(b) Sketch the distribution function of v.
(c) What is the probability that v is between 1 and 2?

4.1.8 Find the mean and variance of the exponential distribution that is given by

$$p(x) \ = \ \frac{1}{m}e^{-x/m}\, U(x)$$

4.1.9 Find the density of $y = |x|$, given that $p_x(x)$ is as shown in Fig. P4.1.9. Also, find the mean and variance of both y and x.

Figure P4.1.9

4.1.10 Find the mean and variance of x, where the density of x is as shown in Fig. P4.1.10.

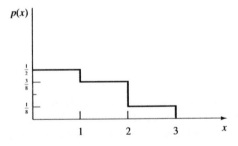

Figure P4.1.10

4.1.11   The density function of x is shown in Fig. P4.1.11. A random variable y is related to x as shown. Determine the density function of y.

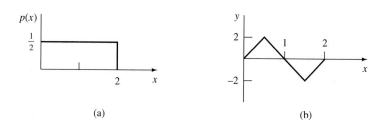

(a)   (b)

Figure P4.1.11

4.1.12   Find the expected value of $y = \sin x$ if x is uniformly distributed between 0 and 2π.

4.1.13   A random variable x has the probability density function

$$p_x(x) = Ae^{-2|x|}$$

(a) What is the value of A?
(b) Find the probability that x is between -3 and $+3$.
(c) Find the variance of x.

4.1.14   Five different symbols, A, B, C, D, and E, are transmitted with equal probability. What is the probability that the message $ABCDE$ will be received?

4.1.15   Binary information is being received, where the probabilities of 0 and 1 are equal. Received messages are divided into 5-bit words. What is the probability that the first received message will contain at least one zero?

4.1.16   A random variable x has the probability density shown in Fig. P4.1.16. A new variable y is defined as the magnitude of x:

$$y = |x|$$

(a) Find the probability density of y, $p(y)$.
(b) Find the mean value of y.
(c) Find the variance of y.

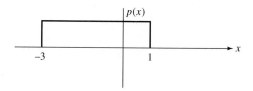

Figure P4.1.16

4.2.1   A Gaussian random variable has a mean value of 2. The probability that the variable lies between 2 and 5 is 0.3. Find the probability that the variable is between 1 and 3.

4.2.2   A Gaussian random variable has a variance of 9. The probability that the variable is greater than 5 is 0.1. Find the mean of the random variable.

4.2.3 A zero-mean Gaussian random variable has a variance of 4. Find x_0 such that

$$\Pr\{|x| > x_0\} < 0.001$$

4.2.4 A random variable is Rayleigh distributed as follows:

$$p(x) = \frac{x}{4} \exp\left(\frac{-x^2}{8}\right) U(x)$$

Find the probability that the random variable is between 1 and 3.

4.2.5 A Gaussian random variable x has zero mean and unit variance. A new variable is defined by $y = |x|$.
   (a) Find the density of y.
   (b) Find the mean value of y.
   (c) Find the variance of y.

4.2.6 A function is defined by

$$y = e^{-2x}$$

Find the density of y if
   (a) x is uniformly distributed between 0 and 3.
   (b) $p(x) = e^{-x}U(x)$

4.2.7 You are told that the mean rainfall per year in California is 10 inches and that the standard deviation in the amount of rain per year is 1 inch. Can you tell, from this information, what is the density of rainfall in California? If so, roughly sketch this density function.

4.2.8 The random variable x is Rayleigh distributed. Find the mean, second moment, and variance.

4.2.9 The random variable x is uniformly distributed between -1 and $+1$. The random variable y is uniformly distributed between 0 and $+5$. x and y are independent of each other. A new variable, z, is formed as the sum of x and y.
   (a) Find the probability density function of z.
   (b) Find the mean and variance of x, y, and z.
   (c) Find a general relationship among the means of x, y, and z.
   (e) Repeat part (c) for the variance.

4.2.10 (a) Find and sketch the density of the sum of two independent random variables as follows: One of the variables is uniformly distributed between -1 and $+1$; the second variable is triangularly distributed between -2 and $+2$.
   (b) Compare the result of part (a) to a Gaussian density function, where the variance should be chosen to make the Gaussian variable as "close" as possible to your answer to part (a).

4.2.11 A voltage is known to be Gaussian distributed with mean value of 4. When this voltage is impressed across a 4-Ω resistor, the average power dissipated is 8 W. Find the probability that the voltage exceeds 2 V at any instant of time.

4.2.12 (a) You are told that the integral of a zero-mean Gaussian density from $x = 10$ to $x = \infty$ is equal to 0.02. Find the variance of this random variable.
   (b) Find the probability that x is between 1 and 3.

4.2.13 A Gaussian random variable x has a mean value of m and a variance of 4. You are told that the probability that the variable is greater than 6 is 0.01. Find the mean value, m.

4.3.1 A random variable k is uniformly distributed in the interval between -1 and $+1$. Sketch several possible samples of the process

$$x(t) = k\sin 2\pi t$$

4.3.2 $x(t)$ is a stationary process with mean value 1 and autocorrelation

$$R(\tau) = e^{-|\tau|}$$

Find the mean and autocorrelation of the process $y(t) = x(t - 1)$.

4.3.3 Find the autocorrelation of the periodic process shown in Fig. P4.3.3. This signal has a Fourier series given by

$$s(t) = \sum_{n-1}^{\infty} \frac{\pi}{n} \sin nt$$

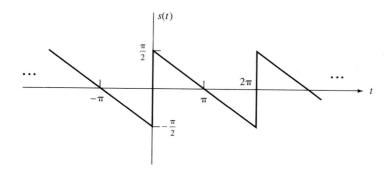

Figure P4.3.3

4.3.4 $x(t)$ is a stationary process with zero mean value. Is the process

$$y(t) = tx(t)$$

stationary? Find the mean and autocorrelation of $y(t)$.

4.3.5 Given a stationary process $x(t)$, find the autocorrelation of

$$y(t) = x(t - 1) + \sin 2\pi t$$

in terms of $R_x(\tau)$.

4.3.6 $x(t)$ is a stationary process with mean value m. Find the mean value of the output $y(t)$ of a linear system with $h(t) = e^{-t}U(t)$ and $x(t)$ as input. That is,

$$y(t) = \int_0^t h(\tau)x(t - \tau)\, d\tau$$

4.3.7 A random process is defined by

$$x(t) = K_1 t + K_2$$

where K_2 is a deterministic constant and K_1 is uniformly distributed between 0 and 1.

(a) Sketch several samples of this process.

(b) Find the mean of the process.

(c) Write an expression for the autocorrelation of the process.

(d) Is the process stationary?

(e) If K_1 is now uniformly distributed between -1 and $+1$, what changes occur in your answers to parts (b), (c), and (d)?

4.3.8 Which of the following could *not* be the autocorrelation function of a process?

$$R_a(\tau) = \begin{cases} 1 - |\tau|, & |\tau| < 1 \\ 0, & |\tau| > 1 \end{cases}$$

$$R_b(\tau) = 5\sin 3\tau$$

$$R_c(\tau) = \begin{cases} 1 & |\tau| < 2 \\ 0, & |\tau| > 2 \end{cases}$$

$$R_d(\tau) = \frac{\sin \tau}{\tau}$$

4.3.9 Find $R_e(\tau)$ in terms of $R_i(\tau)$, where $i(t)$ and $e(t)$ are related by the circuit shown in Fig. P4.3.9.

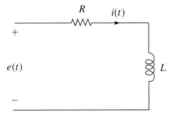

Figure P4.3.9

4.3.10 Given a constant a and a random variable f with density $p_f(f)$, form

$$x(t) = a e^{j2\pi f t}$$

Find $R_x(\tau)$, and show that

$$G_x(f) = |a|^2 p_f(f)$$

4.3.11 Find the autocorrelation and power spectral density of the wave

$$v(t) = A\cos(2\pi f_c t + \theta)$$

where A and f_c are not random and θ is uniformly distributed between 0 and 2π. [*Hint:* Use the definitions of autocorrelation and expected value.]

4.3.12 You are given the RC circuit of Fig. P4.3.12 with input function as shown. You wish to choose the value of R so that the total energy at the output is 50% of the input energy.

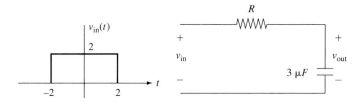

Figure P4.3.12

4.3.13 Use the result of Example 4.8 to find the autocorrelation of the process $y(t) = dx/dt$ in terms of the autocorrelation of $x(t)$. [*Hint:* You can use the definition of the derivative,

$$\frac{dx}{dt} = \lim_{T \to \infty} \frac{x(t) - x(t - T)}{T}$$

and l'Hôpital's rule.]

4.3.14 Use an approach similar to that of Problem 4.3.13 to find

$$E\{x'(t)x(t + \tau)\} \quad \text{and} \quad E\{x'(t + \tau)x(t)\}$$

4.3.15 The random telegraph signal of Example 4.11 forms the input to an ideal lowpass filter with cutoff at f_m. Find the ratio of output to input power as function of f_m and λ.

4.4.1 $x(t)$ is white noise with autocorrelation $R_x(\tau) = \delta(\tau)$. It forms the input to an ideal lowpass filter with cutoff frequency f_m. Referring to Fig. P4.4.1, find the average power of the output signal.

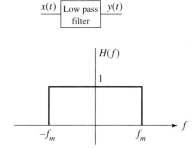

Figure P4.4.1

4.5.1 Prove that the inverse Fourier transform of a Hilbert transform is

$$\mathscr{F}^{-1}(-j\,\text{sgn}(f)) = \frac{1}{\pi t}$$

4.5.2   Find and sketch the Hilbert transform of the function shown in Fig. P4.5.2.

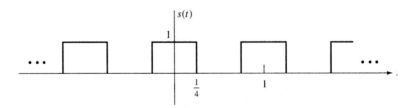

Figure P4.5.2

4.5.3   Express the narrowband noise processes shown in Fig. P4.5.3 in quadrature form. For each process, choose two different center frequencies, and sketch the power spectral density of $x(t)$ and $y(t)$.

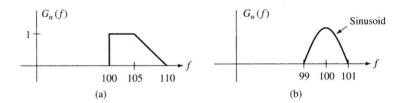

Figure P4.5.3

4.5.4   **(a)** Find the matched filter for the signal

$$s(t) = \begin{cases} 10\cos 2\pi \times 1{,}000t, & 0 < t < 50 \ \mu\text{sec} \\ 0, & \text{otherwise} \end{cases}$$

in white Gaussian noise with power of 10^{-4} watt/Hz.

(b) Find the output SNR for this filter.

4.5.5   Find the cross correlation of narrowband noise with its Hilbert transform. That is, evaluate

$$E\{n(t)\hat{n}(t + \tau)\}$$

in terms of the autocorrelation of $n(t)$.

4.5.6   Show that the cross correlation between the in-phase and quadrature terms in a narrowband noise expansion is given by

$$R_{xy}(\tau) = R_n(\tau)\sin 2\pi f_c \tau - \hat{R}_n(\tau)\cos 2\pi f_c \tau$$

4.5.7   Two narrowband noise processes have power spectral densities as shown in Fig. P4.5.7. Express each of these in quadrature form. For each process, choose two different center frequencies, and sketch the power spectral density of $x(t)$ and $y(t)$.

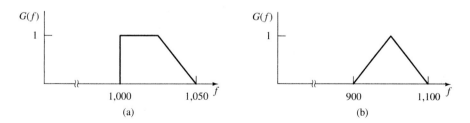

(a)   (b)

Figure P4.5.7

4.6.1   In a given communication system, the signal is $s(t) = 20\cos 2\pi t$. Noise of power spectral density $G_n(f) = e^{-3|f|}$ is added to the signal, and the resultant sum forms the input to a filter.
(**a**) Find the SNR at the input to the filter.
(**b**) If the filter is ideal lowpass with a cutoff of 2 Hz, find the improvement in the SNR, and express it in dB.
(**c**) If the filter is ideal bandpass with a passband from 0.9 to 1.1 Hz, find the improvement in SNR, and express it in dB.

4.6.2   You are given the system shown in Fig. P4.6.2, which is composed of two cascaded filters. $s(t)$ and $n(t)$ are as given in Problem 4.6.1. $H_1(f)$ is a lowpass filter with cutoff of 2 Hz. $H_2(f)$ is a bandpass filter with passband from 0.9 to 1.1 Hz. Find the improvement in SNR of $H_1(f)$ and of $H_2(f)$, and express these in dB. Now find the overall improvement in SNR of the cascaded system, and compare it to the individual improvement in SNR of the two filters.

Figure P4.6.2

4.6.3   Repeat Example 4.18, assuming that the bandpass filter passes from 90 to 210 Hz instead of from 10 to 200 Hz. Compare the results.

4.6.4   A received signal is given by

$$r(t) \;=\; s(t) \;+\; n(t)$$

where

$$s(t) \;=\; 5\cos 2\pi \times 3,000t \;+\; 15\cos 2\pi \times 2,100t$$

The noise $n(t)$ is white with power $N_0 = 0.05$ watt/Hz. The signal $r(t)$ is put through a bandpass filter with a passband between 1,900 and 2,200 Hz. Find the signal to noise ratio at the output of the filter.

4.7.1   (a) Find the matched filter for the signal

$$s(t) = \begin{cases} 5\cos 2\pi \times 2{,}000t, & 0 < 1 < 0.005 \\ \\ 0, & \text{otherwise} \end{cases}$$

in white noise with power spectral density $N_0/2 = 0.1$.

(b)   What is the output signal to noise ratio?

4.7.2   A signal $s(t)$, as shown in Fig. P4.7.2, is added to white noise with a power of 10^{-2} watt/Hz.

(a)   Design a matched filter for this signal.

(b)   What is the signal to noise ratio at the output of the filter?

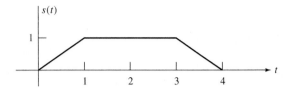

Figure P4.7.2

5

Baseband Transmission

5.0 PREVIEW

What We Will Cover and Why You Should Care

You have finally arrived at the first chapter dealing with communication. The previous chapters have simply laid the necessary groundwork.

In this chapter, we deal with what is known as *baseband transmission*. The term *baseband* refers to the frequency content of the signal. The frequencies used to transmit baseband signals are relatively low. In Chapters 6 and 7, we explore techniques of sending signals in which the transmitted frequencies are high.

Baseband transmission is used in a variety of communication systems, from telephone loops to intercoms. Even if the type of communication used is not baseband, it is important to understand baseband transmission as a stepping-stone to other techniques.

Necessary Background

A working knowledge of Fourier analysis is needed to understand most analog communication systems. This is important because we analyze signals both in the time domain and in the frequency domain.

Analysis of the performance of systems (Section 5.5) requires random process analysis.

5.1 ANALOG BASEBAND

When we use the term *analog baseband*, we are referring to analog signals with Fourier transforms occupying frequencies extending to zero (dc). The reason for using the term *baseband* is that the frequencies are *not* shifted to some other nonzero point on the frequency axis. In later chapters, we explore techniques for shifting frequencies to a range centered around a nonzero frequency.

Since baseband signals occupy relatively low frequencies, they are not suited for transmission through bandpass channels. Baseband signals are typically transmitted through wires or cables.

Because telephones form the backbone of traditional communication systems, the transmission of voice signals represents the most prevalent application of baseband analog communication. We therefore concentrate initially on voice transmission.

The human ear is capable of hearing signals with frequencies in the range of about 20 Hz to 20 kHz. In fact, most people cannot hear the upper portion of this range, and a particular person's upper frequency cutoff decreases with age. It is probably no accident of evolution that signals generated by human beings and their vocal cords fall within the audible range. The magnitude of the Fourier transform of a typical speech waveform is shown in Fig. 5.1. The location of the peak of the waveform depends on the physiology of the speaker (i.e., the resonant frequency of the vocal cavity). It also depends on what the person is saying and what language is being spoken. If, for example, the speaker whistles, the speech waveform is a pure sinusoid and has a Fourier transform consisting of an impulse at the frequency of the whistling. If the person hums at the same frequency, the Fourier transform contains the fundamental frequency plus harmonics at multiples of that frequency.

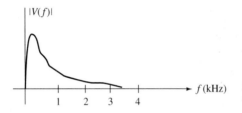

Figure 5.1   Fourier transform of typical voice waveform.

In the early days of telephony, experimentation showed that the portion of the speech waveform between about 300 Hz and 3.3 kHz was sufficient both for intelligibility of speech and for recognition of the speaker. That is, although this range is not considered high fidelity, it permits both understanding of what is being said and identification of the person saying it. High-quality music requires the presence of frequencies higher than 3.3 kHz—perhaps as high as 15 kHz or more. (AM radio transmits frequencies up to 5 kHz, while FM radio transmits frequencies up to 15 kHz. The typical high-quality home entertainment system responds to frequencies above 20 kHz.)

If you speak into a microphone and amplify the resulting waveform using electronic circuitry, you have effectively built a simple analog baseband transmitter. If the output of the transmitter is now connected to a wire channel, and the other end of the channel is connected to a loudspeaker (perhaps through some amplification if the channel contains loss), a complete baseband communication system results. Such systems are used in wire intercoms.

5.2 THE SAMPLING THEOREM

A *discrete* signal is a signal that is *not* continuous in time. That is, it has values only at disconnected points of the time axis. If the discrete signal is analog, its values at any time it is defined lie within a continuum of possible values.

We wish to find a way of converting a continuous (not discrete) analog waveform into a discrete signal. To do this, the time axis must somehow be made discrete. The conversion of the continuous time axis into a discrete axis is accomplished by time sampling. The *sampling theorem* states the following:

> If the Fourier transform of a function of time is zero for $f > f_m$ and the values of the function are known for $t = nT_s$ (for all integer values of n), then the function is known exactly for *all* values of t.

This is remarkable: Knowing the value of the function of time at discrete time points allows us to fill in the curve between these points *precisely and accurately!* Of course, something this remarkable must have some limitations: You couldn't be given two values separated by hours and expect to fill in the curve between these points. Indeed, for the samples to give all of the information, they must be "close enough" to each other. The restriction is that the spacing T_s between samples be less than $1/2f_m$, where f_m is the maximum frequency of the signal. Alternatively, $s(t)$ can be uniquely determined from its values at a sequence of equidistant points in time. The upper limit of T_s, $1/2f_m$, is known as the *Nyquist sampling interval.*

The upper limit on T_s can be expressed in a more meaningful way by taking the reciprocal of T_s to obtain the sampling frequency, denoted $f_s = 1/T_s$ in samples per second. The restriction then becomes

$$T_s < \frac{1}{2f_m}$$

$$\frac{1}{T_s} > 2f_m$$

$$f_s > 2f_m$$

Thus, the sampling frequency must be greater than twice the highest frequency of the signal being sampled. For example, if a voice signal has 4 kHz as a maximum frequency, it must be sampled at least 8,000 times per second to comply with the conditions of the sampling theorem. Twice the highest frequency is known as the *Nyquist frequency.*

Before going further, let us observe that the spacing between the sample points is inversely related to the maximum frequency f_m. This is intuitively satisfying, since the higher f_m, the faster we would expect the function to vary. The faster the function varies, the closer together the sample points should be in order to permit reconstruction of the function.

We present two proofs of the sampling theorem. The first is physical and intuitive, while the second is more mathematical.

Proof 1. Figure 5.2 shows a pulse train multiplying the original signal $s(t)$. If the pulse train consists of narrow pulses, one would say that the output of the multiplier is a *sampled version* of the original waveform. In actuality, the output depends not only on the sample values of the input, but on a range of values around each sample point. The theory does not require these extra values, which represent redundant information. However, practical systems sometimes sample over a small range of time surrounding the actual sample points. As we prove the theorem, it should become obvious that the multiplying

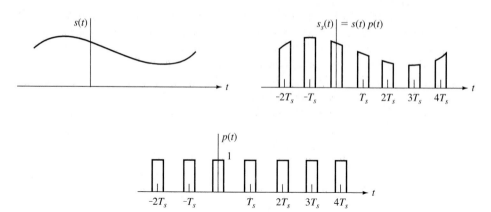

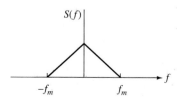

Figure 5.2   Product of pulse train and $s(t)$.

function need not consist of perfect square pulses. In fact, the function can be any periodic signal, and the pulse widths can approach zero (i.e., multiply by a train of impulses).

Multiplying $s(t)$ by $p(t)$ of the type shown in Fig. 5.2 is a form of *time gating*. It can be viewed as the opening and closing of a gate, or switch.

Our goal is to show that the original signal can be recovered from the sampled waveform $s_s(t)$. We do this by examining the Fourier transform of $s_s(t)$. The sampling theorem requires that we assume that $s(t)$ has no frequency components above f_m. The Fourier transform of $s(t)$, $S(f)$, therefore cuts off at f_m. Figure 5.3 shows a representative shape for this transform. While we use this triangular shape throughout the text, we do not mean to restrict the actual transform to that shape.

Figure 5.3   Representative $S(f)$.

Since the multiplying pulse train is assumed to be periodic, it can be expanded in a Fourier series. The $p(t)$ shown is an even function, so we can use a trigonometric series containing only cosine terms (although this is not necessary to prove the theorem). Thus,

$$s_s(t) = s(t)p(t)$$

$$= s(t)\left[a_0 + \sum_{n=1}^{\infty} a_n \cos 2\pi n f_s t\right] \tag{5.1}$$

$$= a_0 s(t) + \sum_{n=1}^{\infty} a_n s(t) \cos 2\pi n f_s t$$

where

$$f_s = \frac{1}{T_s} \tag{5.2}$$

The goal is to isolate the first term in the final expression of Eq. (5.1), which is proportional to the original $s(t)$. We can undo the effects of any constant multiplier with an amplifier or attenuator.

Each of the terms in the summation of Eq. (5.1) is of the form of $s(t)$ multiplied by a sinusoid. When a time signal is multiplied by a sinusoid, the result is a shift of all frequencies of the signal by an amount equal to the frequency of the sinusoid. The frequency content of each term in Eq. (5.1) is then centered around the frequency of the multiplying sinusoid. (When we discuss AM, we will call this the *carrier frequency*.) The Fourier transform of $s_s(t)$ is sketched in Fig. 5.4.

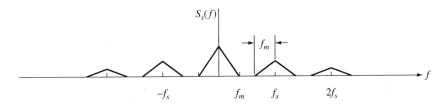

Figure 5.4   Fourier transform of sampled wave.

The shape centered at the origin is the transform of $a_0 s(t)$, and the shifted versions represent the transforms of the various harmonic terms. We see that the various terms do not overlap in frequency, provided that $f_s > 2f_m$. But this is nothing more than the condition given in the sampling theorem. Since the various terms occupy different bands of frequency, they can be separated from each other using linear filters. A lowpass filter with a cutoff frequency of f_m can be used to recover the $a_0 s(t)$ term.

Proof 2. The second proof we present is less intuitive than the first. We take the time to explore it, since the approach supplies an insight into the mathematical principles of sampling.

Since $S(f)$ is nonzero along a finite portion of the f-axis, we can expand it in a Fourier series in the interval

$$-f_m < f < f_m$$

In expanding $S(f)$ in this manner, you should be careful not to let the change in notation confuse the issue. The t used in the Fourier series is an independent functional variable, and any other letter could be substituted for it. Performing the Fourier series expansion, we obtain

$$S(f) = \sum_{n=\infty}^{\infty} c_n e^{jn2\pi t_0 f} \tag{5.3}$$

where

$$t_0 = \frac{1}{2f_m} \tag{5.4}$$

The c_n in Eq. (5.3) are given by

$$c_n = \frac{1}{2f_m} \int_{-f_m}^{f_m} S(f) e^{-jn2\pi t_0 f} df \qquad (5.5)$$

However, the Fourier inversion integral tells us that

$$s(t) = \int_{-\infty}^{\infty} S(f) e^{j2\pi t} df = \int_{-f_m}^{f_m} S(f) e^{j2\pi f t} df \qquad (5.6)$$

In the rightmost expression in Eq. (5.6), the limits of integration have been reduced, since $S(f)$ is equal to zero outside of the interval between $-f_m$ and $+f_m$. Upon comparing Eq. (5.5) with Eq. (5.6), we see that

$$c_n = \frac{1}{2f_m} s(-nt_0) = \frac{1}{2f_m} s\left(-\frac{n}{2f_m}\right) \qquad (5.7)$$

Equation (5.7) is the result we have been seeking. It says that the c_n are specified by the values of $s(t)$ at the points $t = n/2f_m$. Once the c_n are known, $S(f)$ is known, and once $S(f)$ is known, $s(t)$ is known. We have thus proven the sampling theorem.

Although the proof is complete, we will carry the mathematics a step further to actually solve for $s(t)$ in terms of the sample values. We substitute the c_n of Eq. (5.7) into Eq. (5.3) to get

$$S(f) = \frac{1}{2f_m} \sum_{n=-\infty}^{\infty} s\left(-\frac{n}{2f_m}\right) e^{jn\pi ft/f_m} \qquad (5.8)$$

We now find the inverse Fourier transform of $S(f)$:

$$s(t) = \sum_{n=-\infty}^{\infty} \frac{1}{2f_m} \int_{-f_m}^{f_m} s\left(-\frac{n}{2f_m}\right) e^{jn\pi f/f_m} e^{j2\pi ft} df$$

$$\qquad (5.9)$$

$$= \sum_{n=-\infty}^{\infty} s\left(-\frac{n}{2f_m}\right) \left[\frac{\sin(2\pi f_m t + n\pi)}{2\pi f_m t + n\pi} \right]$$

Equation (5.9) is the final statement of the sampling theorem. We can use it to find the value of $s(t)$ at any point in time simply by knowing the sample values of $s(t)$. That is, the only unknowns on the right side of the equation are the sample values.

You can get a feel for the sampling theorem by sketching a few terms in Eq. (5.9). We do this in Figure 5.5 for a representative $s(t)$ and three sample points. Note that only the term centered at each sampling point has non-zero value; all of the other components go to zero at the sampling points. Between sampling points, we must calculate the sum of the various terms from adjacent points.

Computer Exercise

The waveform shown in Fig. 5.5 was generated using computer software. We present the instruction set for both *Mathcad* and MATLAB.

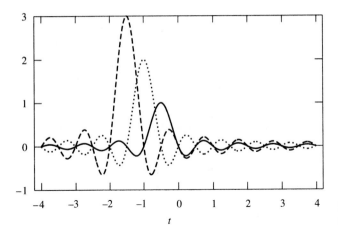

Figure 5.5 Terms from the sampling reconstruction.

We have plotted three waveforms corresponding to samples of 1, 2, and 3 at $t = -0.5$, $t = -1$, and $t = -1.5$, respectively. The Mathcad instruction set is:

```
t:=-4,-3.99..4
s1(t):=sin(2*π*t)+π/2*π*t+π
s2(t):=2*sin(2*π*t+2*π)/2*π*t+2*π
s3(t):=3*sin(2*π*t+3*π)/2*π*t+3*π
```

Notes: Enter ":=" by simply pressing ":". Enter ".." by pressing ";". Enter π by pressing CONTROL+p.

Then press "@" to create a graph. Enter "t" for the abscissa and "s1(t), s2(t), s3(t)" for the ordinate.

The MATLAB instruction set is:

```
t= -4:.01:4;
s1=sin(2*pi*t+pi)./(2*pi*t+pi);
s2=2*sin(2*pi*t+2*pi)./(2*pi*t+2*pi);
s3=3*sin(2*pi*t+3*pi)./(2*pi*t+3*pi);
plot(t,s1,t,s2,t,s3)
```

Notes: The semicolon (;) after each instruction lines stops MATLAB from printing all results immediately after pressing RETURN.

The period in front of the division sign is critical. If it is omitted, MATLAB presupposes that it is dividing two matrices (vectors).

The plot statement superimposes three graphs on the same set of axes.

Example 5.1

A bandlimited signal occupies the frequency range between 990 Hz and 1,010 Hz. A typical Fourier transform is shown in Figure 5.6. Although the sampling theorem indicates that the sampling rate must be higher than 2,020 samples per second, investigate the possibilities of sampling at rates as low as 20 samples per second.

Solution: Figure 5.7 shows the Fourier transform that results from sampling at twice the highest frequency (2,020 samples per second) and also at twice the bandwidth (20 samples per second). We are assuming that all harmonics have equal amplitude (i.e., we assume sampling with an ideal impulse train).

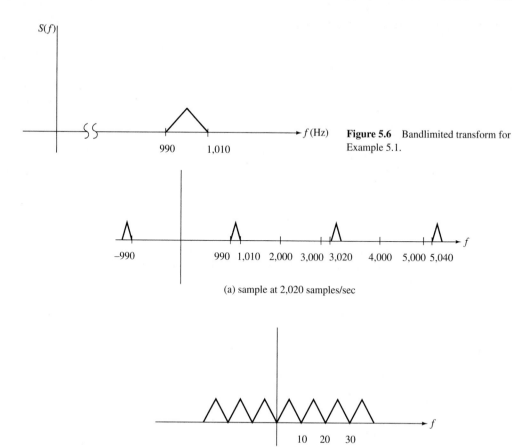

Figure 5.6 Bandlimited transform for Example 5.1.

(a) sample at 2,020 samples/sec

(b) sample at 20 samples/sec

Figure 5.7 Fourier transform of sampled waveforms.

Note that it is possible to recover the original signal when sampling at only 20 samples per second, *provided that* the exact band of frequencies occupied is known. At this lower sampling rate, it is not sufficient to know only the maximum frequency. Note also that if the limits of the band were not multiples of the bandwidth, we would need to sample at a higher rate to avoid aliasing. (We explore this phenomenon in the "Problems" section at the end of this chapter.)

The preceding example shows that it is possible to recover a bandlimited signal by sampling at a rate as low as twice the *bandwidth* of the signal.

Errors in Sampling   The sampling theorem indicates that $s(t)$ can be perfectly recovered from its samples. If sampling is attempted in the real world, however, errors result from three major sources. *Round-off errors* occur when the various sample values are rounded off in the communication system. Rounding off takes place in digital communication, where we send only discrete values. We will later call this error *quantization noise*, and it is examined in detail in Chapter 8. *Truncation errors* occur if the sampling is done over a finite time. That is, the sampling theorem requires that samples be taken for all time in the infinite interval, and every sample is used to reconstruct the value of the original function at any particular time. The theorem *does not* say, "Give me two sample values, and I'll tell you exactly how to draw a line connecting them." In a real system, the signal is observed over a limited time interval. We can define an error as the difference between the reconstructed time function and the original function, and upper bounds can be placed on the magnitude of the error. Such bounds involve sums of the rejected time sample values, and some examples are included in the "Problems" section at the end of the chapter.

A third error results if the sampling rate is not high enough. This situation can be intentional or accidental. For example, if the original time signal has a Fourier transform that asymptotically approaches zero with increasing frequency, a conscious decision is often made to define a maximum frequency beyond which signal energy is negligible. In order to minimize the resulting error, the input signal is often lowpass filtered prior to sampling. Even if we design a system with a high enough sampling rate, an unanticipated high-frequency signal (or noise) may appear at the input. In either case, the error caused by sampling too slowly is known as *aliasing*, a name derived from the fact that the higher frequencies disguise themselves in the form of lower frequencies. This is the same phenomenon that occurs when a rotating device is viewed as a sequence of individual frames, as in a television picture. As the rotation speed of the device increases, a point is reached where the perceived angular velocity starts to decrease. Eventually, a speed is reached (matched to the frame rate) at which the device appears to be standing still. Further increases make the rotation appear to reverse direction.

Analysis of aliasing is most easily performed in the frequency domain. Before doing that, we illustrate a simple example of aliasing in the time domain. Figure 5.8 shows a sinusoid at a frequency of 3 Hz. Suppose we sample this sinusoid at four samples per second. The sampling theorem tells us that the minimum sampling rate for unique recovery is six samples per second, so four samples per second is not fast enough. The samples at the slower rate are indicated in the figure. But alas, these are the same samples that would result from a sinusoid at 1 Hz, as shown by the dashed curve. The 3-Hz signal is thus disguising itself (aliasing) as a 1-Hz signal.

The Fourier transform of the sampled wave is found by periodically shifting and repeating the Fourier transform of the original signal. If the original signal has frequency components above one-half of the sampling rate, these components *fold back* into the frequency band of interest. Thus, in Fig. 5.8, the 3-Hz signal folded back to fall at 1 Hz.

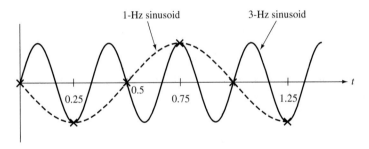

Figure 5.8   Example of aliasing.

Figure 5.9 illustrates the case where a representative signal is sampled by an ideal train of impulses (we use this as the ideal theoretical limit of narrow pulses) at less than the Nyquist rate.

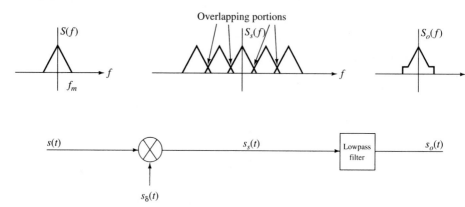

Figure 5.9   Impulse sampling at less than Nyquist rate.

Note that the transform at the output of the lowpass filter is no longer the same as the transform of the original signal. If we denote the filter output as $s_0(t)$, the error is defined as

$$e(t) = s_0(t) - s(t) \tag{5.10}$$

Taking the Fourier transform of both sides of Eq. (5.10) yields

$$E(f) = S_0(f) - S(f)$$
$$= S(f - f_s) + S(f + f_s) \text{ for } f < f_m \tag{5.11}$$

Observe that if $S(f)$ were limited to frequencies below $f_s/2$, the error transform would be zero. Without assuming a specific form for $S(f)$, we cannot carry this example further. In general, various bounds can be placed upon the magnitude of the error function based upon properties of $S(f)$ for $f > f_s/2$. You can explore aliasing in more detail using computer analysis in the "problems" section at the end of the chapter.

Example 5.2

A 100-Hz square wave (assume amplitude levels of 0 and 1) forms the input to the RC filter shown in Figure 5.10. The output of the filter is sampled at 700 samples per second. Find the aliasing error.

Solution: The square wave can be expanded in a Fourier series to yield

$$v_{in}(t) = \frac{1}{2} + \frac{2}{\pi}\cos 2\pi \times 100t - \frac{2}{3\pi}\cos 2\pi \times 300t + \frac{2}{5\pi}\cos 2\pi \times 500t \ldots$$

$$= \frac{1}{2} + \sum_{n=1,n\text{ odd}}^{\infty} (-1)^{\frac{n+3}{2}} \frac{2}{n\pi}\cos 2\pi n \times 100t$$

The filter transfer function is

$$H(f) = \frac{1}{1 + j2\pi f RC} = \frac{1}{1 + j2\pi f (0.00167)}$$

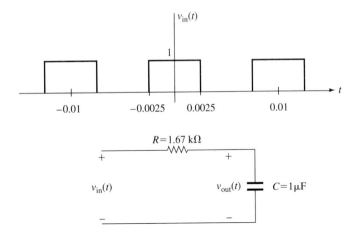

Figure 5.10   Square wave and filter for Example 5.2.

The output of the filter is found by modifying each term in the input Fourier series. The amplitude is multiplied by the magnitude of the transfer function, and the phase is shifted by the phase of the transfer function. The result is

$$v_{out}(t) = \frac{1}{2} + 0.45\cos(2\pi \times 100t - 45°) - 0.067\cos(2\pi \times 300t - 71.6°)$$

$$+ 0.025\cos(2\pi \times 500t - 78.7°) - 0.013\cos(2\pi \times 700t - 81.9°)$$

Let us assume ideal impulse sampling. The result is that the component at 500 Hz appears at 200 Hz in the reconstructed waveform, and the component at 700 Hz appears at dc (zero frequency). We shall ignore the higher harmonics. The reconstructed waveform is therefore given by

$$\frac{1}{2} + 0.45\cos(2\pi \times 100t - 45°) - 0.067\cos(2\pi \times 300t - 71.6°)$$

$$+ 0.025\cos(2\pi \times 200t - 78.7°) - 0.013\cos(-81.9°)$$

The last two terms represent the aliasing error.

Example 5.3

Assume that the bandlimited function

$$s(t) = \frac{\sin 20\pi t}{\pi t}$$

is sampled at 19 samples/sec. The sampling function is a unit-height pulse train with pulse widths of 1 msec. The sampled waveform forms the input to a lowpass filter with cutoff frequency of 10 Hz, as illustrated in Fig. 5.11. Find the output of the lowpass filter, and compare it to the original $s(t)$.

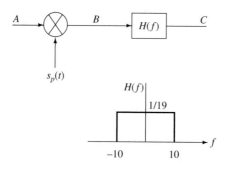

Figure 5.11   Sampling for Example 5.3.

Solution: We need to know only the first two coefficients in the Fourier series expansion of the pulse train. These are

$$a_0 = \frac{0.001}{1/19} = 0.019$$

$$a_1 = 38 \int_{-5\times10^{-4}}^{5\times10^{-4}} \cos 2\pi \times 19t\,dt = \frac{2}{\pi} \sin(19 \times 10^{-3}\pi t) = 0.038$$

The Fourier transforms of the signals at points A, B, and C in Fig. 5.11 are shown in Fig. 5.12. The output function of time is the inverse Fourier transform of $S_c(f)$ and is given by

$$s_0(t) = \frac{0.019\sin 20\pi t}{\pi t} + 0.038\frac{\sin \pi t}{\pi t}\cos 19\pi t$$

The second term in the result represents the aliasing error. Suppose we wish to find the maximum amplitude of this term. It should be obvious that this occurs at $t = 0$, but if you wished to use a simple MATLAB program, the instructions would be as follows:

```
t=-5:.01:=5;
s=.038*sin(pi*t)./(pi*t).*cos(19*pi*t);
MAX=max(s)
```

The maximum amplitude is 0.038 at $t = 0$, at which point the signal portion [the first term of $s_0(t)$] has amplitude 0.019. Do not be tempted to calculate a percentage error by taking the ratio of the amplitude of the error to the amplitude of the desired signal. Since the first term goes to zero at periodic points, and the second term is not necessarily zero at these same points, the percentage error would approach infinity. Errors are often analyzed by looking at the energy of the function of time representing the error. Energy is the area under the square of the function. We could therefore find the energy of the second term in the equation. Once

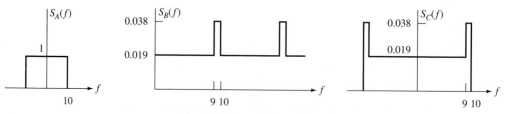

Figure 5.12   Fourier transform of sampling signals.

this energy is found, it is divided by the total energy of the desired signal (the first term) to get a percentage error. As an example, suppose we wanted to perform this operation over two side lobes of the main signal. That is, we compare the signal to the aliasing error over the range of time between -0.2 and $+0.2$ second. A simple MATLAB program calculates the SNR as the ratio of mean square values:

```
t=-.2:.01:.2;
sout=.019*sin(20*pi*t)./(pi*t)
salias=.038*sin*pi*t)./(pi*t).*cos*19*pi*t);
snr=(std(sout)^2+mean(sout)^2)/(std(salias)^2+mean(salias)^2)
snrdb=10*log10(snr)
```

MATLAB returns an SNR of 14.804, or 11.704 dB. Note that we are using the fact that the mean square value is the variance plus the square of the mean.

We conclude this section with the idea that the restriction on $S(f)$ imposed by the sampling theorem is not very severe in practice. All signals of interest in real life do possess Fourier transforms that are approximately zero above some frequency. No physical device can transmit infinitely high frequencies, since all channels contain series inductance and parallel (parasitic) capacitance. The inductance opens and the capacitance shorts as frequencies increase.

5.3 DISCRETE BASEBAND

5.3.1 Pulse Modulation

When a signal is discrete, it can be thought of as a list of numbers representing the sample values of an analog waveform. One way to send such a list through a channel is to send a pulse waveform—one pulse is placed at each sampling point. Each pulse carries information about the corresponding sample values. Each sample value can be conveyed as the amplitude, width, or position of the pulse. If we choose the amplitude of the pulse, the result is known as *pulse amplitude modulation* (PAM). Figure 5.13 illustrates a periodic pulse train $s_c(t)$, a portion of a typical analog signal $s(t)$, and the result $s_m(t)$ of controlling the pulse heights with the sample values.

Note that since the pulse tops are horizontal, the modulated waveform is *not* simply the product of the pulse train and the analog signal. Such a product would appear as in Figure 5.14. It results when $s(t)$ forms the input to a *gating* circuit.

Both of the foregoing waveforms are considered to be pulse amplitude modulated (PAM) waveforms; the waveform of Fig. 5.13 is called *flat-top* or *instantaneous-sampled* PAM, while that of Fig. 5.14 is known as *natural-sampled* PAM. The former is instantaneous sampled because the pulse height depends only upon the value of $s(t)$ at the sampling point, and not on the signal values across the range of the pulse width. Flat-top PAM is generated with a *sample-and-hold* circuit. A simplified sample-and-hold circuit is shown in Fig. 5.15.

In the figure, switch S_1 closes instantaneously at the sampling point, and the capacitor charges to the sample value. The switch is then opened, and the capacitor remains at

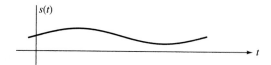

Figure 5.13   Pulse amplitude modulation.

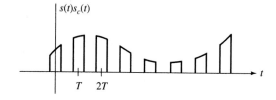

Figure 5.14   Product of pulse train and $s(t)$.

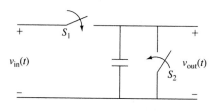

Figure 5.15   Idealized sample-and-hold circuit.

that value until the closing of switch S_2 provides a discharge path. Practical sample-and-hold circuits need additional electronics to provide the energy to charge the capacitor rapidly (i.e., the series resistance is never zero) and to prevent slow discharge (leakage) prior to switch S_2 closing.

We now calculate the Fourier transform of a PAM waveform in order to determine the channel requirements. We begin by evaluating the Fourier transform of the natural sampled waveform. The function $s_c(t)$ is expanded in a Fourier series to obtain

$$s_c(t) = a_0 + \sum_1^\infty a_n \cos 2\pi f_s t \tag{5.12}$$

When this is multiplied by $s(t)$, the result is a summation of products of $s(t)$ with sinusoids:

$$s(t)s_c(t) = a_0 s(t) + \sum_1^\infty a_n s(t) \cos 2\pi f_s t \tag{5.13}$$

The Fourier transform of each term in the summation is the signal transform $S(f)$, shifted up and down by the frequency of the sinusoid (the *modulation theorem*). The transform of $s(t)s_c(t)$ is sketched in Fig. 5.16, where we assume that $s(t)$ has the transform shown in the figure. f_m is the maximum-frequency component of $s(t)$.

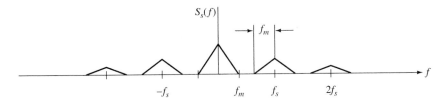

Figure 5.16   Fourier transform of natural-sampled PAM.

The Fourier transform of instantaneous-sampled PAM is more difficult to evaluate. The evaluation is simplified, however, by considering the hypothetical system of Fig. 5.17. We begin by sampling $s(t)$ with an ideal train of impulses. We then shape each impulse into the desired pulse shape—in this case, a square pulse with a flat top.

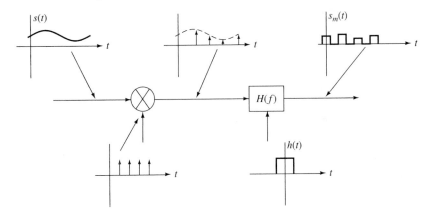

Figure 5.17   Generation of instantaneous-sampled PAM.

The Fourier transform of the sampled signal at the input to the filter is found using the sampling theorem. The Fourier series of the impulse train has equal Fourier coefficient values for all n. The Fourier transform of the impulse-sampled waveform is therefore as shown in Fig. 5.18.

The Fourier transform of the filter output (instantaneous-sampled PAM) is simply the product of the Fourier transform of the impulse-sampled waveform and the transfer

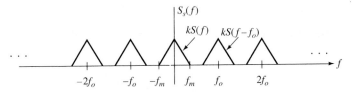

Figure 5.18   Fourier transform of impulse-sampled waveform.

function of the filter. The transfer function of the filter is shown in Fig. 5.19 and is found from a table of Fourier transforms. (See Appendix II.)

Finally, the output transform is as shown in Fig. 5.20. Note that the low-frequency portion of this transform is *not* an undistorted version of $S(f)$.

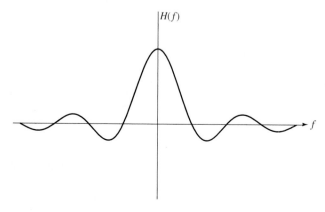

Figure 5.19   Transfer function of filter of Fig. 5.17.

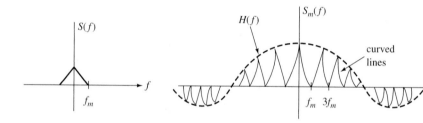

Figure 5.20   Fourier transform of flat-top PAM.

Example 5.4

A signal is of the form

$$s(t) = \frac{\sin \pi t}{\pi t}$$

It is transmitted using PAM. The pulse waveform $s_c(t)$ is a periodic train of triangular pulses, as shown in Fig. 5.21. Find the Fourier transform of the modulated waveform.

Solution: Consider the system of Fig. 5.17. The output of the ideal impulse sampler has the transform

$$S_\delta(f) = \frac{1}{T} \sum_{n=-\infty}^{\infty} S(f - nf_0)$$

where $S(f)$ is the transform of $\sin \pi t / \pi t$. $S(f)$ is a pulse, as shown in Fig. 5.22(a). The filter must change each impulse into a triangular pulse. Its impulse response is therefore a single triangular pulse that has as its transform

$$H(f) = \frac{\sin^2 \pi f \tau}{\tau^2 \pi^2 f^2}$$

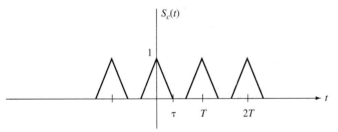

Figure 5.21   Pulse waveform for Example 5.4.

Finally, the transform of the PAM waveform is given by the product $S_\delta(f)H(f)$, as shown in Fig. 5.22(b).

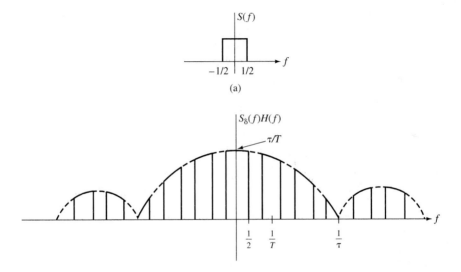

Figure 5.22   Result for Example 5.4.

A significant general observation to make about the transform of a PAM wave is that it occupies all frequencies, from zero to infinity.

5.3.2 Time Division Multiplexing

It would be very impractical to have a separate channel for every signal to be communicated. In telephone communication, this would mean a wire connection for every conversation; if up to 100,000 simultaneous calls between the United States and Europe were anticipated, 100,000 channels (i.e., pairs of wires or satellite channels) would be needed. An even more absurd example is provided by television: If 13 television stations wanted to broadcast simultaneously (forgetting uhf), we would need 13 parallel atmospheric systems—even the science fiction writers would have trouble with that. So we need a way to share a channel among different users.

Signals can be easily separated from each other if they are nonoverlapping in either time or frequency. An example of separation in time occurs in the classroom when students and the professor alternately share the communication channel: When a student talks, the professor stops, and vice versa. An exact dual of this is separation in frequency. Suppose one speaker has unusual vocal cords, and his or her voice occupies frequencies between 1 kHz and 2 kHz. Suppose a second speaker is a baritone, all of whose essential signal components are below 1 kHz. Filters (lowpass and bandpass) can easily separate these two signals even if the two people speak simultaneously. We will deal extensively with separation in frequency beginning with the next chapter. For now, we concentrate on separation in time. Fortunately, over portions of the time axis the PAM waveform is zero, so that separation in time is possible.

Time division multiplexing of signals with identical sampling rates can be viewed as interspersing pulses. Figure 5.23 illustrates this process for two signals. Note that the switches alternate between each of the two positions, making sure to take no longer than one sampling period to complete the entire cycle. That is, two pulses are sent in each sampling period, so the pulse rate on the channel is twice the sampling rate.

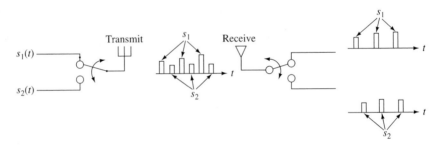

Figure 5.23   Multiplexing of two channels.

Suppose that we now increase the number of channels from 2 to 10. Then the switch becomes a commutator, as shown in Fig. 5.24. The switch must make one complete rotation fast enough so that it arrives at Channel 1 in time for the second sample. The rotation of the receiver switch must be synchronized with that at the transmitter. In practice, this synchronization (known as *frame synchronization*) requires effort. If we knew exactly what was being sent on one of the channels, we could identify its samples at the receiver. Indeed, a common method of synchronization is to sacrifice one of the channels and send a known synchronizing signal in its place. We shall see this in some of the digital transmission systems in later chapters.

The only thing that limits how fast the switch can rotate, and therefore how many channels can be multiplexed, is the fraction of time required for each PAM signal. That fraction is the ratio of the width of each pulse to the spacing between adjacent samples of a single channel. The trade-off design consideration is that the more narrow each pulse, the wider will be the bandwidth of the resulting signal.

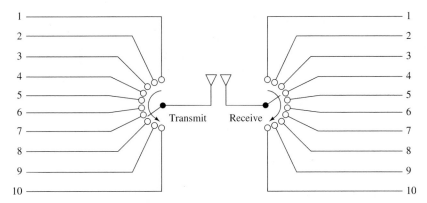

Figure 5.24   Multiplexing of 10 channels.

Multiplexing of Dissimilar Channels   The commutator approach to multiplexing requires that the sampling rate of the various channels be identical. If signals with different sampling rates must be multiplexed, two other general approaches can be taken. One uses a buffer to store sample values and then spit them out at a fixed rate. This buffer approach, which is also effective if the sampling rates are variable, is known as *asynchronous multiplexing*. It is important to design the system so that the buffer always has samples to send when requested by the channel. This might require inserting *stuffing samples* if the buffer gets empty. Alternatively, the buffer must be large enough so that it does not *overflow*.

The buffer approach is also used if the various sources are transmitting asynchronously. That is, suppose they are not always transmitting information. Then setting the size of the buffer requires a probability analysis, and the resulting multiplexer is known as a *statistical multiplexer* (stat MUX). The stat MUX represents an efficient technique for multiplexing channels, since a source has a time slot only when it needs it. On the negative side, since individual messages are not being transmitted at a regular rate, the messages must be *tagged* with a user ID. If the channels are synchronous with samples occurring at a regular and continuous rate, the stat MUX is not the best approach.

The second general approach involves *sub-* and *supercommutation*. This requires that all sampling rates be multiples of some basic rate. Meeting such a requirement might necessitate sampling some of the channels at a rate higher than what you would use without multiplexing. For example, if you have two channels with required sampling rates of 8 kHz and 15.5 kHz, then in order to effect that combination, you might choose to sample the faster channel at 16 kHz.

The concept of sub- and supercommutation is quite simple, and we illustrate it with an example. Figure 5.25 shows a *commutator wheel* with 32 slots. Suppose we wish to multiplex the following 44 channels:

    1 channel sampled at 80 kHz
    1 channel sampled at 40 kHz
    18 channels sampled at 10 kHz
    8 channels sampled at 1,250 Hz
    16 channels sampled at 625 Hz

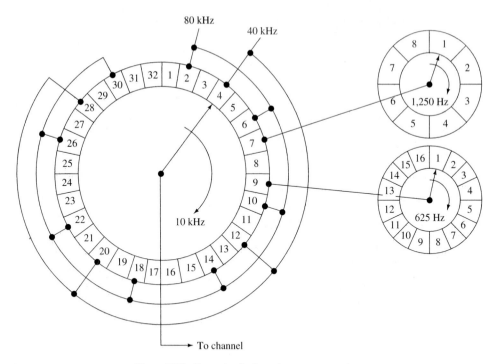

Figure 5.25   Example of sub- and supercommutation.

All of the sampling rates are multiples of 625 Hz. Let us choose the basic rate of the commutator wheel to be 10,000 rotations per second. Therefore, each of the 18 channels that must be sampled at 10 kHz get one slot on the wheel. The channel that must be sampled at 40 kHz needs four equally spaced slots on the wheel, so it is sampled four times during each 0.1-msec rotation of the wheel. Similarly, the 80-kHz channel needs eight equally spaced slots on the wheel. These higher rate channels are multiplexed using *supercommutation*.

The channels sampled at less than 10 kHz need to be sampled only on selected rotations of the wheel. For example, a 1,250-Hz channel needs to be sampled once every 8 rotations of the wheel, while a 625-Hz channel is sampled only once every 16 rotations. We accomplish this using *subcommutation* wheels, as shown in the figure. The eight 1,250-Hz channels are commutated together with a wheel rotating at a rate of 1,250 rotations per second. Each 0.1 msec, one of the channels is connected to a cell on the main commutator wheel. Similarly, the 16 625-Hz channels are commutated with a wheel rotating at 625 rotations per second.

Clearly, a lot of design work had to go into this configuration. We have chosen numbers that work out perfectly. Life is usually not so nice, however, and manipulation is needed to design the commutation system. In some ways, this process resembles finding the lowest common denominator in combining fractions, but of course, it is orders of magnitude more complex than this simple algebraic problem.

5.3.3 Cross Talk

As shown in Figs. 5.19 and 5.20, the envelope of the Fourier transform of the PAM waveform follows a $\sin(f)/f$ curve. In fact, this envelope is the Fourier transform of a single
square pulse. The first zero of the envelope is at $1/\Delta t$, where Δt is the width of each pulse.

  The pulsed waveform may be transmitted through a coaxial cable. The signal is
transmitted with little distortion, provided that the upper frequency cutoff of the channel is
sufficiently high (at least several times $1/\Delta t$).

Example 5.5

  A discrete-time analog signal is created by sampling a speech waveform at 10,000 samples
  per second. Each sampling pulse is 0.01 msec wide and is transmitted through a channel that
  can be approximated by a lowpass filter with cutoff frequency at 100 kHz. Evaluate the effects of channel distortion.

  Solution

  Since the sampling is occurring at 10,000 samples per second, we will assume that the speech
  waveform has a maximum frequency below 5 kHz. The reason for the two-to-one ratio was
  discussed in Section 5.2. The transmitted signal consists of pulses 0.01 msec wide. If one of
  these pulses forms the input to a lowpass filter with cutoff at 100 kHz, the output of the filter

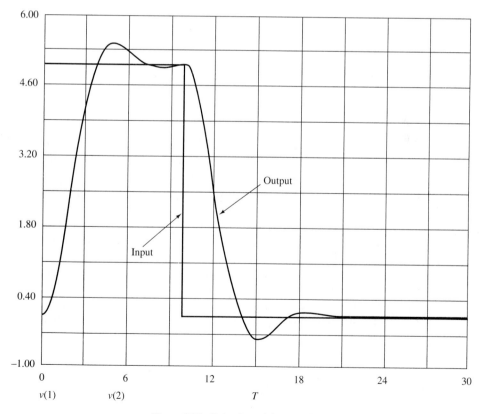

Figure 5.26   Pulse through lowpass filter.

is as shown in Figure 5.26.[1] Thus, although the original pulse may be confined to its assigned time interval, the filtering effects of the channel may widen the pulse to overlap adjacent intervals.

The overlap from one time slot to adjacent time slots is known as *intersymbol interference* or *crosstalk*. The term *crosstalk* also applies to the leakage of signals from one set of wires to an adjacent set of wires, as when multiple wires form part of one cable. We will restrict our discussion to the overlap of time slots in multiplexed systems. There are several ways of reducing the effects of intersymbol interference. Pulse spreading can be decreased by increasing the bandwidth of the system. Unfortunately, this is a luxury that requires a flexibility we don't often have. However, one parameter over which we do have control is the shape of the pulses used to transmit the sample values.

As a start toward the analysis of desirable pulse shapes, suppose we transmit ideal impulse samples. Suppose further that the channel can be modeled as an ideal lowpass filter. This shapes each impulse into a (sin *t*)/*t* type of pulse, as shown in Fig. 5.27.

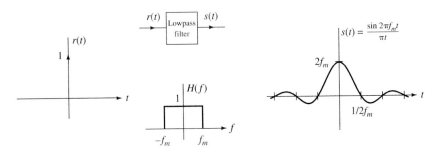

Figure 5.27   Pulse shaping by the channel.

Although the received waveform clearly extends into adjacent time slots, it has the very nice property that interference is zero at the sample points. If the channel passes frequency up to f_m, the zeros of the spread pulse are spaced by $1/2f_m$. Thus, if we sample at the Nyquist rate, the spread pulse goes through zero at all adjacent sample points. Figure 5.28 illustrates this for three adjacent sample values.

Note that while the pulses interact with each other, at each sampling point only one (sin *t*)/*t* curve is nonzero. All of the others go through zero at that point. The sum of the various (sin *t*)/*t* curves reconstructs the original signal *s*(*t*), according to the sampling theorem. Although the ideal impulse transmission–ideal lowpass filter channel represents an unrealistic assumption, the results apply as long as the combination of transmitted pulse shape and channel characteristics results in a (sin *t*)/*t* shape for the received pulse.

The ideal bandlimited pulse shape is impossible to achieve because of the sharp corners on the frequency spectrum. A compromise is the *raised cosine* characteristic. The Fourier transform of this pulse is similar to the square transform of the ideal lowpass filter, except that the transition from maximum to minimum follows a sinusoidal curve. This is shown in Fig. 5.29. The value of the constant K determines the width of the constant por-

---

[1]We have used a circuit simulation program, MICRO-CAP IV, to produce the output waveform. The channel was approximated as a third-order Butterworth lowpass filter with normalized transfer function $1/(s^3 + 2s^2 + 2s + 1)$.

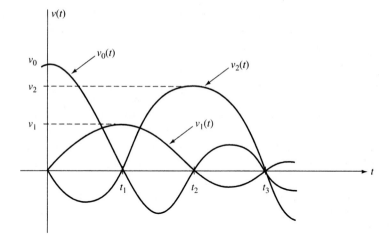

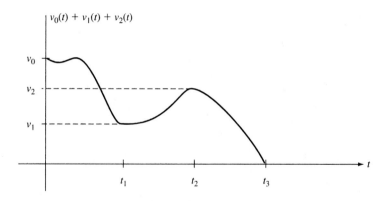

Figure 5.28   Ideal lowpass filter shaping.

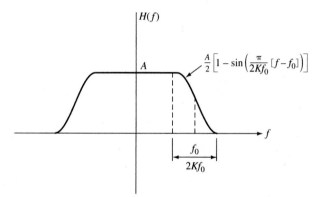

Figure 5.29   Raised cosine frequency characteristic.

tion of the transform. If $K = 0$, the transform is that of the ideal lowpass filter. If $K = 1$, the flat portion is reduced to a point at the origin. The function of time corresponding to the Fourier transform is

$$h(t) = A\frac{\sin 2\pi f_0 t}{2\pi f_0 t}\ \frac{\cos 2\pi K T f_0}{1 - (4K f_0)^2} \tag{5.14}$$

This function is sketched for several representative values of K in Fig. 5.30.

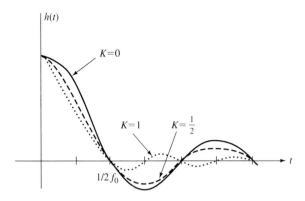

Figure 5.30 Raised cosine impulse response.

Note that for $K = 0$, the function is of the form $(\sin t)/t$. This goes to zero at multiples of $1/2 f_0$. For $K = 1$, the response goes to zero not only at these points, but also at points midway between them (except for the first set of points around the origin). For $K = 1$, the Fourier transform has frequency content up to $2 f_0$. An ideal lowpass filter with cutoff at that frequency has an impulse response with zeros every $1/4 f_0$. Therefore, the difference between the raised cosine with $K = 1$ and the ideal lowpass filter with the same cutoff is that the ideal filter has two additional zeros in its impulse response. Beyond the point $t = 1/2 f_0$, the zeros of both impulse responses coincide. It is much easier to build the raised cosine filter than the ideal lowpass filter; indeed, the latter cannot be built in the real world. The price we pay is intersymbol interference between adjacent samples (no interference more than one sample period away). We can compensate for this by using a technique known as *partial response signaling*. In digital communication, this is called *duobinary*. It is a form of controlled intersymbol interference. Since we know the proportion of one sample value that interacts with the next sample point, we can compensate by giving our receiver memory.

5.3.4 Pulse Width Modulation

Let us again start with a signal that is a periodic train of pulses. Figure 5.31 shows an unmodulated pulse train, a representative information signal $s(t)$, and the resulting *pulse width–modulated* (PWM) waveform. The width of each pulse varies in accordance with the instantaneous sample value of $s(t)$. The larger the sample value, the wider is the corresponding pulse. Since the pulse width is not constant, the power of the waveform is also not constant. Thus, as the amplitudes of the signal increase, the power transmitted also increases.

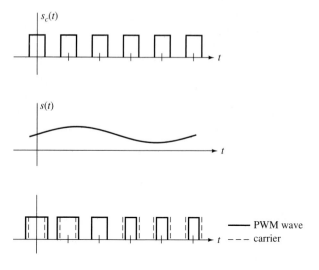

Figure 5.31   Pulse Width Modulation.

Finding the Fourier transform of the PWM waveform is a complex computational task. Part of the reason for this complexity is that PWM is a *nonlinear* form of modulation. A simple example illustrates this. If the information signal is a constant, say, $s(t) = 1$, the PWM wave consists of equal-width pulses. This is so because each sample value is equal to every other sample value. If we now transmit $s(t) = 2$ via PWM, we again get a pulse train of equal-width pulses, but the pulses would be wider than those used to transmit $s(t) = 1$. The principle of linearity dictates that if the modulation is linear, the second modulated waveform should be twice the first. But this is not the case, as is illustrated in Fig. 5.32. (You should stop here and take a moment to convince yourself that, by contrast, PAM is a *linear* form of modulation.)

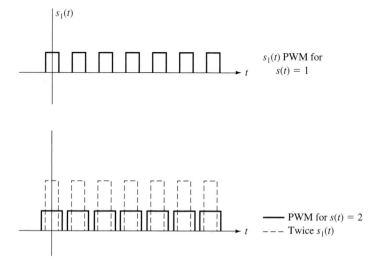

Figure 5.32   PWM is nonlinear.

If one assumes that the information signal is slowly varying (i.e., sampled at a fast rate compared to the Nyquist rate), then adjacent pulses have almost the same width. Under this assumption, an approximate analysis of the modulated waveform is possible: The PWM waveform can be expanded in an approximate Fourier series. We shall not perform the analysis here, but will explore it in the "Problems" section at the end of the chapter.

PAM and PWM are related to each other, and it is possible to construct systems that convert from one form to the other. We can use a sawtooth generator to convert between time and amplitude. The sawtooth waveform we employ is shown in Fig. 5.33.

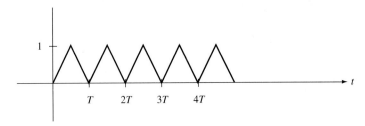

Figure 5.33   Sawtooth waveform for PWM-to-PAM conversion.

The conversion process is illustrated in Fig. 5.34. Figure 5.34(a) shows a block diagram of the generator, and Fig. 5.34(b) shows typical waveforms.

We start with an information signal $s(t)$. This is put through a sample-and-hold circuit to yield $s_1(t)$. The sawtooth is shifted down by one unit in order to form $s_2(t)$. The sum of $s_1(t)$ and $s_2(t)$ is $s_3(t)$. The times for which $s_3(t)$ is positive represent intervals whose widths are proportional to the original sample values. We need only put the shifting sawtooth into a comparator with output of 1 for positive input and 0 for negative input. This results in $s_4(t)$, the PWM waveform. The range of pulse widths can be adjusted by scaling the original function of time. Our illustration assumes that the original $s(t)$ was normalized to lie between 0 and 1.

Since the heights of the pulses in PWM are constant, but the widths depend on $s(t)$, the power of the PWM waveform varies with the amplitude of $s(t)$. For example, if $s(t)$ were a musical selection, less power would be required during soft parts of the music and more power during loud parts. This reduces the efficiency of the communication system, since the pulse amplitudes would have to be chosen to assure that the maximum power does not exceed that permitted by the system.

When we investigate performance later in the chapter, we will find that PWM is disturbed by noise less than PAM is.

5.3.5 Pulse Position Modulation

Pulse position modulation (PPM) possesses the noise advantage of PWM without the problem of a variable power that is a function of the amplitude of the signal. An information signal $s(t)$ and its PPM waveform are illustrated in Fig. 5.35. We see that the larger the sample value, the more the corresponding pulse deviates from its unmodulated position.

A PPM waveform can be derived from a PWM waveform. The relationship between the two is that, while the position of the pulse varies in PPM, the location of the leading

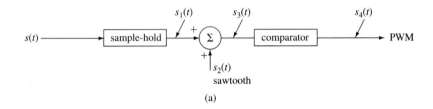

(a)

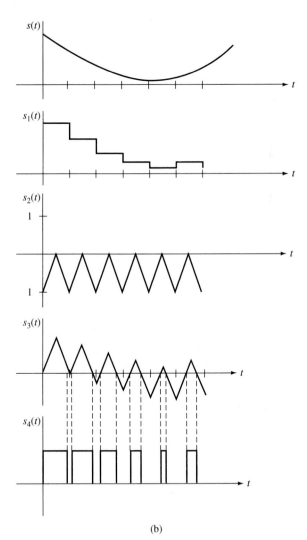

(b)

Figure 5.34 PWM generator.

(or trailing) edge of the pulse varies in PWM. Suppose, for example, that we detect each trailing edge in a PWM waveform. (We differentiate and look for large negative pulses.) If we now place a constant-width pulse at each of these points, the result is PPM. This is illustrated in Fig. 5.36.

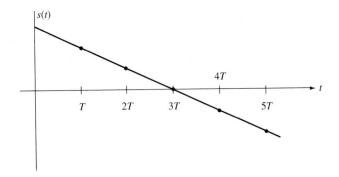

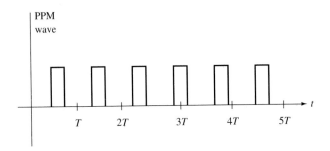

Figure 5.35   Pulse position modulation.

Clearly, both PWM and PPM are more complex than is PAM. The justification for choosing one of these more complex systems is that it provides greater noise immunity than does PAM. In PAM, additive noise directly affects the reconstructed sample value. The disruption is less severe in PPM and PWM, where the additive noise must affect the zero-crossings in order to cause an error. Along with their greater complexity, PWM and PPM have other negative properties. In multiplexed (TDM) systems, one must be sure that adjacent sample pulses do not overlap. If pulses are free to shift around or to get wider, as they are in PPM and PWM, one cannot simply insert other pulses in the spaces and be confident that no interaction will occur. Sufficient spacing must be maintained to allow for the largest possible sample value. This decreases the number of channels that can be multiplexed.

5.4 RECEIVERS

5.4.1 Analog Baseband Reception

Analog baseband reception consists of filtering the received signal, amplifying it, and feeding the result into a transducer (e.g., a speaker). In the case of audio, the receiver is simply an audio amplifier. In designing receivers, one needs to be concerned with filter characteristics and noise. Noise is added in the channel, and additional noise is added by

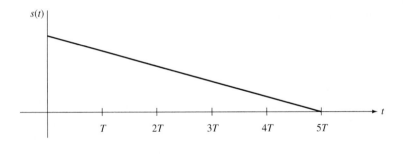

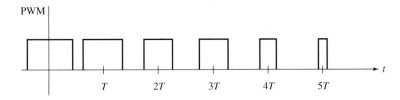

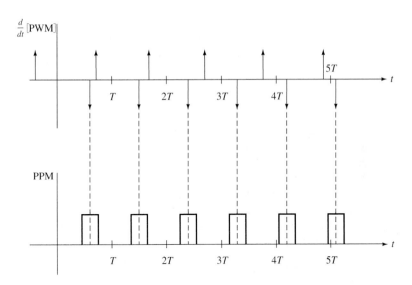

Figure 5.36   Conversion from PWM to PPM.

the electronics in the receiver. We consider the performance of receivers in the next section. Receiver design is normally considered part of a study of electronics.

5.4.2 Discrete Baseband Reception

Reconstruction of the original signal from *natural-sampled PAM* follows directly from the sampling theorem. Indeed, natural-sampled PAM is the type of signal processing we encountered in our first proof of the sampling theorem. Recovery of the original analog signal from its sampled version requires a lowpass filter. The natural-sampled PAM receiver is therefore as shown in Fig. 5.37. The process of converting a PAM waveform to the continuous analog waveform is known as *demodulation*.

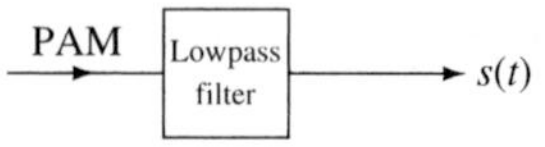

Figure 5.37   Natural-sampled PAM receiver.

Demodulation of *instantaneous-sampled PAM* requires a little more work. We could use a *sample-and-hold* circuit to recover a staircase approximation of the original waveform. The holding time is set equal to the sampling period. The result is shown in Fig. 5.38 for a representative $s(t)$. The resulting staircase function can be lowpass filtered to get a smooth approximation to the original waveform. We ask you to show in Problem 5.4.1 that this form of sample and hold is an approximation to a lowpass filter.

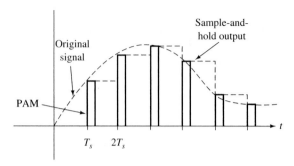

Figure 5.38   Sample and hold for PAM demodulation.

The Fourier transform of instantaneous PAM is shown in Fig. 5.39(a). We derive it by multiplying the sampled Fourier transform by $H(f)$, the transfer function of the filter that changes impulse samples to pulse samples. The baseband portion of the PAM Fourier transform is of the form $S(f)H(f)$. Therefore, $s(t)$ can be recovered from $s_m(t)$ by using a shaped lowpass filter where the transfer function is the reciprocal of $H(f)$. This is shown in Fig. 5.39(b). The filter with transfer function $1/H(f)$ is known as an *equalizer*, since it cancels the effects of the pulse shaping.

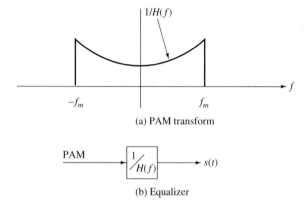

(a) PAM transform

(b) Equalizer

Figure 5.39   Flat-top PAM demodulator.

The equalizer and lowpass filter demodulators do not require that the receiver recover timing information. On the other hand, the sample-and-hold demodulator *does* require such timing information at the receiver. That is, the receiver must "know" when to sample the incoming waveform—at the original sample points. This requires *symbol synchronization*. We consider synchronization in detail when we study digital reception later in the text.

Reception of PWM or PPM can be viewed as a two-step process. We first convert the received waveform into PAM and then use a PAM receiver, as described earlier in this section. The conversion of PWM to PAM is accomplished using an integrator. For PWM, we simply start the integrator at the sample point and integrate the received pulse. Since the height of the pulse is constant, the integral is proportional to the pulse width. The output of the integrator is sampled prior to the next signal sampling point, and the sample generates a PAM waveform. The process is illustrated in Figure 5.40. There are two observations you should make regarding this figure: First, we are using the form of PWM that sets the left edge of each pulse at the sampling point; second, the resulting PAM waveform is delayed by one sampling period.

Conversion of PPM to PAM is also illustrated in Fig. 5.40. Here, we start the integrator at each sampling point and set it to integrate a constant. The integrator stops when the pulse arrives. Since the PPM pulse is at the trailing edge of the PWM pulse, there is no essential difference between PWM and PPM reception.

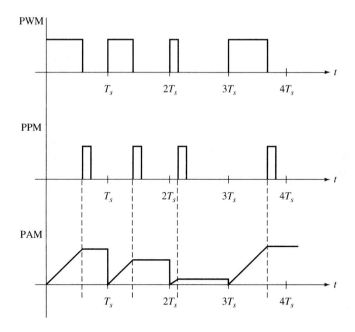

Figure 5.40   Conversion of PWM and PPM to PAM.

5.5 PERFORMANCE

We will learn to design a variety of communication systems. Performance evaluation needs to apply to a wide range of system inputs. In analog communication, we normally wish the output to be as close to the input waveform as possible. The more common measure of such closeness is the S/N power ratio, since the human ear is sensitive to this quantity. In general, the ear can hear additive disturbances if the ratio of S/N power is below a certain threshold.

In digital communication, the normal measure of performance is the rate at which bit errors occur.

5.5.1 Analog Baseband

The output SNR of a baseband analog receiver depends on the input SNR, the filtering characteristic of the receiver, and the noise added by the electronics in the receiver.

The SNR at the input to the receiver depends on the characteristics of the channel and of the noise that intrudes during transmission. We normally consider this additive noise to be white Gaussian, so the total power of that noise is proportional to the system bandwidth.

The noise added by the receiver is characterized by the *noise figure. Thermal noise* is produced by the random motion of electrons in a medium.

If we have a system with a number of noise-generating devices within it, we often refer to the system *noise temperature* T_e, in °K. This is the temperature of a single noise source that would produce the same total noise power at the output.

If the input to the receiver contains noise, the receiver then adds its own noise to produce a larger output noise. The system *noise figure* is the ratio of the noise power at the output to that at the input. It is usually expressed in decibels (dB). For example, a noise figure of 3 dB indicates that the system is adding an amount of noise equal to that which appears at the input, so the output noise power is twice that of the input.

We have concentrated on the noise added by the transmitter and the receiver. We now turn our attention to the channel. The technique we use to analyze the additive noise in the channel is typical of the way we will approach broadcast communication systems in the chapters that follow. Let us assume a channel noise model as shown in Fig. 5.41. In this model, the noise $n(t)$ is added to the signal $s(t)$, and the sum is lowpass filtered. The lowpass filter is normally part of the receiver. We include it in the model because, without it, the noise has significant components outside the frequency band of interest.

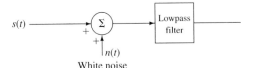

Figure 5.41   Additive noise in baseband channel.

We assume that the additive noise is white with two-sided power spectral density $N_0/2$. That is, the noise has a power of N_0 watts/Hz. The noise power at the output of the filter is then $N_0 f_m$ watts, and

$$\mathrm{SNR} = \frac{P_s}{N_0 f_m} \qquad (5.15)$$

where P_s is the power of the signal.

The performance of a baseband analog receiver also depends on nonlinearities in the electronics. This is expressed in measures such as dynamic range and harmonic distortion. The term *dynamic range* usually refers to the ratio (in decibels) of the strongest to the weakest signal that a receiver can process without noise or distortion exceeding acceptable limits. Although this sounds like a simple concept, application to practical transmission is quite complex. For example, the behavior of a receiver when a single sinusoid forms the input may be quite different from that when the input is a complex sum of many signal components. Keep in mind that we are discussing nonlinear effects, and the actual dynamic range of the receiver may depend upon characteristics of the input signal.

Harmonic distortion is normally measured by setting the receiver input to be a single sinusoid. Nonlinearities in the receiver change this sinusoid to a periodic function with harmonics. The ratio of the power of the harmonics to the power of the fundamental is a measure of harmonic distortion.

5.5.2 Discrete Baseband

The SNR in a PAM signal depends on the form of the receiver. If the receiver simply samples the received waveform at periodic points in time, the sample values are

$$r(nT_s) = s(nT_s) + n(nT_s) \tag{5.16}$$

where $n(t)$ is the additive noise. The SNR is then

$$\frac{S}{N} = \frac{\overline{s^2}}{\overline{n^2}} \tag{5.17}$$

The numerator of Eq. (5.17) is the average signal power, while the denominator is the average noise power. If the channel can be modeled as an ideal lowpass filter, the noise power is simply $N_0 BW$, where N_0 is the noise power per Hz and BW is the system bandwidth.

Calculation of the SNR for PPM and PWM systems is more complex. Since we can easily convert a PWM signal to a PPM signal, we shall calculate the SNR only for PPM. Figure 5.42 shows an *ideal* square pulse and an approximation to a *practical* square pulse.

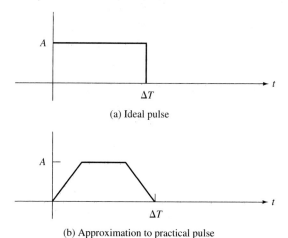

(a) Ideal pulse

(b) Approximation to practical pulse

Figure 5.42   Ideal and practical square pulse.

The job of the PPM receiver is to locate the trailing edge of the latter pulse. One way to do this is with a comparator or threshold detector. A threshold is set, and when the signal breaks through it, we assume that we have located the trailing edge. The threshold value would normally be set at the midpoint, $A/2$, of the pulse amplitude, and this value is known as the *slicing level*. In the case of the *ideal* square pulse, as long as the additive noise waveform never exceeds the slicing level in magnitude, the location of the trailing edge will not be affected.

If we now add noise to the practical approximation to the square pulse, we have the situation shown in Fig. 5.43, where we have focused upon the trailing edge of the pulse.

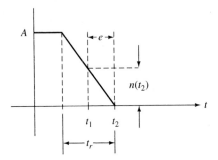

Figure 5.43   Noise affecting location of trailing edge.

Similar triangles can be used to derive the relationship

$$\frac{e}{n(t_2)} = \frac{t_r}{A} \tag{5.18}$$

where A is the amplitude of the pulse, t_r is the pulse rise time, and $n(t_2)$ is the additive noise at the time the perturbed signal crosses the slicing level. We solve for the timing error to obtain

$$e = \left(\frac{t_r}{A}\right) n(t_2) \tag{5.19}$$

The mean square value of the error is then

$$\overline{e^2} = \left(\frac{t_r}{A}\right)^2 \overline{n^2(t_2)} \tag{5.20}$$

Let us assume that the mean square value of the noise is given by the noise power per Hz multiplied by the bandwidth. This presupposes additive white noise. The error is then

$$\overline{e^2} = \left(\frac{t_r}{A}\right)^2 N_0 BW \tag{5.21}$$

The pulse rise time is related to the bandwidth. If we assume the maximum possible slope (a square pulse through a lowpass filter), we get the relationship

$$t_r = \frac{1}{2BW} \tag{5.22}$$

Equation (5.22) is related to the sampling theorem. We also would like to relate the error to the pulse energy instead of the amplitude, since many systems are energy limited. The pulse energy is

$$E_p \approx A^2 \Delta T \tag{5.23}$$

The approximation improves as the rise time decreases. Combining Eqs. (5.21), (5.22), and (5.23) yields

$$\overline{e^2} = \frac{\Delta T}{4BW E_p} N_0 \tag{5.24}$$

Equation (5.24) is the desired result. It shows that the mean square timing error is inversely proportional to the system bandwidth. To convert the PPM waveform to an analog signal, we change the pulse location to a pulse amplitude and then take the ratio of the mean square value of the signal samples to the mean square value of the error. The noise power is therefore related to the mean square timing error of Eq. (5.24). We explore this relationship in the "Problems" section at the back of this chapter.

PROBLEMS

5.2.1 You are given the function

$$s(t) = \frac{\sin t}{t}$$

This function is sampled by the train of impulses, $s_\delta(t)$ as shown in Fig. P5.2.1.

$$s_s(t) = s(t)s_\delta(t)$$

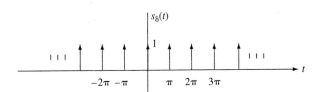

Figure P5.2.1

(a) What is the Fourier transform $S_s(f)$ of the sampled function?

(b) Find $s_s(t)$, the inverse Fourier transform of your answer to part (a).

(c) Should your answer to part (b) have been a train of impulses? Did it turn out that way? Explain any discrepancies.

(d) Design a system to recover the original $s(t)$ from $s_s(t)$, and demonstrate that your system works correctly for this example.

5.2.2 A signal $s(t)$ with $S(f)$ as shown in Fig. P5.2.2 is sampled by two different sampling functions, $s_{\delta 1}(t)$ and $s_{\delta 2}(t)$, where

$$s_{\delta 2}(t) = s_{\delta 1}\left(t - \frac{T}{2}\right)$$

and

$$T = \frac{1}{f_m}$$

Therefore, each of the two sampling waveforms is at one-half of the Nyquist minimum sampling frequency. Find the Fourier transforms of the sampled waveforms, $s_{s1}(t)$ and $s_{s2}(t)$.

Now consider $s(t)$ to be sampled by $s_{\delta3}(t)$, a train of impulses spaced $T/2$ apart (i.e., at the Nyquist rate). This new sampling function is the sum $s_{\delta1}(t) + s_{\delta2}(t)$. Show that the Fourier transform of $s_{s3}(t)$ is equal to the sum of the transforms of $s_{s2}(t)$ and $s_{s2}(t)$. That is, show that, although the transforms of the original two sampled functions contain aliasing error, this error is not present in the sum. [*Hint:* You will have to keep track of phases.]

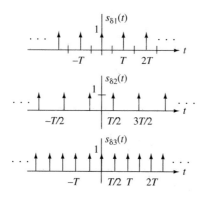

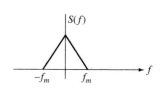

Figure P5.2.2

5.2.3 The function

$$s(t) = \cos 2\pi t$$

is sampled every $\frac{3}{4}$ second. Evaluate the aliasing error.

5.2.4 The signal

$$s(t) = \frac{\sin 2\pi t}{\pi t}$$

is sampled at 1.1 times the Nyquist rate. The sampling is performed for t between -1 and $+1$ second. The signal is reconstructed using a lowpass filter. Find the truncation error.

5.2.5 The function

$$s(t) = \cos 2\pi t$$

is sampled at a rate of 2.5 samples/sec for t between 0 and 10 seconds. The signal is reconstructed using a lowpass filter. Find the difference between the original and reconstructed waveforms at the points $t = 4.9$, $t = 5$, and $t = 5.1$ sec.

5.2.6 You are given a low-frequency bandlimited signal $s(t)$. This signal is multiplied by the pulse train $s_c(t)$, as shown in Fig. P5.2.6. Find the Fourier transform of the product $s(t)s_c(t)$. What restrictions must be imposed so that $s(t)$ can be uniquely recovered from the product waveform?

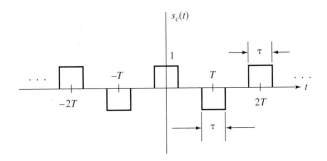

Figure P5.2.6

5.2.7 A signal is given by

$$s(t) = \frac{\sin 5\pi t}{\pi t} + \frac{\sin 10\pi t}{\pi t}$$

This signal is to be sampled with a periodic pulse train consisting of narrow pulses.
(a) Find the Nyquist sampling rate.
(b) Assuming that the sampling is done at the Nyquist rate, sketch the Fourier transform of the sampled waveform.

5.2.8 The signal

$$s(t) = \sin 100\pi t$$

is sampled at the points $t = n/100$. This represents sampling at the Nyquist rate of 100 Hz. However, all of the sample values will be zero, and the original wave cannot be reconstructed. Explain the reason for this situation.

5.2.9 A signal is of the form

$$s(t) = \sin \pi t + 3\sin 3\pi t$$

It is sampled at 1.5 times the Nyquist rate and transmitted using PAM. The pulse waveform, $s_c(t)$, is a periodic train of triangular pulses, as shown in Fig. 5.21. Find the Fourier transform of the modulated waveform.

5.2.10 Derive the dual of the time-sampling theorem—that is, a Fourier transform of a time-limited signal $s(t)$ is completely known from its sample values. Find the minimum frequency spacing between samples to permit reconstruction of the Fourier transform.

5.3.1 An information signal is of the form

$$s(t) = \frac{\sin \pi t}{\pi t}$$

Find the Fourier transform of the waveform that results if each of the two carrier waveforms shown in Fig. P5.3.1 is pulse amplitude modulated with this signal.

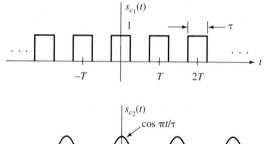

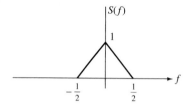

Figure P5.3.1

5.3.2 The signal

$$s(t) = \cos 2\pi t$$

is sampled every 0.4 sec and sent using natural-sampled PAM with pulse widths of 0.1 sec. The channel can be modeled as an ideal lowpass filter with a cutoff at 10 Hz. Find the received waveform. Also, find the reconstructed waveform after the receiver uses a lowpass filter to recover the original signal $s(t)$.

5.3.3 Consider a two-channel TDM PAM system where both channels are used to transmit the same signal $s(t)$ with Fourier transform $S(f)$, as shown in Fig. P5.3.3. The system samples $s(t)$ at the minimum rate. Find the Fourier transform of the TDM waveform, and compare it to the Fourier transform of a single-channel PAM system used to transmit $s(t)$.

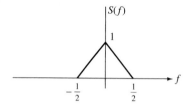

Figure P5.3.3

5.3.4 Three asynchronous sources transmit PAM waveforms to a buffer multiplexer. Each of the sources transmits at a pulse rate that is Gaussian distributed with a mean value of 1,000 pulses/sec and a variance of 9. The channel is capable of transmitting 3,000 pulses per second. How large must the buffer be such that the probability of overload is less than 1%?

5.3.5 Three information signals are to be sent using time-division multiplexed PAM. Suppose that the maximum frequency of each of the first two signals is 5 kHz and that the maximum frequency of the third signal is 10 kHz. Design the multiplex system and draw a block diagram.

5.3.6 Ten signals are to be time division multiplexed and transmitted using PAM. Four of the signals have a maximum frequency of 5 kHz, two have a maximum frequency of 10 kHz, two have a maximum frequency of 15 kHz, and two have a maximum frequency of 20 kHz.

   (a) Design a multiplex system.
   (b) How many pulses per second must be transmitted?

5.3.7 You wish to sample and time division multiplex 36 channels with the following maximum frequencies:

   One channel has a maximum of 10 kHz.

   Three channels have a maximum of 5 kHz.

   Eight channels have a maximum of 2.5 kHz.

   Eight channels have a maximum of 300 Hz.

   Sixteen channels have a maximum of 150 Hz.

   Design a system using sub- and supercommutation. Make any reasonable approximations.

5.3.8 A discrete-time analog signal is created by sampling a speech waveform at 10,000 samples per second. Each sampling pulse is 0.01 msec wide and is transmitted through a channel that can be approximated by a lowpass filter with cutoff frequency at 50 kHz. Evaluate the effects of channel distortion. Compare your answer to that of Example 5.5.

5.3.9 A PWM system multiplexes three signals derived from waveforms with the same maximum frequency f_m. Sample values are normalized (to lie between zero and unity), and the width w of each pulse is related to the normalized sample value $s_i(nT_s)$ by

$$w(nT_s) = 1 + s_i(nT_s) \ \mu\text{sec}$$

   (a) What is the maximum frequency f_m of the baseband signals that would permit this multiplexing to take place?
   (b) What is the minimum bandwidth of the channel?

5.4.1 Show that a sample-and-hold circuit is an approximation to a lowpass filter, provided that the sampling is performed at the Nyquist rate or higher. [*Hint:* You may wish to find the step response of the sample-and-hold circuit and compare it to the step response of a lowpass filter.]

5.4.2 The system shown in Fig. P5.4.2 is similar to a sample-and-hold circuit, where the input can be considered to be an impulse-sampled version of the original waveform.
   (a) Find the impulse response of this system.
   (b) Find the transfer function, and compare it to the transfer function of a lowpass filter.

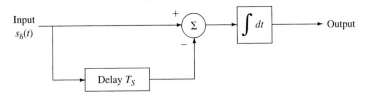

Figure P5.4.2

5.4.3 Two signals,

$$s_1(t) = \cos 2\pi t$$

$$s_2(t) = \cos \pi t + 2\cos 2\pi t$$

are sampled every 0.4 sec and are sent using multiplexed natural-sampled PAM. The channel can be modeled as an ideal lowpass filter with a cutoff at 10 Hz.
  (a) Find the received waveform.
  (b) Find the reconstructed waveforms after the receiver uses a lowpass filter and demultiplexer.
  (c) Repeat part (b), with $s_1(t)$ changed to $\cos 1.9\pi t$.

5.4.4 You wish to investigate the use of a first-order lowpass filter (e.g., an *RC* circuit) to recover the original time signal from a PWM waveform. You can assume that each PWM pulse has its leading edge at the sample point. Analyze this system, and comment on how well it acts as a demodulator.

5.5.1 Consider the system shown in Fig. P5.5.1. We wish to compare $y(t)$ to $x(t)$ in order to evaluate the sample-and-hold circuit as a PAM demodulator. The comparison between $y(t)$ and $x(t)$ is performed by defining an error

$$e = \frac{1}{T}\int_0^T [y(t) - x(t)]^2 dt$$

Find the value of this error term.

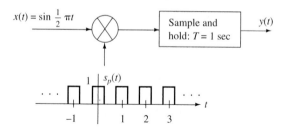

Figure P5.5.1

5.5.2 A sinusoidal signal

$$s(t) = \sin 2\pi t$$

is sampled every 0.4 sec and transmitted using PPM.
  (a) Design the PPM system (i.e., choose a pulse width and a relationship between pulse position and sample value).
  (b) White noise of power $N_0 = 10^{-3}$ watt/Hz adds during transmission. Find the mean square timing error at the receiver.
  (c) Find the approximate SNR after reconstruction at the receiver.

6

Amplitude Modulation

6.0 PREVIEW

What We Will Cover and Why You Should Care

This is the first of two chapters dealing with modulation techniques for analog communication. You will learn the basic concepts of modulation and examine the motivation for using various modulation schemes. After reading the chapter, you will:

- Understand amplitude modulation and the difference between suppressed and transmitted carrier modulation
- Know how to construct modulators
- Know how to construct demodulators
- Know how standard broadcast AM radio works
- Understand the various types of AM stereo
- Know how to perform video transmission (e.g., TV)
- Possess the necessary tools to evaluate and compare the performance of systems.

Necessary Background

The earlier portions of this chapter require that you have an understanding of Fourier transforms. You will need to know systems analysis in order to be in a position to design modulators and demodulators. A working knowledge of random processes and probability is needed to be able to evaluate system performance.

6.1 CONCEPT OF MODULATION

Suppose you were given the job of transmitting either speech or baseband data through a channel. The first question you should ask yourself is whether the signals must be modified before injecting them into the channel. If the answer is no, your job is very simple: You must simply decide how to couple the signal into the channel (i.e., interface the two).

For many channels, the answer will be yes, and the signal will have to be modified. The Fourier transform of a typical speech waveform is sketched in Figure 6.1.

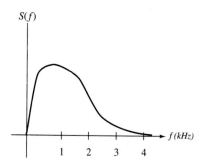

Figure 6.1 Fourier transform of baseband signals.

In the case of short-range transmission, as in the local loop of a telephone circuit or in the path between the pre-amp and amplifier or between the amplifier and speakers of a sound system, these baseband (low-frequency) signals are sent through wires. For longer distances, it is sometimes difficult to use wires, since they require *rights of way*. Additionally, since transmission is *point to point*, one must specify the location of *every* terminal. In the case of television, the wire would have to terminate in the home of every prospective viewer (as in cable television). Mobile communication by wire is almost impossible. (We say *almost* because some missiles actually trail a wire behind them that unwinds as does a fishing line—but this is the exception.) For all of these reasons, *broadcast* communication has been a popular form of transmission.

Suppose we take an audio signal and attempt to transmit it through the air. Let us choose a typical audio frequency of 1 kHz. The wavelength of a 1-kHz signal in air is approximately 300 km (about 180 miles). A quarter-wavelength antenna would then have to be 75 km (45 miles) long, and erecting such antennas in backyards of homes would be a bit impractical! But even if we were willing to erect them, we would still be left with two very serious problems. The first is related to the characteristics of air at audio frequencies: While propagation does occur at frequencies below 10 kHz, these frequencies are not efficiently transmitted through air. Even more serious is the second problem: interference. Often, it is desirable to transmit more than one analog signal at a time. For example, many local radio stations transmit broadcasts simultaneously. If they used quarter-wavelength antennas, they would each have antennas 75 km long on top of their studios (or on mountaintops), and they would pollute the air with many audio signals. The listener would erect an antenna 75 km high and receive a weighted sum of all of the signals (depending on relative distances and antenna patterns from the different transmitting antennas to the receiving antenna). Since the only information the receiver would have about the signals is that they would all be bandlimited to the same upper cutoff frequency, there would be absolutely no way of separating the signal from one station from those from all of the others.[1]

Given the preceding scenario, it is desirable to modify a low-frequency signal before sending it from one point to another. An added bonus arises if the modified signal is less susceptible to noise than is the original signal.

---

[1]Walk into a crowded, noisy room, and try to distinguish one conversation from all of the others. Then record the sounds in the room, and try again to distinguish the sounds, this time by listening to the recording. Ask yourself why there is a difference.

The most common method of accomplishing the modification is to use the low-frequency signal to modulate (i.e., modify the parameters of) another, higher frequency signal. Most commonly, this other signal is a pure sinusoid.

We start, then, with a pure sinusoid $s_c(t)$ called the *carrier* waveform. It is given this name because it is used to *carry* the information signal from the transmitter to the receiver. Mathematically,

$$s_c(t) = A \cos (2\pi f_c t + \theta) \tag{6.1}$$

If f_c is properly chosen, this carrier waveform can be efficiently transmitted. For example, suppose you were told that frequencies in the range between 1 MHz and 3 MHz propagate in a mode that allows them to be reliably sent over distances up to about 200 km. If you chose the frequency f_c to be in this range, then the pure sinusoidal carrier would transmit efficiently. The wavelength of transmission in the range of 1 MHz to 3 MHz is on the order of 100 meters, and antennas of reasonable length can be used.

We now ask the question whether the preceding pure sinusoidal carrier waveform can somehow be altered in a way that (a) the altered waveform still propagates efficiently and (b) the information we wish to send is somehow superimposed on the new waveform in a way that it can be recovered at the receiver. In other words, we are asking whether there is some way that the sinusoid can *carry* the information along. The answer is yes, as we now illustrate.

The right-hand side of Eq. (6.1) contains three parameters that may be varied: the amplitude A, the frequency f_c, and the phase θ. Using the information signal to vary A, f_c, or θ leads to *amplitude modulation, frequency modulation,* and *phase modulation,* respectively.

We will show that efficient transmission is achieved for each of these three cases. We will also show that if more than one signal is simultaneously propagated through the channel, separation of the signals at the receiver is possible. In addition, we will find it critical to illustrate a third property: The information signal $s(t)$ must be uniquely recoverable from the received modulated waveform; it would not be of much use to modify a carrier waveform for efficient transmission and station separability if we could not reproduce $s(t)$ accurately at the receiver.

This chapter concentrates on a thorough treatment of amplitude modulation. Parallel treatments of frequency and phase modulation follow in the next chapter.

6.2 DOUBLE-SIDEBAND SUPPRESSED CARRIER

If we modulate the amplitude of the carrier of Eq. (6.1), the following modulated waveform results:

$$s_m(t) = A(t)\cos (2\pi f_c t + \theta) \tag{6.2}$$

The frequency f_c and the phase θ are constant. The amplitude $A(t)$ varies somehow in accordance with the baseband signal $s(t)$–the signal we want carried through the channel.

We simplify the expression by assuming that $\theta = 0$. This will not affect any of the basic results, since the angle actually corresponds to a time shift of $\theta/2\pi f_c$. A time shift is not considered distortion in a communication system.

If somebody asked you how to vary $A(t)$ in accordance with $s(t)$, the simplest answer you could suggest would be to make $A(t)$ equal to $s(t)$. This would yield a modulated signal of the form

$$s_m(t) = s(t)\cos 2\pi f_c t \qquad (6.3)$$

Such a signal is given the name *double-sideband suppressed carrier* (DSBSC) *amplitude modulation* for reasons that will soon become clear.

This simple equating of $A(t)$ with $s(t)$ does indeed satisfy the criteria demanded of a communication system. The easiest way to illustrate this fact is to express $s_m(t)$ in the frequency domain, that is, to find its Fourier transform.

Suppose that we let $S(f)$ be the Fourier transform of $s(t)$. We require nothing more of $S(f)$ than that it be the Fourier transform of a baseband signal. That is, $S(f)$ must equal zero for frequencies above some cutoff frequency f_m. (The subscript m stands for *maximum*.) Figure 6.2 gives a representative sketch of $S(f)$. We do not mean to imply that $S(f)$ must be of the shape shown; the sketch is meant only to indicate the transform of a general low-frequency bandlimited signal.

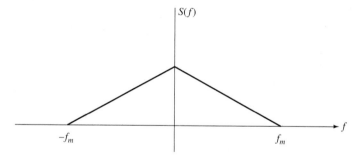

Figure 6.2 General form of baseband $S(f)$.

The modulation theorem is used to find $S_m(f)$:

$$S_m(f) = \mathcal{F}[s(t)\cos 2\pi f_c t] = \frac{1}{2}[S(f + f_c) + S(f - f_c)] \qquad (6.4)$$

This transform is sketched as Figure 6.3. Note that modulation of a carrier with $s(t)$ has shifted the frequencies of $s(t)$ both up and down by the frequency of the carrier. This is

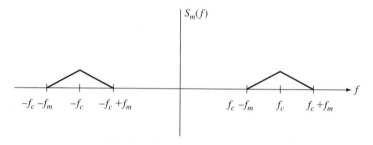

Figure 6.3 $S_m(f)$, the transform of $s_m(t)$.

analogous to the trigonometric result that multiplication of a sinusoid by another sinusoid results in sum and difference frequencies. That is,

$$\cos A \cos B = \frac{1}{2} \cos(A + B) + \frac{1}{2} \cos(A - B) \tag{6.5}$$

If $\cos A$ is replaced by $s(t)$, where $s(t)$ contains a continuum of frequencies between 0 and f_m, the trigonometric identity can be applied term by term to yield a result containing all sums and differences of the frequencies.

Figure 6.3 indicates that the modulated waveform $s_m(t)$ contains components with frequencies between $f_c - f_m$ and $f_c + f_m$. As long as signals in this range of frequencies transmit efficiently and an antenna of reasonable length can be constructed, we have solved the first of the two problems. Let us plug in some typical audio numbers. Let f_m be 5 kHz and f_c be 1 MHz. Then the range of frequencies occupied by the modulated waveform is from 995,000 to 1,005,000 Hz.

The second objective is separation of the signals. We see that if one information signal modulates a sinusoid of frequency f_{c1} and another information signal modulates a sinusoid of frequency f_{c2}, the Fourier transforms of the two modulated carriers do not overlap in frequency, provided that f_{c1} and f_{c2} are separated by at least $2f_m$. This is illustrated in Figure 6.4. Since the signals are "stacked" in frequency, we refer to this situation as *frequency*

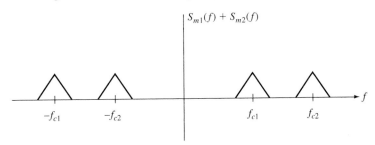

Figure 6.4   Fourier transform of two AM signals.

division multiplexing (FDM). It is the exact dual of time division multiplexing (TDM), which we introduced in Chapter 5.

If the frequencies of the two modulated waveforms are not too widely separated, both signals can even share the same antenna. That is, although the optimum antenna length is not the same for both channels, the total bandwidth can be made relatively small compared to the carrier frequency. In practice, the antenna is usable over a *range* of frequencies rather than just being effective at a single frequency. If this were not true, radio broadcasting would not exist.

As an example, you don't have to readjust the length of your car antenna whenever you tune across the AM dial. The effectiveness of the antenna does not vary greatly from one frequency limit to the other. Instructions accompanying early car antennas suggested that their length be shortened to about 75 centimeters when changing from AM to FM. (Don't try doing this if you have the type of antenna that is sandwiched within the windshield.) Modern receivers have sufficient sensitivity that such tuning is no longer necessary, even with frequency changes of two orders of magnitude.

If signals are nonoverlapping in time, gates or switches can be used to effect their separation. For AM, the signals are nonoverlapping in frequency, and they can be separated from each other by means of frequency gates (bandpass filters). Thus, a system such as that shown in Fig. 6.5 could be used to separate the two modulated carriers of Fig. 6.4 from each other.

The extension of this system to more than two channels should be obvious. Even if many modulated signals were transmitted over the same channel, they could be separated

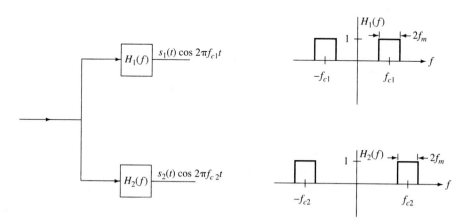

Figure 6.5   Separating two nonoverlapping channels.

at the receiver using bandpass filters that accepted only those frequencies that were present in the desired modulated signal. This is true, provided that the separate carrier frequencies are wide enough apart to prevent overlapping of the Fourier transforms. We see from Fig. 6.4 that the minimum spacing is $2f_m$. In practice, a spacing larger than this is desirable for two reasons: First, even though we may view the information signal as limited to frequencies below f_m, no matter how sharply we lowpass filter it, the signal still has some components above f_m. Second, if the minimum spacing is used, the bandpass filters that separate out the desired channel must be perfect, with flat response in the passband and an infinite roll-off.

Example 6.1

An information signal is of the form

$$s(t) = \frac{\sin 2\pi t}{t}$$

The signal amplitude modulates a carrier of frequency 10 Hz. Sketch the AM waveform and its Fourier transform.

Solution: The AM waveform is given by the equation

$$s_m(t) = \frac{\sin 2\pi t}{t} \cos 20\pi t$$

This function is sketched in Fig. 6.6.

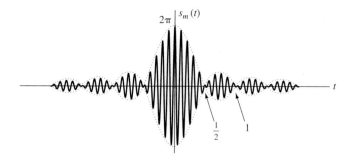

Figure 6.6   AM waveform for Example 6.1.

We note that when the carrier, $\cos 20\pi t$, is equal to 1, $s_m(t) = s(t)$, and when the carrier is equal to -1, $s_m(t) = -s(t)$. In sketching the AM waveform, we start by drawing $s(t)$ and its mirror image, $-s(t)$, as a guide. The AM waveform periodically touches each of these curves and varies smoothly between the periodic points. In this manner, we develop the sketch of the waveform. In most practical situations, the carrier frequency is much higher than that illustrated in this example. In fact, it is so high that if you observed $s_m(t)$ on an oscilloscope, you would not be able to see the back-and-forth oscillations unless you greatly expanded the time axis. Instead, you would see the $s(t)$ and $-s(t)$ outlines and what looks like shading between them.

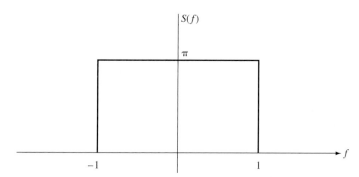

Figure 6.7   Fourier transform of $s(t)$ for Example 6.1.

The Fourier transform of the information signal $s(t)$ is shown in Fig. 6.7. It is found in the table in Appendix II.

The transform of the modulated waveform is given by the following equation, where we have applied the modulation theorem:

$$S_m(f) = \frac{S(f - 10) + S(f + 10)}{2}$$

This is shown in Fig. 6.8.

We have indicated that an AM wave of the type discussed could be transmitted efficiently and that more than one signal could share the channel. A critical property that must still be addressed is whether the information signal, $s(t)$, can be uniquely recovered from the AM waveform.

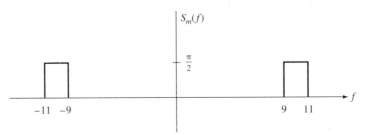

Figure 6.8   Fourier transform of modulated waveform.

Since $S_m(f)$ was derived from $S(f)$ by shifting all of the frequency components of $s(t)$ by f_c, we should be able to recover $s(t)$ from $s_m(t)$ by shifting the frequencies again by the same amount, but this time in the opposite direction.

The *modulation theorem* states that multiplication of a function of time by a sinusoid shifts the Fourier transform of the function both up and down in frequency. Thus, if we *remultiply* $s_m(t)$ by a sinusoid at the carrier frequency, the Fourier transform shifts *back down* to its low-frequency baseband position. The multiplication also shifts the transform up to a position centered about $2f_c$, but this part can easily be rejected using a lowpass filter. The process is illustrated in Fig. 6.9.

The recovery of $s(t)$ is described by the following equations:

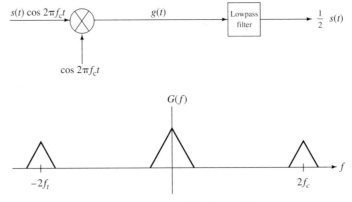

Figure 6.9   Recovery of $s(t)$ from $s_m(t)$.

$$s_m(t)\cos 2\pi f_c t = [s(t)\cos 2\pi f_c t]\cos 2\pi f_c t$$
$$= s(t)\cos^2 2\pi f_c t \tag{6.6}$$
$$= \frac{s(t) + s(t)\cos 4\pi f_c t}{2}$$

In Eq. (6.6), we have used the trigonometric identity

$$\cos^2(A) = \frac{1}{2} + \frac{1}{2}\cos(2A) \tag{6.7}$$

The output of the lowpass filter is therefore $s(t)/2$, which is an undistorted version of $s(t)$.

This process of recovering $s(t)$ from the modulated waveform is known as *demodulation*. We have taken the time to begin our discussion of demodulation now, rather than waiting until Section 6.5, where we will have more to say about it. Indeed, if $s(t)$ could not be recovered from $s_m(t)$, there would be no reason to go on.

6.3 DOUBLE-SIDEBAND TRANSMITTED CARRIER

In the previous section, we studied double-sideband suppressed carrier AM. We found that the waveform resulting from the multiplication of the information signal with a carrier sinusoid possesses desirable properties. In particular, the modulation process shifts frequencies from a band around dc to a band around the carrier frequency. This permits efficient transmission and also allows simultaneous transmission of more than one signal.

We now explore a modification of AM in which we add a portion of the pure sinusoidal carrier to the modulated waveform. We will see in Section 6.5 that this addition greatly simplifies the demodulation process.

Figure 6.10 shows the addition of a pure sinusoidal carrier to the double-sideband suppressed carrier waveform. The resulting waveform is

$$s_m(t) = s(t)\cos 2\pi f_c t + A\cos 2\pi f_c t \tag{6.8}$$

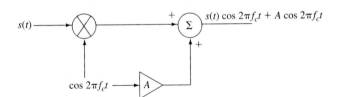

This is known as *double-sideband transmitted carrier* (DSBTC). The type of AM discussed in the previous section did not include an explicit carrier term. That is why it is labeled *suppressed carrier*. We begin by examining the function of time and its Fourier transform.

The Fourier transform of transmitted carrier AM is the sum of the Fourier transform of suppressed carrier AM and the Fourier transform of the pure carrier. The transform of the carrier is a pair of impulses at $\pm f_c$. The complete transform of the AM wave is therefore as shown in Fig. 6.11.

Figure 6.10   Addition of a carrier term.

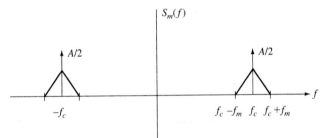

Figure 6.11   Fourier transform of AM transmitted carrier.

The function of time can be sketched if we first combine terms in Eq. (6.8). Doing so, we can rewrite the waveform as

$$s_m(t) = [A + s(t)]\cos 2\pi f_c t \tag{6.9}$$

This function is sketched in the same manner as that used to draw the suppressed carrier waveform. We first draw the outlines at $[A + s(t)]$ and $-[A + s(t)]$. The AM waveform periodically touches these two curves. We then fill in with a smooth, oscillating waveform. This is illustrated for a sinusoidal $s(t)$ [e.g., someone whistling into a microphone] in Figure 6.12.

Figure 6.12(a) shows the sinusoidal $s(t)$, Fig. 6.12(b) shows the AM waveform for a value of A less than the amplitude of $s(t)$, and Fig. 6.12(c) shows the waveform where A is greater than the amplitude of $s(t)$.

Efficiency

We ask you to accept for now the fact that the addition of the carrier makes demodulation easier. The price we pay is in efficiency: A portion of the transmitted power is used to send a pure sinusoid that does not carry any useful information about the signal.

We see from Eq. (6.8) that the carrier power is the power of $A\cos 2\pi f_c t$, or $A^2/2$ watts. The power of the signal portion is the power of $s(t)\cos 2\pi f_c t$, which is the average of $s^2(t)$ divided by 2. The average of $s^2(t)$ is simply the power of $s(t)$, or P_s. Therefore, the signal power is $P_s/2$. The total transmitted power[2] is the sum of this and $A^2/2$.

We define *efficiency*, η, as the ratio of the signal power to the total power. The efficiency is then given by

$$\eta = \frac{P_s/2}{(A^2 + P_s)/2} = \frac{P_s}{A^2 + P_s} \tag{6.10}$$

As an example of the application of Eq. (6.10), suppose we view the AM wave of Fig. 6.12(c) and set A equal to the amplitude of the sinusoid. P_s is then $A^2/2$, and the efficiency is

$$\eta = \frac{A^2/2}{A^2 + A^2/2} = 33\% \tag{6.11}$$

The efficiency depends on the size of the modulated term compared to the size of the pure sinusoidal carrier term. We define a dimensionless quantity m as the ratio of the maximum amplitude of the modulated term to the amplitude of the carrier. That is,

$$m = \frac{\max |s(t)|}{A} \tag{6.12}$$

The quantity m is known as the *index of modulation*. Viewing Fig. 6.12, we see that if the envelope of the waveform extends down to the zero axis, the index of modulation is 1. As the index of modulation decreases, the efficiency also decreases. The index is sometimes expressed as a percentage by multiplying by 100.

---

[2]The square of the sum contains a cross product, but if the average of $s(t)$ is zero, the average of this product is also zero.

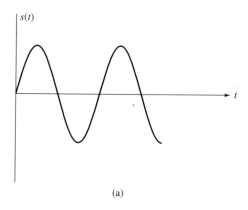

(a)

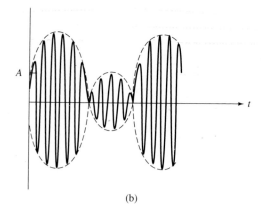

(b)

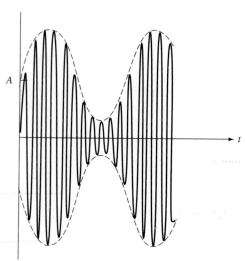

(c)

Figure 6.12 AM transmitted carrier waveform.

6.4 MODULATORS

Figure 6.13 presents the block diagram of an amplitude modulator. The system of Fig.
6.13(a) produces double-sideband suppressed carrier AM, while the systems of Figs.
6.13(b) and 6.13(c) produce double-sideband transmitted carrier AM.

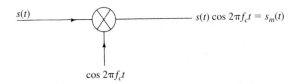

(a) Suppressed carrier

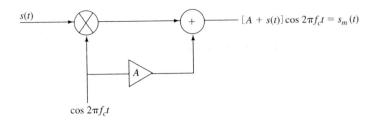

(b) Transmitted carrier

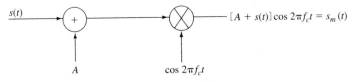

Figure 6.13   Block diagram of amplitude modulator.

You may ask why we devote an entire section of the text to modulators if Fig. 6.13
tells the whole story. Indeed, if we were simply interested in drawing system block diagrams, there would be no need to go any further. However, if you ever intend to implement any system design, you must have some idea of the components that go into each
block in the system block diagram. The diagrams of Fig. 6.13 represent a departure from
the systems considered earlier in the text. The modulator represents the first time we have
come across a system that is *not* linear. This makes its implementation very different from
that of linear filters.

Why is modulation not linear? Any linear system has an output whose Fourier transform is the product of the Fourier transform of the input and $H(f)$, the system function. If
the Fourier transform of the input is zero over some range of frequencies, the output transform must also be zero for this range of frequencies. In other words, a general property of
linear systems is that they cannot generate any output frequency that does not appear in
the input. Since amplitude modulation shifts frequencies to a new range, no linear system
can perform such an operation.

The synthesis of a system that is not linear is, in general, complicated. Fortunately, simplifications are possible in the case of the modulator. We begin with two classes of indirect amplitude modulator: the gated and the square law modulators.

The *gated modulator* uses the fact that multiplication of $s(t)$ by any periodic function produces a series of AM waves at carrier frequencies that are multiples of the fundamental frequency of the periodic function. We illustrate this in Figure 6.14. The output of the multiplier is given by

$$s(t)p(t) = s(t)\left[a_0 + \sum_{n=1}^{\infty} a_n\cos(2\pi nf_c t)\right] \tag{6.13}$$

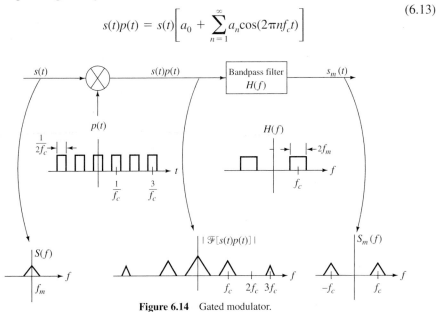

Figure 6.14   Gated modulator.

In Eq. (6.13), f_c is the fundamental frequency of the periodic waveform (the reciprocal of the period), and the a_n are the Fourier series coefficients. We have assumed that $p(t)$ is an even function to avoid having to write the sine terms in the series. The bandpass filter of Fig. 6.14 blocks all but one term in the series, with the result that the output is an AM waveform. We have shown the filter as being tuned to the fundamental frequency, but it could have been tuned to one of the harmonics, thereby resulting in an AM waveform at that higher carrier frequency. In practice, we would favor the lower harmonics, since the Fourier coefficients decrease in magnitude with increasing n. At some point, the output AM waveform would be so small that it would be lost in the circuit noise.

What have we accomplished? If we cannot build a multiplier to take the product of $s(t)$ and a cosine waveform, what makes us think we can build the multiplier of Fig. 6.14? The answer lies in a specific choice of $p(t)$: a periodic pulse train gating function, as shown in Fig. 6.15. Since $p(t)$ is always either 0 or 1, the multiplication can be viewed as a gating operation, where the input is switched on and off.

The output of the bandpass filter is found by expanding $p(t)$ in a Fourier series and finding a_1. The modulator output is then

$$s_m(t) = \frac{2}{\pi}s(t)\cos 2\pi f_c t \tag{6.14}$$

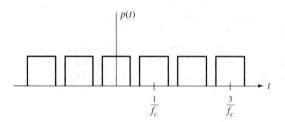

Figure 6.15   Gating with a pulse train.

Equation (6.14) has been written for a gating function that spends half of its time high and half at zero. In fact, an AM wave will be produced for any value of the *duty cycle*.

The gating function can be implemented either passively or actively. Figure 6.16 shows two passive implementations. Figure 6.16(a) shows a simple switch that periodically shorts out the input. When the switch is open, the output equals the input. When the switch is closed, the output is zero. The resistance is the resistance of the source. The disadvantage of mechanical switching is that S must switch at a rate equal to the carrier frequency (or a submultiple of it if we select a harmonic). If our carrier frequency is in the megahertz range, mechanical switching is not practical.

Figure 6.16(b) presents a variation of the switch circuit in which the switching is accomplished using a *diode bridge* circuit. When $\cos 2\pi ft$ is positive, the point labeled B is at a higher potential than the point labeled A. In this condition, all four (ideal and matched) diodes are open circuited, and the circuit is equivalent to that of Fig. 6.16(a) with the switch open. On the other hand, when $\cos 2\pi ft$ is negative, point A is at a

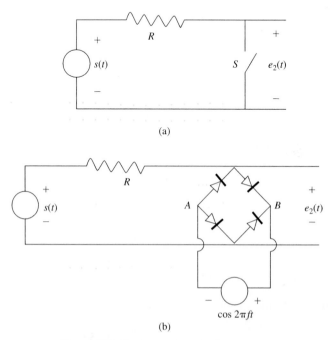

Figure 6.16   Implementation of gating function.

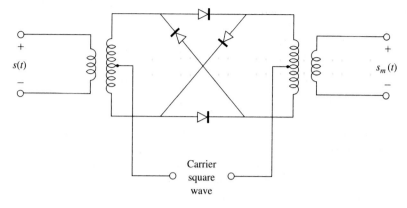

(a) Modulator

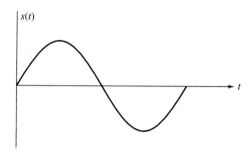

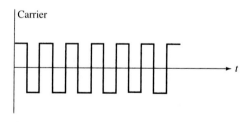

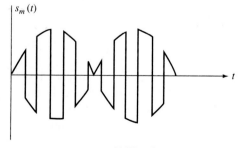

(b) Waveforms

Figure 6.17   Ring Modulator.

higher potential than point *B*, and all four diodes are short circuits. This is equivalent to the switch being closed. The only limit to the rate of switching is imposed by the fact that practical diodes are not ideal (i.e., they have a nonzero capacitance).

The gating can also be accomplished using active electronic devices such as transistors operating between cutoff and saturation. A cutoff transistor acts as an open switch, while a saturated transistor acts as a closed switch.

The *ring modulator* is a variation of the gated modulator. The circuit is shown in Fig. 6.17(a). The carrier, a square wave, is fed into the center taps of the two transformers. The output is a gated version of the input and needs only be filtered to produce AM. We illustrate sample waveforms in Fig. 6.17(b).

The second broad class of modulator we consider is the *square-law modulator*. This modulator takes advantage of the fact that the square of a sum of two functions contains a cross-product term that is the product of the two functions. That is,

$$[s_1(t) + s_2(t)]^2 = s_1^2(t) + s_2^2(t) + 2s_1(t)s_2(t) \tag{6.15}$$

If $s_1(t)$ is the information and $s_2(t)$ is the carrier, we have

$$[s(t) + \cos 2\pi f_c t]^2 = s^2(t) + \cos^2 2\pi f_c t + 2s(t)\cos 2\pi f_c t \tag{6.16}$$

The third term on the right-hand side of Eq. (6.16) is the desired AM waveform. We must find a way of separating it from the other two terms. We know that separation will be simple if the terms are nonoverlapping either in time or in frequency. Clearly, they overlap in time, so our only hope is to look toward frequency.

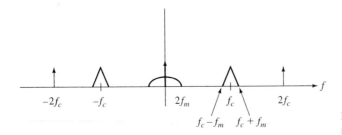

Figure 6.18   Fourier transform of squared-sum signal.

Figure 6.18 shows the Fourier transform of the signal in Eq. (6.16). The impulses at the origin and at $2f_c$ result from expanding the square of the cosine by means of the trigonometric identity

$$\cos^2\theta = \frac{1}{2} + \frac{1}{2}\cos 2\theta \tag{6.17}$$

The continuous shape shown at low frequency represents the Fourier transform of $s^2(t)$. We do not know the exact shape of $s(t)$, but only that its Fourier transform is limited to frequencies below f_m. The Fourier transform of $s^2(t)$ is limited to frequencies below $2f_m$. One way to show this is to observe that the transform of $s^2(t)$ is the convolution of $S(f)$ with itself. Graphical convolution easily shows that this transform goes to zero at $2f_m$. Another way to show it is to consider $s(t)$ as a sum of individual sinusoids at frequencies below f_m. When this sum is squared, the result is all possible cross products of terms. Trigonometric

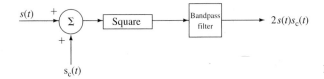

Figure 6.19(a)   Square-law modulator.

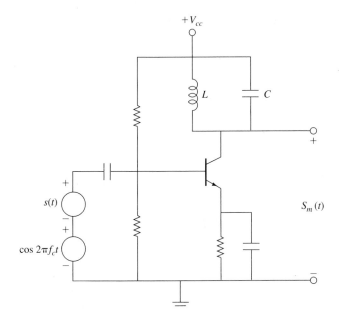

Figure 6.19(b)   Practical square-law modulator.

identities tell us that this leads to sums and differences of the various frequencies. None of these sums or differences can exceed $2f_m$ if the original frequencies do not exceed f_m.

Figure 6.18 indicates that as long as the carrier frequency exceeds $3f_m$, the terms do not overlap in frequency, and the AM waveform can be separated using a bandpass filter. In most practical situations, the carrier frequency is much higher than f_m, so the condition is easily met. Figure 6.19(a) shows the overall block diagram of the square-law modulator. This block diagram contains a summing device, a squarer, and a bandpass filter. You already know how to build a bandpass filter. Summing devices can be active or passive. Any resistive circuit with two sources produces (through superposition) weighted sums of these sources throughout the network. Alternatively, summing op-amp circuits can be used.

Square-law devices are not quite that simple. Any practical nonlinear device has an output-versus-input relationship that can be expanded in a power series. This assumes that no energy storage is taking place; that is, the output at any time depends only on the value of the input at that same time, and not on any past input values. With $y(t)$ as output and $x(t)$ as input, the nonlinear device obeys the relationship

$$y(t) = a_0 + a_1x(t) + a_2x^2(t) + a_3x^3(t) + \cdots \tag{6.18}$$

The term we are interested in is $a_2x^2(t)$. If we could somehow find a way of separating this term from all the others, the nonlinear device could be used as a squarer.

Unfortunately, the various power terms overlap both in time and frequency. (Take the time to verify this! It requires only simple trigonometry and basic Fourier transform theory.)

The nonlinear device must essentially be a squarer; the a_n in Eq. (6.18) must have the property that

$$a_n \ll a_2, \quad \text{for } n > 2$$

There are several things to note about the nonlinearity before we move on. The first is that, if the $n = 1$ and $n = 2$ terms in the series predominate, the result is *transmitted carrier AM*. Further, if the a_n are *not* insignificant for $n > 2$, AM is still possible if $s(t)$ is made very small. Then $s^n(t) \ll s(t)$ for $n > 1$, and the transmitted carrier AM terms will predominate. This is not a desirable situation, because of the small amplitudes that result.

Semiconductor diodes have terminal relationships that are good approximations to square-law devices over limited operating ranges. Figure 6.19(b) shows a simple implementation of the square-law modulator, where the *RC* combination filters out the higher frequency term and the transistor provides the nonlinearity. We will see later that this same circuit configuration can be used to demodulate certain AM waveforms.

In reality, it is not difficult to build a modulator, provided that efficiency is not critical. In fact, most circuits produce modulation products even when these are not wanted. Superposition dictates that sums of signals will appear in a circuit, and practical linear devices always have some inherent nonlinearity. Designers of linear circuits go through a lot of effort to reduce unwanted modulation products.

We can relax the constraints on the nonlinear device by constructing the *balanced modulator*. Figure 6.20 shows a block diagram and one possible implementation of such a modulator. This system adds the carrier to $s(t)$ and places this sum through a nonlinear device. The operation is then repeated using $-s(t)$ as the information signal. The difference of the two outputs is taken, resulting in a cancellation of the terms due to odd powers in the expansion of Eq. (6.18). We illustrate the process by examining the cubic term in the equation. When we expand

$$[s(t) + \cos 2\pi f_c t]^3$$

the term that overlaps the frequency band of the AM waveform is

$$s^2(t) \cos 2\pi f_c t$$

This term remains unchanged when $-s(t)$ is substituted for $s(t)$. The result is that it cancels from the output of the balanced system. The desired term, $s(t)\cos 2\pi f_c t$, changes sign when $-s(t)$ is substituted for $s(t)$. Therefore, the operation of taking the difference has the effect of doubling the desired term. A similar approach is used to show that higher odd powers do not create undesired terms in the output. The balanced modulator is particularly effective if the nonlinearity has strong linear, square, and cubic terms, and all higher terms in the series are negligible. We should note that, since the first-order term is eliminated, the output of the balanced modulator is *suppressed carrier AM*.

A practical implementation of a square-law modulator is shown in Fig. 6.21. This common-emitter transistor circuit uses the nonlinearity of the transistor to produce the product of the signal and the carrier. The tuned circuit in the collector filters out the undesired harmonics.

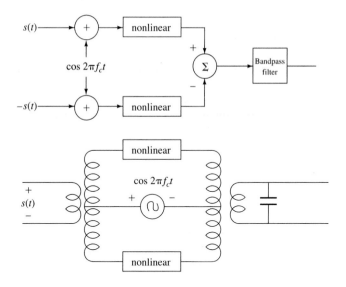

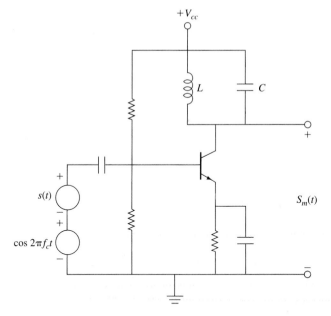

Figure 6.20   Balanced modulator.

Figure 6.21   Implementation of square-law modulator.

The *waveshape modulator* can be thought of as a brute-force device. If you wanted to modulate a flow of water in a hose, you could hold your hand on the valve and keep turning it back and forth. An analogy to this simple system exists in electronics. You can envision building a power amplifier (or oscillator) that produces the carrier. Then simply vary the supply voltage to this amplifier in a manner that follows the information signal. The collector-modulated circuit of Fig. 6.22 does just this. The waveform appearing at the top of the *RLC* tuned collector circuit is the sum of V_{CC} and the information signal. We are therefore essentially varying the supply voltage in accordance with $s(t)$. The output is bandpass filtered to eliminate harmonics created by nonlinear operations of the transistor.

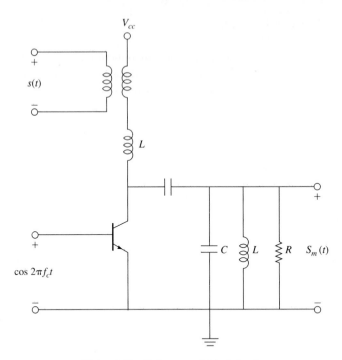

Figure 6.22   Collector-modulated circuit.

6.5 DEMODULATORS

We divide demodulators into two broad classifications: coherent and incoherent. Coherent demodulators must be configured to take advantage of all information received, including the amplitude and timing of the waveform. Incoherent demodulators do not need to establish absolute timing (phase) relationships.

Coherent Demodulation

We have previously observed that $s(t)$ is recovered from $s_m(t)$ by *remodulating* $s_m(t)$ and then passing the result through a lowpass filter. This yields the demodulator system whose block diagram appears in Fig. 6.23. Such a system is known as a *synchronous demodulator*. It gets its name from the observation that the oscillator is *synchronized* in both frequency and phase with the received carrier. Since the multiplier in the figure looks no different from the multiplier used in the modulator, we might expect variations of the gated and square-law modulators to be applicable.

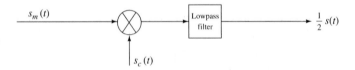

Figure 6.23   AM demodulator.

6.5.1 Gated Demodulator

We first investigate the use of the gated modulator for demodulation. The gated demodulator is shown in Fig. 6.24. The function $p(t)$ is a gating function consisting of a periodic train of unit-amplitude pulses. It can be expressed in a Fourier series as

$$p(t) = a_0 + \sum_{n=1}^{\infty} a_n \cos 2\pi n f_c t \tag{6.19}$$

The input to the lowpass filter is then

$$s_m(t)p(t) = s(t)\cos 2\pi f_c t \left[a_0 + \sum_{n=1}^{\infty} a_n \cos 2\pi n f_c t \right]$$

$$= a_0 s(t) \cos 2\pi f_c t \tag{6.20}$$

$$+ \frac{s(t)}{2} \sum_{n=1}^{\infty} a_n [\cos(n-1)2\pi f_c t + \cos(n+1)2\pi f_c t]$$

The output of the lowpass filter is

$$s_o(t) = \frac{1}{2} a_1 s(t) \tag{6.21}$$

and demodulation is accomplished.

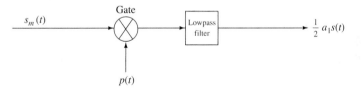

Gate

$s_m(t)$

Lowpass filter

$\frac{1}{2} a_1 s(t)$

$p(t)$

Figure 6.24   Gated demodulator.

We have illustrated the operation of the gated demodulator for suppressed carrier AM. If we substitute $A + s(t)$ for $s(t)$ in Eq. (6.20), we see that the gated demodulator produces an output

$$s_o(t) = \frac{1}{2} a_1 [A + s(t)] \tag{6.22}$$

This represents the original information signal, shifted by an amplitude constant. If the system contains ac-coupled devices, the constant will not appear in the output. If all amplifiers in the system are dc coupled, we may wish to remove the constant using a relatively large series capacitor that charges to the average value of the signal. We are assuming that the average value of the information, $s(t)$, is zero. If this were not true, removal of the constant would also remove some of the signal. Fortunately, most $s(t)$ information signals have zero dc value.

6.5.2 Square-Law Demodulator

We investigate the effect of adding the AM wave to a pure carrier term and then squaring the sum. This yields

$$[s_m(t) + A\cos2\pi f_c t]^2 \tag{6.23}$$

Equation (6.23) can be rewritten as

$$\{[s(t) + A]\cos2\pi f_c t\}^2 = [s(t) + A]^2\cos^2 2\pi f_c t$$

$$= \frac{[s(t) + A]^2 + [s(t) + A]^2\cos4\pi f_c t}{2} \tag{6.24}$$

The second term in Eq. (6.24) is an AM wave with a carrier frequency of $2f_c$ Hz. It can therefore be easily rejected by a lowpass filter. The first term can be expanded as

$$\frac{s^2(t)}{2} + \frac{A^2}{2} + As(t) \tag{6.25}$$

Unfortunately, the frequency content of $s^2(t)$ overlaps that of $s(t)$, and the two terms cannot be separated. However, suppose we used a lowpass filter to isolate the entire term

$$\frac{[s(t) + A]^2}{2} \tag{6.26}$$

from Eq. (6.24). Note that this lowpass filter must pass frequencies up to $2f_m$. We have then recovered the square of the sum of A and $s(t)$. We could subsequently take the square root of this to get

$$0.707|s(t) + A| \tag{6.27}$$

Taking the magnitude of a signal represents a severe form of distortion. As a simple example, suppose the signal were a pure sinusoid. Then the magnitude would be a full-wave rectified sine wave with fundamental frequency twice the original frequency. The rectified signal no longer contains a single frequency, but includes harmonics. If we listened to their sound in a speaker, the original sinusoid would be a pure tone, while the full-wave rectified sine wave would be a raspy tone one octave higher, due to the harmonic content. If the original signal were composed of a mixture of many frequencies, the distortion effect would be far more severe. Indeed, full-wave rectified voice is not intelligible. (Try it in the lab!)

But suppose A is large enough such that $s(t) + A$ never goes negative. In that case, the magnitude of $s(t) + A$ is equal to $s(t) + A$, and we have accomplished demodulation. This means that the added carrier at the receiver must have an amplitude greater than or equal to the maximum negative excursion of $s(t)$.

Effects of Frequency Mismatch

The demodulators we have been discussing require that we generate a replica of the carrier at the receiver. The replica must be synchronized with the received carrier. (Frequency and phase must be matched.) Let us investigate the consequences of frequency and phase mismatches. We illustrate the phenomenon for a suppressed carrier AM wave. Suppose that the local oscillator of Fig. 6.23 is mismatched in frequency by Δf and in phase by $\Delta\theta$. The output of the multiplier is then

$$s_m(t)\cos[2\pi(f_c + \Delta f)t + \Delta\theta]$$

$$= s(t)\cos 2\pi f_c t\cos[2\pi(f_c + \Delta f)t + \Delta\theta] \tag{6.28}$$

$$= s(t)\left[\frac{\cos[2\pi\Delta ft + \Delta\theta]}{2} + \frac{\cos[2\pi(2f_c + \Delta f)t + \Delta\theta]}{2}\right]$$

Since Eq. (6.28) forms the input to the lowpass filter of the synchronous demodulator, the output of this filter is

$$s_o(t) = s(t)\frac{\cos(2\pi\Delta ft + \Delta\theta)}{2} \tag{6.29}$$

This results because the second term of Eq. (6.28) has frequency content around $2f_c + \Delta f$ and is therefore rejected by the lowpass filter. Equation (6.29) represents a signal $s(t)$ multiplied by a sinusoid at Δf Hz. We can assume that Δf is small, since we attempt to make it equal to zero. The modulation theorem then indicates that $s_o(t)$ has a Fourier transform with frequencies ranging up to $f_m + \Delta f$. Thus, even though the lowpass filter is designed to pass frequencies only up to f_m, it is reasonable to assume that this entire term passes through the filter (since $\Delta f << f_m$).

Note that if the phase and frequency are perfectly adjusted, Eq. (6.29) reduces simply to $s(t)/2$, as we already knew for the synchronous demodulator.

Suppose first that we are able to match the frequency precisely, but that the phase is mismatched. Equation (6.29) then reduces to

$$s_o(t) = \frac{s(t)\cos\Delta\theta}{2} \tag{6.30}$$

This is an undistorted version of $s(t)$, so we would normally not be concerned. However, as the phase mismatch approaches 90°, the output goes to zero. If noise is added to the signal, the attenuation presented by the $\cos\Delta\theta$ term could become a significant negative factor. That is, as $\Delta\theta$ deviates from zero, the signal to noise ratio decreases.

One method of making the receiver insensitive to phase variations (i.e., making it robust) is to use the *quadrature receiver,* as shown in Fig. 6.25. We have indicated a phase shift of $\Delta\theta$ on both the sine and cosine multiplier signal. Equivalently, we could have indicated this phase shift on the input carrier.

Using trigonometric identities, we find the outputs of the two lowpass filters to be

$$s_1(t) = \frac{1}{2}s(t)\cos\Delta\theta$$

$$s_2(t) = -\frac{1}{2}s(t)\sin\Delta\theta \tag{6.31}$$

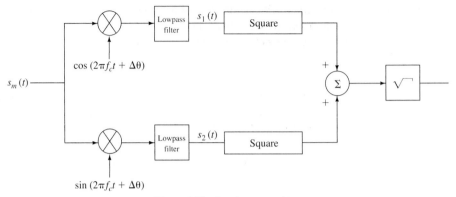

Figure 6.25   Quadrature receiver.

After taking the square root of the sum of the squares, we find that

$$s_o(t) = \frac{1}{2}\sqrt{s^2(t)} = \frac{1}{2}|s(t)| \tag{6.32}$$

As in the case of the square-law demodulator, undistorted demodulation is possible only if $s(t) \geq 0$. This means that the quadrature demodulator works only for transmitted carrier AM. Indeed, for that mode of transmission, we will find much simpler ways to demodulate a signal. We present the quadrature demodulator only to develop this important building block for later application.

Let us return to Eq. (6.29) and assume now that the phase has been perfectly matched, but that the frequency is mismatched. The output of the synchronous demodulator is then

$$s_o(t) = \frac{s(t)\cos 2\pi \Delta f t}{2} \tag{6.33}$$

The frequency mismatch is usually small (we try to make it zero), so the result will be a slowly varying amplitude (beating) of $s(t)$. If, for example, $s(t)$ is an audio signal and the frequency mismatch is 1 Hz, the effect would be to multiply $s(t)$ by a 1-Hz sinusoid. This is like taking the volume control of your radio and smoothly varying it from zero to maximum twice each second! Clearly, it is totally unacceptable. With a carrier frequency of 1 MHz, the 1-Hz mismatch represents only one part in 10^6. But suppose you were an expert at frequency matching, and your mismatch was only 10^{-3} Hz. Then your volume goes from maximum to zero once every 500 seconds. Unless we can derive the exact carrier from the incoming wave, or unless both the transmitter and receiver carriers are derived from the same source, synchronous demodulation is doomed to highly limited use.

6.5.3 Carrier Recovery in Transmitted Carrier AM

We have seen that synchronous demodulation requires *perfect* matching of the frequency and a phase mismatch that does not approach 90°. Frequency matching is possible if the AM waveform contains a periodic component at the carrier frequency. That is, the Fourier transform of the received AM waveform must contain an impulse at the carrier frequency. This is the case with transmitted carrier AM.

We assume that the received signal is of the form

$$s(t)\cos 2\pi f_c t + A\cos 2\pi f_c t \tag{6.34}$$

One way to extract the carrier is with a very narrow bandpass filter tuned to the carrier frequency. In the steady state, all of the carrier term will pass through this filter, while only a portion of the modulated carrier will go through. The Fourier transform of the filter output is

$$S_o(t) = \frac{S(f - f_c) + S(f + f_c) + A\delta(f + f_c) + A\delta(f - f_c)}{2} \tag{6.35}$$

This equation applies for the range of frequencies in the passband of the filter, that is,

$$f_c - \text{BW}/2 < f < f_c + \text{BW}/2$$

The inverse transform is then

$$s_o(t) = A\cos 2\pi f_c t + \int_{f_c - \text{BW}/2}^{f_c + \text{BW}/2} S(f - f_c)\cos 2\pi ft\,df \tag{6.36}$$

The integral in Eq. (6.36) is bounded by

$$\frac{1}{2\pi t} S_{\max}(f)\text{BW} \tag{6.37}$$

The smaller the bandwidth of the filter, the closer is the output to the pure carrier term.

An alternative to the narrow filter is a *phase-lock loop,* illustrated in Fig. 6.26. The phase-lock loop is discussed in detail in the next chapter. For now, we merely indicate that, if properly designed, the loop will lock on to the periodic component in the input to produce a sinusoid at the carrier frequency.

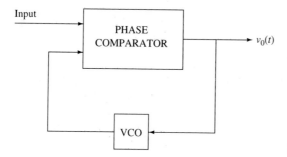

Figure 6.26   The phase-lock loop.

The implementation of the synchronous detector for transmitted carrier AM is shown in Fig. 6.27. Figure 6.27(a) shows the bandpass filter used for carrier recovery, and Fig. 6.27(b) shows the phase-lock loop.

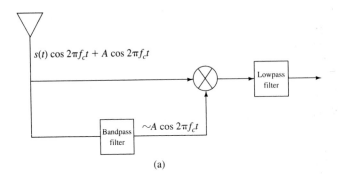

$s(t) \cos 2\pi f_c t + A \cos 2\pi f_c t$

Lowpass filter

$\sim A \cos 2\pi f_c t$

Bandpass filter

(a)

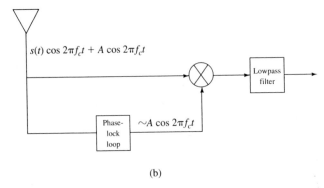

$s(t) \cos 2\pi f_c t + A \cos 2\pi f_c t$

Lowpass filter

Phase-lock loop

$\sim A \cos 2\pi f_c t$

(b)

Figure 6.27 Carrier recovery in transmitted carrier AM.

6.5.4 Incoherent Demodulation

Coherent demodulators (detectors) require reproduction of the carrier at the receiver. Since the exact carrier frequency and phase must be matched at the detector, accurate timing information is needed.

   If the carrier term is sufficiently large in transmitted carrier AM, it is possible to use incoherent detectors that do not have to reproduce the carrier or determine timing information. Let us suppose that the amplitude of the carrier is large enough so that $A + s(t) \geq 0$. We sketch a typical AM waveform in Fig. 6.28.

Square-law Detector

We noted earlier that the square-law demodulator is effective for the waveform presented in Fig. 6.28. We repeat that demodulator as Fig. 6.29. The output of the squarer is

$$[A + s(t)]^2 \cos^2 2\pi f_c t = \frac{[A + s(t)]^2 + [A + s(t)]^2 \cos 4\pi f_c t}{2} \qquad (6.38)$$

The output of the lowpass filter (which passes frequencies up to $2f_m$) is

$$s_1(t) = \frac{[A + s(t)]^2}{2} \qquad (6.39)$$

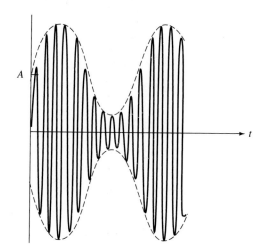

A

Figure 6.28   Transmitted carrier AM where $A + s(t) \geq 0$.

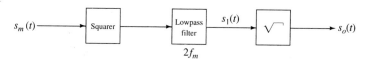

$s_m(t)$ → Squarer → Lowpass filter → $s_1(t)$ → $\sqrt{}$ → $s_o(t)$

$2f_m$

Figure 6.29   Square-law detector for transmitted carrier AM.

If we now assume that A is large enough such that $A + s(t)$ never goes negative, the output of the square rooter is

$$s_o(t) = 0.707[A + s(t)] \tag{6.40}$$

and demodulation is accomplished.

Rectifier detector

The squarer can be replaced by other forms of nonlinearity. In particular, consider the rectifier detector shown in Fig. 6.30. The rectifier can be either half wave or full wave. We will consider the full-wave variety here and ask you to examine the half-wave rectifier in one of the problems at the end of the chapter.

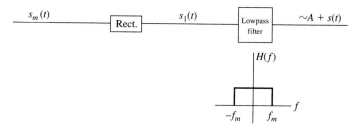

$s_m(t)$ → Rect. → $s_1(t)$ → Lowpass filter → $\sim A + s(t)$

$|H(f)|$

$-f_m \quad f_m$ f

Figure 6.30   Rectifier detector.

Full-wave rectification is equivalent to the mathematical operation of taking the absolute value. The output of the rectifier is then

$$s_1(t) = |A + s(t)||\cos 2\pi f_c t| \qquad (6.41)$$

but since we assume that $A + s(t)$ never goes negative, we can remove one set of absolute value signs to get

$$s_1(t) = \{A + s(t)\}|\cos 2\pi f_c t| \qquad (6.42)$$

The absolute value of the cosine is a periodic wave, as shown in Fig. 6.31. Its fundamental frequency is $2f_c$. We rewrite $s_1(t)$ by expanding the rectified cosine in a Fourier series:

$$s_1(t) = [A + s(t)][a_0 + a_1\cos 4\pi f_c t + a_2\cos 8\pi f_c t + \cdots] \qquad (6.43)$$

The output of the lowpass filter is then

$$s_o(t) = a_0[A + s(t)] \qquad (6.44)$$

and demodulation is accomplished.

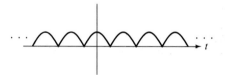

Figure 6.31 Rectified sine wave.

Before leaving the rectifier detector, we will point out the mechanism by which this detector reconstructs the carrier waveform. Figure 6.32 shows that full-wave rectification of the AM wave is equivalent to multiplying the waveform by a square wave at the carrier frequency. That is, the process of taking the absolute value flips around the negative portion of the carrier. This is equivalent to multiplication by -1. Therefore, the rectifier,

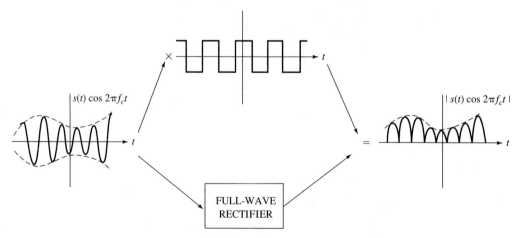

Figure 6.32 Multiplication by square wave vs. full-wave rectification.

which does not need to know the exact carrier frequency, is performing an operation that is equivalent to multiplication by a square wave carrier at the exact frequency and phase of the received carrier. We leave it as an exercise at the end of the chapter to show that a synchronous demodulator can operate by multiplying the wave either by a cosine matching the carrier or by a square wave matching the carrier.

6.5.5 Envelope Detector

The final detector we examine is by far the simplest. Let us observe the transmitted carrier AM waveform of Fig. 6.33. If $A + s(t)$ never goes negative, the upper outline, or *envelope,* of the AM wave is exactly equal to $A + s(t)$. If we can build a circuit that follows this outline, we will have built a demodulator.

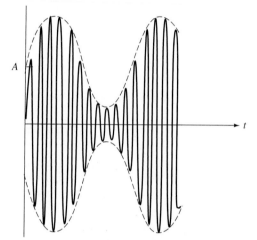

Figure 6.33   Transmitted carrier AM waveform.

It may be helpful to borrow an example from mechanics. Suppose that instead of representing a voltage waveform as a function of time, the curve represented the shape of a wire. One can envisage a cam moving along the top surface. If the cam is attached by means of a shock absorber, or viscous damper device, it will approximately follow the upper outline of the curve. This is shown in Fig. 6.34. The behavior is much the same as that of an automobile suspension system. You want the car to follow the outline of the road, but you do not wish it to track every ripple and bump in the road surface.

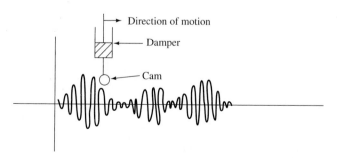

Figure 6.34   Mechanical outline follower.

The higher the carrier frequency, the more smoothly the cam will describe the upper outline of the curve, provided that it can respond fast enough to follow the shape of the outline. (You would probably not want your car to fly through the air between peaks of the road surface.) The outline, or envelope of the waveform, has a maximum frequency of f_m, while the ripples (the carrier) have a frequency of f_c. Intuition tells us that as long as $f_c \gg f_m$, we can design the mechanical system.

We now construct the electrical analogy to this mechanical system. The mass of the cam is represented by a capacitor. ($F = ma = mv'$ is replaced by $i = Cv'$ for a capacitor.) The viscous friction provides a force proportional to velocity, much as a resistor provides a voltage proportional to current. Finally, the cam is not attached to the wire; the wire (road) can push upon the cam, but cannot pull it. This is a mechanical diode. The equivalent circuit is then as shown in Fig. 6.35.

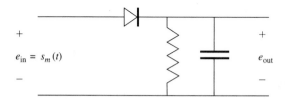

$e_{in} = s_m(t)$

e_{out}

Figure 6.35   Envelope detector.

The circuit of Fig. 6.35 is known as an *envelope detector*. When properly designed, it serves as a demodulator and is clearly far simpler to build than the demodulators we have discussed previously.

We first examine the operation of the envelope detector and then explore the appropriate choice of parameter values. Let us begin by removing the resistor, as shown in Fig. 6.36(a). This circuit is known as a *peak detector*.

The analysis of the peak detector requires only two observations: (a) The input can never be greater than the output (for an ideal diode), and (b) the output can never decrease with time. The first observation is true because, if the input did exceed the output, the diode would be supporting a positive forward voltage. The second observation follows from the fact that the capacitor has no discharge path. Figure 6.36(b) shows a transmitted carrier AM waveform (the carrier frequency has been drawn much lower than it would be in practice, for illustrative purposes) and the output of the peak detector. The output is always equal to the maximum past value of the input.

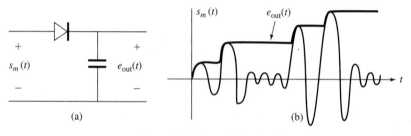

(a)

$s_m(t)$ $e_{out}(t)$

(b)

Figure 6.36   The peak detector.

If a discharging resistor is now added to the circuit, the output follows an exponential curve between peaks of the AM wave. This is shown in Fig. 6.37. If the time constant of the RC circuit is appropriately chosen, the output approximately follows the outline of the input curve, and the circuit acts as a demodulator. The output contains ripple at the carrier frequency (residual radio frequency), but this does not cause a problem, since we are interested only in frequencies below f_m.

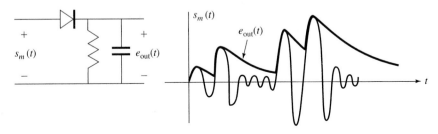

Figure 6.37   Addition of a discharging resistor.

The RC time constant must be short enough that the envelope can track the changes in peak values of the AM waveform. The peaks are spaced at intervals equal to the reciprocal of the carrier frequency, while the heights of these peaks follow the information, $s(t)$. We can consider a worst case where $s(t)$ is a pure sinusoid at a frequency of f_m. This would provide the fastest possible change in peak values. At this frequency, the peaks vary from a maximum to minimum in $1/2f_m$ sec. It takes an exponential function 5 time constants to get within 0.7 percent of its final value. Therefore, if we set the RC time constant to 10 percent of $1/f_m$, the envelope detector can follow even the highest frequency. For example, with an f_m of 5 kHz, the time constant would be set to 1/50 msec, or 20 μsec. This rule of thumb represents a first cut at envelope detector design. We chose the highest envelope frequency and viewed the waveform peaks (maximum and minimum). In reality, a sinusoid has its maximum slope in the middle and minimum slope at the extremes. Therefore, choosing the time constant on the basis of extremes means that our detector will not track all carrier peaks in between these extremes. We are saved by noting that typical information signals spend only a small fraction of their time at the highest frequency. Also, the large difference between carrier and envelope frequency allows a lot of leeway in choosing the time constant. If the information signal $s(t)$ is such that we expect significant periods near the highest frequency (as in the case, for example, of a soprano singing a particularly high-pitched selection), we could safely choose a time constant considerably smaller than that previously indicated.

Example 6.2

Design an envelope detector for use in demodulating a transmitted carrier AM waveform. Suppose the carrier frequency is 1 MHz and the information signal is a voice waveform.

Solution: Once the diode is selected, all that is required in the design of the envelope detector is to choose the values of R and C in the circuit.

The highest frequency of the envelope of the AM waveform is f_m, which we will assume is 5 kHz for voice. The envelope detector must be capable of responding to the fastest possible changes in the signal. The period of a 5-kHz waveform is 0.2 msec, and our guideline calls for choosing an *RC* time constant that is 10 percent of this value, or 20 μsec. This choice will not guarantee that the envelope detector output hits all of the peaks of the carrier. However, since a certain amount of ripple at 1 MHz will not hurt the system, we can afford to shorten the time constant. For example, a time constant of 10 μsec would allow the signal to come within 0.005 percent of the final value in tracking the fastest envelope frequency, and the carrier response would decrease only to 0.975 of the peak (i.e., the exponential decay over one period of the carrier). Even though the envelope is very rarely at the maximum frequency for any sustained period of time, we can certainly afford to play it safe and design for a rather short time constant. The fact that a system is designed for audio does not mean that all transmitted signals will be voicelike. Rather, it means only that they will occupy audio frequencies.

Once the time constant is chosen, it is necessary to specify the type of diode, the resistance and capacitance, and the power rating of the components. The resistance is normally chosen with a view toward input and output impedance matching. For the input to the envelope detector, we would like *R* to be as large as possible to avoid loading the previous circuitry. Let us choose $R = 1$ kΩ. Then, to achieve a time constant of 10 μsec, the capacitor value must be 0.01 μF.

6.5.6 Integrated Circuit Modulators and Demodulators

Several integrated circuit (IC) manufacturers have produced balanced modulators and demodulators. Among these are the Signetic MC1496/MC1596 and the Analog Devices AD630. These ICs contain differential amplifiers either that are driven into saturation or that simulate an electronic commutator (a device that alternately multiplies by positive and negative values). The reader may consult the literature for details of the electronics; we will concentrate upon applications in this text.

Figure 6.38 shows the MC1496 used as a transmitted carrier amplitude modulator. The same circuit can be used to generate suppressed carrier AM by choosing different resistor values in the carrier adjust circuitry.

The MC1496 is also used for demodulation of transmitted carrier AM. The circuit is shown in Fig. 6.39. The carrier for this operation is derived by driving the high-frequency amplifier into saturation, thereby providing an amplified and limited output that resembles a square wave at the carrier frequency. This carrier feeds into one of the MC1496 inputs, along with the AM wave into the other. The output must be lowpass filtered to recover the information signal.

6.6 BROADCAST AM AND SUPERHETERODYNE RECEIVER

The electromagnetic spectrum has been described as a natural resource. The dictionary defines *resource* as "something that lies ready to use or that can be drawn upon for aid or to take care of a need." *Natural resource* is defined as "those actual and potential forms of wealth supplied by nature." The electromagnetic spectrum does indeed lie ready for use to take care of a need—the need to communicate. Also, it is a potential form of wealth, and while it is not expendable in the sense in which oil and natural gas are, it does get used up in the sense that only a limited number of users can employ it at any one time.

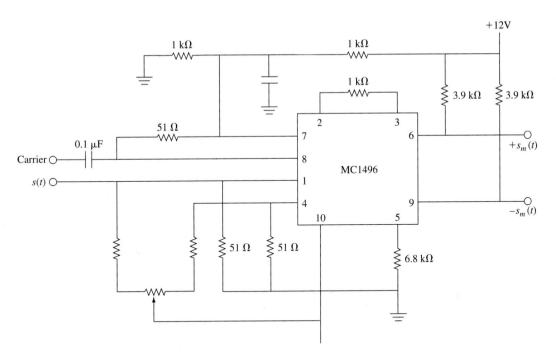

Figure 6.38   AM modulator.

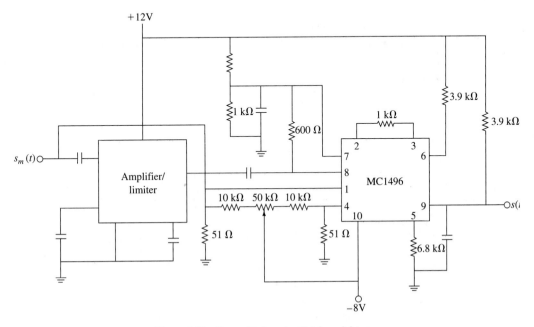

Figure 6.39   Transmitted carrier AM demodulator.

For these reasons, it has become necessary to regulate this valuable resource. Not only the transmission, but also the reception, of electromagnetic waves, is regulated. Interesting legal issues surround the reception of police radio or scrambled satellite entertainment channels. In the early days of satellite entertainment, stories circulated about a man who was sued for erecting a backyard antenna to receive satellite signals countersued the broadcaster for injecting signals onto his private property without his permission! Thus, we provide additional job security for lawyers.

When U.S. regulatory bodies were debating technical solutions to eavesdropping on cellular telephone conversations, they opted instead simply to make it illegal to intercept such signals. (You can draw your own conclusions about the effectiveness of this approach) Nonetheless, the need for regulation is universally acknowledged. The *International Telecommunications Union* (ITU), based in Geneva, Switzerland, is charged with worldwide regulation of the airwaves; its membership extends across most nations throughout the world. The ITU has designated broad bands of frequency and has issued regulations within each band. The bands are as follows:

VLF	3–30	kHz
LF	30–300	kHz
MF	300–3,000	kHz
HF	3–30	MHz
VHF	30–300	MHz
UHF	300–3,000	MHz
SHF	3–30	GHz
EHF	30–300	GHz

VLF transmissions are in the form of surface waves, and they can travel over long distances. Radio-frequency energy in this band is capable of penetrating oceans, so the band is used in submarine communications. The VHF and UHF bands are used for television. Additionally, cellular telephones use portions of the UHF band. We shall concentrate on the MF band around 1 MHz.

The regulatory body within the United States that sets rules (within the ITU general rules) is the *Federal Communications Commission* (FCC). The commission's regulations fill many volumes, which are available in technical libraries.

In the United States, the AM broadcast band extends from 535 to 1,605 kHz. Other parts of the world have a slightly higher upper frequency cutoff (1,606.5 kHz). Within the band, stations are licensed by the FCC to operate commercial broadcast services. Transmission is by transmitted carrier AM, and the maximum information frequency, f_m, is specified as 5 kHz. Assigned carrier frequencies must therefore be separated by 10 kHz. With this spacing, the entire band can support up to 107 stations. Licensing of stations depends on many factors, the most important being the location of the antenna, radiated power, antenna pattern, and times of broadcast.

The location of the *antenna* is important because a low-power station in a rural area may have fewer constraints than a high-power station in a large metropolitan area where many broadcasters are competing for the limited frequency space. If two high-power stations were assigned adjacent carrier frequencies, the filtering demands on the receiver would be excessive. The height and altitude of the antenna, of course, also affect the range of transmission.

The *radiated power* affects the range of transmission as well. Accordingly, the FCC must be careful not to assign identical carrier frequencies to two separate stations whose broadcasts will be received at the same point.

The *antenna pattern* affects the range as a function of bearing from the antenna. For example, a 5-kW station that broadcasts omnidirectionally creates far less interference with a station 500 km away than does a 5-kW station that beams *all* of its power in that particular direction.

The *time of broadcast* affects the range of transmission. The transmission character-istics of air at medium frequencies depend on temperature and humidity, which generally differ from day to night. Indeed, it is not uncommon for a station that is heard at distances up to 150 km during the day to be received 500 km away after dark. Therefore, some sta-tions are licensed to operate only during daylight hours.

With the preceding as an introduction, we are now in a position to understand the operation of the standard broadcast AM receiver. We probably will not be able to repair one (a technician with a familiarity of electronics is needed for that), but we certainly can understand its block diagram.

Several basic operations can be identified in any broadcast receiver. The first is *sta-tion separation*: We must pick out the one desired signal and reject all the other signals. The second operation is *amplification*: The signal intercepted by the radio antenna (on the order of 1 μV) is far too weak to drive the electronics in the receiver without first being amplified. The third operation is *demodulation*: The incoming signal is amplitude modu-lated and contains frequencies centered about the carrier frequency.

Separating one channel from the others requires a very accurate bandpass filter with a sharp frequency cutoff characteristic. Suppose, for example, that we wish to listen to a station with carrier at 1.01 MHz (1010 on the AM dial). That station actually occupies 1.005 to 1.015 MHz in frequency. The adjacent stations occupy 0.995 to 1.005 and 1.015 to 1.025 MHz, so the bandpass filter must be very close to ideal. The FCC deals with this situation by avoiding licensing of powerful stations in adjacent frequency slots. Nonethe-less, the receiver must be capable of distinguishing among closely separated local stations when they do exist. Assuming that the listener would want to tune the receiver to any sta-tion in the band, the filter would have to be adjustable. That is, the band of frequencies that it passes must be capable of variation.

To make a bandpass filter approach the ideal characteristics, we need multiple stages of filtering; a single *RLC* circuit does not have a high enough Q to accomplish the station separation. The tuning of multiple-section filters is no easy task, however. For example, you might have to vary three unequal capacitors in a particular manner to achieve adjustment of the center frequency of a third-order Butterworth filter. Early AM receivers contained such filters and typically had three tuning dials (variable capacitor controls) that had to be simul-taneously adjusted. The family gathered around the *wireless*, and the head of the household (in those days, probably a man) would twiddle the three knobs. It was a major cause for cel-ebration when a station (complete with crackling sounds) appeared at the speaker.

Fortunately, Edwin Armstrong changed all of this when he invented the superhetero-dyne receiver in 1918. This simple device eliminated the need for complex adjustment of the filter and ushered in the radio era.

Recall that multiplication of a signal by a sinusoid shifts all frequencies up and down by the frequency of the sinusoid. Because of this, station selection can be accomplished by building a *fixed* bandpass filter and shifting the input frequencies so that the station of interest falls in the passband of the filter. Looked at another way, we construct a viewing window on the frequency axis. Then, instead of moving this window around to view a particular portion of the axis, we keep the window stationary and shift the entire axis. The shifting process is known as *heterodyning*, and the resulting receiver is the *superheterodyne* receiver. The process of heterodyning is applicable to other forms of modulation (e.g., FM).

A block diagram of an AM broadcast receiver is shown in Fig. 6.40.

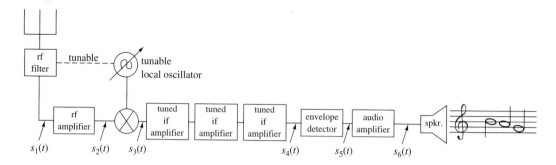

Figure 6.40   AM broadcast receiver.

In the figure, the antenna receives a signal that is a weighted sum of all broadcasted signals. After some filtering, which we will examine in a moment, the incoming signal is amplified in a radio frequency (rf) amplifier. The signal, $s_2(t)$, is then shifted up and down in frequency by multiplying by a sinusoidal generator, called the *local oscillator*. The shifting or heterodyning operation is also known as *mixing*.

The output of the heterodyner is applied to the sharp bandpass filter consisting of multiple filtering stages. This filter is normally combined with amplifiers. The fixed bandpass filter is set to 455 kHz, called the intermediate frequency (if), and has a bandwidth of 10 kHz, matching that of each station. This frequency is not within the AM broadcast band and is specified by the FCC. If stations were authorized to broadcast at this carrier, some of the signal would enter the if portion of the receiver (since every piece of wire acts as an antenna) and would be heard on top of the desired station. In most receivers, the if filter is made up of three tuned circuits that are aligned so as to generate a Butterworth filter characteristic (poles around a semicircle in the *s*-plane). At $s_4(t)$, we have a modulated signal whose carrier frequency has been shifted to 455 kHz and that has already been amplified and separated out from the other signals.

Let us now determine the required frequency of the local oscillator. Suppose you wish to listen to a station at the lower end of the dial, say, a carrier frequency of 540 kHz. To shift this frequency to 455 kHz, you would have to multiply by a sinusoid of either 85 or 995 kHz. Now suppose you wish to listen to a station at the top of the dial, with carrier at 1,600 kHz. Then the local oscillator setting must be either 1,145 or 2,055 kHz. To tune in any station in the band, the oscillator must be tunable over the range from 85 to 1,145 kHz or the range from 995 to 2,055 kHz.

The second of these ranges is selected for practical reasons. The local oscillator is set at the sum of 455 kHz and the desired carrier frequency. The resulting oscillator must tune over a range where the highest frequency is a little more than two times the lowest. If the first range had been selected, the highest frequency would be 13.5 times the lowest. It is much easier to construct variable oscillators for ranges that vary by a factor of 2 to 1 than by a factor of 13.5 to 1. The higher range might require a *range switch*.

The receiver then puts $s_4(t)$ through an envelope detector and then amplifies the signal (usually using push-pull power amplifiers) before applying it to a loudspeaker. After detection, the signal is at audio frequency (af). Some filtering may be done to provide treble and base controls.

One significant problem not mentioned earlier is that heterodyning produces both an upshift and a downshift in frequency (i.e., sums and differences). While one of these shifts moves the desired station into the if window (450 to 460 kHz), the other moves another band of frequencies into the same window. This undesired signal is called an *image*, and eliminating it is not very difficult.

As an example, suppose you wish to listen to a station at a carrier frequency of 600 kHz. Then the local oscillator is set to 600 + 455 = 1,055 kHz. Multiplication by this sinusoid places the desired 600-kHz station right into the if filter passband. But there is another station at a carrier of 1,055 + 455 = 1,510 kHz that, when multiplied by the local oscillator, will produce a component at 455 kHz. This *image station* would be heard right on top of the desired station.

The separation between the image and the desired station is twice the if frequency, or 910 kHz. This places the image 91 frequency slots away from the desired station. The 89 stations between the two are eliminated by the if filter.

A bandpass filter with a bandwidth of less than 1,820 kHz would accomplish the separation. Such a filter must pass the desired station while rejecting the station 910 kHz away. The filter must be tunable, but it need not be a sharp bandpass filter. We do not care what it does to the 89 stations between the image and the desired signal. A single tuned stage is therefore sufficient. This is shown in Fig. 6.41.

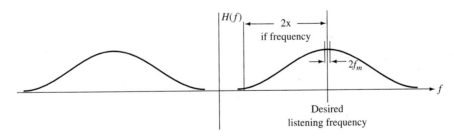

Figure 6.41   Image rejection process.

In practice, when the tuning dial on a receiver is turned, this sloppy RF rejection filter is tuned at the same time that the frequency of the local oscillator is changed. Before electronic tuning, the shaft of the tuning dial connected to two separate variable-capacitor sections. One of these formed part of the image rejection filter, and the other formed part of the tuning circuit of the local oscillator. This is indicated in Fig. 6.40 as a dashed line connecting the two functions.

Integrated Circuit Receiver

Many of the functions of the superheterodyne AM radio receiver have been implemented using integrated circuits. One example is the TEA5550 AM radio circuit from Signetics. This chip contains the balanced mixer, if amplifier, detector, rf amplifier, local oscillator, and automatic gain control seen in the superheterodyne receiver. Its block diagram and pin configuration are shown in Fig. 6.42. We have discussed all of these functions except *automatic gain control* (AGC). As you tune a receiver across the AM dial, the various stations that come in have different voltage levels. The level depends on the transmitted power, the distance from the transmitter, the antenna pattern of both the transmitter and the receiver, and the frequency characteristics of the channel. To minimize the amount of adjustment of the volume control as we change from one station to another, and to decrease time-varying volume effects, a form of feedback is employed in most radio receivers. The output amplitude is sensed (using a lowpass filter with a time constant on the order of several seconds), and the level is fed back to the various amplifier stages to affect the Q points. As the level increase, the Q points are adjusted to reduce amplification, thus providing a form of negative feedback that tends to keep the output level constant.

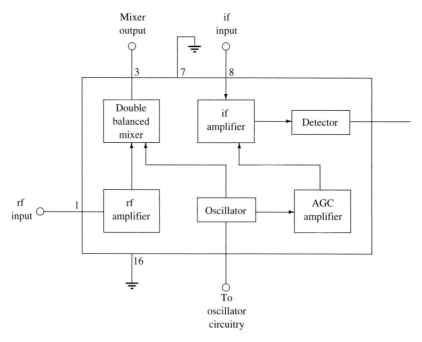

Figure 6.42   Integrated circuit AM receiver.

When the AM radio receiver chip is used as part of a radio, external circuitry is required. In particular, fabrication of capacitors of the size needed is not practical on the IC. We must therefore configure external circuitry for the oscillator and tuning portions of the system. (See application manuals from IC manufacturers for details of the actual circuitry for using the IC in a radio receiver.)

6.7 ENVELOPES AND PRE-ENVELOPES

We have referred to the *envelope* of a waveform as an outline that follows the peaks of the carrier. The envelope detector attempts to follow this outline. Our intuitive definition could be formalized as follows:

> Given $s(t)\cos 2\pi f_c t$, where the frequencies $s(t)$ are much less than f_c, the *envelope* is defined as the absolute value of $s(t)$.

This definition is based on observation of the time waveform.

The foregoing intuitive definition is not very satisfying, since it applies only to a narrow class of signals. We shall now explore a mathematical definition of the envelope that is identical to the intuitive definition for the class of functions to which the intuitive approach applies.

The *pre-envelope* (also known as the *complex envelope* or *analytic function*) of a waveform $r(t)$ is defined as the complex function of time whose Fourier transform is given by

$$Z(f) = 2U(f)R(f) \tag{6.45}$$

where $U(f)$ is the unit step function and $R(f)$ is the transform of $r(t)$. That is, the transform of the pre-envelope is zero for negative f and twice the original transform for positive f. The symbol $z(t)$ is commonly used to represent the pre-envelope function. Note that $z(t)$ cannot be a real function of time, since the magnitude of its transform is never even.

Example 6.3

Find the pre-envelope of $r(t) = \cos 2\pi f_c t$.
Solution: The Fourier transform of $r(t)$ is given by

$$R(f) = \frac{1}{2}[\delta(f - f_c) + \delta(f + f_c)]$$

The transform of the pre-envelope is then

$$Z(f) = \delta(f - f_c)$$

The function of time corresponding to this transform is

$$z(t) = e^{j2\pi f_c t}$$

You have seen this function in sinusoidal steady-state analysis in your circuits course. When you wish to find the response of a circuit to a sinusoid, you replace the sinusoid with a complex exponential. You do this primarily to carry the amplitude and phase in a single operation, rather than having to track sines and cosines through the system. However, in reality, you are substituting the pre-envelope of the sinusoid for the actual function of time.

We now express $z(t)$ in terms of $r(t)$. That is, we translate the Fourier transform truncation process of Eq. (6.45) into an equivalent operation in the time domain. We start with

$$z(t) = 2r(t)*\mathcal{F}^{-1}[U(f)] \tag{6.46}$$

The inverse Fourier transform of $U(f)$ is found from the table in appendix II:

$$U(f) \leftrightarrow \frac{1}{2}\delta(f) - \frac{1}{2\pi jt} \tag{6.47}$$

Finally,

$$z(t) = 2r(t)*\frac{1}{2}\delta(t) + 2r(t)*\frac{-1}{2\pi ft}$$

$$= r(t) + \frac{j}{\pi}\int_{-\infty}^{\infty}\frac{r(\tau)}{t-\tau}d\tau \tag{6.48}$$

Note that the real part of $z(t)$ is the original function of time, $r(t)$, just as the real part of $e^{j2\pi ft}$ is cos $2\pi ft$. The imaginary part of $z(t)$ is given by the convolution of $r(t)$ with $1/\pi t$. This convolution is useful in a number of applications and is known as the *Hilbert transform* of $r(t)$. The symbol for the Hilbert transform of a function of time is the same letter as that denoting the function, with the addition of a caret. Thus, the Hilbert transform of $r(t)$ is $\hat{r}(t)$.

The Fourier transform of $1/\pi t$ is sgn(f). Therefore, taking the Hilbert transform of a function of time is equivalent to taking the mirror image of the portion of the Fourier transform to the left of the origin. Using this fact, we can see clearly how the pre-envelope results from Eq. (6.48). Taking the Fourier transform of that equation yields

$$Z(f) = R(f) + R(f)\text{sgn}(f) = R(f) + \hat{R}(f) \tag{6.49}$$

where $\hat{R}(f)$ is the Fourier transform of $\hat{r}(t)$. For positive f, $Z(f)$ is $2R(f)$, while for negative f, it is zero.

The Hilbert transform arises whenever we perform a truncation of a portion of the frequency axis. We will see it again when we study single sideband in the next section.

The *envelope* is defined as the magnitude of the pre-envelope.

Example 6.4

Find the envelope of $r(t) = \cos 2\pi f_c t$.
Solution: The pre-envelope of $r(t)$ was found in Example 6.3 to be

$$z(t) = e^{j2\pi f_c t}$$

The magnitude of this function is equal to unity. Therefore, the envelope of the function is a constant, 1. We already knew this from our intuitive definition of the envelope, since the function of time represents an unmodulated carrier wave.

Example 6.5

Find the envelope of

$$s_m(t) = s(t)\cos 2\pi f_c t$$

Solution: We must first find the pre-envelope of the function. The Hilbert transform is found by inverting the negative-f portion of the Fourier transform, $S_m(f)$:

$$\hat{S}_m(f) = S_m(f)\text{sgn}(f) = \frac{1}{2}[S(f - f_c) - S(f + f_c)]$$

In writing this equation, we have assumed that $S(f + f_c)$ lies completely to the left of the origin. This result therefore applies as long as $f_m < f_c$. The inverse Fourier transform of $S_m(f)$ is given by

$$\hat{s}_m(t) = js(t)\sin 2\pi f_c t$$

The pre-envelope is given by

$$z_m(t) = s(t)\cos 2\pi f_c t + js(t)\sin 2\pi f_c t$$

The envelope is the magnitude of the pre-envelope, or

$$\left|z_m(t)\right| = \sqrt{s^2(t)[\cos{}^2 2\pi f_c t + \sin{}^2 2\pi f_c t]}$$

$$= \sqrt{s^2(t)} = \left|s(t)\right|$$

This agrees with our intuitive definition. However, note that the intuitive definition requires that $f_c \gg f_m$, while the mathematical definition requires only that $f_c > f_m$.

We have established a definition of envelope that applies to any function of time. A worthwhile, although very difficult, task would be to analyze the envelope detector circuit with general-input functions of time and compare its output to the (mathematical) envelope of the input.

The intuitive definition that applies when the envelope is slowly varying will prove sufficient for our work in AM. The concept of the pre-envelope proves useful in detection theory, which we explore in later chapters of the book dealing with digital communication.

6.8 SINGLE SIDEBAND (SSB)

In the AM systems we have studied, the range of frequencies required to transmit the signal is the band between $f_c - f_m$ and $f_c + f_m$. The frequency f_c is the carrier frequency, and f_m is the maximum frequency of the baseband signal $s(t)$. The total *bandwidth* is then $2f_m$. The frequency spectrum is a "natural resource" whose conservation is critical. The more frequency bandwidth required for each channel, the fewer number of stations can communicate simultaneously. Wouldn't it be lovely if we could find a way to send information using less than $2f_m$ of bandwidth?

Single sideband is a technique that allows transmission in half of the bandwidth required for AM double sideband. In Fig. 6.43, we define that portion of $S_m(f)$ which lies in the band above the carrier as the *upper sideband*. The portion below the carrier is the *lower sideband*. A double-sideband AM wave is composed of a lower and an upper sideband.

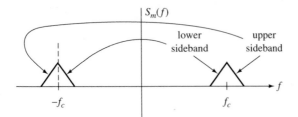

Figure 6.43   Definition of sidebands.

We can use the properties of the Fourier transform to show that the two sidebands are dependent on each other. The transform of the AM wave is formed by shifting $S(f)$, the transform of the signal, up and down in frequency. The lower sideband is formed from the negative-f portion of $S(f)$, and the upper sideband is the positive-f portion of $S(f)$. We assume that the information signal $s(t)$ is a real function of time. Therefore, the magnitude of $S(f)$ is even and the phase is odd. The negative-f portion of $S(f)$ can be derived from the positive-f portion by taking the complex conjugate. Similarly, the lower sideband of $s_m(t)$ can be derived from the upper sideband. Since the sidebands are not independent, it should be possible to transmit all essential information by sending only a single sideband. This is the essence of single-sideband communication.

Figure 6.44 shows the Fourier transforms of the upper and lower sideband versions of the AM wave, denoted $s_{usb}(t)$ and $s_{lsb}(t)$, respectively. The double-sideband AM wave is the sum of the two sidebands:

$$s_m(t) = s_{lsb}(t) + s_{usb}(t) \tag{6.50}$$

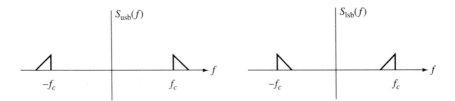

Figure 6.44   Single-sideband Fourier transforms.

Since the single-sideband waveform resides in a subset of the band of frequencies occupied by the double-sideband waveform, it automatically satisfies two of the requirements of a modulation system. That is, by proper choice of the carrier frequency, we can move the modulated waveform into a range of frequencies that transmits efficiently. We can also use different bands for different signals, thereby allowing simultaneous transmission of multiple signals.

The synchronous demodulator can be used to demodulate single sideband. This can be shown either pictorially in the frequency domain or mathematically in the time domain.

Looking first at frequencies, we know that multiplication by a sinusoid shifts the Fourier transform both up and down by the frequency of the sinusoid. Figure 6.45(a) shows the Fourier transform that results when $s_{usb}(t)$ is multiplied by a sinusoid at a frequency of f_c, and Fig. 6.45(b) shows a similar result for the lower sideband signal. In both cases, a low-pass filter would recover a replica of the original information signal.

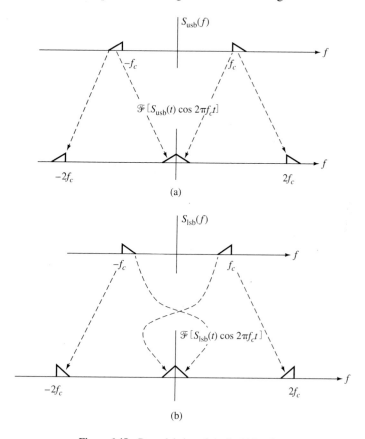

Figure 6.45   Demodulation of single sideband.

We can illustrate this same result in the time domain by multiplying the single-sideband waveform by a cosine at the same frequency as the carrier. Before continuing, however, we need to derive an expression for the single-sideband waveform that is a function of time.

We begin by expressing the lower sideband Fourier transform as a product of the double-sideband transform with a bandpass filter function. The filter passes only the lower sideband:

$$S_{lsb}(f) = S_m(f)H(f)$$

$$= \frac{S(f + f_c)U(f + f_c) - S(f - f_c)U(f - f_c) + S(f - f_c)}{2} \tag{6.51}$$

We next express each unit step function of Eq. (6.51) in an alternate manner, using the *sign* (sgn) function:

$$U(f + f_c) = \frac{1 + \text{sgn}(f + f_c)}{2} \tag{6.52}$$

$$1 - U(f - f_c) = \frac{1 - \text{sgn}(f - f_c)}{2}$$

We can now relate these expressions to the Hilbert transform. Recall that the Fourier transform of the Hilbert transform is given by

$$\hat{S}(f) = \frac{S(f)\text{sgn}(f)}{j} \tag{6.53}$$

Substituting Eqs. (6.52) and (6.53) into (6.51) yields

$$S_{\text{lsb}}(f) = \frac{1}{2}\left[\frac{S(f + f_c) + S(f - f_c)}{2} + \frac{\hat{S}(f - f_c) - \hat{S}(f + f_c)}{2j}\right] \tag{6.54}$$

Both ratios in brackets in Eq. (6.54) should look familiar. The first is in the form of the Fourier transform of an AM wave. The positive and negative frequency shifts represent multiplication of $s(t)$ by a cosine function. The second ratio corresponds to the function of time that is produced when $s(t)$ is multiplied by a sine wave. The result is that the lower sideband signal is given by

$$s_{\text{lsb}}(t) = \frac{1}{2}s(t)\cos 2\pi f_c t + \frac{1}{2}\hat{s}(t)\sin 2\pi f_c t \tag{6.55}$$

The upper sideband can be derived from the lower sideband by observing that the sum of the two sidebands is the double-sideband AM waveform. Therefore,

$$s_{\text{usb}}(t) = s_m(t) - s_{\text{lsb}}(t) \tag{6.56}$$

$$= \frac{1}{2}s(t)\cos 2\pi f_c t - \frac{1}{2}\hat{s}(t)\sin 2\pi f_c t$$

Now that we have functions of time for the single-sideband waveforms, we can return to the analysis of the synchronous demodulator. We use the time domain expression of Eqs. (6.55) and (6.56) for the single-sideband waveforms:

$$s_{\text{ssb}}(t)\cos 2\pi f_c t = \frac{s(t)\cos^2 2\pi f_c t \pm \hat{s}(t)\sin 2\pi f_c t\cos 2\pi f_c t}{2} \tag{6.57}$$

The plus sign applies to the lower sideband and the minus sign to the upper sideband. We use trigonometric expansions to express Eq. (6.57) as

$$s_{\text{ssb}}(t)\cos 2\pi f_c t = \frac{s(t) + s(t)\cos 4\pi f_c t \pm \hat{s}(t)\sin 4\pi f_c t}{4} \tag{6.58}$$

The output of the lowpass filter with this quantity as input is simply $s(t)/4$, so we have accomplished demodulation.

Recall that in double sideband, we did not particularly favor the synchronous demodulator, since a phase mismatch led to attenuation and a frequency mismatch resulted in a very serious form of multiplicative distortion. Such a mismatch in single sideband is also serious, but slightly more forgiving than in the case of double sideband.

If the phase is mismatched, Eq. (6.58) becomes

$$s_{ssb}(t)\cos2\pi(f_c + \Delta\theta)t = \frac{1}{2}s(t)\cos2\pi f_c t\cos2\pi(f_c + \Delta\theta)t$$
$$\pm \frac{1}{2}\hat{s}(t)\sin2\pi f_c t\cos2\pi(f_c + \Delta\theta)t \tag{6.59}$$

The output of the lowpass filter is then

$$\frac{s(t)\cos\Delta\theta}{4} \tag{6.60}$$

As in the case of double-sideband AM, the phase mismatch causes an attenuation.

If the frequency is now mismatched, Eq. (6.58) becomes

$$s_{ssb}(t)\cos 2\pi(f_c + \Delta f)t =$$

$$\frac{s(t)\cos 2\pi\Delta ft}{4} + \frac{s(t)\cos 4\pi(f_c + \Delta f)t}{4} \tag{6.61}$$

$$+ \frac{\hat{s}(t)\sin 2\pi\Delta ft}{4} + \frac{\hat{s}(t)\sin 4\pi(f_c + \Delta f)}{4}$$

It is clear that only the first and third terms go through the lowpass filter. If only the first term appeared in the output, the effect would be the same as that experienced in double sideband. However, the addition of the Hilbert transform term changes the output. The specific form of that change can be seen if we consider the special case of a sinusoidal modulating signal. That is, let $s(t) = \cos2\pi f_m t$. The output of the lowpass filter is then proportional to $\cos2\pi(f_m \pm \Delta f)t$, with the plus sign obtaining for lower sideband and the minus sign for upper sideband. (You can prove this either from Eq. (6.61) or by writing the single-sideband waveform as a single sinusoid at a frequency of either $f_c - f_m$ or $f_c + f_m$, depending upon whether lower or upper sideband is being considered.) The effect of the frequency mismatch is therefore a frequency offset in the demodulated wave. If the information signal is a sum of sinusoids, the demodulated signal will be a sum of shifted sinusoids. Thus, in the general case, we see an overall shifting of frequencies of $s(t)$ by the amount of the frequency mismatch. You might think that this results in a change in pitch of the sound, but such is not the case. For example, suppose you hummed into a microphone. Then the resulting waveform would be periodic with a fundamental frequency equal to the frequency at which you are humming. Harmonics would occur at multiples of this frequency. If each frequency component is shifted by the same amount, the relationship among the harmonics is destroyed, and the resulting sound changes. The effect upon music is generally considered to be unacceptable. The effect upon voice is sometimes acceptable, since the resulting sound is usually intelligible (at least, for small Δf relative to the frequencies that are present). Some have described the effect as a "Donald Duck" voice.

Therefore, while mismatches are to be avoided, the effects may be considered less devastating than in the case of double sideband.

With double sideband, we made demodulation simpler by adding a carrier. We can also add a carrier to the single-sideband waveform, recognizing that the carrier would be at the edge of the band of frequencies occupied by the waveform. The carrier could be extracted using a filter at the receiver, or a phase lock loop could be used.

Can we use an incoherent detector, such as the envelope detector? To answer this question, let us examine transmitted carrier lower sideband, which is given by

$$s_{lsb}(t) + A\cos 2\pi f_c t = \left[A + \frac{s(t)}{2} \right] \cos 2\pi f_c t + \frac{\hat{s}(t)\sin 2\pi f_c t}{2} \tag{6.62}$$

The envelope of this waveform is found by combining the cosine and sine into a single sinusoid with a time-varying amplitude and phase. The envelope is given by

$$\sqrt{\left[A + \frac{1}{2}s(t) \right]^2 + \left[\frac{1}{2}\hat{s}(t) \right]^2} \tag{6.63}$$

In general, this does not look anything like $s(t)$. However, if the constant A is very large, the first term predominates, and the expression is approximately equal to $A + s(t)/2$. Of course, large values of A mean inefficient operation, with most of the transmitted energy going into the carrier. Therefore, a system of this type finds limited application.

6.9 VESTIGIAL SIDEBAND

The advantage of single over double sideband (SSB and DSB) is the former's economy of frequency usage. That is, SSB uses half the corresponding bandwidth required for DSB transmission. The primary disadvantage of SSB is the difficulty in building a transmitter or an effective receiver for it. One problem is that when we attempt to build a sharp filter to remove one of the sidebands, the phase characteristic of the filter develops ripple. The closer one approaches the amplitude characteristic of the ideal filter, the worse becomes the phase characteristic. One area in which frequency conservation becomes critical is television, where bandwidths are orders of magnitude greater than those used for voice transmission. Phase distortion in a video signal causes offset of the resulting scanned image, and this is seen as ghost images on the screen. The eye is much more sensitive to such forms of distortion than is the ear to equivalent forms of voice distortion. We therefore have reason to explore a compromise between SSB and DSB.

Vestigial sideband (VSB) possesses a frequency bandwidth advantage approaching that of single sideband without the disadvantage of difficulty in building a modulator. It is also easier to construct a demodulator for this form of communication.

As the name implies, VSB includes a vestige, or trace, of the second sideband. Thus, instead of completely eliminating the second sideband, as in the case of SSB, we eliminate most, but not all, of it.

Suppose we begin with DSB but filter out one of the sidebands. In contrast to SSB, with VSB we use a filter that does *not* closely approach the ideal infinite roll-off. The result might resemble Fig. 6.45, where we show the double-sided transform, the filter characteristic, and the output transform that is generated.

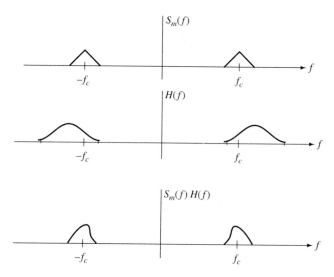

Figure 6.46   Vestigial sideband generator.

Suppose now that this output signal forms the input to a synchronous (coherent) demodulator. Then, when the VSB signal is multiplied by a cosine at the carrier frequency, the Fourier transform shifts both up and down by the carrier frequency. The part that shifts down passes through the lowpass filter. The part that shifts up resides around $2f_c$ and is rejected by the filter. The filter output then has a transform given by

$$S_0(f) = \frac{S(f)[H(f + f_c) + H(f - f_c)]}{4} \tag{6.64}$$

Equation (6.64) can be used to set the conditions on the filter. The bracketed sum is shown in Fig. 6.47 for a typical filter transfer function $H(f)$. This filter transfer function, $H(f)$, must display odd symmetry for frequencies around the carrier such that the sum of the two terms approximates a constant characteristic. The tail of the filter characteristic must be asymmetric about $f = f_c$. That is, the output half of the tail must fold over and fill in any difference between the inner-half values and the value for an ideal filter.

Suppose that we add a carrier term to a vestigial sideband signal. The vestigial sideband transmitted carrier waveform is then of the form

$$s_v(t) + A \cos 2\pi f_c t \tag{6.65}$$

This carrier term can be extracted at the receiver using either a very narrow bandpass filter or a phase lock loop. If the carrier term is large enough, an incoherent detector (e.g., an envelope detector) can be used. We said this same thing in regard to SSB, where the carrier had to be much larger than the signal. In DSB, the carrier need only be of the same order of magnitude as the signal. The required carrier size for VSB is between these two extremes. While the addition of a strong carrier significantly decreases efficiency, the ease of

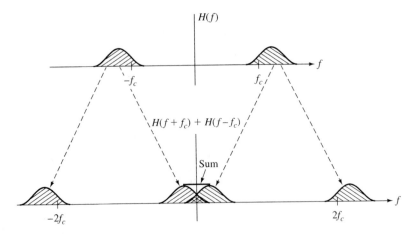

Figure 6.47 Bandpass filter constraint for vestigial sideband.

construction of an envelope detector makes this the system of choice in television, which we discuss in Section 6.12.

6.10 HYBRID SYSTEMS AND AM STEREO

We now investigate a hybrid modulation technique that permits a new form of multiplex-ing of signals. We have repeatedly stated that signals can be separated, provided that they do not overlap in time or in frequency. Double-sideband AM maintains frequency separa-tion in order to keep channels from interfering with each other, but it uses twice the band-width of single sideband. The latter fact, however, hints that it might be possible to send two double-sideband AM signals that overlap in both time and frequency *and yet still be able to separate them* at the receiver. Indeed, *quadrature amplitude modulation* (QAM) accomplishes this.

Suppose we have two information signals $s_1(t)$ and $s_2(t)$, each of which is limited to frequencies below f_m. We now modulate two carriers of exactly the same frequency with these two signals. However, the carriers are in phase quadrature (90° out of phase) with each other. The sum of the two AM waveforms is then

$$s_{m1}(t) + s_{m2}(t) = s_1(t)\cos 2\pi f_c t + s_2(t)\sin 2\pi f_c t \tag{6.66}$$

Even though the two AM waveforms overlap in both frequency and time, they can be sep-arated by the receiver shown in Fig. 6.48. The signal at the input of the upper lowpass filter is

$$s_a(t) = s_1(t)\cos^2 2\pi f_c t + s_2(t)\sin 2\pi f_c t \cos 2\pi f_c t$$

$$= \frac{1}{2}[s_1(t) + s_1(t)\cos 4\pi f_c t + s_2(t)\sin 4\pi f_c t] \tag{6.67}$$

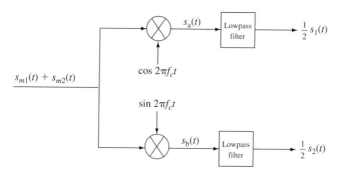

Figure 6.48   QAM receiver.

To derive the rightmost expression of Eq. (6.67), we applied trigonometric identities to the products of sines and cosines in the preceding expression. The output of the filter is then $s_1(t)/2$.

The signal at the input of the lower lowpass filter is

$$s_b(t) = s_1(t)\cos2\pi f_c t \sin2\pi f_c t + s_2(t)\sin^2 2\pi f_c t \qquad (6.68)$$

$$= \frac{1}{2}[s_1(t)\sin4\pi f_c t + s_2(t) - s_2(t)\cos4\pi f_c t]$$

The signal at the output of the filter is then $s_2(t)/2$, and separation is accomplished. Of course, this scheme requires perfect phase control at the receiver to avoid having one signal interfere with the other. Hence, since phase carries some of the information, incoherent detection cannot be used; we must be able to recover the carrier precisely.

Example 6.6

A QAM scheme of the type shown in Fig. 6.48 is used to simultaneously transmit two waveforms in a channel in the frequency range $f_c \pm f_m$. The oscillators in the receiver are in error by $\Delta\theta$. Assuming that the information signals are sinusoids of equal amplitude, find the maximum value of $\Delta\theta$ such that the interference (cross talk) is limited to -20 dB.

Solution: We rederive Eqs. (6.66) and (6.67) for the receiver shown in Fig. 6.49. The output of the upper filter is now

$$\frac{1}{2}[s_1(t)\cos \Delta\theta + s_2(t)\sin \Delta\theta]$$

and the output of the lower filter is

$$\frac{1}{2}[-s_1(t)\sin \Delta\theta - s_2(t)\cos \Delta\theta]$$

The ratio of the amplitude of the undesired term to that of the desired signal is $\sin\Delta\theta/\cos\Delta\theta$. Note that as the phase error approaches zero, this ratio also approaches zero, as we would hope. In order for the interference to be 20 dB below the desired signal, the amplitude ratio must be 0.1. That is,

$$\frac{\sin \Delta\theta}{\cos \Delta\theta} = \tan \Delta\theta = 0.1$$

The phase mismatch must be less than $\tan^{-1}(0.1) = 5.7°$.

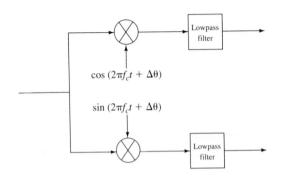

Figure 6.49   QAM receiver with phase mismatch.

AM Stereo

The idea behind AM stereo is to send two independent audio signals within the 10-kHz bandwidth allocated by the FCC to each commercial broadcast station. Additionally, the FCC requires compatibility with monaural receivers. Thus, if the two audio signals represent the left and right channels of a stereo system, a monaural receiver must recover the sum of these two signals.

Quadrature amplitude modulation represents one technique to send two signals simultaneously. If the two signals are designated as $s_L(t)$ and $s_R(t)$, the composite signal can be written as

$$q(t) = s_L(t)\cos 2\pi f_c t + s_R(t)\sin 2\pi f_c t \qquad (6.69)$$

If both $s_L(t)$ and $s_R(t)$ are audio signals with a maximum frequency of 5 kHz, $q(t)$ occupies the band of frequencies between $f_c - 5$ kHz and $f_c + 5$ kHz, a total bandwidth of 10 kHz. The composite signal can then be rewritten as the single sinusoid

$$q(t) = A(t)\cos [2\pi f_c t + \theta(t)] \qquad (6.70\text{a})$$

where

$$A(t) = \sqrt{s_L^2(t) + s_R^2(t)}$$

$$\theta(t) = -\tan^{-1}\left(\frac{s_R(t)}{s_L(t)}\right) \qquad (6.70\text{b})$$

The envelope detector in a monaural AM receiver would produce $A(t)$. This is a distorted version of the sum of the two channels and does not meet the compatibility requirement.

We therefore need to investigate how we might modify the system. The modification does not prove difficult, and we will delay presenting it until we continue with the analysis of this system as if it met the requirements for AM stereo.

Figure 6.50 presents a block diagram of the stereo modulator and demodulator. The dashed block in the demodulator contains a phase lock loop that is used to recover the carrier. In fact, the output of the phase lock loop is $\cos(2\pi f_c t - 45°)$. The various functions of time in the figure are as follows:

$$s_1(t) = \cos(2\pi f_c t - 45°)$$

$$s_2(t) = \cos 2\pi f_c t$$

$$s_3(t) = \sin 2\pi f_c t$$

$$s_4(t) = s_L(t)\cos^2 2\pi f_c t + s_R(t)\sin 2\pi f_c t \cos 2\pi f_c t$$

$$s_5(t) = s_L(t)\sin 2\pi f_c t \cos 2\pi f_c t + s_R(t)\sin^2 2\pi f_c t$$

$$s_6(t) = \frac{1}{2}s_L(t)$$

$$s_7(t) = \frac{1}{2}s_R(t)$$

$$(6.71)$$

Now that we see that the two channels can be separated, we present several specific techniques which assure compatibility. One technique (proposed in two different forms, by Belar and by Magnavox) angle modulates the carrier with one audio signal and amplitude modulates the resulting modulated carrier with the second signal. The angle modulation is narrowband. (We discuss angle modulation in the next chapter.)

Figure 6.51 presents one possible AM/FM configuration: a simplified version of the Belar AM stereo system. The frequency deviation of the FM is $\Delta f = 320$ Hz, which is much less than the maximum audio frequency of 5 kHz. This assures that the resulting modulated signal can be kept within the 10-kHz assigned band. The system meets the compatibility requirement, since an envelope detector will recover the sum signal. The limiter in the receiver removes the amplitude modulation, leaving an FM wave with approximately constant amplitude.

The idealized system of Fig. 6.51 requires some modification to make it practical. Timing is important in a stereo system. If the $L - R$ and $L + R$ channels are not synchronized, the original signals cannot be recovered without distortion. Since the paths traversed by the $L - R$ and $L + R$ signals do not take identical lengths of time, it is necessary to insert a time delay in one of the lines so that the two signals can be properly aligned at the output.

A more complex technique of stereo transmission starts by amplitude modulating a frequency-modulated carrier with the sum signal, $L + R$, as in the preceding system. However, the carrier is not angle modulated with the difference signal. Instead, the angle modulation is performed in such a way that the information in the left channel is carried on the lower sideband and that in the right channel on the upper sideband. To do so requires a relatively complex system involving a 90° phase shift between $L + R$ and $L - R$ signals and also employing automatic gain control (AGC) circuitry. One advantage of this system is

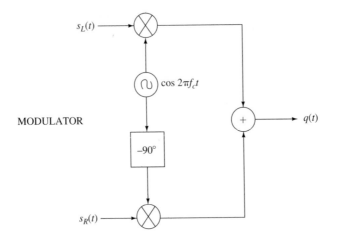

MODULATOR

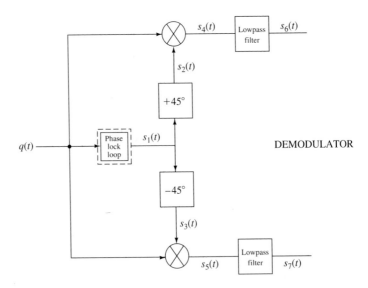

DEMODULATOR

Figure 6.50   Quadrature modulation stereo system.

that it is possible to produce a stereo effect with two monaural receivers merely by tuning one receiver slightly above, and the other slightly below, the carrier.

Yet a third technique is a refinement of the QAM stereo system discussed at the beginning of this section. Recall that the problem with that system was one of compatibility: A monaural receiver would recover the square root of the sum of the squares of the left and right signals, a distorted version of the sum waveform.

TRANSMITTER

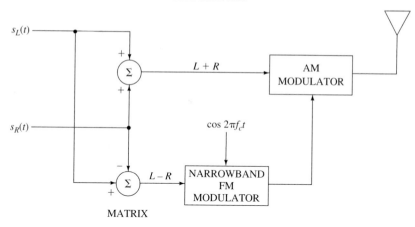

MATRIX

RECEIVER

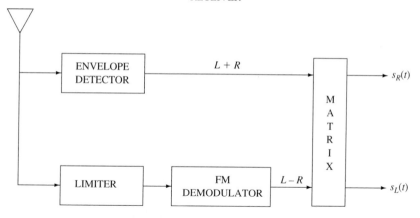

Figure 6.51   AM/FM system for AM stereo.

A *compatible* QAM (C-QAM) system (a simplified version of the Motorola configuration) is shown in Fig. 6.52. The system begins with a QAM signal whose two components are the sum and difference waveforms. The signal at $s_1(t)$ is

$$s_1(t) = \sqrt{[a + s_L(t) + s_R(t)]^2 + [s_L(t) - s_R(t)]^2}$$

$$\times \cos\left(2\pi f_c t - \tan^{-1}\left[\frac{s_L(t) - s_R(t)}{A + s_L(t) + s_R(t)}\right]\right)$$

(6.72)

We now perform an operation to replace the square root by the sum signal. That is, we restore

$$s_2(t) = [A + s_L(t) + s_R(t)] \cos\left(2\pi f_c t - \tan^{-1}\left[\frac{s_L(t) - s_R(t)}{A + s_L(t) + s_R(t)}\right]\right) \quad (6.73)$$

This is done by limiting the amplitude of the QAM waveform [Eq. (6.66)] to a constant and then amplitude modulating with the sum signal.

  The system of Fig. 6.52 produces a signal that meets the compatibility requirement since an envelope detector recovers the sum signal. The problem is that the stereo receiver

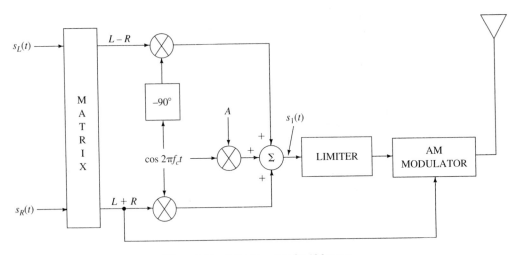

Figure 6.52 C-QAM system for AM stereo.

must perform a complex function. Given the waveform of Eq. (6.73), the receiver must first replace the $L + R$ factor by the square root factor of Eq. (6.72). The information to do this restoration of the QAM waveform is present because the sum signal exists as the envelope and the difference signal can be found by detecting the phase and combining it with the sum signal waveform. The problem is that the receiver must perform these operations with relatively simple circuitry. The receiver uses a phase lock loop to detect the phase and then remodulates the received waveform with that phase.

6.11 PERFORMANCE

In the process of communicating a signal, noise arises in various ways. The information signal $s(t)$ is corrupted by some noise before it even reaches the modulator in the transmitter. This noise is generated by electronic devices in the modulator. Thus, the signal at the output of the modulator is of the form

$$[s(t) + n_1(t)]\cos 2\pi f_c t \quad (6.74)$$

where $n_1(t)$ is the additive noise. Additional noise exists because the carrier sinusoid is not a pure cosine wave, but contains harmonic distortion because of nonlinearities.

The modulated signal is affected by multiplicative noise in the process of being transmitted from transmitter to receiver. This type of noise is due to turbulence in the air and reflection of the signal. The turbulence causes the characteristics of the transmission medium to change with time. The reflections, when recombined with the primary path signal, either add to or subtract from the strength of the signal. With the multiplicative noise taken into account, the signal at the receiving antenna is of the form

$$A[1 + n_2(t)]s_m(t) \tag{6.75}$$

where $n_2(t)$ is the multiplicative factor of n and we assume that the transmitted signal is $s_m(t)$ (i.e., we have neglected the modulator noise).

The AM signal also is affected by additive noise during transmission. This noise is generated by a multitude of sources, including passing automobiles, static electricity, lightning, power transmission lines, and sunspots. If one could listen to it, it would sound like the static one hears over the radio, with occasional crackling sounds added. Assuming that the transmitted signal is $s_m(t)$ (i.e., neglecting the types of noises discussed earlier in this section), the received signal is of the form

$$As_m(t) + n_3(t) \tag{6.76}$$

where $n_3(t)$ is the additive noise.

Additional noise occurs in the receiver. Electronic devices and components are present, thus generating thermal and shot noise. Further, the wires in the receiver act as small antennas, thereby picking up some transmission noise. For purposes of analysis, this receiver noise can be treated as additive and included in $n_3(t)$, as long as it occurs prior to detection in the receiver.

Of the various types of noise introduced, additive transmission noise is generally the most annoying. It normally contains the most power of all. This is not to imply, however, that other types of noise are not critical: Multiplicative transmission noise (turbulence) becomes significant as frequencies approach those of light, so that, to a certain extent, UHF television signals are affected by such noise. Also, very low-frequency multiplicative noise causes fading in microwave systems.

6.11.1 Coherent Detection

Double Sideband

We first examine the case of double-sideband suppressed carrier transmission and synchronous demodulation. We assume that we have been able to match the carrier frequency and phase exactly.

The received waveform at the input to the receiver is

$$r(t) = Ks(t)\cos 2\pi f_c t + n(t) \tag{6.77}$$

where K is a constant that accounts for attenuation during transmission and $n(t)$ is the additive noise. We assume that $n(t)$ is white Gaussian noise with two-sided power spectral density $N_0/2$. That is, the noise power is N_0 watts/Hz.

Let us begin by finding the signal to noise ratio at the *input* to the synchronous de-modulator of Fig. 6.53. The synchronous demodulator contains a bandpass filter that did not appear in our earlier discussions. This filter is known as a *predetection filter*. In a theoretical analysis it is redundant, since any frequencies rejected by it would also be rejected

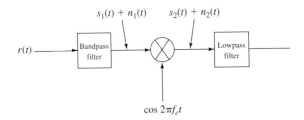

Figure 6.53   Synchronous demodulator.

by the final lowpass filter. It is included both for practical reasons and to simplify the analysis. The electronic device that performs the multiplication could get overloaded and driven into saturation if the input contained too much energy outside of the band of interest. Additionally, the white noise at the input to the system contains (ideally) infinite power. Including it as well would also complicate the analysis.

The signal power at the input to the detector is the average power of $Ks(t)\cos 2\pi f_c t$. This is one half of the average of the square of the cosine amplitude, or

$$\frac{\overline{K^2 s^2(t)}}{2} = \frac{K^2 P_s}{2} \tag{6.78}$$

where P_s is the average power of $s(t)$. The average power of the filtered noise is N_o times the bandwidth of the filter, or $2f_m N_o$. The input signal-to-noise ratio, SNR_i, is then

$$\text{SNR}_i = \frac{K^2 P_s}{4N_o f_m} \tag{6.79}$$

We now wish to derive the signal to noise ratio at the output. The signal at the output of the synchronous demodulator is $Ks(t)/2$, and its average power is $K^2 P_s/4$. To find the noise at the output of the detector, we must turn our attention to the time domain. That is, since the demodulator performs the non-linear operation of multiplication, we can no longer track noise power through the system using the power spectral density.

The bandlimited noise at the detector input can be expanded into its quadrature components thus:

$$n_1(t) = x(t)\cos 2\pi f_c t - y(t)\sin 2\pi f_c t \tag{6.80}$$

The power spectral densities of $x(t)$ and $y(t)$ are calculated as in the example in the Section 4.5. These are shown in Fig. 6.54. The noise at the input to the lowpass filter is

$$n_2(t) = [x(t)\cos 2\pi f_c t - y(t)\sin 2\pi f_c t]\cos 2\pi f_c t$$

$$= \frac{x(t) + x(t)\cos 4\pi f_c t - y(t)\sin 4\pi f_c t}{2} \tag{6.81}$$

where we have used the trigonometric identities

$$\cos^2 2\pi f_c t = \frac{1}{2} + \frac{1}{2} \cos 4\pi f_c t$$

and   (6.82)

$$\sin 2\pi f_c t \cos 2\pi f_c t = \frac{1}{2} \sin 4\pi f_c t$$

to obtain the rightmost expression in Eq. (6.81).

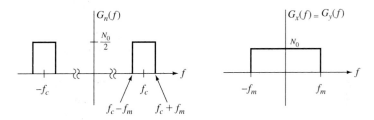

Figure 6.54   Quadrature expansion of noise.

The only term in Eq. (6.81) that passes through the lowpass filter is the first term, so the noise at the output of the detector is $x(t)/2$. The power of this is the power of $x(t)$ divided by 4. The power of $x(t)$ is found by integrating $G_x(f)$, so the total noise power at the detector output is $N_0 f_m/2$. We found the signal power to be $K^2 P_s/4$, so the output signal-to-noise ratio is

$$\text{SNR}_0 = \frac{K^2 P_s}{2 f_m N_0} \tag{6.83}$$

Comparing this to Eq. (6.79), we find that

$$\text{SNR}_0 = 2\text{SNR}_i \tag{6.84}$$

The demodulation process has doubled the signal to noise ratio. Let us try to give an intuitive justification of this result. In double sideband, the two sidebands are related to each other. Thus, knowing one of the sidebands, you can derive the second. The synchronous demodulator essentially realigns the two sidebands to add to each other, so the effect upon the signal is similar to adding a signal to itself. This *coherent* addition doubles the amplitude and therefore multiplies the power by a *factor of four*. On the other hand, the noises in the two sidebands are unrelated (i.e., independent). When these two noise sources are added, it is like adding two independent noise processes. In that case, the mean square values add, and the power *doubles*. The signal power has increased by a factor of four, while the noise power has only doubled. Therefore, the signal to noise ratio doubles.

Single Sideband

We now repeat the foregoing analysis for single sideband. The received signal is of the form

$$r(t) \ = \ \frac{Ks(t)\cos 2\pi f_c t \ \pm \ K\hat{s}(t)\sin 2\pi f_c t}{2} \ + \ n(t) \tag{6.85}$$

where, once again, K is a constant that accounts for attenuation during transmission and $n(t)$ is the noise at the output of the predetection filter of Fig. 6.53.

The signal power at the input to the detector is the average of the square of the signal portion of $r(t)$. This is equal to

$$\frac{\overline{K^2[s^2(t)\cos^2\pi f_c t \ + \ \hat{s}^2(t)\sin^2 2\pi f_c t]}}{4}$$

$$\pm \ \frac{\overline{K^2 2s(t)\hat{s}(t)\cos2\pi f_c t\sin2\pi f_c t}}{4} \tag{6.86}$$

where the bar represents the average of the function over time. The last term is equal to zero, since the average of the cosine multiplied by the sine is zero. (Take the integral over one period.) The squares of the sinusoids have an average value of $\frac{1}{2}$, so the input signal power becomes

$$P \ = \ \frac{K^2(P_s \ + \ P_{\hat{s}})}{8} \tag{6.87}$$

where P_s is the power of $s(t)$ and $P_{\hat{s}}$ is the power of the Hilbert transform $\hat{s}(t)$. The Hilbert transform results from putting $s(t)$ through a filter with $H(f) = -j\text{sgn}(f)$, as shown in Fig. 6.55.

The output power spectral density is given by

$$G_{\hat{s}}(f) \ = \ |H(f)|^2 G_s(f) \tag{6.88}$$

The square of the magnitude of the filter characteristic is unity, so the power of the Hilbert transform is the same as the power of the original signal. Therefore, the input signal power is found from Eq. (6.87) to be $K^2 P_s/4$. The input noise power is N_0 multiplied by the bandwidth of the bandpass filter, or $N_0 f_m$. The signal to noise ratio at the detector input is then

$$\text{SNR}_i \ = \ \frac{K^2 P_s}{4N_0 f_m} \tag{6.89}$$

We now turn our attention to the detector output. The signal at the output is given by $Ks(t)/4$. If we expand the detector input noise in a quadrature expansion, as in Eq. (6.80) for double sideband, the output noise is again given by $x(t)/2$. However, the power spectral density of $x(t)$ is not the same as that given in Fig. 6.54. Figure 6.56 shows the power spectral density of $x(t)$.

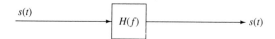

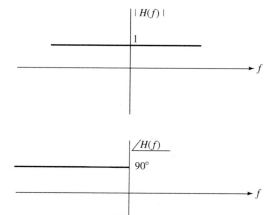

Figure 6.55   The Hilbert transform.

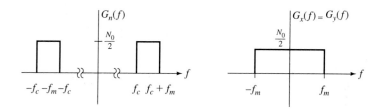

Figure 6.56   Power spectral density of quadrature noise.

The output signal power is $K^2 P_s / 16$, and the output noise power is $N_0 f_m / 4$. The output signal to noise ratio is then

$$\text{SNR}_0 = \frac{K^2 P_s}{4 N_0 f_m} \qquad (6.90)$$

which *is the same* as the signal to noise ratio at the detector input. That is,

$$\text{SNR}_0 = \text{SNR}_i \qquad (6.91)$$

This result distinguishes single sideband from double sideband. Indeed, one would expect to get some benefit from using twice the bandwidth.

Example 6.7

A baseband signal $s(t) = 5\cos 2{,}000\pi t$ is transmitted using DSBSC and demodulated using a synchronous demodulator. Noise with a power of 10^{-4} watt/Hz is added to the signal prior to reception. Find the signal to noise ratio at the output of the receiver.

Solution: We do not have any information about the attenuation due to the channel or due to the antenna patterns. We shall make the unrealistic assumption that the received signal is identical to the transmitted signal. We can then scale the signal to noise ratio according to the square of any attenuation factor.

The signal to noise ratio is found directly from Eq. (6.79):

$$\text{SNR}_0 = \frac{K^2 P_s}{4 N_0 f_m} = \frac{2 \times 25/2}{4 \times 10^{-4} \times 1{,}000} = 62.5 \rightarrow 18 \, dB$$

6.11.2 Incoherent Detection

Transmitted Carrier AM

We now consider the case of transmitted carrier where the carrier amplitude is large enough to permit incoherent demodulation. The demodulator we examine is the envelope detector shown in Fig. 6.57.

The received waveform is of the form

$$r(t) = K[A + s(t)] \cos 2\pi f_c t + n(t) \tag{6.92}$$

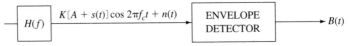

Figure 6.57   Envelope detector for transmitted carrier AM.

The signal power at the input to the detector is $K^2 P_s/2$. Note that A is not considered to be part of the signal, since it carries no information. The power of the noise at the detector input is $2 N_0 f_m$ and the input signal-to-noise ratio is

$$\text{SNR}_i = \frac{K^2 P_s}{4 N_0 f_m} \tag{6.93}$$

Note that this is identical to the input signal-to-noise ratio for double-sideband *synchronous* demodulation [Eq. (6.79)].

To find the output of the detector, we expand the input noise into quadrature form and then combine terms into a single sinusoid. We obtain

$$[KA + Ks(t) + x(t)]\cos 2\pi f_c t - y(t)\sin 2\pi f_c t$$
$$= B(t)\cos[2\pi f_c t + \theta(t)] \tag{6.94}$$

where

$$B(t) = \sqrt{[KA + Ks(t) + x(t)]^2 + y^2(t)} \tag{6.95}$$

and

$$\theta(t) = -\tan^{-1}\left(\frac{y(t)}{KA + Ks(t) + x(t)}\right) \tag{6.96}$$

The output of the envelope detector is $B(t)$. Unfortunately, this contains nonlinear operations, which lead to higher order noise components and cross products between signal and noise. We cannot obtain any general results for the output signal to noise and would have to assume a specific form for $s(t)$ in order to carry the analysis further.

We shall try to gain some insight into the situation by considering the limiting cases of input signal to noise ratio. That is, we will consider the case where the signal is much larger than the noise, and we will also consider the opposite situation.

It is helpful to view Eq. (6.94) in phasor form. Figure 6.58 illustrates this, with the abscissa aligned with the cosine. If $A + s(t)$ is much larger than the noise, $B(t)$ can be approximated by

$$B(t) \approx K[A + s(t)] + x(t) \tag{6.97}$$

This can be seen either from Eq. (6.95) or from Fig. 6.58, where we assume that $y(t) \ll K[A + s(t)] + x(t)$.

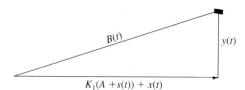

Figure 6.58 Phasor diagram of envelope detector input.

We now need to find the signal to noise ratio at the output of the detector. We can save time by noting that the output of Eq. (6.97) is exactly twice the output of the synchronous demodulator for double-sideband suppressed carrier. Since both the signal and noise ratio are doubled, the signal to noise ratio remains unchanged from that given in Eq. (6.83):

$$\text{SNR}_0 = \frac{K^2 P_s}{2 f_m N_0} \tag{6.98}$$

The ratio of output SNR to input SNR is therefore the same as for double-sideband coherent demodulation. That is,

$$\text{SNR}_0 = 2\text{SNR}_i \tag{6.99}$$

This identical result can be deceiving. We must bear in mind that the price we are paying is lower efficiency. In comparing the various systems, it is important to do so under equivalent conditions. We can get an approximate result by assuming that $s(t)$ is a pure sinusoid, $\cos 2\pi f_m t$. Since this sinusoid has a maximum negative excursion of -1, the minimum value of A is 1. Using this value, we find that $A + s(t)$ is given by

$$1 + \cos 2\pi f_m t \tag{6.100}$$

The signal power at the output is $\frac{1}{2}$ watt, and the power of the dc term in the output is 1 watt. Thus, the true signal to noise ratio at the output is one-third of that found in Eq.

(6.98). In comparing systems, we therefore often use the following expression for incoherent detection:

$$SNR_0 = \frac{2SNR_i}{3} \tag{6.101}$$

We now consider the other extreme in incoherent detection, that is, a very low signal to noise ratio. To analyze this situation, we will redraw Fig. 6.58, but this time referenced to the noise signal. That is, we add the signal vector to the larger noise vector. The realigned diagram is shown in Fig. 6.59. Note that the angle between the signal and noise is

$$\theta(t) = \tan^{-1}\left(\frac{y(t)}{x(t)}\right) \tag{6.102}$$

The resultant vector is approximately given by

$$\sqrt{x^2(t) + y^2(t)} + K[A + s(t)]\cos\theta(t) \tag{6.103}$$

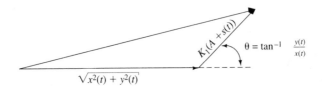

Figure 6.59   Figure 6.58 redrawn for low SNR.

The only place where the signal appears is in the last term, and it is multiplied by a random noise term, $\cos\theta(t)$. Thus, we have *both additive and multiplicative noise*. It can be shown that $\theta(t)$ is uniformly distributed between 0° and 360°. It should therefore not be surprising to observe that the signal *cannot* be recovered from the envelope detector output.

As the signal to noise ratio decreases from a high value, a threshold is reached. For signal to noise ratios above this threshold, the output signal to noise ratio is linearly related to the detector input signal to noise ratio. For signal to noise ratios below this threshold, the dependence approaches a quadratic relationship. That is, for each decrease in input signal to noise ratio by a factor of 2, the output signal to noise ratio decreases approximately by a factor of 4.

6.12 TELEVISION

Public television had its beginnings in England in 1927. In the United States, it started three years later, in 1930. These early forms used *mechanical* scanning of the picture to be transmitted. That is, a picture was changed into an electrical signal by scanning the entire image along a spiral starting at the center. The scanning was accomplished by means of a rapidly rotating wheel with holes cut in it. As the wheel rotated, light from various parts of the total picture passed through the holes.

During this early period, broadcasts did not follow any regular schedule. Such regular scheduling did not begin until 1939, during the opening of the New York World's Fair.

The concepts of television and picture transmission spread into many exciting areas; facsimile transmission, satellite broadcasts, video telephone, video-text, and cable TV represent only a few examples. A cable TV revolution is occurring, with two-way communication links becoming common. We can anticipate that TV will eventually replace the newspaper, supermarket, baby-sitter, theater, and perhaps (heaven forbid!) the university campus. The theory to be presented here, although geared toward broadcast TV, is applicable to most forms of picture transmission.

A Picture Is Worth A Thousand Words?

Nonsense! A picture is equivalent to far more than a thousand words. While we could rigorously define the information content of a picture using concepts from the science of information theory, we will not do that here. Instead, we will subdivide the picture into small components similar to words.

Suppose we divide a picture into squares, where each square is a certain shade. The number of squares in any given area determines the *resolution*. For example, Fig. 6.60(a) shows a picture of the letter A, where 81 squares have been used to define the picture. In Fig. 6.60(b), the same letter is shown with the number of squares increased to 342. The result is improved resolution.

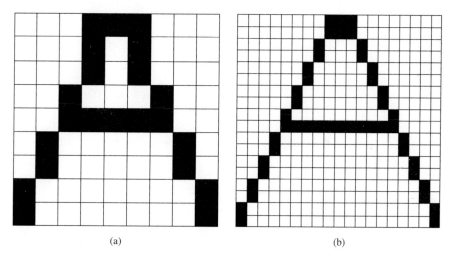

(a)   (b)

Figure 6.60   Definition of picture elements.

The number of squares used in U.S. television is set by the FCC at 211,000 picture elements (*pixels*). This is divided into 426 elements in each horizontal line and 495 visible horizontal lines in each picture. The job of television transmitters is then to step through the 211,000 picture elements and send an intensity value for each one. The receiver interprets these transmissions and reconstructs the picture from the 211,000 intensity levels. This information must be updated rapidly to simulate motion. (We confine our attention here to monochrome (black-and-white) television.)

A conventional TV receiver is not much different from a conventional analog laboratory oscilloscope. A beam of electrons is shot toward a screen and bent by deflection plates. When a negative charge is placed on a plate, the electron beam is repelled. In the oscilloscope, we apply a sawtooth waveform to the horizontal deflection plates to sweep the beam from left to right (creating a time axis) and then more rapidly back to the left. This traces a line. In TV receivers, we add a second dimension to the sweep. While the beam is rapidly sweeping from left to right on the screen, it is less rapidly sweeping from top to bottom. The net result is a series of almost horizontal lines on the screen, as sketched in Fig. 6.61. This is known as the TV *raster*. We are describing traditional analog scanning TV. Digital television has the option of controlling the choice of picture elements using digital counting circuitry. Picture elements are not necessarily bombarded with an aimed electron beam, but could be implemented as individual LED or LCD elements arranged in a matrix and digitally scanned. Nonetheless, conventional TV remains the norm for the present. Even in the future, it will represent an educational historical study of how a significant problem was solved.

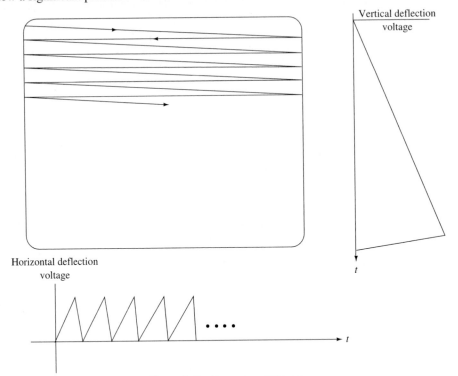

Figure 6.61   Generation of TV raster.

The screen has 495 horizontal lines in order to comply with the FCC regulation. After scanning the entire screen from top to bottom, the beam returns to the top of the screen, taking the equivalent time of an additional 30 lines to do so. We can therefore think of the picture as having 525 lines, 30 of which occur during the *vertical retrace* time.

We now assign timing to this process. The human eye requires a certain picture rate (the number of times the entire screen is traced each second) to avoid seeing flicker, as in early motion pictures. The minimum number is somewhere near 40 per second. United States TV uses 60 frames per second. (Color TV actually uses a number slightly less, approximately 59.94 Hz.) This matches the frequency of household current and is chosen so as to minimize the effects of the video equivalent of 60-Hz hum. That is, if the 60-Hz power signal is not completely filtered out of the video, it causes a slight gradation of brightness over the height of the picture. If this gradation is stationary, the eye probably will not notice it. However, with a frequency mismatch, the gradation will exhibit a migration in the vertical direction (a rolling), and the likelihood of detecting it increases. We emphasize here that the numbers being presented are the U.S. standard (this applies to the United States, Canada, the Netherlands, Brazil, Colombia, Cuba, Japan, Mexico, Peru, Surinam, and Venezuela). Many of the standards in use in other parts of the world include 625 lines and a frame frequency of 50 Hz.

At 525 lines/frame and 60 frames/sec, the product of these is 31,500 lines/sec. The reciprocal of this is the time per line, 31.75 μsec/line. Of this 5.1 μsec are used for horizontal retrace, leaving 26.65 μsec for the visible part of each horizontal line. Dividing this by the 426 elements in a line yields the time per element of 0.0625 μsec/element, or 16 million elements/sec. The system therefore must be capable of transmitting 16 million independent shades (black, white, gray, and so on) per second. In the worst case, we may wish to display a perfect checkerboard design of alternating black and white squares. The system would then jump from the darkest to lightest shades and back again 16 million times a second.

If we now think of this light intensity information as a signal, we see from Fig. 6.62 that it has a fundamental frequency of 8 MHz. The square wave actually has a maximum frequency of infinity, but if we round it to a sinusoid, the human eye would still see the checkerboard design. Thus, no matter what modulation scheme we use to transmit this pattern, at least 8 MHz of bandwidth is required.

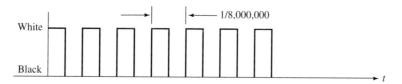

Figure 6.62   Intensity signal for checkerboard pattern.

Here the law steps in again, and the FCC mandated a maximum video signal bandwidth of 4.2 MHz. Alas, how can these seemingly contradictory specifications (60 frames/sec to avoid flicker, 211,000 picture elements for proper resolution, and a 4.2-MHz maximum bandwidth) be met?

Engineers had observed many decades of the development of motion pictures in which a similar predicament occurred. In standard motion pictures, only 24 different pictures are shown each second. But 24 flashes/sec on the screen would appear to flicker considerably. Contemporary motion picture projectors flash each image twice. (The film moves into position and the shutter opens and closes twice before the film moves again.) Thus, the frame rate is 48/sec, although the rate of presenting new pictures is 24/sec.

Television's founders decided to play a similar trick. They cut the signal frequency in half by cutting the number of lines per second in half, from 31,500 to 15,750 lines/sec. However, it was necessary to fill the entire screen each $\frac{1}{60}$ sec to avoid flicker. The technique for cutting the line frequency in half without changing the frame frequency is known as *interlaced scanning*. During the first $\frac{1}{60}$ sec, the odd-numbered lines are traced (ending with $\frac{1}{2}$ line). The beam then returns to the top center of the screen to trace the even-numbered lines in the next $\frac{1}{6}$ sec. Thus, while it takes $\frac{1}{30}$ sec to send the frame consisting of all 211,000 picture elements, the screen is scanned twice (each scan is called a *field*) during this period. The eye fills in the missing rows and detects no flicker. There is some loss of resolution on fast-moving objects, but this was deemed appropriate for conventional TV.

Signal Design and Transmission

We now translate the foregoing information into an electrical signal format. If we plot light intensity as a function of time, a staircase function results. Figure 6.63 shows an example of the letter T in dark black, followed by a period in light gray. The associated signal is also shown; for simplicity, interlaced scanning has not been included, and the number of lines has been drastically reduced.

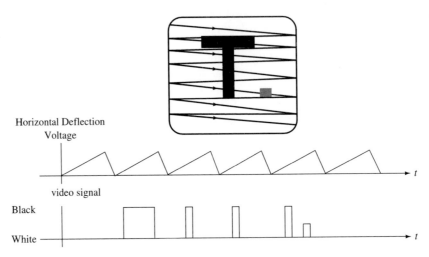

Figure 6.63   Video signal for particular message.

In broadcast TV, the information (video) signal would be a similar staircase function with minimum step width 0.125 μsec. In the actual video signal, the voltage corresponds to the light intensity, and the receiver uses this voltage to control the electron gun. The higher the voltage applied to a grid placed between the gun and screen, the fewer is the number of electrons that hit the screen and the darker is the spot. As an additional modification, the staircase function is smoothed to reduce the bandwidth. The eye cannot tell the difference between a smooth or rapid transition in the signal during 0.125 μsec. After all, this corresponds to only $\frac{1}{426}$ of the width of the TV screen.

There is an additional aspect to this electrical video signal known as *blanking*: While the beam on the cathode ray tube is retracing, it is desirable that the electron stream be turned off so that the retrace is not seen as a line on the screen.

Taking all this into account, the video signal corresponding to the picture shown in Fig. 6.63 is redrawn as Fig. 6.64.

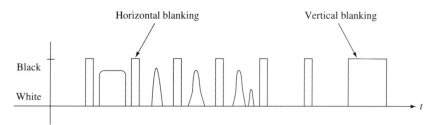

Figure 6.64   Video signal with addition of blanking.

Synchronization

The transmitter rapidly traces from left to right and from top to bottom many times each second. It sends a record of light intensity as a function of time. The receiver must be sure that it is placing the transmitted intensity element in the same location on the screen as intended in transmission. If the beam in the receiver does not start a scan at the same instant that the received waveform does, the picture will appear split at best and totally scrambled at worst. A method is thus required for synchronizing the two sweeping operations. This is done by means of synchronization pulses added to the video signal. The pulses are added during the blanking intervals, thereby not affecting what is seen on the screen. Figure 6.65 shows the signal of Fig. 6.64 modified with the addition of synchronization pulses. Two types of synchronizing pulses are shown in Fig. 6.65. The narrow pulses are horizontal synchronizing pulses, and the wide pulses are vertical synchronizing pulses.

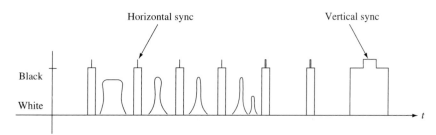

Figure 6.65   Video signal with addition of synchronization.

The receiver separates these pulses from the remaining signal through the use of a threshold circuit. The horizontal and vertical pulses are then distinguished by means of a single *RC* integrator circuit. The integral of the wider vertical sync pulses is larger than that of the narrower horizontal sync pulses. The separate pulses are then used to synchronize (trigger) the horizontal and vertical oscillators.

Modulation Techniques

The video signal has a maximum frequency of about 4 MHz. The FCC allocates 6 MHz of bandwidth to each television channel, and this space must contain both the video and audio sections of the transmitted signal. Obviously, the use of double-sideband AM must be rejected, since this would require over 8 MHz of bandwidth for each channel.

Single-sideband transmission is also rejected. Its generation requires a very sharp filtering of the double-sideband signal to remove one of the sidebands. However, it is difficult to control the phase characteristic of a filter that has a very sharp amplitude characteristic. In designing practical filters, we can approach either the phase or amplitude characteristics as closely as desired, but to achieve both simultaneously is extremely difficult. In audio applications, phase deviations are not very serious. They represent varying delays of the frequency components of the message, and the human ear is not sensitive to such variations. In a video signal, these varying delays would be manifest as shifts in position on the screen. These are commonly referred to as *ghost images* and are highly undesirable.

The video portion of the TV signal is sent using vestigial sideband. A carrier is added that is large enough to allow the use of an envelope detector for demodulation. The entire upper sideband and a portion of the lower sideband are sent.

Figure 6.66 shows the frequency composition of a TV signal. Note that the audio and video are frequency multiplexed, and their carriers are separated by 4.5 MHz. The audio is sent using FM, which is described in Chapter 7.

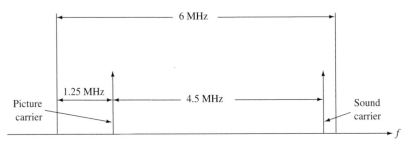

Figure 6.66   Frequency composition of TV signal.

Functional Block Diagram of TV Receiver

We are now in a position to examine the overall block diagram of a black-and-white TV receiver, which is shown in Fig. 6.67. The composite (video plus audio) signal is received by the antenna and is amplified by an RF amplifier. It then enters the tuner, which includes a mixer and an IF filter, just as in the case of the superheterodyne radio receiver. The frequency band allocated to commercial TV in the United States is shown in Table 6.1.

Most large cities with many active TV stations use channels 2, 4, 5, 7, 9, 11, and 13. Examination of the table of frequency allocations shows that this choice leaves a frequency separation between adjacent active stations, thus easing the requirements for design of the IF filters.

The IF frequency used for TV is 40 MHz. After IF amplification and filtering, the signal enters the video detector, which is simply an envelope detector. During this stage of processing, the sound signal is separated. Since the sound carrier is 4.5 MHz above the picture carrier, a filter (called a *sound trap*) is used to separate the sound signal from the video signal. The sound is sent by FM, so we defer discussion of the demodulating and processing of it until Chapter 7.

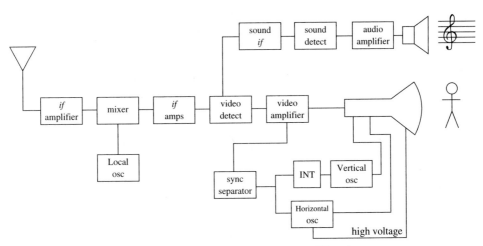

Figure 6.67   Block diagram of monochrome TV receiver.

The synchronizing pulses are separated from the video signal by means of a threshold circuit (clipper) known as the *sync separator*. The vertical sync pulses are then distinguished from the horizontal by an integrator. Both sets of pulses are used to trigger the corresponding sweep oscillators. As an added bonus, the retrace portion of the horizontal deflection voltage (with a very large slope due to a 5.1-μsec retrace time) is used to generate the very high voltage (over 20 kV) required on the CRT anode to pull the electrons in a straight line toward the face of the tube. This high voltage is generated by differentiating the retrace ramp using what is known as a *flyback transformer* (an inductor, so voltage is proportional to the derivative of current).

TABLE 6.1   FREQUENCY ALLOCATIONS FOR BROADCAST TV

Channel	Frequency Range (MHz)	Comments
2	54–60	
3	60–66	
4	66–72	
5	76–82	Note gap between 4 and 5
6	82–88	Between 6 and 7 is FM radio, aircraft, government, railroad, and police
7	174–180	
8	180–186	
9	186–192	
10	192–198	
11	198–204	
12	204–210	
13	210–216	
14–83	470–890	

Color Television

Color TV takes advantage of the theory behind the color wheel. All colors, including black and white, can be formed as combinations of the three primary colors. It is interesting to note that the theory of resolving colors into components dates back to Sir Isaac Newton (a scientist), Johann Wolfgang von Goethe (a poet), Thomas Young (a physicist), and Hermann von Helmholtz (a physicist). The term *primary colors* can be interpreted in a couple of ways. The *primary colors of the spectrum* are red, green, and blue, while the *primary colors of paints* are red, yellow, and blue. Television chose the primary colors of the spectrum, since phosphors that glow these colors when bombarded with an electron beam are readily available.

The color TV receiver can be thought of as three separate TVs, one generating red, one generating blue, and one generating green. By mixing these in varying strengths, any color can be formed.

Traditional color CRTs actually contain three separate electron guns, one for each color. The transmitted signal must therefore be capable of generating three separate video signals to control each of the guns.

But certainly, you are asking where the two additional signals can be squeezed. Monochrome transmission already uses all of the bandwidth allowed for TV. We need a system to transmit all three video signals without increasing the 6-MHz bandwidth, yet at the same time allowing compatible transmission. That is, a black-and-white TV receiver should receive the correct composite image, not just one of the colors.

The answer to this dilemma lies in a frequency analysis of the video signal. If we examine the Fourier transform of a black-and-white video signal, we find that it resembles a train of impulses. An actual signal for one frame contains 525 horizontal traces. Since most pictures contain some form of *vertical continuity* (that is, the content of one horizontal line closely resembles that of the next horizontal line), the video signal is almost periodic with a fundamental frequency of 15,750 Hz (the number of lines per second).

If the video signal were exactly periodic, its Fourier transform would be a train of impulses at multiples of the fundamental frequency. Since it is almost periodic, the Fourier transform consists of pulses (not impulses) centered around multiples of the line frequency. The more periodic the time signal, the sharper are the pulses in frequency.

Since the Fourier transform approaches zero between multiples of the line frequency, additional information can be placed in these spaces through the use of a form of frequency multiplexing. The additional information needed for color transmission modulates a carrier that is midway between two multiples of the line frequency. This is assured by using a carrier with a frequency that is an odd multiple of half of the line frequency. The figure used is 3.579545 MHz, which is the 455th harmonic of half of the line frequency. The two additional signals are combined and transmitted using quadrature AM.

To make the signal compatible with monochrome receivers, the three signals are *not* the three primary colors. Instead, the basic system transmits brightness (known as *luminance*), position in the color spectrum (*hue*), and how close the color is to a pure single frequency (*saturation*). The hue and saturation are combined with the luminance in a matrix operation that has the goal of making it easy to reproduce the three separate colors at the receiver and, at the same time, matching characteristics of our human color perception. The system does not provide perfect separation, and if the picture lacks vertical continuity over a span of time, the colors will interact. The combination scheme used in the United States is known as the *National Television Systems Committee* (NTSC) *standard*. This standard is used by 32 countries in the world. Some other parts of the world combine the colors according to different standards, notably the *phase-alternation line* (PAL), which is used in 63 countries, and *sequential color and memory* (SECAM), used in 42 countries. Sixteen countries have only black-and-white television, and 9 have no television at all (1988 figures). The bandwidth of the luminance signal is limited to 4.2 MHz, while the bandwidths of the other two signals are 1.5 MHz and 500 kHz. Again, this is a compromise and results in some loss of performance. For example, if you ever watched a movie on TV that used red for the titles, you probably noticed a decrease in resolution. The letters became fuzzy. The sharp corners of the letters represent frequency components above the cutoff of the signal sending this information.

High-definition TV

At the time that standards were developed for television, few people dreamed of its evolution into a type of universal communication terminal. While these standards are acceptable for entertainment video, they are not sufficient for many evolving applications, such as videotext. For those, we require a high-resolution standard. *High-definition TV* (HDTV) is a term applied to a broad class of new systems. These systems have received worldwide attention. Indeed, the U.S. government put seed money into this consumer-related development in the hope of generating global competition and to be a catalyst in spurring the growth of new systems.

Of course, if we want to start from scratch, we can set the bandwidth of each channel to a number greater than 6 MHz, thereby achieving higher resolution. In fact, the Japan Broadcasting Corporation has done just that by assigning 10 MHz per channel and using compression techniques to achieve further improvement. The Japanese system permits 1,125 lines per frame, with 30 frames per second as 60 fields per second. (We will discuss compression techniques in Chapter 9.)

In the United States, the FCC has ruled that any new HDTV system must permit continuation of service to contemporary NTSC receivers. This significant constraint applies to terrestrial broadcasting (as opposed to videodisc, videotape, and cable television).

Developments in digital signal processing and high-speed RAMs have opened up interesting possibilities for increasing resolution while staying within the 6-MHz allocation per channel. For example, the number of lines could be increased if the frame rate were decreased. If the parameters for each pixel are stored, processing can be performed between frames. In the simplest example, the system could interpolate values between frames and create estimates of intermediate values. Indeed, more sophisticated compression algorithms can now be implemented, including variations of the powerful techniques currently applied to facsimile.

Other HDTV systems relax the 6-MHz constraint. For example, a VHF channel could be combined with a UHF channel, thus providing a total bandwidth of 12 MHz. Other systems involve transmission from direct-broadcast satellites. The one common thread among the various proposals is that the number of lines per frame is generally twice the current rate.

As of this writing, none of the competing systems has received anything near uniform support. The situation is such that the technical aspects of the problem are merged with social and economic issues. We can only hope that we do not end up with a multitude of incompatible systems.

PROBLEMS

6.2.1   Given an information signal $r(t)$, with

$$R(f) = A(f)e^{j\theta(f)}$$

(i.e., $R(f)$ is complex), find the Fourier transform of

$$r(t)\cos 2\pi f_c t$$

Also, find the Fourier transform of

$$r(t)\cos\left(\frac{2\pi f_c t + \pi}{4}\right)$$

6.3.1   The signal shown in Fig. P6.3.1 amplitude modulates a carrier of frequency 10^6 Hz.
   (a) If the modulation is DSBSC, sketch the modulated waveform.
   (b) The modulated wave of part (a) forms the input to an envelope detector. Sketch the output of the detector.
   (c) A carrier term is now added to the DSBSC waveform. What is the minimum amplitude of the carrier such that envelope detection can be used?
   (d) For the modulated signal of part (c), sketch the output of an envelope detector.
   (e) Draw a block diagram of a synchronous detector that could be used to recover $s(t)$ from the modulated waveform of part (a).
   (f) Sketch the output of the synchronous demodulator if the waveform of part (c) forms the input.

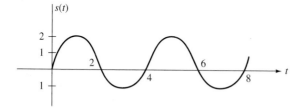

Fig. P6.3.1

6.4.1   You are given the voltage signals $s(t)$ and $\cos 2\pi f_c t$, and you wish to produce the AM wave. Discuss two practical methods of generating this AM waveform.

6.4.2   A system is as shown in Fig. P6.4.2. Note that the system resembles a gated modulator, except that the gating function goes between $+1$ and -1 instead of between $+1$ and 0, and the bandpass filter has been replaced with a lowpass filter.
   Can this system still produce an AM waveform? If your answer is yes, find the minimum and maximum values of f_{LPF} in order for the system to act as a modulator. If your answer is no, show all work that made you arrive at this conclusion.

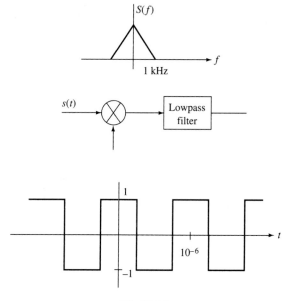

Fig. P6.4.2

6.5.1   The signal

$$s(t) = \frac{2\sin 2\pi t}{t}$$

modulates a carrier of frequency 100 Hz. A signal of the form

$$n(t) = \frac{\sin 199 \, \pi t}{t}$$

adds to the AM waveform, and the sum forms the input to a synchronous demodulator. Find the output of the demodulator.

6.5.2   The waveform $v_{in}(t)$ shown in Fig. P6.5.2 forms the input to an envelope detector. Sketch the output waveform.

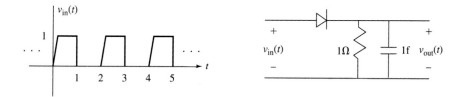

Fig. P6.5.2

6.5.3   You are given the system shown in Fig. P6.5.3. An AMTC waveform forms the input. $p(t)$ is the periodic function, and $S(f)$ is as sketched. $X(f)$, $Y(f)$, and $Z(f)$ are the Fourier transforms of $x(t)$, $y(t)$, and $z(t)$, respectively. Assume that $f_c \gg f_m$. Assume also that $s(t)$ never goes negative. (The impulse at the origin in $S(f)$ indicates a dc value.)

(a) Sketch $|X(f)|$.
(b) Sketch $|Y(f)|$.
(c) Sketch $|Z(f)|$.

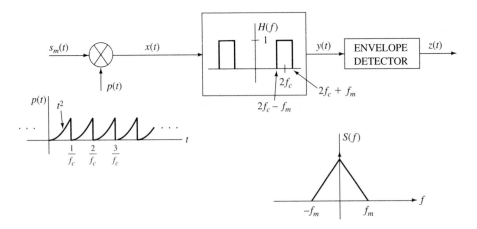

Fig. P6.5.3

6.5.4 The input to an envelope detector is

$$r(t)\cos 2\pi f_c t$$

where $r(t)$ is always greater than zero.

(a) What is the output of the envelope detector?
(b) What is the average power of the input in terms of the average power of $r(t)$?
(c) What is the average power of the output?
(d) Discuss any apparent discrepancies.

6.5.5 Replace the local carrier in a synchronous demodulator with a square wave at a fundamental frequency of f_c. Will the system still operate as a demodulator? Will the same be true if periodic signals other than the square wave are substituted for the oscillator?

6.5.6 An AMTC signal $s_m(t)$ is given by

$$s_m(t) = [A + s(t)]\cos(2\pi f_c t + \theta)$$

This AM waveform is applied to both of the systems shown in Fig. P6.5.6. The maximum frequency of $s(t)$ is f_m, which is also the cutoff frequency of the lowpass filters. Show that the two systems yield the same output. Also, comment on whether the two lowpass filters in the system of part (b) can be replaced by a single filter either following or preceding the square root operation.

6.5.7 A synchronous demodulator is used to detect an amplitude-modulated suppressed carrier double-sideband waveform. In designing the detector, the frequency is matched perfectly, but the phase differs from that of the received carrier by $\Delta\theta$, as shown in Fig. P6.5.7. The phase difference is random and Gaussian distributed with a mean of zero and variance of σ^2. When the phases are matched, the output is $s(t)/2$. What is the maximum variance of the phase error such that the output amplitude is at least 50% of this optimum value 99% of the time?

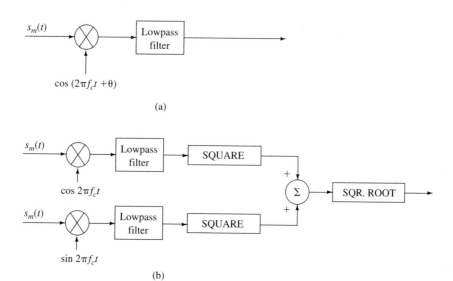

(a)

(b)

Fig. P6.5.6

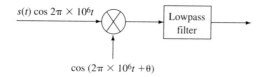

Fig. P6.5.7

6.5.8   Show that the rectifier detector of Fig. 6.32 demodulates transmitted carrier AM when the rectifier is half wave.

6.5.9   The system illustrated in Fig. P6.5.9 performs a simple scrambling operation: reversing frequencies. (That is, dc switches to the highest frequency, while the highest frequency switches to dc; frequencies near f_m flip to a location near zero.)

(a) Sketch the Fourier transform of the output signal $Y(f)$.

(b) Find the output time signal if

$$r(t) = 5 \cos 100 \, \pi t + 10 \cos 200 \, \pi t + 3 \cos 1{,}000 \, \pi t$$

(c) Design a system that can recover the original signal $r(t)$ from the scrambled output of the system.

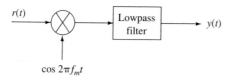

Fig. P6.5.9

6.5.10 Consider the carrier injection system shown in Fig. 6.10. The information signal is

$$r(t) = \cos 4\pi t + \cos 20\pi t + \cos 40\pi t + \cos 60\pi t$$

Suppose that the bandpass filter in the receiver is tuned to the carrier frequency with a bandwidth that passes 3 Hz on either side of this frequency. Thus, the reconstructed carrier is perturbed by a 2-Hz signal. Find the output of the system. Now define the error as the difference between this output and an appropriately scaled version of $r(t)$. Find the percentage of error.

6.5.11 A quadrature receiver is shown in Fig. P6.5.11. The local oscillators have shifted by 30°, as indicated. Find the output of the receiver if an AMTC signal forms in the input as shown.

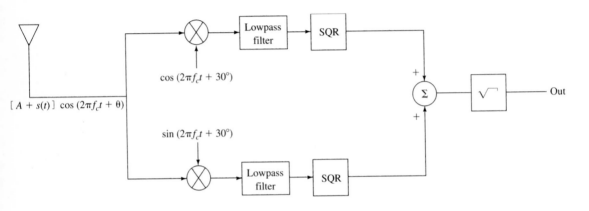

Fig. P6.5.11

6.5.12 You have a friend who is a guitar player. Your friend asks you to design a system that will help in the tuning of a guitar. Specifically, you are asked to design a system that accepts two inputs,

$$\cos(2\pi f_0 t); \quad \cos[2\pi(f_0 + \Delta f)t + \theta]$$

and produces an output which is a dc signal that is proportional to the frequency difference Δf. The phase angle θ is unknown. Draw a block diagram of your system.

6.5.13 You wish to design a *Doppler radar system*. A sinusoidal generator continuously generates the signal,

$$s(t) = A\cos 2\pi f_c t$$

This signal is transmitted to a speeding car, and the reflected signal is of the form

$$r(t) = B\cos[2\pi(f_c + \Delta f)t]$$

The situation is shown in Fig. P6.5.13. The frequency difference is

$$\Delta f = 10s$$

where s is the speed of the car in miles per hour.

You wish to display the speed of the car on a voltmeter that reads from 0 to 100 volts, and you wish the voltage reading to be the same as the speed. That is, if the car is traveling at 50 miles per hour, you want the meter to read 50 V. You have a laboratory full of equipment, including filters, multipliers, and any other type of device you need. Design the system.

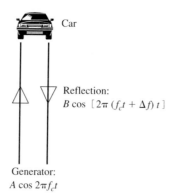

Car

Reflection:
$B \cos [2\pi (f_c t + \Delta f) t]$

Generator:
$A \cos 2\pi f_c t$

Fig. P6.5.13

6.6.1 Design a superheterodyne receiver that operates on AM signals in the band between 1.7 and 2 MHz. As a minimum, your design must include selection of an IF and design of the heterodyne circuitry.

6.6.2 Figure P6.6.2 shows a *dual conversion receiver*. Assume that the first IF frequency is 30 MHz and the second is 10 MHz. Assume further that the receiver is designed to demodulate a band of channels between 135 and 136 MHz, each of which is 100 kHz in bandwidth.
(a) Suggest the range of frequencies for the local oscillators.
(b) Determine all possible image station frequencies.
(c) How would you remove the unwanted image stations?

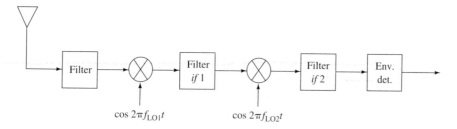

Fig. P6.6.2

6.6.3 Twenty-five radio stations are broadcasting in the band between 3 MHz and 3.5 MHz. You wish to modify an AM broadcast receiver to receive the broadcasts. Each audio signal has a maximum frequency $f_m = 10$ kHz. Describe, in detail, the changes you would have to make to the standard broadcast superheterodyne receiver in order to receive the broadcast.

6.7.1 Determine the envelope of the waveform

$$s(t) = \cos 10\pi t + 17\cos 30\pi t \cos 1000\pi t$$

6.8.1 Starting with the transform of a lower sideband single-sideband wave

$$S_{lsb}(f) = \frac{1}{2}H(f)[S(f - f_c) + S(f + f_c)]$$

where $H(f)$ is as shown in Fig. P6.8.1, prove that synchronous demodulation can be used to recover $s(t)$ from $s_{lsb}(t)$.

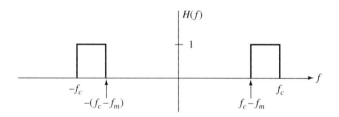

Fig. P6.8.1

6.8.2 Show that the system of Fig. P6.8.2 produces a single-sideband (lower sideband) waveform. What changes would you make to the system to produce the upper sideband waveform?

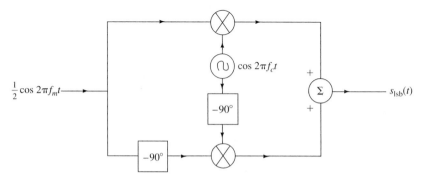

Fig. 6.8.2

6.9.1 A vestigial sideband (VSB) signal is formed by amplitude modulating a carrier with the signal $s(t)$ shown in Fig. P6.9.1. A pure carrier term of amplitude 2 is added to the result. The double-sideband signal is then filtered with the system function shown. Assume that f_m is large enough to pass all significant harmonics of the upper sideband. An envelope detector is used for demodulation. Find the minimum value of f_v such that the error is less than 10%. Define the error as the difference between the demodulated signal and $s(t)$. The percent error is the ratio of error power to signal power.

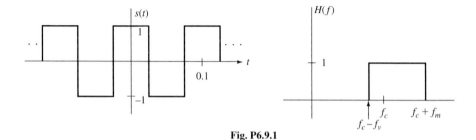

Fig. P6.9.1

6.11.1 An information signal $s(t) = 5\cos 1,000\pi t$ is transmitted using single-sideband suppressed carrier and is demodulated using a synchronous demodulator. Noise with power spectral density $G_n(f) = N_o/2 = 10^{-4}$ adds to the signal during transmission. Find the SNR at the output of the receiver, in dB.

6.11.2 A signal

$$s(t) = 20 \cos 1,000 \, \pi t + 10 \cos 2,000 \, \pi t$$

is transmitted using single sideband. Noise of power spectral density $G_n(f) = N_0/2 = 10^{-3}$ adds during transmission.
(a) Sketch a block diagram of the required receiver.
(b) Find the SNR at the output of the receiver.
(c) Suppose a bandpass filter with passband between 400 and 1,100 Hz is added to the output of the receiver. Find the improvement in SNR of this filter.

6.11.3 A signal

$$s(t) = 4 \sin (200 \, \pi t + 10°)$$

is transmitted using double-sideband transmitted carrier, double-sideband suppressed carrier, and single sideband. Noise of power spectral density $G_n(f) = N_0/2 = 10^{-2}$ adds during transmission. Find the SNR at the output of the appropriate receiver for each case. (Assume that an envelope detector is used for double-sideband transmitted carrier.)

6.11.4 A signal $s(t)$ is transmitted using single-sideband AM. The power spectral density of $s(t)$ is as shown in Fig. P6.11.4. White noise of spectral density $N_0/2$ adds during transmission. Find the SNR at the output of a synchronous demodulator.

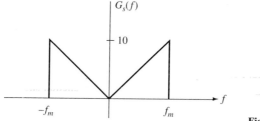

Fig. P6.11.4

6.11.5 A DSBSC waveform with $f_m = 5$ kHz and $f_c = 1$ MHz is transmitted. Nonwhite noise with power spectral density as shown in Fig. P6.11.5 adds to the signal prior to detection with a synchronous demodulator. Find the SNR at the output of the detector, assuming that the power of the signal, $s(t)$, is 1 watt.

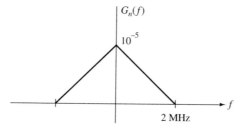

Fig. P6.11.5

6.12.1 Discuss the trade-off decision required to increase the vertical resolution of commercial TV by 10%.

7

Angle Modulation

7.0 PREVIEW

What We Will Cover and Why You Should Care

Chapter 6 introduced a technique for sending a signal by superimposing the waveform on a *carrier*. In that analysis, we varied the carrier amplitude using the information signal. This technique yielded desirable properties that allowed transmission over the channel. It also permitted sharing of the channel among multiple sources.

Angle modulation is a second method of superimposing a signal on a carrier. Just as with AM, we start with an unmodulated carrier waveform

$$s_c(t) = A\cos(2\pi f_c t + \theta)$$

If f_c is varied in accordance with the information we wish to transmit, the carrier is said to be *frequency modulated*. If θ is varied, the carrier is *phase modulated*. Both of these techniques are known as *angle modulation*. They have potential advantages over AM and are used in a wide variety of applications, from broadcast radio to the sound portion of broadcast television.

Necessary Background

To understand the various properties of angle-modulated waveforms, you need a working knowledge of Fourier transform theory. We will also use the Bessel function, but will present the theory behind it as it is needed. Finally, the actual systems for modulating and demodulating the waveforms require a knowledge of linear systems. It is also helpful if you have some familiarity with electronics, including oscillator and phase-locked loops.

7.1 INSTANTANEOUS FREQUENCY

When f_c is varied with time, the carrier, $s_c(t)$, is no longer a sinusoid, and the definition of frequency that we have used all of our technical lives must be modified accordingly.

Let us examine three functions of time:

$$s_1(t) = A\cos 6\pi t$$

$$s_2(t) = A\cos(6\pi t + 5^o) \tag{7.1}$$

$$s_3(t) = A\cos(2\pi t e^{-t})$$

The frequencies of $s_1(t)$ and $s_2(t)$ are clearly 3 Hz. The frequency of $s_3(t)$ is, at present, un-defined: <u>Our traditional definition of frequency</u> does not apply to this type of waveform. Some thought indicates that the traditional, intuitive definition of frequency can be stated as follows:

> If $s(t)$ can be put into the form $A\cos(2\pi ft + \theta)$, where A, f, and θ are constants, then the frequency is defined as f Hz.

Since $s_3(t)$ cannot be put into this form, we need to broaden our concept of frequency to apply to the case where the frequency is not constant. We will define *instantaneous frequency* in a manner that can be applied to general waveforms. The definition will be structured so that it agrees with the long-standing concept of frequency for those waveforms which have constant frequency.

We express the given function in the form

$$s(t) = A\cos\theta(t) \tag{7.2}$$

In Eq. (7.2), A is a constant. Any general $s(t)$ can be expressed in this manner by first nor-malizing and then defining $\theta(t)$ as the inverse cosine. That is,

$$s(t) = |s_{max}| \cos\left(\cos^{-1}\frac{s(t)}{|s_{max}|}\right) \tag{7.3}$$

The reason we must divide by $|s_{max}|$ is that the inverse cosine is defined only for arguments with magnitudes bounded by unity.

The instantaneous frequency is defined as the rate of change of the phase. That is,

$$2\pi f_i(t) = \frac{d\theta}{dt} \tag{7.4}$$

where $f_i(t)$ is the instantaneous frequency in Hz. Note that both sides of Eq. (7.4) have units of radians per second.

You should convince yourself that the instantaneous frequencies of the signals given in Eq. (7.1) are 3 Hz, 3 Hz, and $e^{-t}(1 - t)$ Hz, respectively.

Example 7.1

Find the instantaneous frequency of the waveform

$$s(t) = \begin{cases} \cos 2\pi t, & t < 1 \\ \cos 4\pi t, & 1 < t < 2 \\ \cos 6\pi t, & 2 < t \end{cases}$$

Solution: The wave is of the form

$$s(t) = \cos [2\pi tg(t)]$$

where $g(t)$ is as shown in Fig. 7.1. The instantaneous frequency is then

$$f_i(t) = \frac{d}{dt}[tg(t)] = g(t) + t\frac{dg}{dt}$$

This is shown in Fig. 7.2.

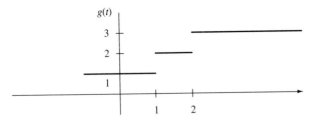

Figure 7.1   Function $g(t)$ for Example 7.1.

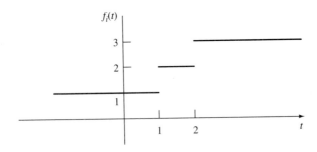

Figure 7.2   Instantaneous frequency for Example 7.1.

Example 7.2

Find the instantaneous frequency of the function

$$s(t) = 10 \cos 2\pi (1,000\, t + \sin 10\pi\, t)$$

Solution: We apply the definition to this waveform to find

$$f_i(t) = \frac{1}{2\pi}\frac{d\theta}{dt} = 1,000 + 10\,\pi \cos 10\pi t$$

This is shown in Fig. 7.3.

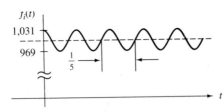

Figure 7.3   Instantaneous frequency for Example 7.2.

7.2 FREQUENCY MODULATION

Frequency modulation was invented by Edwin Armstrong (the same engineer who invented the superheterodyne receiver) in 1933. In frequency modulation we modulate the instantaneous frequency $f_i(t)$, with the signal $s(t)$, just as in amplitude modulation we modulate the amplitude A with the signal. We still want efficient transmission and the ability to separate stations, so we must shift the frequencies of $s(t)$ up to some carrier frequency. Therefore, rather than vary the frequency around zero, we vary it around a positive value f_c. For these reasons, we *define* frequency modulation (FM) as a waveform with the instantaneous frequency

$$f_i(t) = f_c + k_f s(t) \tag{7.5}$$

where f_c is the (constant) carrier frequency and k_f is a proportionality constant relating frequency changes to amplitude values of $s(t)$. If $s(t)$ is in volts, k_f has units of Hz/volt, or (volt-sec)$^{-1}$.

You might visualize Eq. (7.5) by thinking of operating a laboratory oscillator—the type with a rotating frequency adjustment. You set the frequency to f_c and then start wiggling the dial back and forth to follow the waveform $s(t)$. As $s(t)$ increases, the frequency increases. The constant k_f determines the magnitude of your wiggling of the frequency dial. Carrying this example further, suppose $s(t)$ is the result of your whistling into a microphone. Then if you whistle louder, your hand will have to swing further in each rotational direction of the frequency adjustment dial. If you whistle at a higher pitch, your hand will wiggle back and forth faster. It may help to refer back to this example from time to time, since FM is conceptually more difficult to understand than is AM.

Equation (7.5) describes the instantaneous frequency of a waveform. Let us now develop the formula for that waveform. We know that frequency is the derivative of phase, so the phase is of the form

$$\theta(t) = 2\pi \int_0^t f_i(\tau)\,d\tau = 2\pi\left[f_c t + k_f \int_0^t s(\tau)\,d\tau \right] \tag{7.6}$$

We have assumed a zero initial condition on the phase. The modulated waveform is then given by

$$\lambda_{fm}(t) = A\cos\theta(t)$$

$$= A\cos 2\pi\left[f_c t + k_f \int_0^t s(\tau)\,d\tau \right] \tag{7.7}$$

Note that if we set $s(t) = 0$, Eq. (7.7) becomes a pure carrier waveform, as we desire. Note also that the average power is $A^2/2$, which is independent of $s(t)$.

Example 7.3

Sketch the FM and suppressed carrier AM waveforms for the information signals of Fig. 7.4.
Solution: The suppressed carrier AM and FM waveforms corresponding to the information signals of Fig. 7.4 are illustrated in Fig. 7.5.

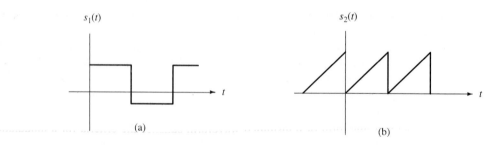

Figure 7.4   Waveforms for Example 7.3.

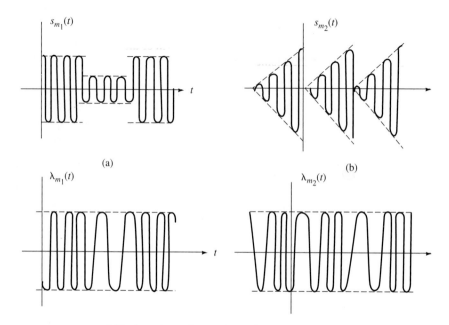

Figure 7.5   Modulated waveforms for Example 7.3.

The frequency of $\lambda_{\text{fm}}(t)$ varies from $f_c + k_f[\min s(t)]$ to $f_c + k_f[\max s(t)]$. You might find it useful to refer back to the laboratory oscillator scenario: As you wiggle the frequency adjustment dial back and forth, the preceding expressions represent the maximum excursions from the rest position of f_c.

You can observe from Eq. (7.5) that by making k_f arbitrarily small, the frequency of $\lambda_{\text{fm}}(t)$ can be kept arbitrarily close to f_c. You might reason that this could result in a great bandwidth savings. Unfortunately, this reasoning is flawed: The Fourier transform of a waveform is the result of evaluating an integral over all time, and instantaneous frequency does not take the intended time interval into account. This observation becomes clearer by looking at a constant function $s(t) = A$, which has a frequency of zero. A square pulse consists of piecewise constant waveforms, so the instantaneous frequency of the pulse is zero (except at the transitions). However, the bandwidth of the pulse is inversely proportional to its width (i.e., the Fourier transform is of the form $\sin (f)/f$). Clearly, just because a function's instantaneous frequency is confined to a certain band does not mean that the function occupies that particular bandwidth.

Returning to FM, we shall see in Section 7.4 that no matter how small k_f is made, the bandwidth can never be less than that of AM. Thus, FM cannot be justified on the basis of bandwidth savings. It does, however, possess other advantages over AM, primarily in the area of immunity to noise.

Proceeding in the manner we have established for modulation systems, we must now show that this form of transmitted waveform satisfies the criteria for a modulation scheme. That is, it must be capable of efficient transmission through the air; it must be in a form that allows multiplexing, and $s(t)$ must be uniquely recoverable from the modulated waveform. The demonstration of these properties is much more difficult than it was for AM, because the modulation is not linear in $s(t)$. If $s(t)$ in Eq. (7.5) is replaced with a sum of signals, the resulting FM wave is *not* the sum of the individual FM waveforms. The reason for this is that the cosine is nonlinear. That is,

$$\cos (A + B) \neq \cos A + \cos B \qquad (7.8)$$

Because of this nonlinearity, we will be forced to make many approximations. In the end, we will be satisfied with a far less detailed analysis than was possible for AM.

To show that FM is efficient and can be adapted to frequency multiplexing, the Fourier transform of the FM wave must be found. We will find it easier to first divide FM into two classes, depending on the size of the constant k_f. A relatively simple approximation is possible for very small values of k_f. We call this case *narrowband FM* for reasons that will become obvious in Section 7.4.

7.3 PHASE MODULATION

There is no basic difference between phase modulation (PM) and frequency modulation. In fact, the terms are often used interchangeably. Modulating a phase with a particular signal also modulates the derivative of that phase, frequency, with a related function of time. We therefore need not spend a great deal of time developing PM. Indeed, we will be able to extend the results we obtained for FM to PM with little difficulty.

We again have a modulated waveform of the form

$$\lambda_{pm}(t) = A\cos \theta(t) \qquad (7.9)$$

where $\theta(t)$ (instead of $d\theta/dt$) is modulated with $s(t)$. Therefore,

$$\theta(t) = 2\pi[f_c t + k_p s(t)] \qquad (7.10)$$

The proportionality constant, k_p, has units of volts^{-1}. The PM wave is then given by

$$\lambda_{pm}(t) = A\cos 2\pi[f_c t + k_p s(t)] \qquad (7.11)$$

We have included the factor $2\pi f_c t$ in the phase definition of Eq. (7.10) so that when $s(t) = 0$, the PM wave is a pure carrier. Had we not included this term, the PM wave would become dc when the signal went to zero.

We can relate the PM wave to the FM wave by finding the instantaneous frequency of the signal of Eq. (7.11):

$$f_i(t) = f_c + k_p \frac{ds}{dt} \tag{7.12}$$

This looks very similar to the FM case, where

$$f_i(t) = f_c + k_f s(t) \tag{7.13}$$

In fact, there is no difference between frequency modulating a carrier with $s(t)$ and phase modulating that same carrier with the integral of $s(t)$. Alternatively, there is no difference between phase modulating a carrier with $s(t)$ and frequency modulating the same carrier with the derivative of $s(t)$.

For this reason, all of the results of the following sections are easily translatable between the two forms of modulation.

7.4 NARROWBAND ANGLE MODULATION

Narrowband FM

If k_f is very small, we can use approximations to simplify the equation for the FM waveform. We start with the general form given in Eq. (7.7) which we repeat here:

$$\lambda_{fm}(t) = A\cos 2\pi\left(f_c t + k_f \int_0^t s(\tau)d\tau\right) \tag{7.14}$$

To avoid having to rewrite the integral many times, we define

$$g(t) = \int_0^t s(\tau)d\tau \tag{7.15}$$

Equation (7.14) then becomes

$$\lambda_{fm}(t) = A\cos 2\pi[f_c t + k_f g(t)] \tag{7.16}$$

We now expand the cosine using trigonometric identities, to obtain

$$\lambda_{fm}(t) = A\cos 2\pi f_c t\cos 2\pi k_f g(t) - A\sin 2\pi f_c t\sin 2\pi k_f g(t) \tag{7.17}$$

The cosine of a very small angle is unity, while the sine of a very small angle is the angle itself (i.e., we retain the first term in a power series expansion). Therefore, if k_f is small enough such that $2\pi k_f g(t)$ represents a very small angle, we can make these approximations, and Eq. (7.17) becomes

$$\lambda_{fm}(t) \approx A\cos 2\pi f_c t - 2\pi A g(t)k_f\sin 2\pi f_c t \tag{7.18}$$

This approximation is linear in $g(t)$, and is therefore linear in $s(t)$. We can find its Fourier transform with little difficulty as follows.

The Fourier transform of $g(t)$ is related to $S(f)$ by

$$G(f) = \frac{S(f)}{j2\pi f} \qquad (7.19)$$

This is true because an integrator is a linear system with transfer function $H(f) = 1/j2\pi f$. If we now assume that $S(f)$ is limited to frequencies below f_m, we find that $G(f)$ is also limited to the same range of frequencies. Integration is a process that shifts phase by 90° and divides amplitudes by the frequency; it does not create any new frequencies. (This is simply a statement that it is a linear operation.)

We can now take the Fourier transform of the expression in Eq. (7.18) to find

$$\Lambda_{\mathrm{fm}}(f) = \frac{A}{2}[\delta(f - f_c) + \delta(f + f_c)] + \frac{2\pi A k_f}{4\pi}\left[\frac{S(f - f_c)}{f - f_c} - \frac{S(f + f_c)}{f + f_c}\right] \qquad (7.20)$$

Figure 7.6 shows a representative $S(f)$ and the magnitude of the transform of the resulting narrowband FM waveform. Note that although the shape of $S(f)$ is meant only to be representative of a signal limited to f_m, we have attempted to show the transform of $g(t)$ as being proportional to $S(f)/f$.

We see from the figure that narrowband FM meets the first two objectives of a modulating system. That is, the frequencies that are present can be made as high as necessary

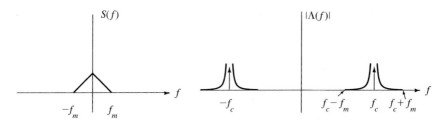

Figure 7.6   Magnitude of transform of FM waveform.

for efficient transmission by adjusting f_c to any desired value. If adjacent carrier frequencies are separated by at least $2f_m$, multiple signals may be transmitted simultaneously on the same channel. As for the third requirement, being able to recover $s(t)$ from the modulated waveform, we shall defer that consideration to Section 7.7, where we discuss demodulation. The same demodulators can be used regardless of the size of k_f, so we will cover the cases of small and large k_f together.

The bandwidth of the FM waveform is $2f_m$, just as for double-sideband AM. This is true regardless of how small k_f is made. Returning to the laboratory oscillator case, suppose a speech waveform is frequency modulating a carrier. The speech waveform has a maximum frequency of 5 kHz. Therefore, you will have to wiggle the frequency control dial back and forth up to 5,000 times per second. Suppose k_f is so small that the maximum deviation from f_c is 1 Hz. You will then be wiggling the dial over a swing of only 2 Hz, from $f_c - 1$ to $f_c + 1$. Yet the Fourier transform of the resulting waveform will occupy the band between $f_c - 5,000$ and $f_c + 5,000$. Clearly, the instantaneous frequency and the manner in which it changes *both* contribute to the FM bandwidth.

That we call the small k_f *narrowband* should hint at the fact that as k_f is raised, the bandwidth increases from its minimum of $2f_m$. This is true, as discussed in the next section.

Narrowband Phase Modulation

We have been concentrating on FM, but as mentioned earlier, the results are easily extendable to PM. Phase modulating with $s(t)$ is exactly the same as frequency modulating with the derivative of $s(t)$. Since the derivative of $s(t)$ contains the same range of frequencies as does $s(t)$, the derivation leading to Fig. 7.6 would reach the same conclusion as regards the bandwidth for PM. That is, a narrowband PM waveform occupies the range of frequencies between $f_c - f_m$ and $f_c + f_m$, with a bandwidth of $2f_m$.

In order for FM to be narrowband, the maximum value of $2\pi k_f g(t)$, where $g(t)$ is the integral of $s(t)$, had to be a very small angle. In order for *PM* to be narrowband, $2\pi k_p s(t)$ must be a very small angle. This permits approximating the cosine and sine by the first term in a series expansion.

Before leaving narrowband angle modulation, we present several vector plots to gain more insight into the modulating process. We want to compare vector plots from AM with those from FM. The transmitted carrier AM waveform for a sinusoidal information signal is

$$s_m(t) \; = \; A\cos 2\pi f_c t \; + \; \cos 2\pi f_m t\cos 2\pi f_c t \tag{7.21}$$

We can write this as the real part of a complex exponential:

$$s_m(t) \; = \; \mathrm{Re}\left\{ e^{j2\pi f_c t}\left[A \; + \; \frac{1}{2}e^{j2\pi f_m t} \; + \; \frac{1}{2}e^{-j2\pi f_m t} \right] \right\} \tag{7.22}$$

The phase of the term in braces has a large angular term due to the phase factor $2\pi f_c t$. Recall that f_c is usually much larger than f_m. We use the common technique of assuming that a "stroboscopic" picture of this phasor is taken with a flash every $1/f_c$ seconds. Hence, we plot only the phasor of the terms in square brackets in Eq. (7.22). This is shown as Fig. 7.7. Note that the resultant is in phase with the carrier term. This is as expected, since there is no phase variation in AM.

We now repeat the foregoing approach for narrowband FM. Equation (7.18) can be rewritten in the following form, where we have let $s(t)$ be $\cos 2\pi f_m t$:

$$\lambda_{fm}(t) \; = \; \mathrm{Re}\left\{ e^{j2\pi f_c t}\left[A \; - \; \frac{Ak_f}{2jf_m}e^{-j2\pi f_m t} \; + \; \frac{Ak_f}{2jf_m}e^{+j2\pi f_m t} \right] \right\} \tag{7.23}$$

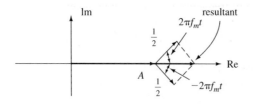

Figure 7.7 Phasor plot of portion of Eq. (7.22).

The quantity in square brackets in Eq. (7.23) is plotted in Fig. 7.8. Note that the resultant has almost the same amplitude as A. The reason that the amplitudes are not exactly the same is that Eq. (7.18) is an approximation. Had we not used an approximation, the amplitude would be constant.

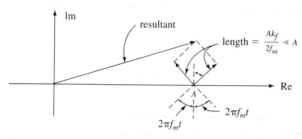

Figure 7.8   Phasor plot of portion of Eq. (7.23).

7.5 WIDEBAND FM

If k_f is not small enough to permit the approximations of the previous section, we have what is known as *wideband FM*. The transmitted signal is of the form

$$\lambda_{fm}(t) = A\cos 2\pi[f_c t + k_f g(t)] \tag{7.24}$$

where $g(t)$ is the integral over time of the information signal $s(t)$.

If $g(t)$ were a known function, the Fourier transform of this FM waveform could be evaluated. We would simply use the defining integral for the Fourier transform, and although numerical techniques might be required to evaluate the integral, it nevertheless could be solved. However, we do not wish to restrict ourselves to a particular $s(t)$: we merely want to constrain $s(t)$ to frequencies below f_m. With this as the only information given, it is not possible to find the Fourier transform of the FM waveform, because of the nonlinear relationship between $s(t)$ and the modulated waveform. Hence, since we cannot find the Fourier transform of the FM wave in the general case, let us see how much we can say about the modulated signal.

In order to show that FM can be efficiently transmitted and that several channels can be multiplexed, we must only gain some feeling for the range of frequencies occupied by the modulated waveform. We have already done this for narrowband angle modulation, so we must now concentrate on wideband.

This is a significant difference from AM, where we found the Fourier transform of $s_m(t)$ and used it throughout the analysis, including when we investigated the operation of modulators and demodulators. In the case of angle modulation, we will use the Fourier transform for only one thing: to determine the range of frequencies occupied by the modulated waveform. All of our other analysis will be performed in the time domain. For this reason, we do not care about the exact shape of the transform.

Even finding the range of frequencies is impossible in general, so we shall begin by restricting ourselves to a specific type of information signal: a pure sinusoid. This will allow us to use trigonometry in the analysis. Once that is completed, we will generalize the results.

We let the information signal be

$$s(t) = a\cos 2\pi f_m t \tag{7.25}$$

where a is a constant amplitude. The instantaneous frequency of the FM wave is then

$$f_i(t) = f_c + ak_f \cos 2\pi f_m t \tag{7.26}$$

The FM waveform is of the form

$$\lambda_{fm}(t) = A\cos\left(2\pi f_c t + \frac{ak_f}{f_m} \sin 2\pi f_m t\right) \tag{7.27}$$

Let us define the *modulation index* as

$$\beta = \frac{ak_f}{f_m} \tag{7.28}$$

Note that the modulation index is a dimensionless quantity; it is the ratio of the maximum value of the time-varying term in Eq. (7.26) to the frequency of that term.

Using this definition of the modulation index, we can simplify Eq. (7.27) to

$$\lambda_{fm}(t) = A\cos(2\pi f_c t + \beta\sin 2\pi f_m t) \tag{7.29}$$

We now want to expand this expression in an attempt to separate the information portion of the signal from the carrier. One way to do this would be to use trigonometric identities, tracking two terms throughout the analysis. However, we can simplify the mathematics and work with only one expression by converting to exponential notation:

$$\lambda_{fm}(t) = \mathrm{Re}\{A\exp(j2\pi f_c t + j\beta\sin 2\pi f_m t)\} \tag{7.30}$$

The exponential terms in Eq. (7.30) can be split into a product where the second term is the one that contains the information signal and is equal to

$$\exp(j\beta\sin 2\pi f_m t) \tag{7.31}$$

The real part of Eq. (7.31) is in the form of a cosine of a sine and is generally difficult to work with. However, we note that this particular exponential is a periodic function with period $1/f_m$. That is, if you substitute $t + 1/f_m$ for t, you do not change the expression. Note that this periodic function is complex. We can expand the function into a complex Fourier series with fundamental frequency f_m as follows:

$$e^{j\beta\sin 2\pi f_m t} = \sum_{n=-\infty}^{\infty} c_n e^{jn2\pi f_m t} \tag{7.32}$$

The Fourier coefficients of Eq. (7.32) are

$$c_n = f_m \int_{-\frac{1}{2f_m}}^{\frac{1}{2f_m}} e^{j\beta\sin 2\pi f_m t} e^{-jn2\pi f_m t}\, dt \tag{7.33}$$

We first observe that the coefficients c_n do not depend on f_m. This can be seen by making a change of variables in Eq. (7.33). Suppose we let $x = f_m t$. Equation (7.33) then becomes

$$c_n = \int_{-\frac{1}{2}}^{\frac{1}{2}} e^{j\beta\sin 2\pi x} e^{-jn2\pi x} \, dx \tag{7.34}$$

The integral in Eq. (7.34) cannot be evaluated in closed form. It does, however, converge to a real value. (See Problem 7.5.1.) That real value is a function of n and β. The integral is tabulated under the name *Bessel function of the first kind*, and its symbol is $J_n(\beta)$.

Bessel Functions

The Bessel function of the first kind is generated as solutions of the differential equation

$$x^2 \frac{d^2y}{dx^2} + x\frac{dy}{dx} + (x^2 - n^2)y(x) = 0 \tag{7.35}$$

Although the Bessel function is defined for all values of n, we will be concerned only with the positive and negative real integers. It can be shown that, for integer values of n,

$$J_{-n}(x) = (-1)^n J_n(x) \tag{7.36}$$

Figure 7.9 shows $J_n(x)$ for $n = 0$, 1, and 2. Note that for very small x, $J_0(x)$ approaches unity, while $J_1(x)$ and $J_2(x)$ approach zero.

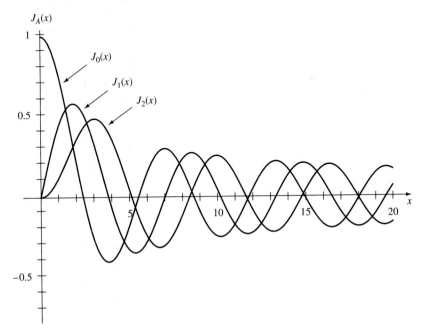

Figure 7.9   Bessel functions for $n = 0$, 1, and 2.

It will be necessary for us to observe the behavior of the Bessel function as n becomes large. To gain insight into this behavior, we will examine a particular point on the curves of Fig. 7.9. Figure 7.10 is a plot of $J_n(10)$ as a function of n. The function appears to be in underdamped oscillation for negative n, but note again that we are concerned only with integer values of n. For these integer values, the symmetry of Eq. (7.36) holds, and we can focus attention on the positive n-axis.

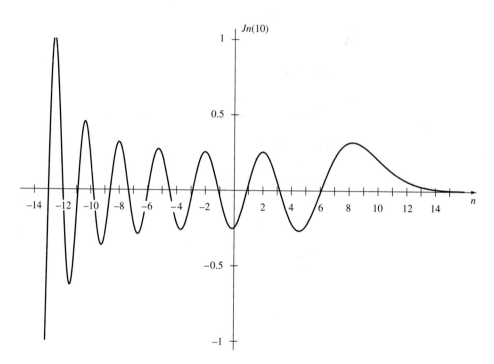

Figure 7.10 $J_n(10)$ as a function of n.

The important observation to make is that for n greater than about 9, the Bessel function asymptotically approaches zero. In fact, for fixed n and large β, the Bessel function can be approximated by

$$J_n(\beta) \approx \frac{(\beta/2)^n}{\Gamma(n+1)} \tag{7.37}$$

In Eq. (7.37), $\Gamma(n+1)$ is the *Gamma function*. Gamma functions approach infinity for arguments greater than 2. For example, the values of the Gamma function for arguments of 2,3,4,5, and 6 are 1,2,6,24, and 120, respectively. Clearly, since the Gamma function is in the denominator, the Bessel function gets small rapidly with increasing n beyond the value of the argument. This is a critical property for finding the bandwidth of the FM waveform.

Returning now to Eq. (7.34), we see that the Fourier coefficients are

$$c_n = J_n(\beta) \tag{7.38}$$

and the FM waveform becomes

$$\lambda_{\mathrm{fm}}(t) = \mathrm{Re}\left\{ Ae^{j2\pi f_c t} \sum_{n=-\infty}^{\infty} J_n(\beta)\, e^{jn2\pi f_m t} \right\} \qquad (7.39)$$

Since $\exp(j2\pi f_c t)$ is not a function of n, we can bring it into the summation to get

$$\lambda_{\mathrm{fm}}(t) = \mathrm{Re}\left\{ A \sum_{n=-\infty}^{\infty} J_n(\beta)\, e^{j2\pi\, t(f_c + nf_m)} \right\} \qquad (7.40)$$

Now, taking the real part of the expression, we obtain

$$\lambda_{\mathrm{fm}}(t) = A \sum_{n=-\infty}^{\infty} J_n(\beta) \cos 2\pi(f_c + nf_m)t \qquad (7.41)$$

Equation (7.41) is the desired result: We have reduced the FM waveform to a sum of sinusoids. The Fourier transform of this sum of sinusoids is a train of impulses and is sketched in Fig. 7.11.

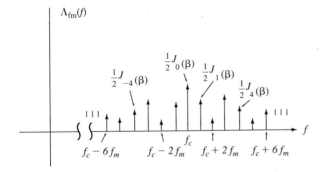

Figure 7.11 Fourier transform of FM with sinusoidal information signal.

We seem to be in big trouble: The transform extends forever in both directions away from the carrier; it would appear to have an infinite bandwidth. Indeed, unless $J_n(\beta)$ goes to zero for n above some value, the bandwidth is *not limited*, and we can neither transmit efficiently nor multiplex channels.

Our previous discussion of Bessel functions shows that, for fixed β, the functions $J_n(\beta)$ eventually do approach zero as n increases. Note that the Bessel functions do not decrease monotonically for n less than β. In fact, for certain choices of β, the $J_0(\beta)$ term goes to zero, and the carrier term is eliminated from the FM waveform. In the case of AM, elimination of the carrier increases efficiency. In FM, elimination of the carrier does not gain anything, since the total power remains constant.

In order to approximate the bandwidth of the FM waveform, we must examine the strengths of the impulses in Fig. 7.11. Let us first choose a small value of β. We see from Fig. 7.9 that if $\beta < 0.5$, $J_2(\beta) < 0.03$. The higher order Bessel functions ($n > 2$) are even smaller. At $\beta = 0.5$, J_1 is 0.24. For these small values of β, it is therefore reasonable to include only the three impulses near the carrier in Fig. 7.11. That is, there is the component at the carrier and two additional components spaced $\pm f_m$ away from the carrier. This yields a bandwidth of $2f_m$. But we already knew that, since very small values of β (ak_f/f_m) correspond to the narrowband condition.

Now suppose that β is not small. For example, suppose it is equal to 10. The properties that we have previously discussed would indicate that $J_n(10)$ attenuates rapidly for $n > 10$. Referring to Fig. 7.11, we then consider the significant components to be the carrier term and 10 harmonics on each side of the carrier. In general, for large β, the number of terms that must be included on each side of the carrier is β (rounded to an integer). This yields a bandwidth of $2\beta f_m$. Note that this bandwidth is an approximation since the Fourier transform of the FM wave is not identically zero beyond any frequency point. However, in practice, if the energy outside the band is rejected with a bandpass filter, the distortion is not noticeable.

In many instances, we will be working with modulation indexes between the two extremes. In the very early days of FM studies, John Carson proposed a rule of thumb that has been widely adopted. This rule gives the approximate bandwidth of FM as a function of the frequency of the information signal and the modulation index. The reason the rule has been widely accepted is that the amount of energy outside this band has been shown to be negligible for most applications. Carson's rule says that the bandwidth is

$$\text{BW} \approx 2(\beta f_m + f_m) \tag{7.42}$$

This agrees with our two limiting cases. For very small β, the bandwidth is approximately $2f_m$, while for large β, it is approximately $2\beta f_m$.

Carson's approximation can be written in a more meaningful way by substituting ak_f/f_m for β. Equation (7.42) then becomes

$$\text{BW} \approx 2(ak_f + f_m) \tag{7.43}$$

Recall that the instantaneous frequency is

$$f_i(t) = f_c + ak_f \cos 2\pi f_m t \tag{7.44}$$

Let us try to give a physical interpretation to the terms in Eq. (7.43). We see that f_m is the rate at which the instantaneous frequency varies, while ak_f is the maximum amount that it deviates from the carrier. Returning to our laboratory oscillator example, f_m is the rate at which your hand must wiggle back and forth, while ak_f is the maximum amount it must twist away from the rest position. It makes intuitive sense that both of these quantities should contribute to the bandwidth of the FM wave.

Example 7.4

Find the approximate band of frequencies occupied by an FM wave with carrier frequency of 5 kHz, $k_f = 10$ Hz/v, and

(a) $s(t) = 10\cos 10\pi t$ volts
(b) $s(t) = 5\cos 20\pi t$ volts
(c) $s(t) = 100\cos 2,000\pi t$ volts

Solution: The bandwidths are

(a) $\text{BW} \approx 2(ak_f + f_m) = 2[10(10) + 5] = 210$ Hz
(b) $\text{BW} \approx 2(ak_f + f_m) = 2[5(10) + 10] = 120$ Hz
(c) $\text{BW} \approx 2(ak_f + f_m) = 2[100(10) + 1,000] = 4$ kHz

The bands of frequencies occupied are therefore as follows:
  (a) 4,895 Hz to 5,105 Hz
  (b) 4,940 Hz to 5,060 Hz
  (c) 3 kHz to 7 kHz

Equation (7.43) was developed for the special case of a sinusoidal information signal. If the modulation were linear in the information signal, we could simply apply the formula to the highest frequency component of $s(t)$ to find the bandwidth. However, FM is nonlinear, so this approach is not correct.

It is time for a rather bold generalization. Let us stare at Eq. (7.43) and attempt to develop a similar formula for the general case. Figure 7.12 shows the instantaneous fre-

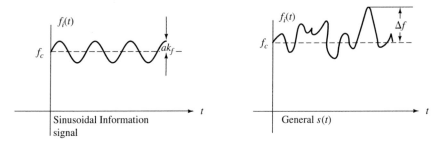

Figure 7.12   Instantaneous frequency.

quency for the special case of a sinusoidal information signal and also for the more general case. Looking first at the special case, we see that the two parameters which feed into Eq. (7.43) are the frequency $f_i(t)$ and the *maximum frequency deviation*, as marked on the figure.

In the general case, $f_i(t)$ is no longer at a single frequency, but contains a continuum of frequencies up to a maximum of f_m. Since we are looking for the maximum bandwidth, it would seem intuitive to replace the "frequency of $f_i(t)$" term in Eq. (7.43) with the "maximum frequency of $f_i(t)$." Because we call each of these f_m, there is no need for any change in this portion of the equation.

For the case of the sinusoidal information signal, ak_f is the maximum deviation of the frequency from f_c. Figure 7.12 shows the equivalent *maximum frequency deviation* for the general case. We assign the symbol Δf to this frequency deviation. The general form of Eq. (7.43) is then

$$BW \approx 2(\Delta f + f_m) \tag{7.45}$$

and we have accomplished our goal.

Let us attempt to attach an intuitive meaning to the wideband case. If Δf is much larger than f_m, we have wideband FM, and the frequency of the carrier is varying by a large amount, but slowly. That is, the instantaneous frequency of the carrier is going from $f_c - \Delta f$ to $f_c + \Delta f$ very slowly. The FM wave therefore approximates a pure sinusoid over long lengths of time. We can think of it as a sum of many sinusoids with frequencies between the two limits. The Fourier transform is then approximately equal to a superposition of the transforms of each of these many sinusoids, all lying between the frequency limits. It is therefore reasonable to assume that the bandwidth is approximately the width of this frequency interval, or $2\Delta f$. Once again, looking at our laboratory oscillator, if we move the dial back and forth very slowly, the Fourier transform of the resulting waveform occupies approximately the same range of frequencies as that over which the dial is varied.

On the other hand, for very small Δf, we have a carrier that is varying over a very small range of frequencies, but doing so relatively rapidly. We can approximate this with two oscillators at the frequency limits, each being gated on for half of the total time. This is shown in Fig. 7.13. The approximate band of frequencies occupied by the output signal

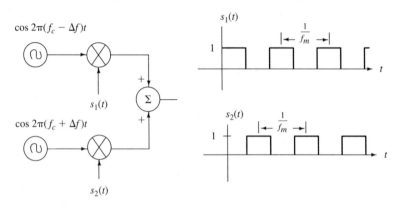

Figure 7.13　Approximation of narrowband FM.

is from $f_c - \Delta f - f_m$ to $f_c + \Delta f + f_m$. For small Δf, this yields a bandwidth of $2f_m$. The details of this derivation are left to Problem 7.5.8.

It should be clear that the bandwidth of an FM waveform increases with increasing values of k_f. At this point, there appears to be no reason to use other than narrowband FM, which has the minimum bandwidth of $2f_m$. However, we will see that wideband FM possesses a noise advantage over both narrowband FM and AM.

Example 7.5

A 10-MHz carrier is frequency modulated by a sinusoidal signal of 5-kHz frequency such that the maximum frequency deviation of the FM wave is 500 kHz. Find the approximate band of frequencies occupied by the FM waveform.

Solution:　We must first find the approximate bandwidth. This is

$$BW \approx 2(\Delta f + f_m)$$

We are given the value of Δf, and the maximum frequency component of the information is 5 kHz. Indeed, this is the *only* frequency component of the information and, therefore, certainly is the maximum. The bandwidth is then

$$BW \approx 2(500 \text{ kHz} + 5 \text{ kHz}) = 1{,}010 \text{ kHz}$$

Thus, the band of frequencies occupied is centered around the carrier frequency and ranges from 9,495 kHz to 10,505 kHz.

The FM signal of this example is wideband. If it were narrowband, the bandwidth would be only 10 kHz.

Example 7.6

A 100-MHz carrier is frequency modulated by a sinusoidal signal of 1-volt amplitude. k_f is set at 100 Hz/volt. Find the approximate bandwidth of the FM waveform if the modulating signal has a frequency of 10 kHz.

Solution: Again, we use Carson's approximation. That is,

$$BW \approx 2(\Delta f + f_m)$$

Since the information $s(t)$ has unit amplitude, the maximum frequency deviation Δf is given by k_f, or 100 Hz. f_m is simply 10 kHz, the frequency of the modulating signal. Therefore,

$$BW \approx 2(100 \text{ Hz} + 10 \text{ kHz}) = 20,200 \text{ Hz}$$

Since f_m is much greater than Δf, this is a narrowband FM signal. The bandwidth necessary to transmit the same information waveform using double-sideband AM would be 20 kHz, which is approximately the same as the bandwidth of the FM wave.

Example 7.7

An angle-modulated waveform is described by

$$\lambda(t) = 10\cos(2 \times 10^7 \pi t + 20\cos 1,000 \pi t)$$

Find the approximate bandwidth of this waveform.

Solution: f_m is equal to 500 Hz. To compute Δf, we first find the instantaneous frequency $f_i(t)$:

$$f_i(t) = \frac{1}{2\pi} \frac{d}{dt} (2 \times 10^7 \pi t + 20\cos 1,000 \pi t)$$

$$= 10^7 - 10,000 \sin 1,000 \pi t$$

The maximum frequency deviation is the maximum value of $10,000\sin 1,000\pi t$, which is simply 10 kHz. The bandwidth is therefore

$$BW \approx 2(10,000 + 500) = 21 \text{ kHz}$$

This is clearly a wideband waveform, since Δf is much greater than f_m. Note that we do not have to know whether the waveform has been frequency or phase modulated in order to find the bandwidth.

7.6 MODULATORS

We have shown that FM waveforms are bandlimited to a range of frequencies around f_c, the carrier frequency. The first two criteria of a useful modulation system are therefore satisfied. We can transmit efficiently by choosing f_c in the proper range, and we can frequency multiplex many separate signals by making sure that the adjacent carrier frequencies are separated by a sufficient amount such that the transforms of the FM waveforms do not overlap in frequency.

It remains for us to show that $s(t)$ can be recovered from the angle-modulated waveform and that modulators and demodulators can be constructed simply.

We start by reexamining narrowband FM. The waveform is expressed by the following approximation:

$$\lambda_{fm}(t) \ = \ A\cos 2\pi f_c t \ - \ 2\pi A g(t) k_f \sin 2\pi f_c t \tag{7.46}$$

This immediately leads to the block diagram of Fig. 7.14. The equivalent expression for narrowband PM is

$$\lambda_{pm}(t) \ = \ A\cos 2\pi f_c t \ - \ 2\pi A k_p s(t)\sin 2\pi f_c t \tag{7.47}$$

Figure 7.14 is therefore modified by replacing $2\pi k_f s(t)$ with $2\pi k_p s(t)$ and eliminating the integrator.

The instantaneous frequency of the output of the system of Fig. 7.14 is

$$f_i(t) \ = \ f_c \ + \ k_f s(t) \tag{7.48}$$

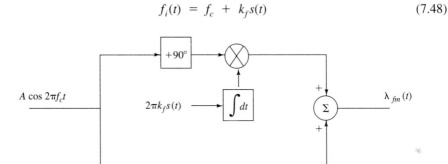

Figure 7.14   Narrowband FM modulator.

The reason this represents a narrowband signal is that the maximum value of $k_f s(t)$ [i.e., the frequency deviation] is small compared to the frequencies present in $s(t)$. Suppose we put the output narrowband FM waveform through a nonlinear device that multiplies all frequencies by a constant C. Then the resulting waveform has an instantaneous frequency

$$f_i(t) \ = \ Cf_c \ + \ Ck_f s(t) \tag{7.49}$$

The frequency deviation of this new waveform is C times that of the old, while the rate at which the instantaneous frequency varies has not changed. This is shown in Fig. 7.15. Therefore, for high enough values of C, frequency multiplication changes narrowband FM into wideband FM. It also moves the carrier frequency, but the carrier has no effect upon whether an FM wave is narrowband or wideband. Looked at another way, if the bandwidth of the FM wave is significantly larger than $2f_m$, the signal is wideband. If the new, higher carrier is not desired, we can shift (heterodyne) the result to any part of the axis without affecting the bandwidth. The resulting FM modulator is shown in Fig. 7.16.

There are more direct ways to generate wideband FM. An electronic oscillator has an output frequency that depends on energy storage devices. There are a wide variety of oscillators whose frequencies depend upon a particular capacitor value. By varying the capacitor value, the frequency of oscillation varies.

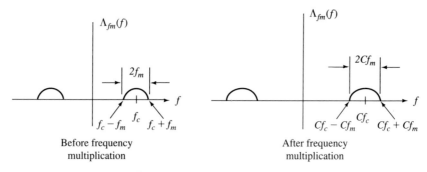

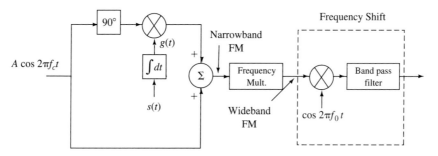

Figure 7.15   Frequency multiplication.

Figure 7.16   Wideband FM modulator.

If the capacitor variations are controlled by $s(t)$, the result is an FM waveform. One way to accomplish this is with *varactor diodes*. When this type of diode is back biased, it acts as a capacitance with a value that depends upon the magnitude of the back biasing voltage.

What we are describing is the *voltage-controlled oscillator* (VCO). A wide variety of VCOs are available in integrated circuit form. The 566 is a representative device that is used for frequencies below 1 MHz. The block diagram of the 566 VCO is shown in Fig. 7.17. The timing capacitor, C_T, receives a constant charging or discharging current from the current source/sink. The voltage on the modulation input controls the amount of current produced by the source/sink. The timing capacitor output forms the input to a Schmitt trigger. When the Schmitt trigger output changes level, the current source/sink reverses its operating mode. Thus, the slope of the voltage ramp on the timing capacitor is proportional to the output frequency of the IC.

PM Modulators

Phase modulators are the same as frequency modulators, except that we first differentiate $s(t)$ before feeding it into the VCO.

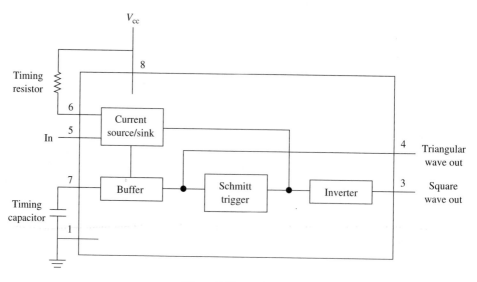

Figure 7.17   566 VCO.

7.7 DEMODULATORS

The problem of FM demodulation can be stated as follows: Given

$$\lambda_{\text{fm}}(t) \; = \; A\cos 2\pi \left(f_c t \; + \; k_f \int_0^t s(\tau)d\tau \right) \tag{7.50}$$

recover the information signal $s(t)$.

Demodulators fall into one of two broad classifications. The first of these employ discriminators, which are devices that discriminate one frequency from another by transforming changes in frequency into changes in amplitude. The amplitude changes are detected just as was done in AM. The second category of demodulator uses a phase lock loop to match a local oscillator to the modulated carrier frequency.

Discriminator Detector

Since differentiation of a sinusoid is a process that multiplies the sinusoid by its instantaneous frequency, we start by investigating the derivative of the FM waveform:

$$\frac{d\lambda}{dt} \; = \; - \, 2\pi A[f_c \; + \; k_f s(t)]\sin 2\pi \left(f_c t \; + \; k_f \int_0^t s(\tau)d\tau \right) \tag{7.51}$$

This derivative is sketched in Fig. 7.18 for a representative $s(t)$.

If we now assume that the instantaneous frequency of the sinusoidal part of Eq. (7.51) is always much greater than f_m (a reasonable assumption in real life), this carrier term fills in the area between the amplitude and its mirror image. We have exaggerated the carrier in sketching Fig. 7.18. Actually, the area between the upper and lower outline

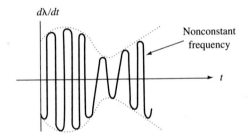

Nonconstant
frequency

Figure 7.18 Derivative of FM waveform.

should be shaded due to the extremely high carrier frequency. Thus, even though the carrier frequency is not constant, the envelope of the waveform is still clearly defined by

$$2\pi|A[f_c + k_f s(t)]| \tag{7.52}$$

The slight variation in the frequency of the carrier would not even be noticed by an envelope detector.

Example 7.8

Use the pre-envelope to prove that the envelope of the expression in Eq. (7.51) is that given in Eq. (7.52).

Solution: We find the pre-envelope using the techniques of Chapter 6:

$$z(t) = -2\pi A[f_c + k_f s(t)]\sin 2\pi\left(f_c t + k_f \int_0^t s(\tau)\,d\tau\right)$$

$$-2\pi jA[f_c + k_f s(t)]\cos 2\pi\left(f_c t + k_f \int_0^t s(\tau)d\tau\right)$$

The envelope of $r(t)$ is the magnitude of $z(t)$:

$$|z(t)| = 2\pi\sqrt{A^2[f_c + k_f s(t)]^2\left\{\sin^2 2\pi\left[f_c t + k_f \int_0^t s(\tau)\,d\tau\right]\right.}$$

$$\overline{\left. + \cos^2 2\pi\left[f_c t + k_f \int_0^t s(\tau)d\tau\right]\right\}}$$

$$= 2\pi|A[f_c + k_f s(t)]|$$

In a practical system, f_c is much larger than $k_f s(t)$. Therefore, the quantity in brackets in Eq. (7.52) is positive, and we can eliminate the absolute value sign. A differentiator followed by an envelope detector can thus be used to recover $s(t)$ from the FM waveform. This is shown in Fig. 7.19.

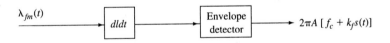

Figure 7.19 FM demodulator.

We note that the preceding analysis did not assume anything about the size of k_f. The demodulator of Fig. 7.19 therefore works for either wideband or narrowband FM signals.

If the transmission is PM rather than FM, the output of the system of Fig. 7.19 is the derivative of $s(t)$. Therefore, to change the system to a PM demodulator, we must simply add an integrator to the output of the system.

The occurrence of the envelope detector is a clue that AM is somehow appearing in the system. Indeed, this is the case, as the following analysis reveals.

The transfer function of a differentiator is given by

$$H(f) = 2\pi jf \tag{7.53}$$

The magnitude characteristic is sketched in Fig. 7.20. The magnitude of the output of the differentiator is linearly related to the frequency of the device's input. The differentiator therefore changes FM into AM. When a differentiator is used in this manner, it is called a *discriminator*.

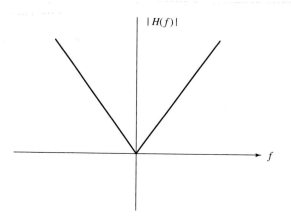

Figure 7.20   Magnitude characteristic of differentiator.

There are other types of discriminator besides differentiators. In order for a system to change FM to AM, the magnitude of its transfer function must be linear throughout the range of frequencies of the FM wave. Even a sloppy bandpass filter will work as a discriminator if we operate over a limited range relative to the filter bandwidth. This is shown in Fig. 7.21.

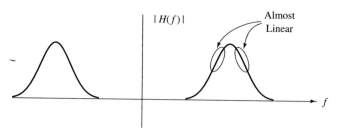

Figure 7.21   Bandpass filter as a discriminator.

We can improve the linearity of the bandpass filter discriminator in a manner similar to that applied to the balanced modulator. That is, we subtract the characteristic from a shifted version of itself, as shown in Fig. 7.22. In other words, we take the difference between the output of two bandpass filters with center frequencies separated as shown.

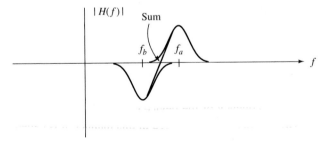

Figure 7.22   Improved linearity discriminator.

If the system is properly designed, we effectively eliminate even powers in the series expansion of the filter roll-off function. This is accomplished by the circuit of Fig. 7.23. The tuned circuit consisting of the upper half of the output winding of the transformer and C_1 is tuned to f_a, and the tuned circuit consisting of the other half of the output winding and C_2 is tuned to f_b. This circuit is known as a *slope demodulator*, since it uses the sloped portion of the filter characteristic as part of the demodulation process.

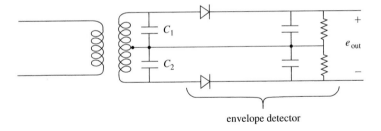

Figure 7.23   Slope demodulator.

If we now return to the original implementation using a differentiator, we can see one other approach: We can approximate the derivative by the difference between two adjacent sample values of the waveform. That is,

$$\lambda(t) - \lambda(t - t_0) \approx t_0 \frac{d\lambda}{dt} \tag{7.54}$$

This leads to the demodulator of Fig. 7.24. Because a time shift is equivalent to a phase shift, this is known as a *phase shift demodulator*.

Phase Lock Loop

The *phase lock loop* is a feedback circuit that can be used to demodulate angle-modulated waveforms. Feedback circuits are often used to reduce an error term toward zero. In the case of the phase lock loop, the error term is the difference in phase between the input sig-

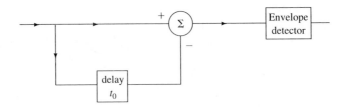

Figure 7.24   Phase shift demodulator.

nal and a reference sinusoid. This important circuit finds application in both analog and digital communication.

The phase lock loop incorporates a VCO in its feedback loop. A typical loop configuration is shown in Fig. 7.25. The loop compares the phase of the input signal to the phase of the signal at the output of the VCO. If the difference in phase is anything other than zero, the output frequency of the VCO changes in a manner that forces the difference toward zero.

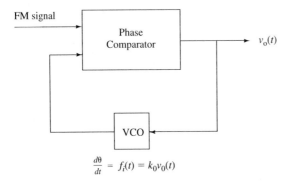

$$\frac{d\theta}{dt} = f_i(t) = k_0 v_0(t)$$

Figure 7.25   Phase lock loop.

The output of the phase comparator, $v_0(t)$, forms the input to the VCO. The output of the VCO is an FM waveform with an instantaneous frequency $f_i(t)$ that is proportional to the phase difference between the input and the VCO output. If, for example, the input frequency is higher than the VCO frequency, the difference in phase between the two will increase linearly with time, causing the VCO frequency to increase until it matches that of the input. Thus, the VCO attempts to follow the input frequency. Since the frequency of the VCO output is proportional to the voltage at its input, this input voltage also tries to follow the frequency of the loop input signal. Thus, monitoring the input to the VCO yields a demodulated version of the FM waveform. If an integrator is added to the output of the phase lock loop, the result is a demodulator for PM waveforms.

Now let us examine the phase comparator, which is also a part of the phase lock loop. The simplest method of phase comparison consists of a multiplier followed by a lowpass filter, as shown in Fig. 7.26. Suppose that the two inputs to the comparator are $\cos(2\pi f_c t + \theta_1)$ and $\cos(2\pi f_c t + \theta_2)$. The output of the lowpass filter is then $\cos(\theta_1 - \theta_2)/2$. This output is zero when the phase difference is 90°. Since the phase comparator generates the loop error function (the quantity that changes the frequency of the

why?

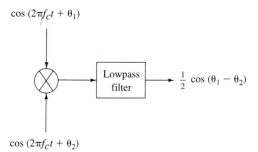

$$\cos (2\pi f_c t + \theta_1)$$

Lowpass filter

$\frac{1}{2} \cos (\theta_1 - \theta_2)$

$$\cos (2\pi f_c t + \theta_2)$$

Figure 7.26   Phase comparator.

VCO), the loop is in lock when the input and the VCO output are at the same frequency and 90° out of phase.

In some cases, it is desirable to have a comparator output that varies *linearly* with the phase difference, rather than sinusoidally. If the input cosine and sine signals are severely clipped (or amplified to the point where they severely saturate the electronics), they can be thought of as square waveforms. A representative situation is illustrated in Fig. 7.27. The product of the two square waves is now averaged by the low-pass filter. This average is proportional to the fraction of time that the square waves are equal. Therefore, the output amplitude is linearly related to the phase difference.

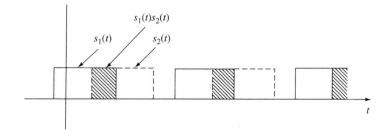

Figure 7.27   Severe clipping in the linear comparator.

Using the multiplier as a phase comparator, we redraw the phase lock loop with FM input and a general *loop filter* in place of the low-pass filter. This is shown in Fig. 7.28.

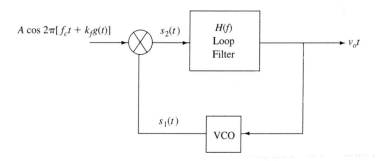

$A \cos 2\pi[f_c t + k_f g(t)]$ $s_2(t)$ $H(f)$ Loop Filter $v_o t$

$s_1(t)$ VCO

Figure 7.28   Phase lock loop with multiplier as phase comparator.

When the loop is in lock, the output of the VCO follows the input FM wave, but is 90° out of phase with it. The VCO output has an amplitude we will call B. The output frequency varies from the nominal setting of f_c by an amount proportional to $v_o(t)$, the input to the VCO. We shall call the VCO proportionality constant k_0. Then,

$$s_1(t) = B\sin 2\pi\left(f_c t + k_0\int_0^t v_o(\tau)d\tau\right)$$
(7.55)

It is not enough to know the lock condition of the loop; it is also important to know the loop's behavior with time, so that we know whether the loop is capable of tracking the input. One way to analyze the loop is to find its step response. That is, we suppose the loop is in lock and the input phase changes instantaneously.

We shall perform this transient analysis for the *first-order loop*, where the loop filter is a lowpass filter. The signal at the output of the multiplier is

$$s_2(t) = AB\cos 2\pi[f_c t + k_f g(t)]\sin 2\pi\left(f_c t + k_0\int_0^t v_o(\tau)d\tau\right)$$

$$= \frac{AB}{2}\sin 2\pi\left(k_f g(t) - k_0\int_0^t v_o(\tau)d\tau\right)$$
(7.56)

$$+ \text{ higher frequency terms}$$

We do not bother writing the higher order terms in Eq. (7.56), since they will not get through the lowpass filter.

Let us define the following two phase factors:

$$\theta_{fm}(t) = 2\pi k_f g(t)$$
(7.57)
$$\theta_0(t) = 2\pi k_0\int_0^t v_o(\tau)d\tau$$

In terms of these factors, the output of the lowpass filter is

$$v_o(t) = \frac{1}{2}AB\sin[\theta_{fm}(t) - \theta_0(t)]$$
(7.58)

If a linear phase detector is used (or if the phase factors are small enough to approximate the sine by its angle), Eq. (7.58) becomes

$$v_o(t) = \frac{1}{2}AB[\theta_{fm}(t) - \theta_0(t)]$$
(7.59)

To find the transient loop response, we take the derivative of both sides of Eq. 7.59 to get

$$\frac{dv_o}{dt} = \frac{AB}{2}\left(\frac{d\theta_{fm}}{dt} - \frac{d\theta_0}{dt}\right) = \pi AB[k_f s(t) - k_0 v_o(t)]$$
(7.60)

Finally, the differential equation is

$$\frac{dv_o}{dt} + \pi k_0 AB v_o(t) = \pi k_f AB s(t) \tag{7.61}$$

The steady-state solution of Eq. (7.61) is found by setting the derivative equal to zero. It follows that

$$v_o(t) = \frac{k_f}{k_0} s(t) \tag{7.62}$$

The transient response of the first-order differential equation is an exponential function. The time constant of the transient response is $1/\pi k_0 AB$. This determines the response time of the loop.

Integrated Circuit Phase Lock Loops

Phase lock loops are available as integrated circuits. A typical device is the LM565, which is shown in Fig. 7.29 configured for frequency demodulation. The IC contains the same

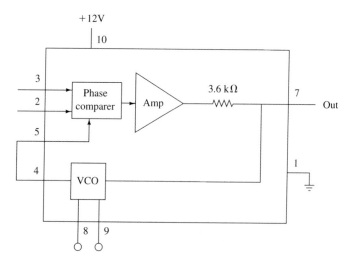

Figure 7.29   LM565 IC phase lock loop.

blocks as in Fig. 7.28. The amplifier coupled with the internal resistor and an external capacitor form a lowpass filter. The lowpass filter output is the demodulated output of the circuit. The 565 operates in a range of frequencies up to 500 kHz. Operation at higher frequencies requires a more complex IC. For example, the NE564 is capable of operation up to 50 MHz.

In selecting the proper phase lock loop for a particular application, one runs across a variety of descriptive terms. We shall spend a moment now defining these terms, for we will be using them again when we use phase lock loops in digital circuitry.

The *free-running frequency*, or center frequency, is the frequency of the VCO without any input.

The *lock range* is the range of frequencies over which the loop is capable of staying in lock. It is normally centered around the free-running frequency. This is similar to the *tracking range,* except for a factor of 2. The lock range is the total change in frequency between the highest and lowest operating frequencies. The tracking range is the maximum deviation permitted from the free-running frequency.

The *capture range* is the range of frequencies over which the loop is capable of locking in. It is normally smaller than the lock range. That is, once the loop is in locked, the frequency can deviate a certain maximum amount without losing lock. However, if the system is not yet in the locked and is searching for the correct frequency, the allowable deviation is less.

The *lock-up* time is the duration of the transient response.

7.8 BROADCAST FM AND STEREO

If you pick up any FM program guide, you will notice that adjacent stations are separated by 200 kHz. The FCC has assigned carrier frequencies of the type 101.1 MHz, 101.3 MHz, 101.5 MHz, etc., to the various transmitting stations. Thus, while the bandwidth allocated to each station in AM is only 10 kHz, it is an impressive 200 kHz for FM stations. This gives each broadcaster lots of room to work with.

The maximum frequency f_m of the information signal is set at 15 kHz. This is three times the figure specified for AM, accounting in part for the fact that FM sounds much better than AM. The higher frequencies provide much fuller musical sounds, and even voice signals sound much clearer.

To transmit this signal using narrowband FM would require a bandwidth of $2f_m$, or 30 kHz. Since 200 kHz is available, we see that broadcast FM is wideband.

Using the approximate formula developed for the bandwidth of an FM signal, we see that a maximum frequency deviation of about 85 kHz is possible. The actual figure used is a conservative 75 kHz.

The FM broadcast receiver looks very much like the superheterodyne AM receiver. The only differences are the range of frequencies of the image rejection filter and local oscillator (the FM band extends from 88 MHz to 108 MHz), the if frequency (which is set at 10.7 MHz), and the addition of the discriminator prior to the envelope detector. This is illustrated in the block diagram of Fig. 7.30.

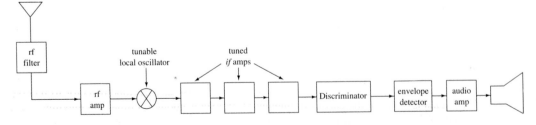

Figure 7.30   Block diagram of FM receiver.

FM Stereo

FM stereo is the process of sending two independent audio signals simultaneously within the same FM channel. Unlike AM, we do not have to look very hard to find room for additional channels. We have already noted that only 30 kHz of bandwidth is needed to send the signal by narrowband FM. Since 200 kHz is allocated, there is plenty of room, although we will have to use narrowband instead of wideband.

We begin with a system that multiplexes the two individual audio channels. This system is illustrated in the left portion of Fig. 7.31. We use two different shapes to represent the two Fourier transforms, $S_1(f)$ and $S_2(f)$. Each of these is meant to represent a

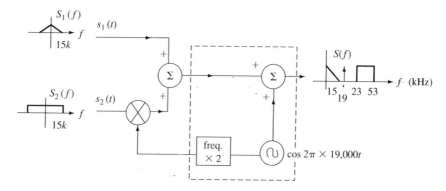

Figure 7.31   Multiplexing stereo signals.

general low-frequency limited signal. The reason for the two different shapes is to make it easier to track the signals through the system. We amplitude modulate a carrier at 38 kHz with $s_2(t)$, thus shifting the signal into the range between 23 kHz and 53 kHz. This does not overlap the frequencies of $s_1(t)$, so we can now add the two signals together confident in the thought that we can pull them apart later. The Fourier transform of the sum is shown on the right side of the figure. Do not worry about the impulse in the transform and the circuit in the dashed box: we will return to them later.

The composite signal

$$s_1(t) \; + \; s_2(t)\cos(2\pi \times 3.8 \times 10^4 t) \tag{7.63}$$

is a function of time with an upper frequency of 53 kHz. We can frequency modulate our carrier with this function. If we use narrowband FM, the resulting waveform uses only 106 kHz of the available 200 kHz.

At the receiver, we demodulate the FM waveform to recover the composite signal. This is shown on the left side of Fig. 7.32. A lowpass filter recovers $s_1(t)$ while rejecting the second term. A bandpass filter would separate the second term from the first, but we then must recover $s_2(t)$ from the modulated waveform. Even if we chose to add a carrier to make this transmitted carrier AM, we could still not use an envelope detector to receive $s_2(t)$. This is because the carrier frequency of 38 kHz is about 2.5 times the maximum frequency of $s_2(t)$. Proper operation of the envelope detector requires that the envelope be at a much lower frequency than the carrier. Accordingly, we *must* use synchronous demodula-

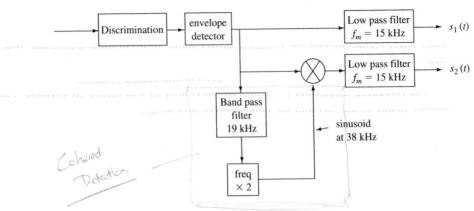

Coherent
Detection

Figure 7.32   FM stereo receiver.

tion. This is shown in Fig. 7.32, where the composite signal is multiplied by a carrier at 38 kHz and then lowpass filtered to recover $s_2(t)$.

How do we assure that the 38-kHz sinusoid in the receiver is perfectly synchronized with the received carrier? We could, of course, transmit some of the carrier and use a phase lock loop to recover it at the receiver. However, a much simpler option exists. Returning to the transmitter of Fig. 7.31, note that the 38-kHz carrier is developed by doubling the frequency of a 19-kHz oscillator. The 19-kHz signal is added to the composite signal. Thus, the actual composite signal that goes to the narrowband FM modulator is

$$s_1(t) \; + \; s_2(t)\cos(2\pi \times 3.8 \times 10^4 t) \; + \; A\cos(2\pi \times 1.9 \times 10^4 t) \qquad (7.64)$$

This composite signal has the Fourier transform shown on the right of Fig. 7.31. Note that the impulse has appeared due to the 19-kHz sinusoid.

At the receiver, the output of the envelope detector of Fig. 7.32 contains this 19-kHz sinusoid. This sinusoid is separated using a bandpass filter, which need not be too sharp, since no signal energy lies within 4 kHz on either side of the 19-kHz sinusoid.[1] The 19 kHz reconstructed signal is doubled in frequency and used to synchronously demodulate the AM waveform. We are assured of perfect synchronization, since the two 38-kHz sinusoids (that in the transmitter and that in the receiver) are derived from the same 19-kHz source.

Only one problem remains: When the FCC set standards for FM stereo, it made *compatibility* a requirement. That is, a monaural receiver should not receive just the left (or right) channel.

The upper leg of Fig. 7.32 represents a monaural receiver whose output is $s_1(t)$. We do not want $s_1(t)$ and $s_2(t)$ to represent the left and right channels, respectively. Instead, we let $s_1(t)$ be the sum of the left and right signals and $s_2(t)$ be their difference. The monaural receiver then recovers the sum of the left and right channels. The stereo receiver must do an additional linear operation: add $s_1(t) + s_2(t)$ to get one channel and take the difference of these signals to get the other channel. This is known as a *matrix* operation.

---

[1]This filter provides us with a bonus. If monaural FM forms the input to the receiver, the output of the 19-kHz bandpass filter is zero (i.e., all of the information in the signal is limited to 15 kHz). The output of the filter can therefore be used to turn on a "stereo" indicator light on the front panel of the receiver.

The unused portion of the allotted band between a 53-kHz and a 100-kHz deviation from f_c is sometimes used for so-called subsidiary communications authorization (SCA) signals. These signals include the commercial-free music heard in some restuarants, transmitted on a subcarrier of 67 kHz. This portion of the frequency band is also licensed for special applications, such us utility load management and paging systems.

The foregoing discussion illustrates the flexibility of FM broadcast systems. This flexibility is due to the fact that the bandwidth of an FM signal depends upon more than just the bandwidth of the information signal.

7.9 PERFORMANCE

Since FM is a nonlinear form of modulation, analyzing noisy FM systems proves much more difficult than analyzing noisy AM systems. We have to be content with some approximations and generalizations.

Let us first assume that the noise-free FM waveform

$$\lambda_{fm}(t) = A\cos 2\pi[f_c t + k_f g(t)] \tag{7.65}$$

is transmitted. In Eq. (7.65), $g(t)$ is the integral with respect to time of the information signal. Noise is added to this FM signal along the path between the transmitter and receiver. The additive noise is assumed to be white with a power of N_0 watts/Hz. The combination of noise and signal, $r(t)$, enters the receiver antenna, as shown in Fig. 7.33. Although the figure indicates that the noise is added prior to the receiver input, this model can also be used for the case of front end receiver noise.

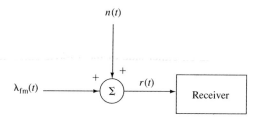

Figure 7.33   FM plus noise into receiver.

FM Receivers

An FM waveform carries its information in the form of frequency; the amplitude of the FM wave is constant. Another way to view this is that the information in the signal is contained in the zero-crossings of the wave. The FM waveform can be clipped at a low level (i.e., it can effectively be changed from a continuous waveform to a square wave) without loss of signal information. Additive noise can significantly affect the amplitude of the FM wave, but it perturbs the zero-crossing to a lesser degree. Receivers therefore often clip, or limit, the amplitude of the received waveform prior to frequency detection. This provides a constant-amplitude waveform as input to the discriminator. One effect of clipping is to introduce higher harmonic terms. A *postdetection* lowpass filter is used to reject these higher harmonics.

A block diagram of a simplified FM receiver is shown in Fig. 7.34. This receiver does not include the heterodyning and *IF* strip, since these do not change the signal to noise ratio.

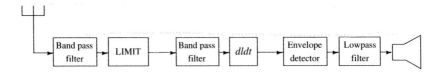

Figure 7.34   Simplified FM receiver.

Signal-to-Noise Ratio

The analysis of noise in FM systems is facilitated by a phasor presentation. Figure 7.35 presents a phasor diagram of the received FM signal plus additive noise. The signal phasor has length A and angle $2\pi k_f g(t)$. The phasor wiggles around according to $2\pi k_f g(t)$, while its amplitude remains constant. The noise phasor can be added to the signal if we expand the narrowband noise in quadrature components:

$$n(t) = x(t)\cos 2\pi f_c t - y(t)\sin 2\pi f_c t \tag{7.66}$$

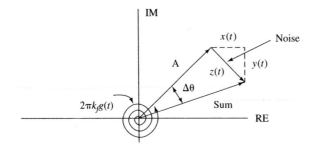

Figure 7.35   Phasor diagram of FM signal plus noise.

Here, $x(t)$ and $y(t)$ are low-frequency random processes. It is the bandpass filter in the receiver that allows us to use this bandlimited quadrature representation for the noise.

The angle of the resultant phasor contains all of the useful information. The length of the vector can be disregarded, since the limiter eliminates any variations in amplitude.

As long as the magnitude of the noise vector does not exceed A, the maximum angular difference $\Delta\theta$ between the resultant and signal vectors is 45°. The limiting value of 45° would occur only when the noise amplitude reached A, and the vectors were at right angles. As the noise gets smaller, the angular variations around the signal vector are limited to decreasing amounts less than 45°.

The angle of the signal vector is $2\pi k_f g(t)$. For wideband FM, this angle can be considerably larger than 360°. Thus, the proportional degradation caused by the noise can be made arbitrarily small by increasing k_f.

As a first step in quantifying these observations, we shall analyze the special case of a large carrier-to-signal ratio. That is, assume that A is much larger than $g(t)$. Thus, in the limit, a pure carrier is transmitted. The signal plus noise entering the limiter is then given by

$$r(t) = A\cos 2\pi f_c t + x(t)\cos 2\pi f_c t - y(t)\sin 2\pi f_c t \tag{7.67}$$

To trace this through the limiter, we combine terms and then eliminate the variations in amplitude. We rewrite $r(t)$ as

$$r(t) = \sqrt{[A + x(t)]^2 + y^2(t)} \ \cos\left[2\pi f_c t - \tan^{-1}\left(\frac{y(t)}{A + x(t)}\right)\right] \tag{7.68}$$

Since the output of the limiter has constant amplitude, we eliminate the amplitude term in Eq. (7.68). The limiter output can then be written as

$$\cos\left[2\pi f_c t - \tan^{-1}\left(\frac{y(t)}{A + x(t)}\right)\right] + \text{ higher harmonics} \tag{7.69}$$

The higher harmonics are rejected by the bandpass filter. The input to the discriminator is therefore the first term in Eq. (7.69). The output of the discriminator–envelope detector combination is

$$\frac{d\theta}{dt} = \frac{[x(t) + A]dy/dt - y(t)dx/dt}{y^2(t) + [x(t) + A]^2} \tag{7.70}$$

where we have taken the derivative of the inverse tangent and have also eliminated the constant term arising from the carrier frequency.

The signal of Eq. (7.70) enters the final lowpass filter of Fig. 7.34. To find the power at the output of this filter, we must examine the frequency content of Eq. (7.70). We know the power spectral densities of $x(t)$ and $y(t)$, and differentiation operations can be modeled as linear systems with $H(f) = j2\pi f$. However, before doing the mathematics, we can simplify Eq. (7.70) considerably under the assumption of high carrier-to-noise ratios (CNRs). That is, we assume that A is much larger than $x(t)$ and $y(t)$. Equation (7.70) therefore becomes

$$\frac{d\theta}{dt} \approx \frac{dy/dt}{A} \tag{7.71}$$

Our goal is to find the output noise power, so we investigate the power spectral density of the output noise, as given in Eq. (7.71). Note that the signal is not present, since we assumed that the received waveform was a pure carrier plus noise. The power spectral density of $y(t)$ is N_0 for frequencies between $-\text{BW}/2$ and $+\text{BW}/2$, where BW is the bandwidth of the receiver input filter. (Review the narrowband noise expansion in Chapter 5 if you have trouble verifying this.) Even though the bandwidth can be many times f_m for wideband FM, the final lowpass filter will cut off any components above f_m. The power spectral density of dy/dt is simply $(2\pi f)^2$ times the density of $y(t)$. Therefore, the power spectral density of the output noise of Eq. (7.71) is as shown in Fig. 7.36.

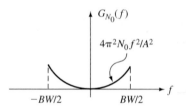

Figure 7.36 Spectral density of output noise.

The output noise power is found by integrating the power spectral density over the frequency range of the lowpass filter:

$$P_N = 2\int_0^{f_m} G_{N_0}(f)df = \frac{8\pi^2}{3}\frac{N_0}{A^2}f_m^3 \tag{7.72}$$

Since the signal waveform at the output of the FM receiver is $2\pi k_f s(t)$, the output signal power is $4\pi^2 k_f^2 P_s$, where P_s is the power of the information signal $s(t)$. The output SNR for high CNR is then

$$\text{SNR}_o = \frac{4\pi^2 K_f^2 P_s}{P_N} = \frac{3A^2 k_f^2 P_s}{2N_0 f_m^3} \tag{7.73}$$

We can give an intuitive interpretation to Eq. (7.73) by considering the special case of a sinusoidal information signal. That is, assume that $s(t) = a\cos 2\pi f_m t$. Then $P_s = a^2/2$, and Eq. (7.73) becomes

$$\text{SNR}_o = \frac{3\beta A^2/2}{2\beta f_m N_0}\left(\frac{a k_f^2}{f_m}\right) = 3\beta^3 \text{SNR}_i \tag{7.74}$$

In Eq. (7.74), we define SNR_i as the power of the received carrier divided by the power of the received noise in the FM bandwidth of $2\beta f_m$. The equation clearly shows the effect of wideband FM: As β increases, the bandwidth increases, but so does the output SNR. The equation makes it look as if the improvement follows the third power of β. This is misleading, however, since the input noise power to the receiver is linearly proportional to β. Therefore, if we referenced output SNR to an SNR that used noise in a fixed bandwidth, the relationship would follow the square of β. This leads to a trade-off decision in the design of systems.

Threshold Effect

Let us now return to Fig. 7.35 to explore more fully the effect of noise on wideband FM. We made the assumption that the magnitude A of the signal vector was larger than the magnitude of the noise vector. If the reverse is true, the diagram could be redrawn as in Fig. 7.37. Alas, the receiver doesn't know the difference between signal and noise. The wideband FM receiver improves noise performance only because the signal is larger than the noise. If the noise gets larger than the signal, the receiver locks onto the noise and suppresses the signal. This phenomenon is known as *noise capture*. It is clearly seen in FM broadcast radio when one listens to a distant station (e.g., in an automobile as you ap-

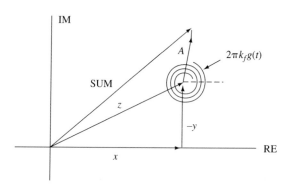

Figure 7.37   Noise larger than signal.

proach the distance limits of reception). The station occasionally drops out and is replaced by static.

The amplitude of the signal remains constant at A. As the average length of the noise vector approaches A, we can expect this random quantity to be sometimes larger and sometimes smaller than A. Figure 7.38 shows a possible trajectory of the noise vector and the resultant sum as the noise vector hovers around a magnitude of A. A counterclockwise rotation around the origin is illustrated. Also in the figure are the resulting angle and its derivative, the frequency. Note that the pulse in frequency is positive and has an area of 2π. If the encirclement of the origin had been in the clockwise direction, the pulse would have been negative. In cases where the origin is not encircled, the average value (total area) of the angular variations is zero. The spike in the frequency waveform would result in an audible *click* in the receiver audio output. Since the noise vector is sweeping around as the quadrature components $x(t)$ and $y(t)$ change, we can expect occasional clicks as $|z|$ approaches A. The larger the average value of $|z|$ gets, the more frequent will be the clicks, until noise capture takes effect. At that point, the audio output would simply be approximately that of the random noise—a rushing sound in the speaker.

Since the noise is random, the clicks cannot be characterized deterministically; we can only derive probabilistic averages. One quantity of interest is the average number of clicks per second.

For simplicity, we will rotate the phasor diagram to align the horizontal axis with the signal vector. The new diagram is redrawn as Fig. 7.39.

In order for the trajectory to cross the horizontal axis at point P, $y(t)$ must have a zero-crossing, and at the same time, $x(t)$ must be less than $-A$. Since the noise quadrature components, $x(t)$ and $y(t)$, are independent of each other, the probability of this joint event is the product of the two individual probabilities. Because $x(t)$ and $y(t)$ are assumed to be Gaussian random processes, the probability that $x(t)$ is less than $-A$ is given by a complementary error function. The probability of a zero-crossing can be found from the joint density of the function and its derivative. That is, if the function is negative at a particular point in time with a sufficiently positive derivative, a zero-crossing results. The probability of a trajectory circling the origin within a time interval t is given by

$$\text{Pr(click)} = \frac{\text{BW }\Delta t}{2\sqrt{3}} \text{ erfc} \left(\sqrt{\frac{A^2}{2N_0\text{BW}}} \right) \qquad (7.75)$$

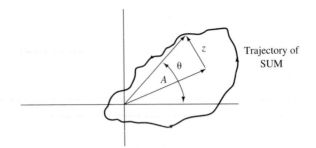

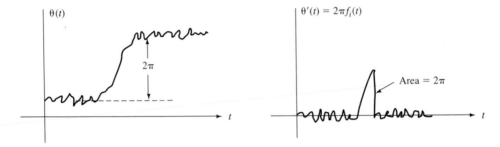

Figure 7.38   Trajectory of sum vector.

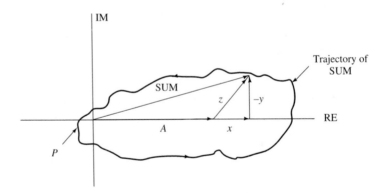

Figure 7.39   Realigned phasor diagram.

where BW is the bandwidth of the FM waveform. The average number of clicks per second is then

$$\text{CPS}_{av} = \frac{\text{BW}}{2\sqrt{3}} \, \text{erfc}\left(\sqrt{\frac{A^2}{2N_0\text{BW}}}\right) \tag{7.76}$$

Note that the argument of the complementary error function is the square root of the signal to noise ratio. Thus, if we define the signal to noise ratio as

$$\text{SNR} = \frac{A^2}{2N_0\text{BW}} \tag{7.77}$$

then the average number of clicks per second becomes

$$CPS_{av} = \frac{BW}{2\sqrt{3}} \operatorname{erfc}\left(\sqrt{SNR}\right) \tag{7.78}$$

As the SNR increases, the complementary error function decreases, and the number of clicks per second also decreases. As the bandwidth of the FM wave increases, the number of clicks per second increases, since additional noise is being admitted into the system.

The area under each pulse in Fig. 7.38 is 2π. We can therefore get an approximation to the output signal to noise ratio near the threshold by adding the power of the clicks to the noise power found in the high-CNR case. The power of the clicks is estimated in a manner similar to that used for impulse noise analysis. The final result is

$$SNR = \frac{3BW\,(SNR)k_f^2\,P_s/f_m^3}{1 + (BW)^3\,4\sqrt{3}\,\operatorname{erfc}\left(\sqrt{SNR}\right)f_m^3} \tag{7.79}$$

As the average number of clicks per second approaches zero, the denominator of Eq. (7.79) approaches unity, and the SNR becomes that of Eq. (7.73) for the high-CNR case.

Pre-emphasis and De-emphasis

Equation (7.73) indicates that the signal to noise ratio of an FM waveform increases with increasing k_f. Thus, a noise advantage is realized by increasing the bandwidth of the FM signal.

An even greater noise advantage can be realized by observing that the noise at the output of the FM receiver is not white, even though the input noise is white. Figure 7.36 shows that the output noise power spectral density increases parabolically with increasing frequency (i.e., when a sinusoid is differentiated by the discriminator, the amplitude is multiplied by the frequency). Therefore, the high information frequencies are more se-verely affected by noise than are the lower frequencies. A more efficient use of the band of frequencies would occur if these higher signal frequencies were *emphasized* and the lower frequencies were *de-emphasized* in a manner that kept the total power constant. The signal could then be shaped by a filter with $H(f)$, as shown in Fig. 7.40. The shaping would be done prior to transmission. All signal frequencies would then be equally affected by the noise. The filter, of course, represents a signal distortion. The filtering operation must therefore be *undone* at the receiver, using an inverse filter known as the *de-emphasis filter*. This filter has a transfer function $1/H(f)$. The composite transfer function of the two filters is unity.

The de-emphasis filter changes the noise power spectral density from that shown in Fig. 7.36 to a density which is approximately white. This decreases the output noise power. It also assures equal noise disturbance over all signal frequencies.

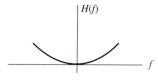

Figure 7.40   Pre-emphasis filter.

Total noise power is not the only consideration. If the noise power is concentrated at high frequencies (as is the case without de-emphasis), the overall SNR may seem acceptable while high-frequency reception is very poor. In real life, many information signals drop off as frequency increases. It may therefore be desirable to overemphasize the signal with a filter characteristic which increases more rapidly with frequency than that shown in Fig. 7.40.

Commercial broadcast FM uses pre-emphasis, but in the interest of economy, not the exact function shown in Fig. 7.40. A simple *RC* lowpass filter in the receiver provides the de-emphasis. The inverse filter is used in the transmitting station for pre-emphasis.

7.10 COMPARISON OF SYSTEMS

We have discussed narrowband and wideband FM and narrowband and wideband PM. Narrowband systems have the same bandwidth as double-sideband AM, so we need to develop guidelines to assist us in choosing the best candidate system. Also, FM and PM have similar characteristics, so again we need guidelines to help choose. In this section, we summarize the essential characteristics of each of the four angle-modulated systems. Included as well is a summary of the AM techniques of the previous chapter.

Narrowband FM can be generated with a system consisting of a multiplier, an integrator, and a phase shifter. It is demodulated with a discriminator, followed by an envelope detector or by a phase lock loop. These are relatively simple implementations. The bandwidth of narrowband FM is $2f_m$, where f_m is the maximum frequency of the information signal. Although the modulation process makes this look very much like a variation of AM, it is definitely different. The modulated waveform has constant amplitude, permitting us to add a limiter to the receiver and thereby cutting down on additive noise. This is particularly effective if the additive noise is in the form of spikes. The process of discriminating in the receiver changes the noise from white to nonwhite, accentuating high frequencies while attenuating low frequencies. This yields a noise advantage over AM if critical signal-related information resides at low frequencies or if the additive noise contains a nonwhite component at a low frequency.

Narrowband PM is very similar to narrowband FM. The integrator in the modulator is traded for an integrator in the demodulator. The bandwidth is $2f_m$. The amplitude of the PM wave is constant, thus affording the noise advantage realized by adding a limiter to the receiver. The final integrating process in the receiver attenuates the high frequencies. This could be advantageous if critical signal-related information resides at high frequencies or if the additive noise contains a nonwhite component at a high frequency.

Wideband FM is generated either indirectly from narrowband FM (through frequency multiplication) or by a VCO. It is demodulated in the same way as narrowband FM. The bandwidth is approximately $2\beta f_m$ which can be considerably greater than that of AM or narrowband angle modulation. The major advantage of wideband FM is capability of reducing the effect of noise. The output signal to noise ratio is approximately proportional to β^2.

Wideband PM bears the same relationship to wideband FM as does narrowband PM to narrowband FM. However, unlike wideband FM, the modulation index cannot be increased without limit. The maximum phase deviation is limited to 180°. Beyond that, there is a phase ambiguity, and the original signal cannot be uniquely recovered.

Double-sideband suppressed carrier AM has a bandwidth of $2f_m$ and an efficiency of 100 percent (i.e., no power is wasted in sending a pure carrier). Modulation is performed by a multiplier, while demodulation requires coherent circuits with the accompanying difficulty of reconstructing the carrier at the receiver.

Double-sideband transmitted carrier AM has a bandwidth of $2f_m$ and an efficiency of less than 50 percent because power is being wasted in sending a pure carrier. It is the easiest to demodulate (envelope detector) of all schemes of modulation. It cannot support a signal with a nonzero *dc* level, since that information would be lost in demodulation.

Single-sideband AM has the smallest bandwidth of all of the systems examined in this text, namely, f_m. It is 100 percent efficient, since no power is wasted in sending a pure carrier. It is not easy to implement the modulator or demodulators, because of the filtering required in the transmitter and the carrier recovery and coherent detection required in the receiver.

Vestigial-sideband suppressed carrier AM has a bandwidth larger than f_m, but less than $2f_m$. The modulator is easier to construct than that of single sideband, but the demodulator requires carrier recovery and also requires a carefully controlled filter shape to combine the sidebands properly.

Vestigial-sideband transmitted carrier AM has a bandwidth larger than f_m, but less than $2f_m$. The modulator is easier to construct than that of single sideband, and if the carrier is large enough, an envelope detector can be used, thus making demodulation extremely simple.

PROBLEMS

7.1.1 Find the instantaneous frequency of

$$s(t) = 10[\cos(10t)\cos(30t^2) - \sin(10t)\sin(30t^2)]$$

7.1.2 Find the instantaneous frequency of

$$s(t) = 2e^{-t}U(t)$$

7.2.1 An information signal is as shown in Fig. P7.2.1. The carrier frequency is 1 MHz, $k_f = 10^4$, and $k_p = 20$. Sketch the FM and PM waveforms, and label significant frequencies and values.

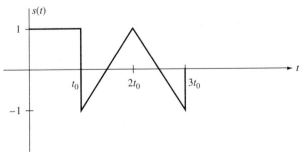

Fig P7.2.1

7.5.1 In Eq. (7.33), we found that

$$c_n = \frac{1}{T} \int_{-\frac{T}{2}}^{\frac{T}{2}} e^{j\beta \sin 2\pi f_m t} e^{-jn2\pi f_m t} \, dt$$

where $T = 1/f_m$. Show that the imaginary part of c_n is zero.

7.5.2 Show that c_n, as given in Eq. (7.33) [see Problem 7.5.1], is not a function of f_m or T. That is, shown that c_n depends only on β and n. [*Hint:* Let $x = 2\pi f_m t$ and make a change of variables.]

7.5.3 Find the approximate Fourier transform of

$$s(t) = \cos\left(\frac{ak_f}{f_m} \sin 2\pi f_m t\right)$$

Do this by expanding the cosine in a Taylor series and retaining the first three terms.

7.5.4 Repeat Problem 7.5.3 for

$$s(t) = \sin\left(\frac{ak_f}{f_m} \sin 2\pi f_m t\right)$$

7.5.5 Use the results of Problems 7.5.3 and 7.5.4 to find the approximate Fourier transform of the FM waveform

$$\lambda_{\text{fm}}(t) = \cos\left(2\pi f_c t + \frac{ak_f}{f_m} \sin 2\pi f_m t\right)$$

7.5.6 Find the approximate band of frequencies occupied by an FM waveform of carrier frequency 2 MHz, where $k_f = 100$ Hz/V and:
(a) $s(t) = 100\cos 2\pi \times 150t$ volts.
(b) $s(t) = 200\cos 2\pi \times 300t$ volts.

7.5.7 Find the approximate band of frequencies occupied by an FM waveform of carrier frequency 2 MHz, where $k_f = 100$ Hz/V and

$$s(t) = 100 \cos 2\pi \times 150t + 200 \cos 2\pi \times 300t \text{ volts}$$

Compare this with your answers to Problem 7.5.6. Contrast the result with AM.

7.5.8 Consider the system shown in Fig. P7.5.8, where the output is alternately switched between a source at frequency $f_c - \Delta f$ and a source at frequency $f_c + \Delta f$. The switching is done at a frequency f_m. Find and sketch the Fourier transform of the output $y(t)$. You may assume that the switch takes zero time to go from one position to another.

7.5.9 A 10-MHz carrier is frequency modulated by a sinusoid of unit amplitude and $k_f = 10$ Hz/V. Find the approximate bandwidth of the FM waveform if the modulating signal has a frequency of 10 kHz.

7.5.10 A 100-MHz carrier is frequency modulated by a sinusoid of frequency 75 kHz such that $\Delta f = 500$ kHz. Find the approximate band of frequencies occupied by the FM waveform.

7.5.11 Find the approximate band of frequencies occupied by the waveform

$$\lambda(t) = 100 \cos (2\pi \times 10^5 t + 35 \cos 100 \pi t)$$

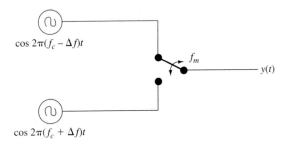

Fig. P7.5.8

7.5.12 An angle-modulated waveform is described by

$$\lambda(t) = 50 \cos (2\pi \times 10^6 t + 0.001 \cos 2\pi \times 500t)$$

  (a) What is $f_i(t)$, the instantaneous frequency of $\lambda(t)$?
  (b) What is the approximate bandwidth of $\lambda(t)$?
  (c) Is λ a narrowband or wideband angle-modulated signal?
  (d) If λ represents an FM signal, what is $s(t)$, the information signal?
  (e) If λ represents a PM signal, what is $s(t)$?

7.7.1 Find the approximate pre-envelope and envelope of

$$s(t) \cos \left(2\pi f_c t + k_f \int_0^t s(\tau)d\tau \right)$$

where the maximum frequency of $s(t)$ is much less than f_c.

7.7.2 A phase lock loop is used to demodulate an FM waveform. The information signal is a voice waveform with maximum frequency 5 kHz. The comparator has an output that varies linearly with input phase difference. An input phase difference of 90° causes a 25-V change in the output. The VCO output varies linearly and changes by 10 kHz for each 1-V change in input. Will the loop act as an effective demodulator?

7.7.3 Show that the system of Fig. P7.7.3 can be used to demodulate an FM waveform. Analyze the performance of this demodulator as a function of the time delay T.

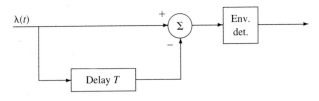

Fig. P7.7.3

7.8.1 You are given the sum $(L + R)$ and the difference $(L - R)$ signals in a stereo receiver system. Design a simple resistive circuit that could be used to produce the individual left and right signals from the sum and difference waveforms.

7.8.2 An FM stereo system is being designed for planet X, where the people have hearing that goes up to 25 kHz. The XCC (equivalent to FCC on the earth) has specified the band of frequencies between 100 MHz and 200 MHz for their broadcast FM. Each channel is allocated 200 kHz within this band. You are asked to design the FM stereo system. Describe, in detail, the changes you would make to the system we use here on earth.

7.9.1 An information signal

$$s(t) = 5 \cos 2{,}000 \, \pi t$$

frequency modulates a carrier of frequency 10^6 Hz. $A = 100$ and $k_f = 100$ Hz/V. Noise of power spectral density 10^{-3} is added during transmission.

(a) Find the SNR at the output of an FM receiver, in dB.

(b) If the output is put through a bandpass filter with passband from 900 to 1,100 Hz, find the improvement in SNR of the filter.

(c) Repeat parts (a) and (b) for AM double-sided transmitted carrier. Compare the results.

7.9.2 Repeat Problem 7.9.1 for

$$s(t) = 5 \cos 2{,}000 \, \pi t + 3 \cos 4{,}000 \, \pi t$$

7.9.3 Repeat Problem 7.9.1 for

$$s(t) = \frac{100 \sin 2{,}000 \, \pi t}{\pi t}$$

7.9.4 Repeat the analysis of Problem 7.9.1 if PM is used instead of FM. (Give answers in terms of k_p.) Does a range exist for k_p such that PM gives better performance than FM? If so, give that range.

7.9.5 You are given the FM signal

$$\lambda(t) = 100 \cos (2\pi \times 10^5 t + 35 \cos 100 \, \pi t)$$

This signal is received along with additive white noise of power spectral density $N_0/2 = 0.1$. Find the average number of clicks per second in the output of the detector.

7.9.6 An FM signal is given by

$$\lambda(t) = 50 \cos (2\pi \times 10^6 t + 0.001 \cos 2\pi \times 500 t)$$

Noise of power spectral density $N_0/2 = 0.05$ is added to the signal. The sum of the signal and noise enters the detector. Is the detector operating near threshold? If your answer is yes, find the average number of clicks per second at the output of the detector.

7.9.7 An FM signal is given by

$$\lambda(t) = A \cos (2\pi \times 10^5 t + 35 \cos 100 \, \pi t)$$

White noise with power of $N_0 = 0.01$ watt/Hz is added to the signal, and the sum forms the input to a detector.

(a) Find the value of A so that the average number of clicks per second is 5.

(b) For the value of A in part (a), find the signal to noise ratio at the detector output.

7.9.8 You receive the phase modulated waveform

$$\lambda(t) = 10^{-3}\cos(2\pi \times 10^7 t + 10K\cos 100\,\pi t + 15K\cos 200\,\pi t)$$

plus additive white Gaussian noise with a power of 10^{-8} watt/Hz. Find the signal to noise ratio at the output of the demodulator if:

(a) $K = 0.01$
(b) $K = 1$
(c) $K = 10$
(d) $K = 100$

8

Source Encoding

8.0 PREVIEW

What We Will Cover and Why You Should Care

This is the first of three chapters devoted exclusively to digital communication. If you refer back to the block diagram of a communication system (Fig. 1.13), you note that the first major operation is source encoding. In the case of digital communication, this is the process of turning the information signal into a sequence of ones and zeros. There is no sense in examining transmission techniques for digital signals until you learn how to create those signals from real-world information sources.

Once you finish studying this chapter, you will know the principal methods for converting information into digital formats. You will also know about and be able to analyze the various errors that occur during encoding. In addition, we introduce the basics of configuring digital signals for transmission through a channel.

Necessary Background

Most of this chapter requires very little background. You need a basic familiarity with binary operations and binary numbers. When you get to the analysis of errors (quantization noise), you will need to know probability and random analysis.

8.1 ANALOG-TO-DIGITAL CONVERSION

Digital communication possesses certain important advantages over analog communication—in particular, improved immunity from noise and simplicity of processing. For this reason, we often send analog information signals using digital techniques. Accordingly, we now discuss techniques of converting an analog signal into a digital signal.

The first step in changing a continuous analog time signal into digital form is to convert the signal into a list of numbers. This is accomplished by sampling the signal. The result is a list of numbers, each of which represents an infinite decimal number. That is, although a particular sample may be stated as a rounded-off number (e.g., 5.758 volts), in actuality it should be continued as an infinite decimal. We call this a *discrete signal*, and it formed the basis for our discussion in Chapter 5.

To turn the discrete signal into a *digital* signal, the list of analog numbers must be *coded* into discrete *code words*. The first approach toward accomplishing this might be to round off each number in the list. Thus, if the samples range, for example, from 0 to 10 V, each sample could be rounded to the nearest integer. This would result in code words drawn from the 11 integers between 0 and 10.

In many communication systems, the form chosen for code words is a *binary number*—that is, a sequence of 1's and 0's. The reasons for this choice will become clear when we discuss specific transmission techniques. With the restriction to binary numbers, we can envision one simple form of *analog-to-digital* (A/D) converter for the example of 0- to 10-V signals. The converter would operate on the samples by first rounding each one to the nearest volt. It would then convert the resulting integer into a four-bit binary number (a *binary-coded decimal*, or BCD).

While practical A/D converters perform an operation similar to that just outlined, an appreciation of this operation is best introduced by starting from the beginning.

Analog-to-digital conversion is known as *quantizing*. The goal is to change a continuous variable into a variable with discrete values. In *uniform quantization*, the continuum of functional values is divided into regions of uniform width, and an integer code is assigned to each region. Thus, all functional values within a particular region are coded into the same number.

Figure 8.1 illustrates the concept of three-bit quantization in two different ways. Figure 8.1(a) shows the range of functional values divided into eight regions. Each of these regions is assigned a three-bit binary number. We have chosen eight regions because 8 is a power of 2. Thus, all three-bit binary combinations are used, leading to greater efficiency. Note that the range of values is given as that between zero and unity. While this may seem restrictive, any function can be *normalized* to fall within this region through the addition of a constant and scaling.

Before continuing, we shall use Fig. 8.1 to make a general observation about binary counting. As you examine the binary numbers along the ordinate, note that the first bit (the *most significant bit*) is equal to 1 for the top half of the range and 0 for the bottom half. It oscillates with a period equal to the total range. The next bit oscillates with a period equal to half of the range and is equal to 1 for the top half *of each half* and 0 for the bottom half *of each half*. This pattern continues, with each successive bit subdividing the region by two and indicating which half of the new subregion the sample value is in.

Figure 8.1(b) illustrates quantization by use of an input-output relationship. While the input is continuous, the output can take on only discrete values. The width of each step is constant, since the quantization is uniform.

Figure 8.2 shows a representative $s(t)$ and the resulting digital form of the signal for both a two-bit and a three-bit A/D converter.

Quantizers

There are three generic types of quantizer:

1. *Serial quantizers*, which generate a code word, bit by bit. That is, they start with the most significant bit and work their way to the least significant bit.

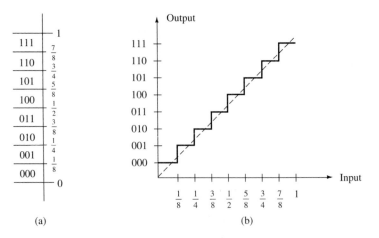

Figure 8.1 Concept of quantization.

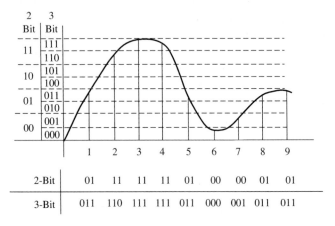

Figure 8.2 Result of A/D conversion for representative $s(t)$.

2. *Counting quantizers*, which sequentially count through each quantizing level.
3. *Parallel quantizers*, which generate all bits of a complete code word simultaneously.

Serial Quantizers

The serial quantizer (also known as the *successive approximation quantizer*) successively divides the ordinate into two regions. First, it divides the range in half and observes whether the sample is in the upper or lower half-range. The result of this observation generates the most significant bit in the code word.

The half-range in which the sample lies is then subdivided into two regions, and a comparison is again performed. This generates the next bit. The process continues a number of times equal to the number of bits in the A/D conversion. Thus, each bit increases the resolution by a factor of 2.

Figure 8.3 shows a flow diagram of the serial quantizer for three bits of encoding and for inputs in the range 0 to 1. The diamond-shaped boxes are *comparators*. They com-

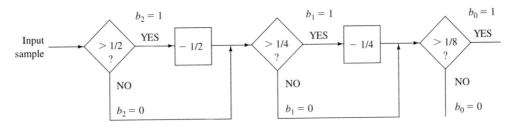

Figure 8.3   The serial quantizer.

pare the input with some fixed value and give one output if the input exceeds the fixed value and another output if the reverse is true. The block diagram indicates these two alternatives as two possible output paths labeled "YES" and "NO." The diagram is shown for three-bit code words and for a range of input values between 0 and 1 V. If the range were not 0 to 1, the sampled signal could be normalized (shifted and then amplified or attenuated) to achieve values within this range. If more (or fewer) bits are required, the appropriate comparison blocks can be added to (or removed from) the diagram.

Bit b_2, the first bit of the coded sample value, is known as the *most significant bit.* Bit b_0, the third and final bit of the coded sample, is known as the *least significant bit.* The reason for this terminology is that the weight associated with b_2 is 2^2, or 4, while the weight associated with b_0 is 2^0, or 1.

Example 8.1

Illustrate the operation of the system of Fig. 8.3 for the input sample values of 0.2 V and 0.8 V.

Solution: For 0.2 V, the first comparison, with $\frac{1}{2}$, yields a "No" answer. Therefore, $b_2 = 0$. The second comparison, with $\frac{1}{4}$, yields a "No" answer also, so $b_1 = 0$. The third comparison, with $\frac{1}{8}$, yields a "Yes" answer, so $b_0 = 1$. The binary code for 0.2 V is therefore 001.

For the 0.8-V input, the first comparison, with $\frac{1}{2}$, yields a "Yes" answer, so $b_2 = 1$. We then subtract $\frac{1}{2}$, leaving 0.3. The second comparison, with $\frac{1}{4}$, results in a "Yes" answer, so $b_1 = 1$, and we subtract $\frac{1}{4}$, leaving 0.05. The third comparison, with $\frac{1}{8}$, yields "No," so $b_0 = 0$. The code for 0.8 V is thus 110.

A simplified system can be realized if, at the output of the block marked "$-\frac{1}{2}$" in Fig. 8.4, a multiplication by 2 is performed and the result is fed back into the comparison with $\frac{1}{2}$. All blocks to the right can then be eliminated, as in Fig. 8.4.

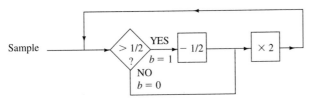

Figure 8.4   Simplified serial quantizer.

The sample of the signal can be cycled through as many times as desired to achieve any number of bits for the length of the code word. You can draw an analogy to using a microscope to view a small sample. You position the sample in the middle of the field of view and then double the magnification. The process is repeated as many times as desired.

Counting Quantizers

Figure 8.5 illustrates a counting quantizer. The ramp generator starts at each sampling point, and a binary counter is simultaneously started. The output of the *sample-and-hold* system is a staircase approximation to the original function, with steps that stay at the previous sample value throughout each sampling interval. A typical waveform is shown in the figure. The duration of the ramp, and therefore the duration of the count, T_s, is proportional to the sample value. This is true because the slope of the ramp is kept constant. The clock frequency is such that the counter has enough time to count to its highest value (all 1's) for a ramp duration corresponding to the maximum possible sample. The ending count on the counter corresponds to the quantizing level.

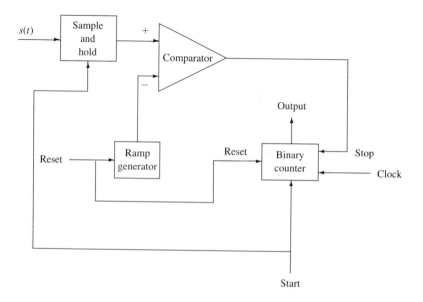

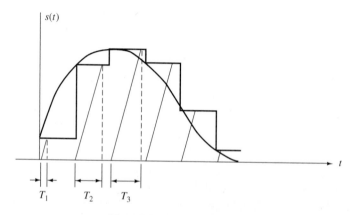

Figure 8.5   Counting quantizer.

Example 8.2

Design a counting quantizer for a voice signal with a maximum frequency of 3 kHz. The slope of the ramp is specified as 10^6 volts/sec, and the amplitudes of the signal range from 0 to 10 volts. Find the required clock frequency if a four-bit counter is used.

Solution: The only reason for considering the maximum frequency of the signal is to see whether the slope of the ramp is sufficient to reach the maximum possible sample value within one sampling period. With a maximum signal frequency of 3 kHz, the minimum sampling rate for recovery is 6 kHz, so the maximum sampling period is $\frac{1}{6}$ msec. Since the ramp can reach the 10-V maximum in 0.01 msec, it is sufficiently fast to avoid overload. The counter must be capable of counting from 0000 to 1111 in 0.01 msec. The clock frequency must be 1.6 MHz, since up to 16 counts are required in this sampling period.

Example 8.3

Design a counting A/D converter to convert $s(t) = \sin 2,000\pi t$ into a four-bit digital signal.

Solution: To comply with the sampling theorem, the sampling rate must be greater than 2,000 samples/sec. Let us choose a rate 25 percent above this value, or 2,500 samples/sec. There are many practical trade-off considerations that go into this choice of sampling rate. If we use a rate close to the minimum, precise lowpass filtering is required at the receiver to reconstruct the original signal. On the other hand, if we use a much higher rate, the bandwidth of the transmitted waveform increases.

The individual sample values range between -1 V and $+1$ V. The counting quantizer discussed in this section operates upon positive samples. We therefore shift the signal by 1 V to assure that samples never go negative. The shifted samples range from 0 V to 2 V. The ramp must be capable of reaching the maximum sample value within the sampling period, 0.4 msec. Hence, the slope must be at least $2/0.4 \times 10^{-3} = 5,000$ volts/sec. In practice, we would choose a value larger than this to account for slight jitter in the timing of the system and also to give the ramp time to return to zero prior to arriving at the next sampling point. We might choose a much larger value for the slope if we wanted to convert the sample in a small fraction of the period. This would apply if a converter is being shared among a number of multiplexed signals. At the minimum slope, it takes the ramp function 0.4 msec to reach the maximum sample value. The counter should therefore count from 0000 to 1111 in 0.4 msec. This requires a counting rate of 40,000 counts/sec.

Parallel Quantizers

The parallel quantizer (or *flash coder*) is the fastest quantizer in operation, developing all bits of the code word simultaneously. It is also the most complex quantizer, requiring a number of comparators that is only one less than the number of levels of quantization. We illustrate the device with the three-bit encoder of Fig. 8.6.

The block labeled "Coder" observes the output of the seven comparators. The coder is a simple combinational logic circuit. If all seven outputs are 1 (YES), the coder output is 111, since the sample value must be greater than $\frac{7}{8}$. If comparator outputs 1 through 6 are 1 and output 7 is 0, the coder output is 110, since the sample must be between $\frac{6}{8}$ and $\frac{7}{8}$. We continue through all levels, and finally, if all comparator outputs are low, the sample is less than $\frac{1}{8}$, so the coder output is 000. The truth table for the coder is shown in Table 8.1. Only 8 of the 128 (2^8) possible coder inputs are valid, the other 120 inputs representing invalid possibilities (e.g., there is no way a sample could be greater than $\frac{7}{8}$ and not greater than $\frac{5}{8}$). Thus, the combinational logic circuit contains a high percentage of "don't care" conditions, and the design is simplified accordingly. Alternatively, the logic can be configured to give an error signal for any invalid input combination.

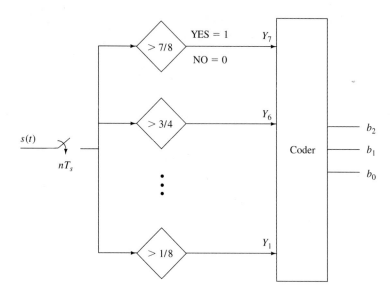

Figure 8.6   The parallel quantizer.

TABLE 8.1   TRUTH TABLE FOR PARALLEL QUANTIZER.

Y_1	Y_2	Y_3	Y_4	Y_5	Y_6	Y_7	b_2	b_1	b_0
0	0	0	0	0	0	0	0	0	0
1	0	0	0	0	0	0	0	0	1
1	1	0	0	0	0	0	0	1	0
1	1	1	0	0	0	0	0	1	1
1	1	1	1	0	0	0	1	0	0
1	1	1	1	1	0	0	1	0	1
1	1	1	1	1	1	0	1	1	0
1	1	1	1	1	1	1	1	1	1

While the serial quantizer takes advantage of the structure of binary numbers when they are counted in sequence, the parallel quantizer does not require such structure. In fact, the code for the quantization regions can be assigned in any useful manner. A problem with sequential assignment is that transmission bit errors cause nonuniform reconstruction errors. A bit error in the most significant bit has a much greater effect than one in the least significant bit.

In 1947, F. Gray, who was working with electronic coding devices, invented a *reflected binary code* (also known as a *Gray code*) in which adjacent numbers differ in only one bit position. We illustrate a four-bit version of the Gray code in Table 8.2. A reversal of one bit in the code word changes the digit by only one. The Gray code is easily implemented in the flash coder logic. It can also be used in the other types of quantizers. In the counting quantizer, we simply vary the count sequence. In the serial quantizer, we follow the decision operations with a simple combinational logic circuit to convert the sequential code to the Gray code.

TABLE 8.2 THE GRAY CODE

Digit	Binary	Gray
0	0000	0000
1	0001	0001
2	0010	0011
3	0011	0010
4	0100	0110
5	0101	0111
6	0110	0101
7	0111	0100
8	1000	1100
9	1001	1101
10	1010	1111
11	1011	1110
12	1100	1010
13	1101	1011
14	1110	1001
15	1111	1000

Practical Quantizers

Practical quantizers are constructed in accordance with the block diagrams presented in this section. Most forms are packaged as single integrated circuits. We discuss one practical quantizer for each class of conversion.

Counting quantizers are also known as *dual-slope* A/D converters. First, the input sample is applied to an integrator for a fixed length of time, thus yielding an integrated output that is proportional to the sample value. Then the input to the integrator is switched to a reference voltage (which is opposite in sign from the voltage of the signal sample), the counter is started, and the output of the integrator is compared to zero. The counter is stopped when the integrator output ramp reaches zero.

The *ICL7126* is a CMOS integrated circuit that simulates a counting quantizer. The IC package contains 40 pins and is illustrated in Fig. 8.7. Pins 2 through 25 are used for the output display. The IC is configured to directly drive liquid crystal displays (LCDs), as it includes seven-segment decoders and LCD drivers. The display is $3\frac{1}{2}$ digit, which means that it can indicate numbers with magnitudes as high as 1,999. The seven-segment outputs for the *units* display are indicated as A1 to G1. Those for the *tens* display use a suffix of 2, and the *hundreds* display uses 3. The *thousands* display is indicated as AB4, and only one lead is needed, since this digit is either 0 or 1 (for a $3\frac{1}{2}$ digit display).

The analog input is applied to pins 30 and 31. The actual operation of the IC proceeds in three phases. The first is *auto-zero*, in which the analog inputs are disconnected and internally shorted to ground (common), pin 32. The output of the comparator is shorted to the inverting input of the integrator.

The second phase occurs when the input signal is integrated for a time corresponding to 1,000 clock pulses. Finally, in the third phase, the reference voltage stored on a capacitor that is externally connected between pins 34 and 35 is used to initiate the second ramp. The range of input values determines the required value of the reference, which is

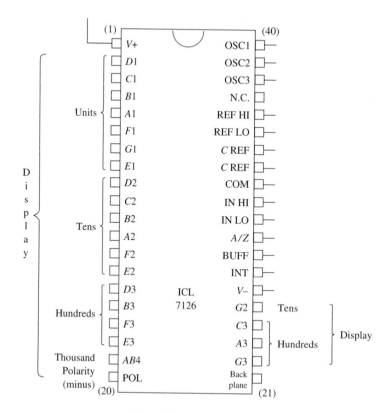

Figure 8.7   ICL7126 quantizer.

input to the REF HI pin, pin 36. If this input is 1 V, the chip is capable of converting voltages with magnitudes as high as 1.999. The clock can be derived from pins 38, 39, and 40. We can use either an external oscillator, a crystal connected between pins 39 and 40, or an *RC* circuit configured across these same two pins. A complete A/D conversion of a single sample requires 4,000 counts. The signal is integrated for one-fourth of this period, or 1,000 counts. The second integration and the auto-zero require the remaining 3,000 counts. The internal clock is developed by dividing the oscillator input by 4. Thus, if, for example, we wish to perform 10 conversions per second, the oscillator input must be 160 kHz. The ICL7126 is not capable of high-speed conversion and should be used for slowly varying signals (low sampling rates) or dc inputs.

The ADC0804 is an example of an integrated circuit that performs *serial A/D conversion*, sometimes known as *successive approximation* conversion. The IC is illustrated in Fig. 8.8. The ADC0804 is an eight-bit device. Its internal construction consists of a number of flip-flops, shift registers, a decoder, and a comparator. The full conversion takes eight internal clock pulses. The internal clock is provided by dividing the clock signal at pins 4 and 19 by 8. Thus, with a 64-kHz signal on these pins, the IC can perform one conversion in 1 msec. The ADC0804 is capable of converting a sample in about 120 microseconds, so we still cannot use it for high-speed sampling.

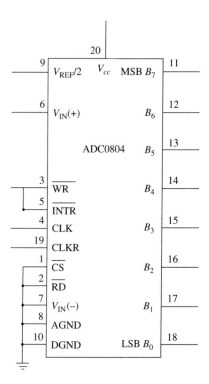

Figure 8.8   ADC0804 A/D converter.

The digital outputs, B_0 through B_7, appear on pins 11 through 18. The IC is compatible with a microprocessor, which is the reason for much of the terminology used in labeling the pins. The pin lines are specified as follows:

	PIN LABEL	Function
1	CS (chip select)	Set LOW to initialize, HIGH to start conversion
2	RD (read)	Goes LOW to indicate that microprocessor is ready to receive data
3	WR (write)	LOW to initialize, HIGH to start conversion
4	CLK	Input external oscillator or connect resistor between 4 and 19 to set oscillation frequency
5	INTR (interrupt)	Goes LOW to tell microprocessor that data are available
6	V_{IN}	Part of differential input (with pin 7 for negative input)
9	$V_{REF}/2$	Reference voltage (one-half of full-scale voltage)

The CA3308 is an example of an integrated circuit that accomplishes *parallel* (or *flash*) A/D conversion. The layout of this 24-pin IC is shown in Fig. 8.9. The circuit is capable of converting a sample in 66.7 nsec. It contains a bank of comparators. The analog input is on pins 16 and 21, and reference voltages are applied to pins 10, 15, 20, 22, and 23. The digital output is read from pins 1 to 8.

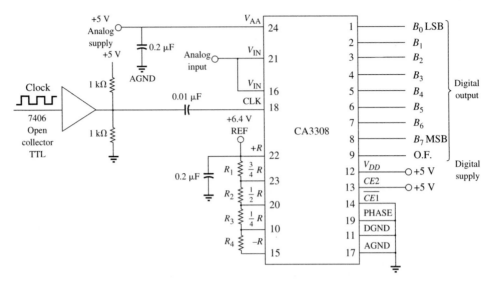

Figure 8.9   CA3308 flash A/D converter.

8.2 DIGITAL-TO-ANALOG CONVERSION

We now shift our attention to the conversion of a digital to an analog signal. This is performed by a *digital-to-analog (D/A) converter*. To perform such a conversion, we need
simply associate a level with each binary code word. Since the code word represents a
range of sample values, the actual value chosen for the conversion is usually the center
point of the region. If A/D conversion is performed as previously described, the reverse
operation is equivalent to assigning a weight to each bit position.

Let us illustrate the procedure for a four-bit binary word. We assume that the analog
sample is normalized (i.e., it falls in the range between 0 and 1 V) and that sequential coding (as opposed to Gray coding) is used. Conversion to the analog sample value is accomplished by converting the binary number to decimal, dividing by 16, and adding $\frac{1}{32}$. Thus,
the code 1101 represents the decimal number 13, so we convert this to $\frac{13}{16} + \frac{1}{32} = \frac{27}{32}$. The addition of the $\frac{1}{32}$ takes us from the bottom to the middle of the $\frac{1}{16}$-wide region.

Figure 8.10 illustrates the conversion. If a 1 appears in the position of the most sig-
nificant bit, a $\frac{1}{2}$-V battery is switched into the circuit. The second bit controls a $\frac{1}{4}$-V battery
and so on.

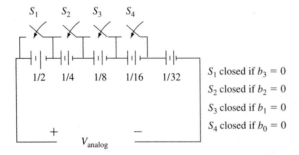

S_1 closed if $b_3 = 0$
S_2 closed if $b_2 = 0$
S_3 closed if $b_1 = 0$
S_4 closed if $b_0 = 0$

Figure 8.10   Digital-to-analog conversion.

The ideal decoder of Fig. 8.10 is analogous to the serial quantizer, since each bit is associated with a particular component of the sample value. A more complex decoder results when an analogy to the counting operation is attempted. Figure 8.11 shows the counter decoder. A clock feeds a staircase generator and, simultaneously, a binary counter.

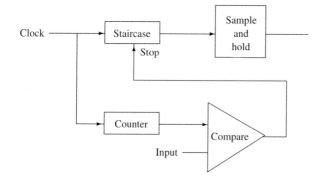

Figure 8.11   Counter D/A converter.

The output of the binary counter is compared to the binary digitized input. When a match occurs, the staircase generator is stopped. The output of the generator is sampled and held until the next sample value is achieved. The final staircase approximation is smoothed by a lowpass filter to recover an approximation to the original signal.

Practical Digital-to-Analog Converters

Figure 8.12 shows a practical version of the ideal decoder of Fig. 8.10. The gain of the operational amplifier is $R/5R_{in}$, where R_{in} is the parallel combination of the resistors included in the input circuit (i.e., those whose associated switches are closed). In terms of input conductance, the gain is $RG_{in}/5$, where G_{in} is the sum of the conductances associated with closed switches. Therefore, closing S_1 contributes $\frac{8}{5}$ to the gain. Closing S_2 contributes $\frac{4}{5}$ to the gain, closing S_3 contributes $\frac{2}{5}$, and closing S_4 $\frac{1}{5}$. To calculate the gain due to more than one switch being closed, we simply add the associated gains. Therefore, if the input voltage source, V, is equal to 5 volts, then S_1 contributes 8 volts to the output, S_2 contributes 4 volts, S_3 contributes 2 volts, and S_4 contributes 1 volt.

The output voltage is an integer voltage between 0 and 15 corresponding to the decimal equivalent of the binary number controlling the switches. To convert the system into a D/A converter analogous to that of Fig. 8.10, we need simply divide the output by 16 and

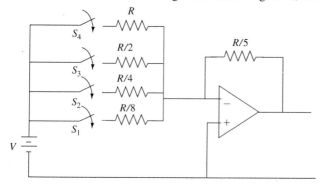

Figure 8.12   Practical D/A converter.

add $\frac{1}{32}$ volt. The division can be performed by scaling V or by scaling the resistor values. The addition of $\frac{1}{32}$ volt can be accomplished by adding a fifth resistor (unswitched) to the input. The value of this resistor would be $2R$.

8.3 DIGITAL BASEBAND

We began our study of analog communication with baseband in Chapter 5. We then saw reasons to shift frequencies away from dc. We take a parallel approach to digital communication.

8.3.1 Signal Formats

We begin this section by presenting various time domain signal formats used in digital communication. One of the simplest ways of transmitting a sequence of 1's and 0's is to transmit a $+V$-volt signal for a digital 1 and a 0-volt signal for a digital 0. Such a scheme is known as *unipolar* transmission, since the signal deflects from zero in only one direction.

A modification of this natural choice is often made. We send a constant of $+V$ volts for a digital 1 and $-V$ volts for a digital 0. This is known as *bipolar* transmission. The voltage amplitude, V, is chosen on the basis of available power and other physical considerations. We use $+V$ and $-V$ rather than $+V$ and 0 because, as we shall see later, it is advantageous to make the two signals as different as possible. Figure 8.13 shows the unipolar and bipolar signal waveforms that would be used to send the digital sequence 11000101.

The specific signaling format illustrated in Fig. 8.13 is *NRZ*, or *nonreturn to zero*. This particular type of transmission is known as NRZ-L, where the voltage level corresponds to the logic level. Alternatively, we could have represented a logic 1 with $-V$ volts and a logic 0 with $+V$ volts. (This modified version is still called NRZ-L.)

Before exploring other signaling formats, we take a few minutes to examine the Fourier transform of the NRZ signal. This Fourier transform is important because it gives us a measure of the required channel bandwidth needed to transmit the waveform. The actual waveform depends on the binary sequence being transmitted. We consider two cases: periodic data and random data.

Suppose that the data comprise an alternating train of ones and zeros. That is, the digital sequence is described by

$$a_n = (-1)^n \tag{8.1}$$

The NRZ-L waveform is then simply a square wave with frequency equal to half the bit rate, as shown in Fig. 8.14. The Fourier series of the square wave is given by

$$s_{\mathrm{NRZ}}(t) = \sum_1^\infty b_n \sin \pi n R_b t \tag{8.2}$$

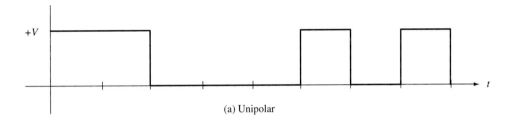

(a) Unipolar

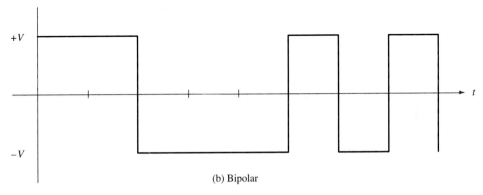

(b) Bipolar

Figure 8.13   Waveforms for sending 11000101.

where R_b is the bit rate in *bits per second* and the Fourier coefficients are

$$a_n = 0, \quad \text{all } n$$

$$b_n = \begin{cases} (-1)^{\frac{n-1}{2}} \dfrac{4V}{n\pi}, & n \text{ odd} \\ \\ 0, & n \text{ even} \end{cases} \qquad (8.3)$$

We are interested in the power spectral density of the NRZ signal. The power of each sinusoid is the square of the amplitude divided by 2. The power spectral density of the periodic NRZ-L waveform is therefore as sketched in Figure 8.14. Note that the outline of the impulses is a $(\sin^2 f)/f^2$ type of curve.

While the result for periodic data gives us a starting point, it is not relevant to the real world: You wouldn't go through a lot of trouble to transmit something as predictable as a periodic data train. A more meaningful spectrum results if we assume a random sequence of 1's and 0's. The power spectral density of a function of time, $s(t)$, can be found by taking the ratio of energy to time. That is, power is the rate of change of energy. We do this operation over a limited time interval and then take the limit as the length of the inter-

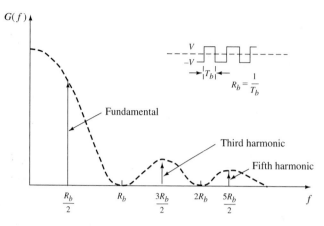

Figure 8.14   Power spectrum of periodic NRZ.

val becomes infinite. Setting this length to ΔT, we find that the power spectral density is given by

$$G(f) = \left| \left(\lim_{\Delta T \to \infty} \frac{1}{\Delta T} \int_{-\Delta T/2}^{\Delta T/2} s(t)e^{-j2\pi ft}dt \right)^2 \right| \tag{8.4}$$

The term in parentheses in Eq. (8.4) is the Fourier transform of the truncated signal. Since $s(t)$ is NRZ bipolar, the signal is always either $+1V$ or $-V$ and stays constant within an interval. As ΔT increases by jumps equal to the bit intervals, we keep adding identical terms, since the squaring operation makes the contribution of each period identical. We can therefore perform the averaging operation over one period to get

$$G(f) = \frac{1}{T_b} \left| \int_{-T_b/2}^{T_b/2} Ve^{-j2\pi ft}dt \right|^2$$

$$= V^2 T_b \left(\frac{\sin\pi f T_b}{\pi f T_b} \right)^2 \tag{8.5}$$

where T_b is the bit period. This power spectral density is sketched in Figure 8.15.

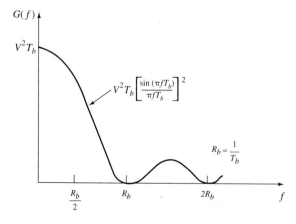

Figure 8.15   Power spectrum of random NRZ.

The total power of the NRZ waveform is V^2, since the square of the NRZ waveform is a function with this constant value. The integral of the power spectral density yields the power in any frequency band. One can show (see the problems at the end of this chapter) that 91 percent of the total power of an NRZ signal is contained in frequencies below the bit rate R_b.

Figure 8.16 illustrates NRZ and RZ (*return to zero*) signaling for the sequence 11000101. Note that in RZ, the pulse returns to the zero state before reaching the end of the bit interval. (For illustrative purposes, we shall assume that the return occurs at the midpoint of the interval.) Thus, the pulses are not as wide as in the NRZ case, and this change increases the bandwidth.

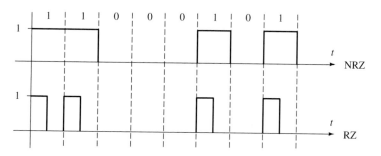

Figure 8.16   NRZ and RZ signaling.

The power spectral density of a (unipolar) RZ signal is found in much the same way as we found it for the NRZ signal. We again separate the analysis into periodic data and random data. The periodic situation gives a function of time, as shown in Fig. 8.17(a). Since we are interested in power, shifting this function does not change the result. We therefore can shift the function $T_b/4$ to the left to yield an even function of time whose Fourier series is given by

$$s_{RZ}(t) = \sum_{n=1}^{\infty} a_n \cos \pi n R_b t \tag{8.6}$$

The resulting power spectral density is sketched in Fig. 8.17(b). Note that the outline of the impulses goes to zero at multiples of $2R_b$, twice that of NRZ signaling.

The analysis for random data is more complex. The signal can be decomposed into a periodic component at the bit rate added to a random component. This decomposition is shown in Fig. 8.18. If we assume that 1's and 0's are equally likely (i.e., each occurs 50 percent of the time), the power spectral density is given by

$$G(f) = \frac{V^2}{16R_b} \frac{\sin^2(\pi f/2R_b)}{(\pi f/2R_b)^2} + \frac{\pi V^2}{8} \sum_{-\infty}^{\infty} \frac{\sin^2(n\pi/2)}{(n\pi/2)^2} \delta(f - nR_b) \tag{8.7}$$

This power spectral density is sketched in Figure 8.19.

One problem associated with the transmission schemes we have presented is that the average value of the waveform is a function of the fraction of 1's in the transmitted signal.

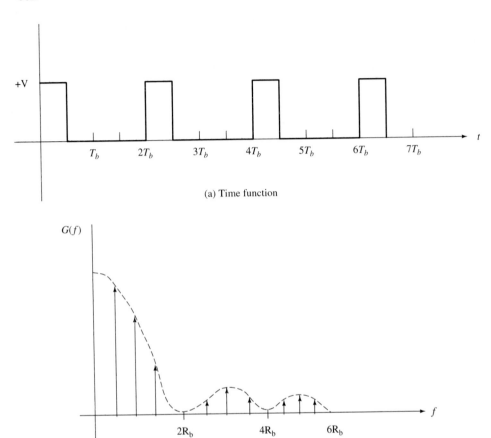

(a) Time function

(b) Power spectral density

Figure 8.17   Power spectrum of periodic RZ.

This causes problems if the channel does not transmit dc (i.e., if it has ac coupling). A variation that overcomes these problems is *alternate mark inversion* (AMI). This is the same as unipolar RZ, except that every other nonzero pulse is inverted. Figure 8.20 illustrates the NRZ-AMI signal for the binary sequence 11000101.

   Two particular problems are associated with NRZ-L transmissions. First, when the data are *static* (i.e., there is no change from one bit interval to the next), there is no transition in the transmitted waveform. This causes timing problems when we try to establish bit synchronization (as we discuss in Chapter 10). The second problem occurs with data inversion. If the levels are reversed during transmission (i.e., if $+V$ is interpreted as $-V$ at the receiver), the entire data train is inverted, and every bit is in error. Inversion can occur in several ways. The most common form occurs when the information is being sent by varying the phase of a sinusoid. A time delay corresponding to a 180° phase shift results in a data inversion. Additionally, some systems consist of numerous electronic devices (e.g., op-amps), each of which inverts the signal. It may sometimes prove difficult to keep track of the number of inversions, or indeed, different signals may encounter different numbers of inverters.

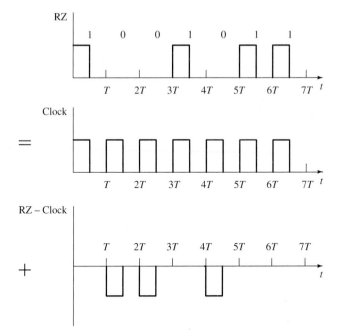

Figure 8.18 Unipolar RZ and sum of periodic and random signal.

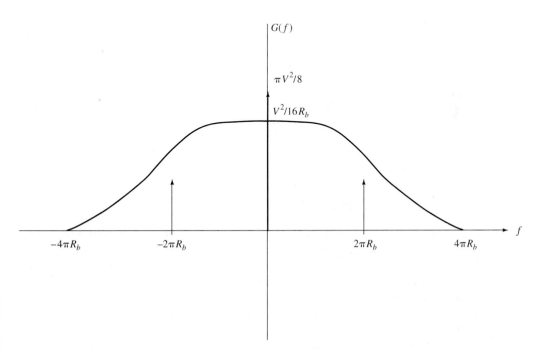

Figure 8.19 Power spectral density of random RZ.

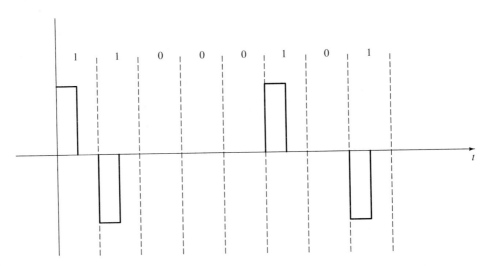

Figure 8.20   NRZ-AMI signal for 11000101.

For the foregoing reasons, we often choose *differential* forms of encoding. In such techniques, the data are represented as *changes in levels*, rather than particular levels, of the signal. We investigate the *NRZ-M* and *NRZ-S* systems. The terminology of M and S stems from MARK and SPACE, a carry-over from the days of the telegraph.

In the NRZ-M system, the datum "1" is represented by a change in level between two consecutive bit times, while the datum "0" is represented by no change in the level. Figure 8.21 illustrates an NRZ-L waveform and the corresponding NRZ-M waveform. Note that we have started the NRZ-M signal at $+V$ volts. We could have started the signal at $-V$ volts.

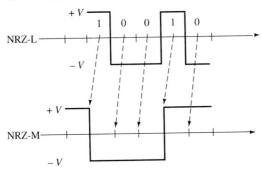

Figure 8.21   NRZ-M differential format.

We can implement an NRZ-M encoder with the use of an exclusive-OR gate and a time delay. This is shown in Fig. 8.22. We start with the NRZ-L signal and exclusive-OR it with the delayed NRZ-M output. If the input bit is a 1, the output represents an inversion of the previous output. That is, if we exclusive-OR anything with the bit 1, the result is the inverse of that thing. In other words,

$$1 \oplus 1 = 0$$
$$1 \oplus 0 = 1$$

where the symbol $\oplus$ denotes the exclusive-OR operation.

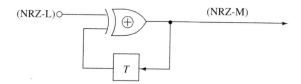

Figure 8.22   NRZ-M encoder.

The decoder for NRZ-M compares the NRZ-M signal to a delayed version of itself. If the two are identical in an interval, we know that a 0 is being sent; if the two are different, a 1 is being sent. This decision rule describes the operation of an exclusive-OR gate, so the decoder can be implemented as in Fig. 8.23.

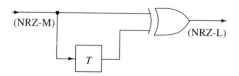

Figure 8.23   NRZ-M decoder.

The NRZ-S system is similar to the NRZ-M, except that the two outputs are reversed: The datum "1" is represented by no change in level between two consecutive bit times, while the datum "0" is represented by a change. Figure 8.24 shows the NRZ-L, NRZ-M, and NRZ-S representations for the same data signal, 10010. The encoders and decoders for NRZ-S are the same as those for NRZ-M, except for the addition of an inverter. A derivation of these systems is left to the problems at the end of the chapter.

While the NRZ-S and NRZ-M differential systems solve the problem of waveform inversion, they do not solve the problem of loss of timing information. For example, in the NRZ-M system, a train of data 0's results in a transmitted waveform without transitions. Similarly, in the NRZ-S system, a train of data 1's results in a loss of bit-timing information.

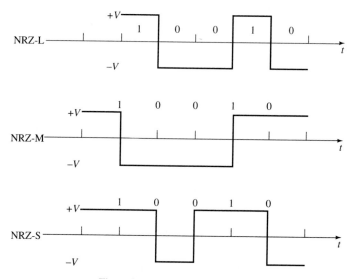

Figure 8.24   NRZ-S and NRZ-M.

The *biphase* (bi-ɸ), or *Manchester*, format overcomes this *static data* and timing problem. Figure 8.25 shows the NRZ-L and biphase-L waveforms for the same data train, 10010, that we have been examining. Also shown in the figure are the basic waveforms used to send a 1 and a 0. Note that a 1 is sent by transmitting $-V$ volts for the first half of the bit interval and $+V$ volts for the second half. A 0 is sent with the inverse signal. Thus, a transition always occurs at the midpoint of each bit interval. An additional transition occurs at the beginning of the interval when two adjacent bits are identical.

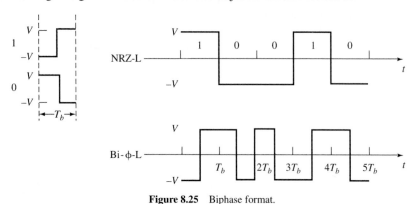

Figure 8.25   Biphase format.

The power spectral density of a random biphase signal is found in a manner similar to that of NRZ and is given by

$$G(f) = \lim_{\Delta T \to \infty} \left| \int_{-\Delta T/2}^{0} \pm\, V e^{-j2\pi ft} dt \;+\; \int_{0}^{\Delta T/2} \mp\, V e^{-j2\pi ft} dt \right|^2$$

$$= \lim_{\Delta T \to \infty} V^2 \frac{1}{\Delta T} \left| \frac{-2 + e^{j\pi f\Delta T} + e^{-j\pi f\Delta T}}{j2\pi f} \right|^2$$

$$= \lim_{\Delta T \to \infty} \frac{V^2}{\Delta T} \left| \frac{1 - \cos \pi f\Delta T}{\pi f} \right|^2 \qquad (8.8)$$

$$= \lim_{\Delta T \to \infty} \left| \frac{\sin^2(\pi f\Delta T/2)}{\pi f} \right|^2$$

We now let ΔT go to infinity in jumps of the bit period to get

$$G(f) = \frac{V^2 T_b}{2} \left[\frac{\sin^2(\pi f T_b/2)}{\pi f T_b/2} \right]^2 \qquad (8.9)$$

This curve is sketched in Fig. 8.26. Note that the first zero occurs at $2R_b$, just as in the case of RZ signaling. However, note also that the spectrum goes to zero as frequencies approach zero. Therefore, a channel that transmits a biphase signal need not be capable of transmitting frequencies down to dc (i.e., ac coupling is permitted).

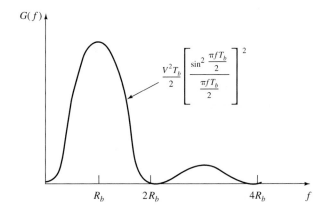

Figure 8.26   Spectral density of random biphase signal.

Power is found by integrating the power spectral density. The total power of a biphase signal is V^2, as expected, since the square of the signal is a dc waveform at V^2. One can demonstrate that (see the problems at the end of the chapter) 65 percent of the total power lies in frequencies below R_b and 86 percent of the total power lies in frequencies below $2R_b$.

Figure 8.27 shows an encoder for biphase-L. We start with NRZ-L and feed both this and a double-frequency clock signal into an exclusive-OR gate. When the data logic level is 0, the signal output of the gate is equal to the clock waveform. When the data logic level is 1, the output of the gate is the inverse of the clock waveform. We have thus solved the timing problem, but inversion of the signal during transmission or reception still results in bit reversal, causing unacceptable bit errors.

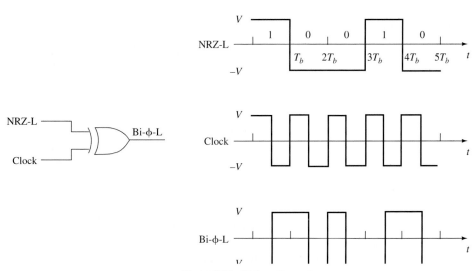

Figure 8.27   Biphase-L encoder.

We can combine differential encoding with biphase encoding to solve both timing and inversion problems. The *biphase-M* and *biphase-S* formats are differential. A transition occurs at the *beginning* of every bit period, unlike the biphase-L format, where transitions occur in the middle of the intervals. In the biphase-M system, a data 1 is represented by an additional transition in the midperiod, while a 0 is represented by no midperiod transition. In the biphase-S system, the data 1 results in no midperiod transition, while the data 0 results in a midperiod transition. These formats are illustrated in Fig. 8.28.

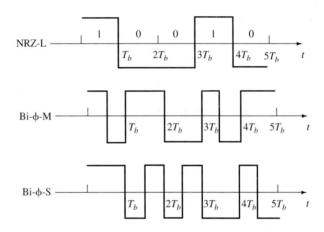

Figure 8.28   Biphase-M and biphase-S formats.

The encoder for biphase-S is shown in Fig. 8.29. We start with NRZ-M, delay it by one-half of the bit interval, and exclusive-OR the delayed signal with the double-frequency clock. The result is to invert every other segment of the NRZ-M signal. If we want biphase-M instead of biphase-S, we start with NRZ-S.

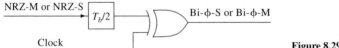

Figure 8.29   Biphase-S encoder.

8.3.2 Pulse Code Modulation

Pulse code modulation (PCM) is a direct application of the A/D converter. Rather than simply present it as such, however, we rederive the results using a different approach. We begin with a pulse amplitude modulation system.

Suppose that the amplitude of each pulse in a PAM system is rounded off to one of several possible levels. Alternatively, suppose that the original continuous function of time is first rounded off to several possible levels, yielding a staircase type of waveform, as shown in Fig. 8.30. We then sample the staircase function and transmit the samples using PAM. The rounding-off process is known as *quantization*, and it introduces an error known as *quantization noise*. That is, the staircase approximation is not identical to the original function, and the difference between the two represents an error. We discuss this error in detail in Section 8.4.

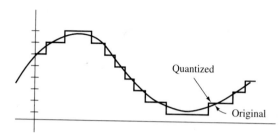

Figure 8.30   Rounding off to discrete levels.

We have reduced the *dictionary* of possible PAM pulse heights to include only specified quantization levels. A received pulse is compared to the possible transmitted pulses, and it is decoded into the dictionary entry that most closely resembles the received signal. In this way, small errors (up to one-half of the quantization step size) are corrected.

Performing *error correction* is the major reason for quantizing. Suppose, for example, you wanted to transmit a signal from California to New York over coaxial cable. If the signal were transmitted via PAM, the resulting additive noise could make the received signal unintelligible. That is, noise would be added along the entire transmission path, with additional noise added in each amplifier. Many amplifiers would have to be located along the path to cancel attenuation.

If the same signal were now sent using quantized PAM, under certain conditions most errors could be corrected. If the amplifiers were spaced in such a way that the noise introduced between any two amplifiers was less than one-half of the quantization step size, each amplifier could restore the function to its original staircase form before amplifying it and sending it on its way. That is, each amplifier would round each received pulse to the nearest acceptable level and then retransmit the reconstructed signal, thereby eliminating all but the strongest noise. Since the amplifiers do more than simply enlarge the received signal, they are given the name *repeater*.

The concept of having a repeater *reconstruct* a signal is crucial to digital communication. To make certain that it is clear, we will take a moment for a simple example. Suppose you want to speak a sentence and have it heard by a friend on the other side of the campus. You could yell loudly, but if the campus is large, the message would not be received. You could place amplifiers along the way, each of which rebroadcasts the received signal through a loudspeaker, but these amplifiers would also rebroadcast all of the received background noise (including other people talking). Once again, your friend could not distinguish your sentence. Suppose now that you position classmates every 25 meters in a line stretching across campus. You yell the sentence to the nearest person. That person turns and yells the same sentence to the next person in line, and so on, until the sentence reaches the destination. Since each "repeater" here is retransmitting a "cleaned-up" version of the signal, it will most likely be correctly received across the campus.

The more quantization round-off levels used, the closer the staircase function resembles the desired signal. The number of levels then determines the *resolution* of the signal, that is, how small a change in signal level can be detected by looking at the quantized version of the signal.

If high resolution is required, the number of quantizing levels must increase. At the same time, the spacing between levels decreases. As the dictionary words get closer and closer together, the noise advantages decrease.

If resolution could be improved without increasing the size of the dictionary (i.e., without moving dictionary words closer together), the error-correcting advantages could be maintained. *Pulse code modulation (PCM)* is a method of accomplishing this.

In a binary PCM system, the dictionary of possible transmitted signals contains only two entries, "0" and "1." The quantization levels are coded into binary numbers. Thus, if there are eight quantization levels, the values are coded into three-bit binary numbers. Three pulses would then be required to send each quantization value. Each of the three pulses represents either a 0 or a 1. This is identical to the concept of A/D conversion.

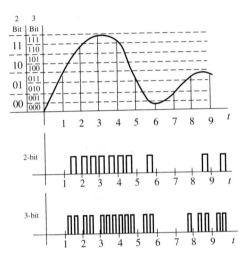

Figure 8.31   Illustration of PCM.

Figure 8.31 shows the two-bit and three-bit PCM waveforms for a representative $s(t)$. For illustrative purposes, a positive pulse is used to represent a binary 1, and a zero pulse represents a 0. We will generalize this scheme later in the chapter, where we discuss using different waveforms to send the binary information. Even when the actual waveforms are not in the shape of pulses, we still call this PCM.

A PCM demodulator is simply a binary D/A converter. The modulator and demodulator are available as large scale integrated circuits (LSI) and are collectively given the name *CODEC* (for coder-decoder).

Nonuniform Quantization

Figure 8.32(a) illustrates uniform quantization as an input-vs.-output function. The range of sample values is divided into quantization regions, each of which is the same size as every other region. Thus, with three-bit quantization, we divide the entire range of sample values into eight equally sized regions.

We often find it advantageous to use quantization intervals that are *not* all the same size. That is, we replace the quantization function in Fig. 8.32(a) with that of Fig. 8.32(b). The function shown in Fig. 8.32(b) has the property that the spacing between quantization levels is not uniform and the output levels are not in the center of each interval.

Let us first intuitively justify nonuniform quantization. Consider a musical piece for which the voltage waveform ranges from -2 to $+2$ volts. Suppose that three-bit uniform quantization is used. Therefore, all voltages between 0 and $\frac{1}{2}$ volt are coded into the same code word, 100, corresponding to the reconstructed output value of $\frac{1}{4}$ volt. Likewise, all samples between 1.5 and 2 volts are coded into a single code word, 111, corresponding to

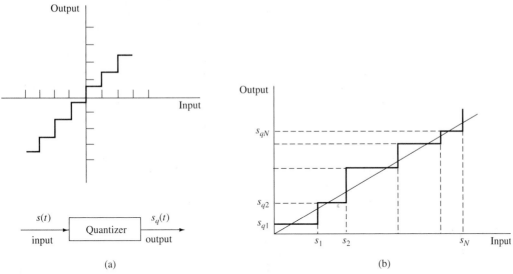

Figure 8.32   Quantization.

a reconstructed output value of $\frac{7}{4}$ volt. During soft music passages, where the signal may not exceed $\frac{1}{2}$ volt for long periods, a great deal of musical definition would be lost. The quantization provides the *same resolution* at high levels as at low, even though the human ear is less sensitive to changes at higher levels. The response of the human ear is nonlinear. It would therefore be desirable to use small quantization steps at the lower levels and larger steps at the higher levels. Another way to look at this is to examine the ratio of the sample value to the maximum round-off error. This ratio is low for small samples, and it increases as the sample size increases.

An alternative justification for examining nonuniform quantization applies if the signal spends a far greater percentage of time at the low levels than at the higher levels. It would be preferable to provide higher resolution at these lower levels at the expense of lower resolution at the high levels. Since the signal spends much less time at the higher levels, the average quantization error may well decrease using this approach.

The *quantization error*, or quantization noise, is a measure of the effectiveness of a quantization scheme. We examine it in detail in Section 8.4; for now, we take a qualitative approach.

The average quantization error is a function of the edge locations of the quantization regions (the s_i's in Fig. 8.32), the round-off values (the s_{qi}'s in Fig. 8.32), and the probability density function of the sample values. We will show later that once quantization regions have been selected, the round-off values are chosen to be the center of gravity of the corresponding portion of the probability density. Figure 8.33 shows a representative example of a probability density function that resembles a Gaussian density. We have divided this function into eight equal regions (this will prove to be a poor choice) and indicated these with boundaries of s_0 to s_8 on the figure. Given this choice of regions, the round-off levels are approximately as shown in the figure. That is, the s_{qi} are at the approximate center of gravity of each quantization region.

If, in addition to the specific quantization levels, we also make the actual intervals variable, a complicated optimization problem results, and a complex system would be re-

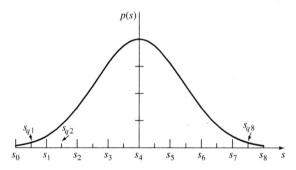

$$p(s)$$

Figure 8.33   Signal probability density.

quired to implement its solution. For this reason, and also because systems should be applicable to a variety of input signals, suboptimal systems are almost always used. This leads us to a study of companding.

Companding

The most common form of nonuniform quantization is known as *companding*. The name is derived from the words "compressing-expanding." The process is illustrated in Fig. 8.34.

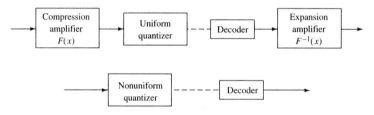

Figure 8.34   Companding.

The original signal is compressed using a memoryless nonlinear device. The compressed signal is then uniformly quantized. Following transmission, the decoded waveform must be expanded using a nonlinear function which is the inverse of that used in compression. This is similar to pre-emphasis and de-emphasis, as used in FM. As a simple example, suppose that you were in a classroom trying to listen to the professor, but the light fixtures were buzzing loudly at 60 Hz. You could hang a *notch filter* on your ear, with the transfer function shown in Fig. 8.35(a). In this manner, you would significantly decrease the annoying hum. However, you would also lose portions of the professor's voice signal around 60 Hz. The professor could compensate for this by using an amplifier with the characteristic shown in Fig. 8.35(b). This amplifier enhances portions of the audio around 60 Hz. If the product of the two transfer functions is a constant, you would hear an undistorted version of the lecture. The additive hum would be attenuated, and the overall transmission would improve.

We begin our examination of companding by analyzing the compression process. Prior to quantization, the signal is distorted by a function similar to that shown in Fig. 8.36. This operation compresses the extreme values of the waveform while enhancing the smaller values, much as the logarithm is used to permit viewing of very large and very small values on the same set of axes. If an analog signal forms the input to the compressor

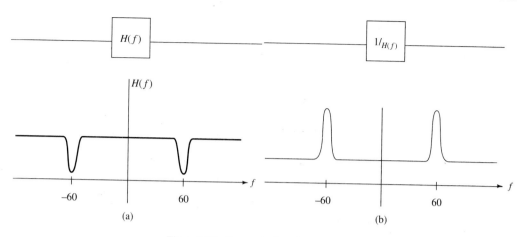

Figure 8.35 Pre-emphasis and de-emphasis.

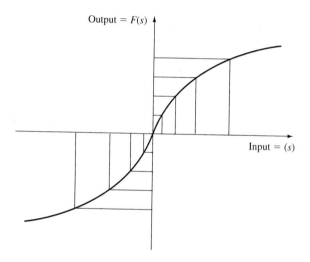

Figure 8.36 Compression function.

and the output is uniformly quantized, the result is equivalent to quantizing with steps that start out small and get larger for higher signal levels.

We have divided the output of the compressor into eight equal quantization regions. The function is used to translate the boundaries of these regions to the abscissa, which represents the uncompressed input signal. Note that the regions on the s-axis start out small and get larger with increasing values of s.

It would be nice if we could agree to some standards for companding in order to realize hardware efficiencies and to permit sending and receiving by a variety of users. If you are to have any hope of purchasing an off-the-shelf PCM encoder with companding, there is a need to agree to a limited number of standard compression formulas.

The most common application of companding is in voice transmission. North America and Japan have adopted a standard compression curve known as μ-*law companding*. Europe has adopted a different, but similar, standard known as A-*law companding*.

The μ-law compression formula is

$$F(s) = \text{sgn}(s)\frac{\ln(1 + \mu|s|)}{\ln(1 + \mu)} \qquad (8.10)$$

The A-law compression formula is

$$F(s) = \text{sgn}(s)\frac{A|s|}{1 + \ln(A)} \qquad \text{for } 0 \le |s| < \frac{1}{A}$$

$$\qquad (8.11)$$

$$F(s) = \text{sgn}(s)\frac{1 + \ln A|s|}{1 + \ln(A)} \qquad \text{for } \frac{1}{A} \le |s| \le 1$$

The μ-law function is sketched for selected values of μ in Fig. 8.37. If we attempted to sketch the A-law on the same set of axes, you would not notice any difference in the curves because they are similar to those generated by the μ-law.

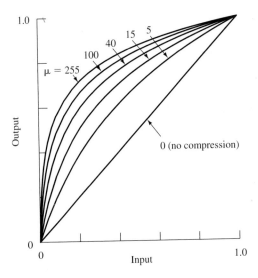

Figure 8.37   μ-law companding.

The parameter μ defines the degree of curvature of the function. A commonly used value is μ = 255. For A-law companding, the commonly used value of the parameter is A = 87.6.

One way to implement the μ-255 compander is to simulate a nonlinear system that follows the μ-255 input/output curve. Then place the sample values through this system, and uniformly quantize the output using an eight-bit A/D converter. An alternative way is to approximate the μ-255 curve by a piecewise linear curve, as shown in Fig. 8.38. We illustrate the positive input portion only; the curve is an odd function.

We have approximated the positive portion of the curve by eight straight-line segments. We divide the positive output region into eight equal segments, which effectively divides the input region into eight *unequal* segments. Within each of these segments, we uniformly quantize the value of samples, using four bits of quantization. Thus, each of these eight regions is divided into 16 subregions, for a total of 128 regions on each side of

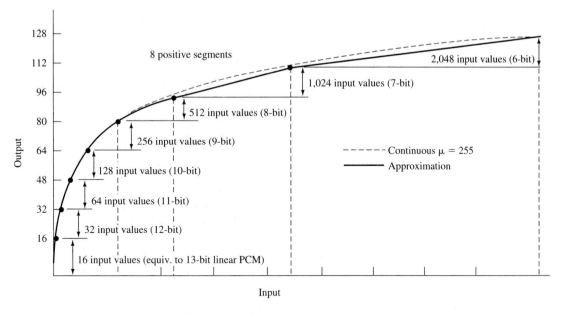

Figure 8.38   μ-255 piecewise linear approximation.

the axis. We therefore have a total of 256, or 2^8, regions, which corresponds to eight bits of quantization.

The specific technique for sending a sample value is to send eight bits coded as follows:

> one bit is used to give the polarity of the sample: 1 for positive, 0 for negative.
> three bits are used to identify which piecewise segment the sample lies in.
> four bits are used to identify the quantization level within each sample region.

The logarithmic relationship of the μ-255 law leads to an interesting interdependence among the eight segments: Each segment on the input axis is twice as wide as the segment to its left. The resolution of the first segment to the right of the origin is therefore twice that of the next segment, and so on. The sixth region to the right of the origin covers an interval on the input axis that is one-sixteenth of the total swing. Thus, the resolution of samples in this particular interval is the same as that of uniform quantization using eight-bits of A/D conversion. The resolution of the region just to the left of the interval in question is the same as that of nine-bit uniform quantization. Similarly, as we move to the left, each region has the resolution of a uniform quantizer with one more bit of quantization than the region to its right. These levels are marked on the figure.

8.3.3 Time Division Multiplexing

The concept of TDM was explored in Section 5.3.2, where we discussed pulse techniques for discrete signals. Here, we need only modify that discussion, since each sample, instead of requiring one pulse for transmission, now requires a number of pulses equal to the number of bits of quantization. Thus, for six-bit PCM, six pulses must be transmitted during each sampling period.

T-1 Carrier System

In the early 1960s, the need arose to provide capability for commercial and public digital communication. This need first manifested itself with regard voice channels. The *Bell System* (which no longer exists) already had an elaborate carrier system in operation, with cables that could handle bandwidths up to several MHz. In addition to the advantage inherent in all forms of digital communication, an additional advantage was that the signaling information required to control telephone switching operations could be economically transmitted in digital form.

The *Bell System T-1 digital communication system* was formulated to be compatible with existing communication systems. The existing equipment was designed primarily for interoffice trunks within an exchange and therefore is suited for relatively short distances of about 15 to 65 km.

The system develops a 1.544-megabit/sec pulsed digital signal for transmission down the cable. The signal is generated by multiplexing 24 voice channels. First, each channel is sampled at a rate of 8,000 samples/second. Then the samples are quantized using eight-bit μ-law companding. The least significant bit is periodically (typically, once every six samples) devoted to *signaling* information. This includes the necessary bookkeeping to set up the call and keep track of billing. The 24 channels are then time division multiplexed, as shown in Fig. 8.39. The interleaved order of channels is selected because of the specific method of combining, which was designed to use existing multiplex equipment. The quantized samples from the 24 channels form what is known as a *frame*, and since sampling is done at an 8-kHz rate, each frame occupies 1/8,000 sec, or 125 μsec.

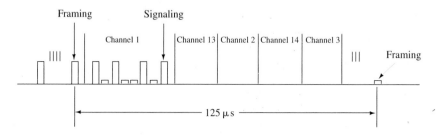

Figure 8.39   TDM in T-1 system.

There are 192 information and signaling bits in each frame (24 × 8), and a 193rd bit is added for *frame synchronization*.This framing signal is a fixed pattern of 1's and 0's in every 193rd pulse position.

The product of 8,000 frames per second and 193 bits per frame yields a transmission rate of 1.544 Mbits/sec.

The T-1 system can be used to send data as well as PCM, since the basic system is simply a data communication system capable of transmitting 1.544 Mbits/sec. When used for data transmission, the T-1 requires effort to interface properly with the hardware (which is expecting a structured frame of coded voice samples). Data may have to be structured into eight-bit blocks, and framing pulses must be included.

As just described, T-1 has been adopted by the United States, Canada, and Japan. The structure of T-1 for Europe is different, however, and follows the recommendation of the Comité Consultatif Internationale de Télégraphique et Téléphonique (CCITT)—the organization that sets voluntary international communication standards. This version of the system multiplexes 32 channels instead of 24, and the bit rate is 2.048 Mbits/sec instead of 1.544 Mbits/sec.

8.3.4 Delta Modulation

In the PCM method of source encoding, each sample value is coded into a binary number. This number must be capable of providing a measure of the sample value over the entire dynamic range. For example, if we start with a signal that ranges from -5 V to $+5$ V, the digital code must be capable of indicating sample values over a 10-volt range. The resulting resolution (and therefore round-off error) is dependent on this dynamic range.

If we could somehow reduce the dynamic range of the numbers communicated, the resolution would improve (without having to increase the number of bits). Various alternative forms of source encoding operate on this principle.

Delta modulation is a simple technique for reducing the dynamic range of the numbers to be coded. Instead of sending each sample value independently, we send the *difference* between a sample and the previous sample. If sampling is performed at the Nyquist rate, this difference may have a dynamic range as large as twice that of the original samples. That is, at the Nyquist rate, each sample is independent of the previous sample, so that two adjacent samples could be at the minimum and maximum signal amplitudes. However, if we sample at a rate higher than the Nyquist rate, the samples are not independent, and the dynamic range of the difference between two samples is less than that of the samples themselves (i.e., adjacent samples cannot vary by much). Sampling at higher rates means that we must send bits more often. But if this process reduces the dynamic range, we may be able to reduce the number of bits sent at each sampling point. Thus, the overall process could result in an improvement.

Delta modulation quantizes the difference between adjacent samples using only one bit of quantization. Thus, a 1 is sent if the difference is positive, and a 0 is sent if the difference is negative. Since there is only one bit of quantization, the differences are coded into only one of two levels. We refer to these two possibilities as $+\Delta$ and $-\Delta$; at every sample point, the quantized waveform can increase or decrease only by Δ.

Figure 8.40 shows a typical analog waveform and its quantized version. Since the quantized waveform can increase or decrease only by Δ at each sample point, we attempt to fit a staircase approximation to the analog waveform. This approximation is indicated in the figure. We shall examine the choice of sampling rate and step size in a few moments. In fitting the staircase to the function, we need only make a simple decision at each sample point. If the staircase is below the analog sample value, the decision is to increment positively (i.e., an *up* step). If the staircase is above the sample value, we increment negatively (i.e., a *down* step). The transmitted bit train for the example shown in Fig. 8.40 is

1 1 1 1 1 1 0 0 0 0 1 1 1 1 1 1 1 0 0 0 0 0

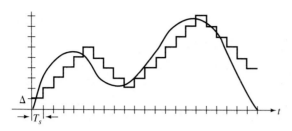

Figure 8.40   Delta modulation.

The receiver reconstructs the staircase approximation directly from the binary information it receives. If a 1 is detected, the receiver increments with a positive step. If a 0 is detected, a negative step occurs.

The preceding description leads to a simple implementation of the quantizer, using a comparator and a staircase generator. The resulting A/D converter and D/A converter are shown in Fig. 8.41.

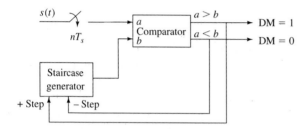

Figure 8.41   Delta modulator and demodulator.

The key to the effective use of delta modulation is the intelligent choice of the *step size* and *sampling rate*. These parameters must be chosen such that the staircase signal is a close approximation to the actual analog waveform. Since the signal has a definable upper frequency, we know the fastest rate at which it can change. However, to account for the fastest possible change in the signal, the sampling frequency and/or the step size must be increased. Increasing the sampling frequency results in the delta-modulated waveform requiring a larger bandwidth. Increasing the step size increases the quantization error.

Figure 8.42 shows the consequences of choosing an incorrect step size. If the steps are too small, we experience a *slope overload*, wherein the staircase cannot track rapid

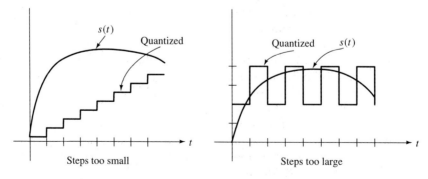

Figure 8.42   Consequences of choosing an incorrect step size.

changes in the analog signal. That is, the maximum slope the staircase can track is Δ/T_s. If, on the other hand, the steps are too large, considerable overshoot occurs during periods when the signal is not changing rapidly. In that case, we have significant quantization noise, known as *granular noise*.

Delta modulation uses a source-encoding process that employs memory to reduce dynamic range. Under certain circumstances, it is possible to achieve the same level of performance as PCM with fewer bits transmitted each second. However, because the system has memory, bit transmission errors propagate. In PCM, a single transmission bit error causes an error in reconstructing the associated sample value, but the error affects *only that single reconstructed sample*. If a bit error occurs in delta modulation, the D/A converter in the receiver goes up instead of down (or vice versa), and *all later values contain an offset error* of twice the step size. If a subsequent bit error occurs in the opposite direction, the offset error is cancelled. If the offset error poses a significant problem, the system can be periodically restarted from a reference level (usually zero). However, this introduces a transient associated with each restart.

Adaptive Delta Modulation

We have shown that the appropriate step size to use in delta modulation depends on how rapidly the signal changes from one sample to the next. When the signal is changing rapidly, a larger step size helps avoid overload. When the signal is changing slowly, a smaller step size reduces the amount of overshoot and, therefore, the quantization noise.

Adaptive delta modulation is a scheme that permits adjustment of the step size depending on the characteristics of the analog signal. It is, of course, critical that the receiver be able to adapt step sizes in exactly the same manner as the transmitter. If this were not the case, the receiver could not uniquely recover the original quantized signal (the staircase function). Since the transmission consists of a series of bits, the step size must be derived from this bit train (unless we are willing to send a separate control signal as *overhead*).

In fact, the bit sequence does contain the necessary information. Let us focus on a group of bits in the sequence. If a string of bits of a particular length contains an almost equal number of 1's and 0's, we can assume that the staircase is oscillating about a slowly varying analog signal. In such cases, we should reduce the step size. On the other hand, an excess of either 1's or 0's within a string of bits indicates that the staircase is "trying to catch up with" the function. In such cases, we should increase the step size.

In one practical implementation, the step size control is performed by a digital integrator that sums the bits over some fixed period. If the sum deviates from what it would be for an equal number of 1's and 0's, the step size is increased. In practice, the sum is translated into a voltage, which is then fed into a variable-gain amplifier that controls the step size. Amplification is a minimum when the input voltage corresponds to an equal number of 1's and 0's in the period.

Several other adaptive delta modulation algorithms are simpler to implement than the one just discussed. Two of the simpler ones are the Song algorithm and the space shuttle algorithm.

The *Song algorithm* compares the transmitted bit with the previous bit. If the two are the same, the step size is *increased* by a fixed amount Δ. If the two bits are different, the step size is *reduced* by Δ. Thus, the step size always changes, and it can get larger and larger, without limit if necessary. We illustrate this for a step input function in Fig. 8.43(a). Such a function represents an extreme case and does not occur in the real world, since a step function possesses infinite frequency. Note that a damped oscillation occurs following the rapid change in the signal.

If an analog signal is expected to have many abrupt transitions resembling the step function, the damped oscillations observed after the transition in the Song algorithm might be troublesome. Photographs of distinct and detailed objects could have many such transitions (i.e., rapid changes from white to black) as they are scanned for transmission. The *space shuttle algorithm* is a modification of the Song algorithm that eliminates the damped oscillations. When the present bit is the same as the previous bit, the step size increases by a fixed amount Δ. This, of course, is the same as for the Song algorithm. However, when the bits disagree, the step size reverts immediately to its minimum size, Δ. This is in contrast to the Song algorithm, in which the step size decreases toward zero at a rate of Δ every sampling period. The space shuttle algorithm is illustrated in Fig. 8.43(b) for the step function input.

Adaptive delta modulation techniques propagate errors in a manner similar to that of delta modulation. A bit error causes the step approximation to move in the wrong direction, and this in turn causes an offset for the portion of the waveform that follows. However, the error depends on the actual sequence of bits. For example, a bit error in a long string of transmitted 1's can make the space shuttle algorithm reset the step size to the minimum. This affects not only the direction of movement, but also the size of the succeeding steps. In addition to an offset, there is a distortion of the waveform that follows the bit error.

8.3.5 Other Techniques

In addition to PCM, delta modulation, and adaptive delta modulation, there are numerous other methods for coding analog information into a digital format. The goal of each system is to send the information with maximum reliability and minimum bandwidth. In this section, we introduce three of these methods: *delta pulse code modulation, differential pulse code modulation, and adaptive differential pulse code modulation.*

Delta modulation can be viewed in a manner different from that which we set forth previously. This new approach will point the way to variations in its implementation. In delta modulation, we approximate a continuous waveform with a staircase wave. At each sampling point, we develop an error term that is the difference between the signal and the staircase function. We quantize this error to develop a *correction term*, which is then added to the staircase function. In the case of basic delta modulation, the quantization is done in units of 1 bit. In *delta pulse code modulation*, we code the error into more than 1 bit of quantization and add this term to the previous staircase value, as shown in Fig. 8.44. Therefore, instead of the stair steps having only one possible magnitude, they can now have a magnitude that is one of any number of choices that is a power of 2. At each sampling point, we must of course send more than 1 bit of information; the various bits represent the PCM code for the error term. The advantage of delta PCM over ordinary PCM is that, with the proper choice of sampling interval, the error being quantized has a smaller dynamic range than that of the original signal. Therefore, for the same number of bits of

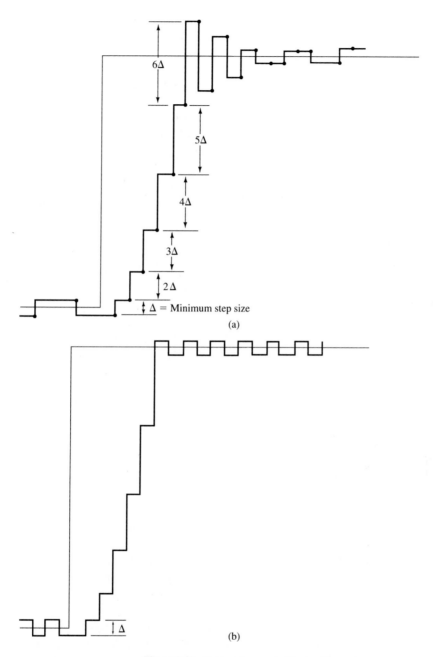

Figure 8.43   Song and space shuttle algorithms.

quantization, we can achieve better resolution. The price we pay is complexity of the modulator. If the signal were always at its maximum frequency (the one that determines the sampling rate in PCM), delta PCM would be essentially the same as PCM. However, since

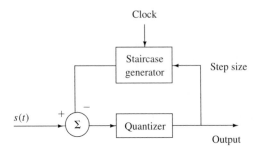

Figure 8.44 Delta pulse code modulation.

the frequencies of the signal are usually distributed over a range, adjacent samples are often correlated, and it is possible to get better performance from this system than from a PCM system with the same rate of bit transmission.

Differential pulse code modulation (DPCM) is another technique for sending information about changes in the samples rather than about the sample values themselves. The approach includes an additional step that is not part of delta PCM. The modulator does not send the difference between adjacent samples, but instead uses the difference between a sample and its *predicted* value. The prediction is made on the basis of previous samples. Differential PCM is illustrated in Fig. 8.45. The symbol $\hat{s}(nT_s)$ is used to denote the predicted value of $s(nT_s)$. The predictions can be as simple as a linear extension of the waveform (i.e., estimate the next sample using a straight line tangent to the curve at the previous sample point).

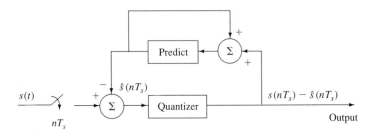

Figure 8.45 Differential pulse code modulation.

In one simple form of prediction, an estimate is made that is a linear function of the previous sample values. For the case of only one sample, we have

$$\hat{s}(nT_s) = As[(n - 1)T_s] \tag{8.12}$$

where A is a constant. The prediction block in Fig. 8.45 is therefore a multiplier with the value A. The challenge is to choose A so as to make the prediction as "good" as possible. We get a measure of the *goodness* of a prediction by defining a prediction error as the difference between the sample and its estimate:

$$e(nT_s) = s(nT_s) - \hat{s}(nT_s)$$
$$= s(nT_s) - As[(n - 1)T_s] \tag{8.13}$$

The mean square value of this error is

$$\text{mse} = E\{e^2(nT_s)\}$$

$$= E\{s^2(nT_s)\} + A^2E\{[s^2([n-1]T_s)]\} - 2AE\{s(nT_s)s([n-1]T_s)\} \quad (8.14)$$

$$= R(0)(1 + A^2) - 2AR(T_s)$$

where $R(t)$ is the autocorrelation of $s(t)$. The error can be minimized with respect to A by setting its derivative equal to zero:

$$\frac{d(\text{mse})}{dA} = 2AR(0) - 2R(T_s) = 0 \quad (8.15)$$

or

$$E\{s([n-1]T_s)[s(nT_s) - As([n-1]T_s]\} = 0 \quad (8.16)$$

Solving for A, we obtain

$$A = \frac{R(T_s)}{R(0)} \quad (8.17)$$

Equation (8.16) has an interesting intuitive interpretation. It states that the expected value of the product of the error and the measured sample is zero. That is, the error has no component in the direction of the observation. In other words, the two quantities are *orthogonal*. This makes sense because if the error *did* have a component in the direction of the observation, we could reduce that component to zero by readjusting the constant A.

The predictor of Fig. 8.45 takes the most recent sample value (which it forms by adding the prediction to the difference term) and weights it by $R(T_s)/R(0)$. We assume that the input process has been observed sufficiently long to be able to estimate its autocorrelation.

Example 8.4

Find the weights associated with a predictor operating on the two most recent samples. Also, evaluate the performance of the predictor.

Solution: The prediction is given by

$$s(nT_s) = As[(n-1)T_s] + Bs[(n-2)T_s]$$

where the object is to make the best choice of A and B. For this choice, the error has no component in the direction of the measured quantities. Thus,

$$E\{(s(nT_s) - As[(n-1)T_s] - Bs[(n-2)T_s])s[(n-1)T_s]\} = 0$$

$$E\{(s(nT_s) - As[(n-1)T_s] - Bs[(n-2)T_s])s[(n-2)T_s]\} = 0$$

Expanding these equations, we find that

$$R(T_s) - AR(0) - BR(T_s) = 0$$

$$R(2T_s) - AR(T_s) - BR(0) = 0$$

and solving for A and B yields

$$A = \frac{R(T_s)[R(0) - R(2T_s)]}{R^2(0) - R^2(T_s)}$$

$$B = \frac{R(0)R(2T_s) - R^2(T_s)}{R^2(0) - R^2(T_s)}$$

The mean square error is then

$$\text{mse} = E\{[s(nT_s) - \hat{s}(nT_s)]^2\}$$
$$= E\{s^2(nT_s)\} - E\{\hat{s}(nT_s)s(T_s)\}$$
$$= R(0) - \frac{R(0)R^2(T_s) + R^2(2T_s) - 2R(2T_s)R^2(T_s)}{R^2(0) - R^2(T_s)}$$

As a specific example, suppose that the autocorrelation of $s(t)$ is as shown in Fig. 8.46 and that the sampling period is 1 sec. The equation for mse then yields

$$\text{mse} = 1.895$$

For comparison, if we had forgotten to measure $s[(n-1)T_s]$ and $s[(n-2)T_s]$ and simply guessed at the mean value, or zero, the mean square error would have been $R(0)$, or 10.

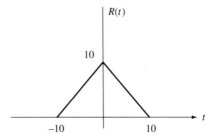

Figure 8.46 Autocorrelation for Example 8.4.

For a speech signal, it has been shown experimentally that a differential PCM system using a predictor operating on the most recent sample can save one bit per sample over ordinary PCM. That is, a differential PCM system can achieve approximately the same error performance as the PCM system with one fewer bit per sample. Thus, if we think of a single voice channel as requiring eight bits of PCM quantization, the transmission rate is 64 kbps (assuming a sampling rate of 8,000 samples/sec). Differential PCM then needs only seven bits per sample, thus reducing the transmission rate to 56 kbps and freeing the channel for other uses.

In *adaptive differential pulse code modulation*, the coefficients of the predictor do not remain constant over the entire transmission. For each group of a given length of samples—say, n—we compute a covariance matrix $[R_{ij}]$. We then use this matrix to solve for the coefficients of the predictor. These are then changed for the next n samples. Since the coefficients of the predictor are no longer constant, there must be some way to assure that the receiver uses the same coefficients. The most common method for accomplishing this is to send the updated coefficients as *overhead*, usually multiplexed with the information in the sample.

The three common alternative techniques that we have presented encode analog information into a digital format. There are other possibilities for doing so as well, many of which represent perturbations and combinations of the basic techniques. For example, the three techniques described can be coupled with companding to improve performance. In each case, we must weigh the improvement in performance (i.e., the decreased quantization error) against the increasing complexity of the hardware.

8.4 QUANTIZATION NOISE

We begin our study of quantization noise in PCM by re-examining the quantization input-output relationship of Fig. 8.47. Quantization noise, or error, is defined as that function of time which is the difference between $s_q(t)$, the quantized waveform, and $s(t)$. That is,

$$e(nT_s) = s(nT_s) - s_q(nT_s) \qquad (8.18)$$

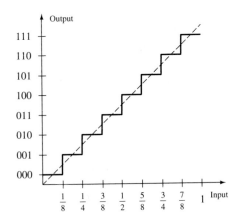

Figure 8.47   Input-output relationship of quantizer.

Figure 8.48(a) illustrates a representative function of time, $s(t)$, and a quantized version, $s_q(t)$, of this function of time.

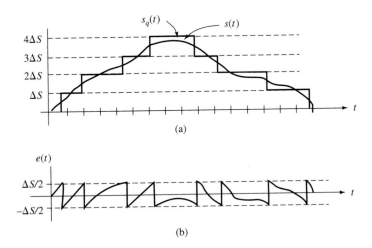

Figure 8.48   Quantization error.

While we are illustrating functions of time, it is important to note that it is the *sample values* that are rounded off, not the analog function. Therefore, the only significant values of $s_q(t)$ are the values at the sample points, nT_s. Figure 8.48(b) shows the quantization error $e(t)$ as the difference between $s(t)$ and $s_q(t)$. Again, note that we are interested in the values of this error function only at the sample points. The magnitude of the error term never exceeds one-half of the spacing between quantization levels.

We wish to find the average statistics of the error. To do so, we must first find the probability density function of the error. Figure 8.49 illustrates the error as a function of the input sample value.

The error curve starts at $-\Delta S/2$ at the lower edge of each quantization interval and increases linearly to reach $+\Delta S/2$ at the top edge. If we now know the probability density

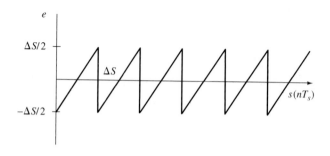

Figure 8.49   Error as a function of sample value.

function of the sample values, it becomes a simple matter to find the probability density function of the error e. This is an application of functions of a random variable. The result is

$$p(e) = \sum_i \frac{p(s_i)}{|de_i/ds_i|} \tag{8.19}$$

The s_i are the various values of s corresponding to e. That is, if we set e to a certain value in Fig. 8.49, there are a number of values of s (equal to the number of quantization regions) that yield the specified value of e. The magnitude of the slope of the function is always unity, so Eq. (8.19) can be simplified to

$$p(e) = \sum_i p(s_i) \tag{8.20}$$

The first value of s_i to the right of the origin is

$$s_i = \frac{e + \Delta S}{2} \tag{8.21}$$

All of the other s_i can be found by successively adding and subtracting ΔS from this value.

If the samples were uniformly distributed over the range of values, each term in the summation of Eq. (8.20) would be the same—the height of the original density function. The result is a uniform error density, as shown in Fig. 8.50.

Now suppose that the samples have a triangular density, as shown in Fig. 8.51. The result is still the uniform error density of Fig. 8.50. You can convince yourself of this by

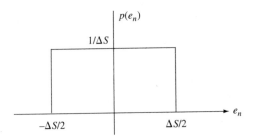

Figure 8.50   Probability density of error.

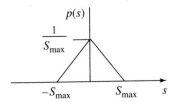

Figure 8.51   Triangular error density.

taking some representative values. In fact, as e increases, all of the s_i of Eq. (8.21) increase by the same amount. In a triangular density, for every value in the sum of Eq. (8.20) that is decreasing, another value is increasing by the same amount. A similar argument can be used for any density that is approximately linear over the range of a single quantization region. For this reason, the uniform density of Fig. 8.50 is assumed to be approximately correct over a wide range of input signals.

Now that we know the density of the error, we can find the average of its square, or the *mean square error*:

$$\text{mse} = \int_{-\infty}^{\infty} p(e_n) e_n^2 de_n$$

(8.22)

$$= \frac{1}{\Delta S} \int_{-\Delta S/2}^{\Delta S/2} e_n^2 de_n = \frac{\Delta S^2}{12}$$

This result gives the average of the square of the error in a single sample of the function of time. To assess the "annoyance" caused by the error, it is necessary to compare this value to the average of the square of the unperturbed time sample. Doing so yields the very important *signal-to-quantization noise ratio*

$$\text{SNR} = \frac{\overline{s^2}}{\overline{e_n^2}} = \frac{12\overline{s^2}}{\Delta S^2} = \frac{12P_s}{\Delta S^2}$$

(8.23)

where P_s is the average signal power. Equation (8.23) is an extremely important result that will be used many times in the work that follows.

Example 8.5

Consider an audio signal comprised of the sinusoidal term

$$s(t) = 3\cos 500 \, \pi t$$

(a) Find the signal-to–quantization noise ratio when the audio signal is quantized using 10-bit PCM.

(b) How many bits of quantization are needed to achieve a signal-to–quantization noise ratio of at least 40 dB?

(c) Repeat part (a) for a signal twice that of the given signal, that is,

$$s(t) = 6\cos 500 \, \pi t$$

Solution: **(a)** Equation (8.23) is used to find the signal-to–quantization noise ratio. The only parameters that need to be evaluated are the average signal power and the quantization region size. The total swing of the signal is 6 volts, so the size of each interval is $6/2^{10} = 5.86 \times 10^{-3}$. The average signal power is $3^2/2 = 4.5$ watts. The signal-to–quantization noise ratio is then given by

$$\text{SNR} = \frac{12 \times 4.5}{(5.86 \times 10^{-3})^2}$$

$$= 1.57 \times 10^6$$

If we wish to express this in dB, we take its logarithm and multiply by 10. Therefore,

$$\text{SNR}_{\text{dB}} = 10 \times \log (1.57 \times 10^6) = 62 \text{ dB}$$

(b) The minimum SNR is specified as 40 dB. This corresponds to a ratio of 10^4. We use Eq. (8.22), where ΔS is unknown. Thus,

$$\frac{12 \times 4.5}{\Delta S^2} > 10^4$$

and solving for ΔS yields

$$\Delta S < 7.35 \times 10^{-2}$$

We next note that the step size is

$$\Delta S = \frac{6}{2^N}$$

where N is the number of bits of quantization. We need to choose an N such that ΔS is no larger than 7.35×10^2. Therefore,

$$\frac{6}{2^N} < 7.35 \times 10^{-2}$$

and

$$2^N > 81.6$$

We could use logarithms to solve this equation, but that really should not be necessary. If N is 6, the left side of the inequality is 64. If N is 7, the left side is 128. Therefore, we require seven bits of quantization to achieve an SNR of at least 40 dB. (You might think that the answer is $N = 6.35$, which you would obtain using a calculator. But keep in mind that N is the number of bits of quantization, and as such, it must be an integer.)

It should be clear that each additional bit of quantization reduces ΔS by a factor of 2. This increases the signal-to-quantization noise ratio by a factor of 4, which corresponds to 6 dB, since

$$10 \log 4 \approx 6$$

Hence, each additional bit of quantization increases the SNR by 6 dB.

(c) The total swing increases from 6 to 12 volts, so ΔS increases from 5.86×10^{-3} to 1.172×10^{-2}. The power is $6^2/2$, or 18, watts. The resulting signal-to-noise ratio is

$$SNR = \frac{12 \times 18}{(1.172 \times 10^{-2})^2} = 1.57 \times 10^6$$

This answer is the same as that found in part (a). It should not be surprising, since the numerator and denominator of the equation for SNR changed by the same factor.

While Eq. (8.23) clearly shows how to find the signal-to-quantization noise ratio as a function of the signal and the quantization step size, it is useful to get one general result to use as a starting point in system design. Suppose the signal $s(t)$ is uniformly distributed between $-S_{max}$ and $+S_{max}$, as shown in Fig. 8.52. For this particular case, Eq. (8.23) reduces to a very simple form. We need only two quantities to solve the equation: the average signal power and the step size. The power is found using results from basic probability theory:

$$P_s = \int_{-\infty}^{\infty} s^2 p(s)\,ds = \frac{1}{2S_{max}} \int_{-S_{max}}^{S_{max}} s^2\,ds = \frac{S_{max}^2}{3} \tag{8.24}$$

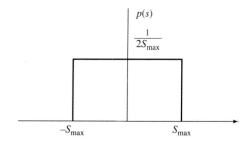

Figure 8.52   Uniformly distributed signal.

The step size is

$$\Delta S = \frac{2S_{max}}{2^N} = \frac{S_{max}}{2^{N-1}} \tag{8.25}$$

Equation (8.23) then becomes

$$SNR = \frac{4S_{max}^2}{S_{max}^2/2^{2N-2}} = 2^{2N} \tag{8.26}$$

Note that, as in the previous example, the specific value of S_{max} does not affect the SNR: As S_{max} changes, both the signal power and the quantization noise power change by the same factor.

We can convert the SNR of Eq. (8.26) into decibels as follows:

$$\text{SNR}_{dB} = 10 \log (2^{2N}) = 20N \log (2) \approx 6N \text{ dB} \qquad (8.27)$$

This result represents a good starting point even when the signal is not uniformly distributed. In part (b) of Example 8.5, we were asked to specify the number of bits of quantization to achieve an SNR of at least 40 dB. In Eq. (8.27), in order for $6N$ to be greater than 40, N must be at least 7, which is the same answer we got in the example. But be cautious in applying Eqs. (8.24) and (8.25): Signals in real life are not uniformly distributed, and these equations apply *only* to the uniformly distributed case. If you incorrectly apply Eq. (8.25) to a nonuniform signal, you will be in danger of designing a system with the wrong value of N. If you use a value smaller than needed, you will not meet the specifications for the SNR. If you use a value too large, you can cost your employer large sums of money by requiring more bits per second than are needed to meet the specifications.

Quantization Noise: Nonuniform Quantization

When the input samples are *not* uniformly distributed, it is possible to improve signal-to–quantization noise ratios by using *nonuniform quantization*. We begin by assuming that the samples are distributed according to a probability density $p(s)$, as shown in Fig. 8.53. Although this distribution appears to be approximately Gaussian, it is meant to be a representative probability density function, and the results we obtain will not depend upon any specific shape. We have illustrated three bits of quantization, with the eight regions marked by their boundaries s_i and by the round-off levels s_{qi}. The mean square quantization error of the samples is

$$\text{mse} = E\{[s(nT_s) - s_q(nT_s)]^2\}$$

$$= \int_{-\infty}^{\infty} (s - s_q)^2 p(s) ds \qquad (8.28)$$

$$= \sum_i \int_{s_i}^{s_{i+1}} (s - s_{q_i})^2 p(s) ds$$

In Eq. (8.28), the s_{qi} are the various quantization round-off levels, and $p(s)$ is the probability density function of the signal samples. We shall return to this equation in a few

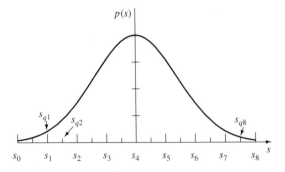

Figure 8.53 Probability density of samples.

moments, when we examine companded systems. For now, we use the expression to prove a statement made earlier about the best location for the round-off levels. Let us assume that the regions have been specified (the s_i are given) and that we wish to find the optimum location of the round-off values s_{qi}. We interpret the word *optimum* to mean those values that minimize the mean square error. To do so, we differentiate Eq. (8.28) with respect to the s_{qi} and set the derivative to zero. This yields

$$\int_{s_i}^{s_{i+1}} (s - s_{q_i})p(s)ds = 0 \quad \text{for all } i \tag{8.29}$$

Equation (8.29) indicates that once the quantization regions have been chosen, the round-off levels are selected to be the center of gravity of the corresponding portion of the probability density. Thus, rather than being in the center of the interval, each quantization level is skewed toward the higher probability end of the interval. This is intuitively satisfying, since more samples can be expected to fall into this region.

Example 8.6

Assume that the density function of $s(t)$ is a zero-mean Gaussian density with variance σ^2 equal to 1/9. Since the probability of a sample exceeding a magnitude of 1 is less than 1 percent (i.e., the 3σ point), suppose that you quantize the region between -1 and $+1$ (that is, values above a magnitude of 1 will saturate at either 000 or 111). Three-bit quantization is used.

(a) Find the mean square quantization error, assuming that uniform quantization is used.

(b) A scheme is proposed wherein the quantization regions are chosen to yield *equal area* under the probability density function over each region. That is, the probability of a sample falling within any particular interval is the same as that for any other interval. Choose the best location for the round-off values, and find the corresponding mean square error.

Solution: (a) We use the formula of Eq. (8.22) to find the mean square error in the uniform case. The size of each interval is $\frac{2}{8} = \frac{1}{4}$. The error is therefore

$$\text{mse} = \Delta S^2/12 = 1/192 = 5.2 \times 10^{-3}$$

(b) We must first find the boundaries of the quantization regions. We are dividing the latter into eight equal-area segments, so the area under the density within each region must be $\frac{1}{8}$. Reference to a table of error functions indicates that the s_i are as follows:

$$-1, -0.38, -0.22, -0.1, 0, 0.1, 0.22, 0.38, 1$$

Equation (8.29) is now used to find the round-off levels s_{qi}. The equation reduces to

$$s_{q_i} = \frac{\displaystyle\int_{s_i}^{s_{i+1}} sp(s)ds}{\displaystyle\int_{s_i}^{s_i=1} p(s)ds}$$

$$= 8\int_{s_i}^{s_{i+1}} sp(s)ds \quad \text{for all } i$$

This can be evaluated in closed form or approximated. The resulting s_{qi} are:

$$-0.54, -0.3, -0.16, -0.05, 0.05, 0.16, 0.3, 0.54$$

Finally, the mean square error is found from Eq. (8.27) to be

$$\text{mse} = 5.3 \times 10^{-3}$$

It appears that the uniform quantizer is better than this particular nonuniform quantizer. However, with the Gaussian density and only three bits of quantization, Eq. (8.22) is not a very good approximation for the mean square error. Recall that that equation required the density to be approximately piecewise linear over the various regions. The exact answer to part (a) can be found by applying Eq. (8.28). The result is 6.2×10^{-3}, and therefore, the nonuniform quantizer does provide an improvement in performance.

Example 8.6 suggested one possible algorithm for selecting quantization regions. In fact, it is not the best algorithm, and it even causes degradation of performance, compared to uniform quantization, for some probability densities.

The equation for the mean square error weights the probability by the square of the deflection from the quantized value before integration. In general, the problem is to minimize the error of Eq. (8.28) as a function of two variables, s_i and s_{qi}. The s_{qi} are constrained to satisfy Eq. (8.29). Unless the probability density can be expressed in closed form, the problem is computationally difficult.

The *mean value theorem* can be used in Eq. (8.28) to get an approximation that improves as the number of bits of quantization increases. When we use this approximation, the following rule for choosing the quantization regions results:

Choose the quantization regions so as to satisfy the uniform moment constraint

$$(s_{i+1} - s_i)^2 p(\text{midpoint}) = \text{constant} \tag{8.30}$$

We explore this approach more fully in the problems at the end of the chapter.

Companded Systems

We derive an approximate expression for the improvement (or degradation) presented by a particular compression function, compared to uniform quantization. The result is approximate, and the approximation improves as the number of bits of quantization increases— that is, as the quantization regions get smaller. We begin by assuming that the round-off values are in the middle of each interval. This is the best choice only if the density can be assumed to be constant over the width of the interval. Using this assumption, and further assuming that the density function can be approximated over the interval by its value at the round-off level, we find that Eq. (8.28) reduces to

$$\text{mse} = \sum_i \int_{s_i}^{s_{i+1}} (s - s_{q_i})^2 p(s)\, ds$$

$$= \sum_i p(s_{q_i}) \frac{(s - s_{q_i})^3}{3} \Bigg|_{s_i}^{s_{i+1}} \tag{8.31}$$

We now use the approximation that s_{qi} is in the middle of the interval:

$$s_{qi} = \frac{s_i + s_{i+1}}{2} \tag{8.32}$$

Equation (8.31) then becomes

$$\text{mse} = \sum_i p(s_{qi}) \frac{(s_{i+1} - s_i)^3}{12} \tag{8.33}$$

It is comforting to check our work occasionally against the known result for uniform quantization. In fact, if uniform steps of size ΔS are substituted into Eq. (8.33), the result reduces to $\Delta S^2/12$, as we found earlier for uniform quantization. (If you simply looked at Eq. (8.33) and said, "Oh sure, he is right," you are fooling yourself. I am not always right. Please take a moment now to prove this statement.) If this were not the case, we would have to recheck the derivation to find the error(s).

We can relate the interval size, $s_{i+1} - s_i$, to the slope of the compression curve. That is, if the compressed output is uniformly quantized with a step size of ΔS, the corresponding step sizes of the uncompressed waveform are given approximately by (see Fig. 8.35)

$$s_{i+1} - s_i \approx \frac{\Delta S}{F'(s_i)} \tag{8.34}$$

We need to take the limit of the summation of Eq. (8.33) as the intervals get smaller and smaller. To do so, we separate the square of the interval from the cubic term and rewrite the squared term using the compression function relationship of Eq. (8.34). The remaining interval multiplier becomes the differential term. Thus,

$$\begin{aligned}
\text{mse} &= \frac{1}{12} \sum_i p(s_{qi})(s_{i+1} - s_i)^2 (s_{i+1} - s_i) \\
&= \frac{1}{12} \sum_i \frac{p(s_{q_i}) \Delta S^2}{[F'(s_i)]^2}(s_{i+1} - s_i)
\end{aligned} \tag{8.35}$$

In the limit as the interval widths approach zero, this becomes

$$\text{mse} = \frac{\Delta S^2}{12} \int_s \frac{p(s)}{[F'(s)]^2} ds \tag{8.36}$$

The mean square error for a uniform quantizer, $\Delta S^2/12$, appears explicitly in this equation. If the integral in this equation is less than unity, the compander has provided an improvement over the uniform quantizer.

Equation (8.36) can be used to evaluate a companding scheme, given any particular probability density of the signal samples. You will have ample opportunity to fine tune your analysis skills in the problems at the end of the chapter. For now, we will present one set of results in a slightly different way.

We wish to compare the companded and uniform systems. For this comparison, we will choose eight-bit quantization, since that is common in voice transmission. If we assume that the signal samples are uniformly distributed, we achieve a signal-to-quantization noise ratio of 48 dB ($6N$ dB).

Suppose now that the signal power decreases, but the quantization levels are not changed (i.e., we do not redesign the A/D converter). Then, as long as the signal fills at least one quantization region ($S_{max}/128$), the quantization noise varies over the entire range of the signal ($-\Delta S/2$ to $+\Delta S/2$), and the average noise power remains unchanged. Therefore, as the signal power decreases, the SNR decreases in the same proportion. We can plot the SNR as a function of the input signal power as shown in the linear curve of Fig. 8.54.

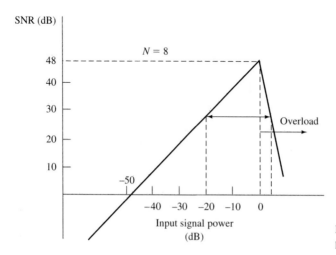

Figure 8.54   SNR as a function of signal power.

As soon as the signal increases beyond the range of the quantization levels (i.e., it overloads), the noise power increases rapidly. This is true because larger samples saturate the system, and the error is no longer limited in magnitude to $\Delta S/2$.

For any specified SNR, the portion of the curve above that level represents the *dynamic range* of the quantizer. For example, if we desire an SNR of at least 28 dB, the dynamic range is from -20 dB to about $+3$ dB relative to full load, as indicated in the figure.

The companded system performs better than the uniform quantization system for small signals. The reason is that the intervals get smaller as the sample size decreases.

We can evaluate the performance of the companded system and compare it to that of the uniform quantizer. Figure 8.55 does this for a uniform signal density and μ-law companding (for various values of μ, including μ-255). The curve of Fig. 8.54 is repeated in that figure for comparison. Note that, as expected, the companded system outperforms the uniform quantizer for low signal levels. As the signal starts to fill the entire range of quantization intervals, the uniform quantizer performs better than the companded quantizer. Thus, the performance of the companded system improves in the presence of low-level signals, as expected. The dynamic range also increases. For example, if we desire an SNR of at least 28 dB, the dynamic range is from -50 dB to about $+3$ dB relative to full load, as indicated on the diagram.

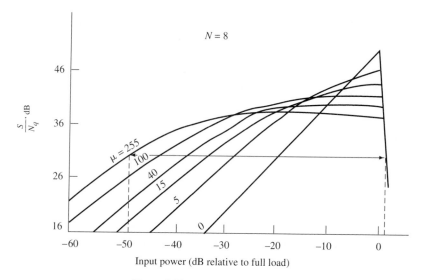

Figure 8.55 Performance of companded systems.

Oversampling

We have been examining the signal-to-quantization noise ratio where the quantization noise power is the mean square value of the quantization error. In companded systems, we found that the signal to quantization noise ratio could sometimes be reduced by quantizing with steps of nonuniform size. We now explore whether there are other techniques for improving the signal-to-quantization noise ratio.

One item to which we can look for help is the lowpass filter, which is used to reconstruct $s(t)$ from its samples. Suppose that filter were to cut off some of the quantization noise. If so, it would improve the SNR.

To analyze the effects of the filter, we need to know something about the frequency distribution of the quantization noise. The noise consists of samples at the sampling points, and we can assume that each sample is independent of all of the others. That is, knowing the value of the quantization noise at one sampling point tells us nothing about the value at the next point. This assumption is valid as long as the sample values of the signal change across many quantization regions between adjacent sample points. In contrast, if there were very few quantization regions (e.g., one bit of quantization), and we sampled much higher than the Nyquist rate, we would expect the quantization noise values to change relatively slowly from sample to sample.

If the sample values of the noise are independent of each other, the noise power spectral density is constant up to one-half of the sampling rate. Intuitively, you can think of it as a reverse application of the sampling theorem: *A signal specified by independent samples at a rate of $2f_m$ samples/sec describes a waveform with a maximum frequency of f_m Hz.*

Now returning to the original problem, we see that if we sample at the Nyquist rate, the noise power is limited to frequencies below f_m, and all of the noise goes through the lowpass filter. The signal to noise ratio is therefore calculated as in Eq. (8.23):

$$\text{SNR} = \frac{12P_s}{\Delta S^2} \tag{8.37}$$

Suppose we now sample at a rate higher than the Nyquist rate—say, $2kf_m$—where k is greater than 1. The noise now has frequencies ranging up to kf_m Hz, and the lowpass filter reduces the noise power by the ratio of

$$\frac{f_m}{kf_m} = \frac{1}{k} \tag{8.38}$$

The signal-to-quantization noise ratio is now

$$\text{SNR} = \frac{12P_s}{k(\Delta S)^2} \tag{8.39}$$

As an example, suppose a speech waveform with maximum frequency of 4 kHz is sampled at 8 kHz and quantized using eight bits. If we assume that the speech is uniformly distributed, the signal to noise ratio is $6N = 48$ dB. Now suppose that we raise the sampling rate to 32 kHz, a factor of *four times oversampling* (or, as the manufacturers of compact disc player advertise to a totally uninformed public, "4 × oversampling"). The SNR then improves by a factor of four, to 54 dB.

Quantization Noise in Delta Modulation

Once again, we define the quantization error as the difference between the original signal and the quantized (staircase) approximation:

$$e(t) = s(t) - s_q(t) \tag{8.40}$$

We assume that the sampling rate and step size are chosen so as to avoid overload. Under this assumption, the magnitude of the quantization noise never exceeds the step size. If we make the simplifying assumption that all signal amplitudes are equally likely, we conclude that the error is uniformly distributed over the range between $-\Delta$ and $+\Delta$, as shown in Fig. 8.56. The mean square value of the quantization noise

$$\text{mse} = \frac{1}{2\Delta} \int_{-\Delta}^{\Delta} e^2 de$$
$$= \frac{\Delta^2}{3} \tag{8.41}$$

In designing digital communication systems, a logical question to ask would be whether to use pulse code modulation or delta modulation as the source-encoding technique. As a designer, you would be concerned with many factors: complexity (i.e., in terms of cost), transmission bit rate (i.e., the required system bandwidth), reliability, quantization noise, the effects of transmission errors, and others. It would be very helpful if we could derive a simple formula that relates the SNR of pulse code modulation to that of delta modulation. Unfortunately, many specific assumptions must be made to arrive at such a result. The bottom line is that, under certain circumstances, delta modulation will

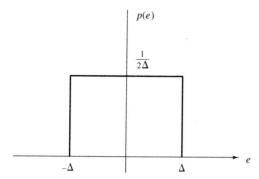

Figure 8.56 Uniform quantization error distribution in delta modulation.

provide the same SNR as pulse code modulation with a lower transmission bit rate. Under other circumstances, the reverse is true. Adaptive delta modulation adds yet another dimension to the analysis. This is a good place for computer simulations.

We start our analysis by solving for the mean square quantization error at the output of the delta modulation receiver. The demodulator includes a lowpass filter to smooth the staircase approximation into a continuous curve. To find the quantization noise power at the output of the filter, we must first find the frequency characteristics of the quantization noise. This is not a simple analytical task, and it requires that a specific form be assumed for $s(t)$.

Let us assume that the original $s(t)$ is a sawtooth-type waveform. This is the simplest example of a waveform that is uniformly distributed. The waveform, its quantized version, and the resulting quantization noise are shown in Fig. 8.57. Notice that the noise function is *almost* periodic with period T_s, the sampling period. The noise would be *exactly* periodic with a period equal to that of the sawtooth waveform if that period were an integral multiple of T_s. We are assuming that the step size and sampling period are chosen to avoid overload and, in this case, to give perfect symmetry. The power spectral density of $s_q(t)$ can be found precisely. It is of the form $\sin^4 f / f^4$, since the Fourier transform of a sawtooth follows a $\sin^2 f / f^2$ shape. The first zero of the power spectral density of the triangular waveform is at $f = 1/T_s$. The lobes beyond this point are attenuated as the fourth power of $1/f$. Therefore, we can assume that there is relatively little power outside of the main lobe, so we suppose that all of the power is concentrated in a low-frequency band stretching to $f = 1/T_s$. Now, since we are assuming that delta modulation sampling takes place at a rate well above the Nyquist rate (typically, about seven times the Nyquist rate), the first zero of the spectrum, at $f = 1/T_s$, is at a much higher frequency than f_m. Thus, the lowpass filter, which cuts off at f_m, passes only a relatively small portion of the main lobe of the noise power spectrum. The situation is shown in Figure 8.58.

To get an approximate result, we will assume that the spectrum is essentially flat over the range between 0 and f_s. The total power of the noise is the mean square error we found earlier, $\Delta^2/3$. Since we are assuming a flat spectrum, the fraction of the power that gets through the lowpass filter is $T_s f_m$, or f_m / f_s. The output noise power is then given by

$$N_q = \frac{\Delta^2}{3} \frac{f_m}{f_s} \tag{8.42}$$

where f_s is the number of samples per second.

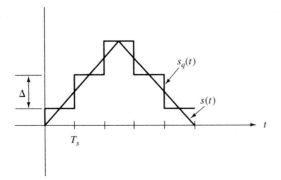

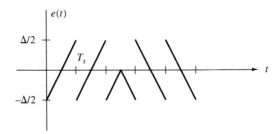

Figure 8.57   Delta modulation of sawtooth waveform.

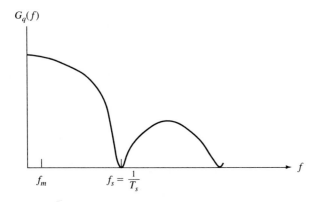

Figure 8.58   Quantization noise power spectrum.

Now that we have an expression for the quantization noise power, we can find the signal to noise ratio as

$$\frac{S}{N_q} = \frac{3f_s}{\Delta^2 f_m}\overline{s^2(t)}$$
(8.43)

Equation (8.43) implies that the signal to noise ratio doubles (a 3-dB increase) for each doubling of the sampling frequency. However, the result is not that simple, since the step size is related to the sampling frequency.

We will now apply Eq. (8.43) to a single sinusoid in order to get a more comprehensive result. Suppose we wish to quantize the signal

$$s(t) = a\cos 2\pi f_m t$$
(8.44)

The maximum slope of the function is

$$\left|\frac{ds}{dt}\right|_{max} = 2\pi f_m a \tag{8.45}$$

The maximum slope of the delta modulation staircase approximation is

$$slope_{max} = \frac{\Delta}{T_s} = \Delta f_s \tag{8.46}$$

To avoid slope overload, the maximum slope of the staircase approximation must be equal to or greater than the maximum slope of the signal.[1] Therefore,

$$\Delta f_s \geq 2\pi f_m a$$

$$f_s \geq \frac{2\pi f_m a}{\Delta} \tag{8.47}$$

Now we put all the pieces together, using the minimum slope in the calculation of Eq. (8.43). This yields

$$\frac{S}{N} = \frac{3f_s}{\Delta^2 f_m}\overline{s^2(t)}$$

$$= \frac{6\pi a a^2}{\Delta^3 2} = 3\pi\left(\frac{a}{\Delta}\right)^3 \tag{8.48}$$

We can obtain an alternative form of this relationship in terms of the frequencies by solving for a/Δ in Eq. (8.47) and using the result in Eq. (8.48). This yields

$$\frac{S}{N} = \frac{3}{8\pi^2}\left(\frac{f_s}{f_m}\right)^3 \tag{8.49}$$

Equation (8.49) shows that, for every doubling of the sampling frequency, the signal to noise ratio increases by a factor of eight (9 dB). To avoid incorrectly applying Eq. (8.49), it is important that you understand the assumptions made in its derivation. The most critical of these is the assumption that Δf_s is the minimum required to avoid slope overload.

Example 8.7

An audio signal of the form

$$s(t) = 3\cos 1000\ \pi t$$

is quantized using delta modulation. Find the signal-to-quantization noise ratio.
Solution: We must first choose a step size and a sampling frequency for the given waveform. The Nyquist rate is $f_s = 1,000$ samples/sec. Suppose, for purposes of illustration, that

---

[1]In real life, signals are rarely pure sinusoids. It would not make sense to go through a lot of trouble to transmit such a predictable signal. When a signal is composed of many different frequencies, we usually allow a limited amount of overshoot so that the sampling frequency does not have to be unrealistically high.

we choose eight times this rate, that is, $f_s = 8,000$ samples/sec. The maximum amount that the function can change in $\frac{1}{8}$ msec is approximately 1 volt. If a step size of 1 volt is chosen, the staircase will not overload. The quantization noise power is

$$M_q = \frac{\Delta^2}{3}\frac{f_m}{f_s} = 41.7 \text{ mW}$$

The signal power is $3^2/2$ or 4.5 watts. Finally, the SNR is

$$\text{SNR} = \frac{4.5}{0.042} = 107 \quad \text{or} \quad 20.3 \text{ dB}$$

This is far less than what is achieved using PCM in this example. Note that, had you applied Eq. (8.49), you would have found an SNR of only 19.45. The reason for the discrepancy is that the step size Δ is assumed to be 1 volt instead of the 2.36 volts specified by Eq. (8.47).

PROBLEMS

8.1.1 Design a four-bit serial quantizer that could be used for a voice signal with a maximum frequency of 5 kHz and sample values between -3 V and $+4$ V. Illustrate the operation of the quantizer for sample values of -2 V and $+3.4$ V.

8.1.2 (a) Design a counting quantizer for five-bit PCM and a voice signal as the input. Assume that the voice signal has a maximum frequency of 5 kHz. Choose all parameter values and justify your choices.
(b) Repeat the design in part (a) for six-bit PCM.

8.1.3 Design a counting analog-to-digital converter to operate on the signal

$$s(t) = 2\sin \pi t$$

The design requires four-bit analog-to-digital conversion.

8.1.4 Design a flash analog-to-digital converter to operate on the signal

$$s(t) = \sin 2 \pi t + \frac{\sin 2 \pi t}{t}$$

8.3.1 A 1-kHz sine wave is sampled 3,000 times each second and encoded using four-bit PCM. (You may choose three bits plus a sign or four-bit encoding).
(a) Sketch the PCM waveform for the first 3 msec, assuming that the NRZ-L format is used.
(b) Repeat part (a) for the NRZ-M differential format.
(c) Repeat part (a) for the biphase-L format.

8.3.2 You are given the NRZ-M signal shown in Fig. P8.3.2. Sketch the NRZ-L signal for the same binary information.

8.3.3 Modify the NRZ-M encoder and decoder to apply to NRZ-S. Do this by adding only one inverter to the system. Investigate whether there is more than one possible location for the inverter.

8.3.4 You are given the function

$$s(t) = 2\sin \pi t$$

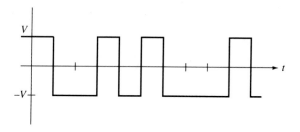

Figure P8.3.2

The signal is to be transmitted using four-bit PCM.
(a) Find the PCM waveform the first 5 sec.
(b) Find the mean square quantization noise power.

8.3.5 Repeat Problem 8.3.4 for a Gray code.

8.3.6 A PCM wave is as shown in Fig. P8.3.6, where voltages of $+1$ and -1 are used to send a 1 and a 0, respectively. Two-bit quantization has been used. Sketch a possible analog information signal $s(t)$ that could have resulted in this PCM wave.

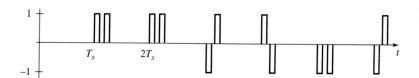

Figure P8.3.6

8.3.7 Two voice signals with 5-kHz maximum frequency are transmitted using PCM with eight levels of quantization. How many pulses per second are required to be transmitted?

8.3.8 The input to a delta modulator is $s(t) = 5t + 1$. The sampler operates at 20 samples per sec. The step size is 1 V. Sketch the output of the delta modulator for the first 2 sec.

8.3.9 Delta modulation is used to transmit a voice signal with a maximum frequency of 3 kHz. The sampling rate is set at 20 kHz. The maximum amplitude of the analog voice signal is 1 V. Design the modulator, and discuss your reasons for selecting a particular step size.

8.3.10 An adaptive delta modulator is used to transmit a signal $s(t) = 5t + 1.1$. The sampler operates at 10 samples/sec. The step size is either 0.5, 1, or 1.5 V, depending on the number of 1's in a sequence of four bits. For zero or four 1's, the maximum step size is chosen. For one or three 1's, the step size is 1 V, and for two 1's, the minimum step size is chosen. Sketch the staircase approximation to $s(t)$ and the transmitted digital signal for the first 3 sec.

8.3.11 Repeat Problem 8.3.10, using the Song algorithm.

8.3.12 Repeat Problem 8.3.10, using the space shuttle algorithm.

8.3.13 You are given the function

$$s(t) = \sin 2 \pi t + 4\cos 4\pi t$$

(a) Find the binary sequence that results if this signal is to be transmitted using delta modulation with a sampling rate of 20 samples/sec. Choose an appropriate step size.

(b) Repeat part (a) for the Song algorithm.

8.3.14 Design a delta-PCM modulator for the signal

$$s(t) = \sin 2\pi t + 4\cos 4\pi t$$

Samples are taken at the rate of 10 samples/sec, and the bit transmission rate is 20 bps. (That
is, the correction factor is coded into two-bit words.)

8.3.15 Find the weights for a predictor that operates on the three most recent samples in order to
predict the next sample value. The autocorrelation of the signal is shown in Fig. P8.3.15.
Evaluate the performance of the predictor, and compare the resulting error to that given in
Example 8.4.

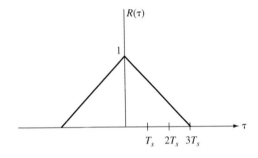

$R(\tau)$

$T_s \quad 2T_s \quad 3T_s$

Figure P8.3.15

8.3.16 (a) Design a linear predictor to approximate the signal $s(t_0 + T)$ as a linear combination of
$s(t_0)$ and its derivative $s'(t_0)$. You may either derive a general result in terms of autocorre-
lations or use the autocorrelation given in Problem 8.3.15 to arrive at a numerical answer.

(b) Evaluate the mean square error.

8.3.17 Design a linear predictor to estimate the value of $s(nT + \Delta)$ as a linear combination of $s(nT)$
for all n. Find the mean square error. Show that if the conditions of the sampling theorem are
met, the mean square error goes to zero.

8.3.18 Design a PCM system that could be used to transmit the video portion of a black-and-white
television signal. The video signal contains frequencies up to 4 MHz. A study of viewers has
indicated that at least 70 gray-scale levels are required for a pleasing picture.

8.3.19 Prove that about 91% of the total power of a random NRZ-L signal is contained in frequencies below the bit rate. [*Hint:* Integrate the expression within the scope of the integral sign in
Eq. (8.5)].

8.3.20 Prove that about 65% of the total power of a random biphase-L signal is contained in frequencies below the bit rate and that about 86% of the power is contained in frequencies below twice the bit rate. [*Hint:* Integrate the expression within the scope of the integral sign in
Eq. (8.22)].

8.4.1 The signal $s(t) = \sin 2\pi t$ is sampled at a rate of 4 samples/sec. The samples are converted to
a binary signal using a four-bit analog-to-digital converter. Compute the actual value of
quantization noise at each sampling point, and find the mean square average noise. (You
need do this only over one complete period of the waveform.) Compare your answer to that
given by Eq. (8.11), and explain any discrepancies.

8.4.2 A signal $s(t)$ is uniformly distributed between -5 V and $+5$ V. How many bits of uniform quantization are needed to achieve a signal-to-quantization noise ratio greater than 40 dB?

8.4.3 A signal is Rayleigh distributed with a mean value of 1. How many bits of quantization are needed such that the signal-to-quantization noise ratio is greater than 40 dB? (Make any reasonable approximations.)

8.4.4 Verify the numerical answers given in Example 8.6. (You probably will want to use a computer program.)

8.4.5 Find the mean square quantization error if three-bit PCM is used for an input signal that is Gaussian distributed. Assume that the quantization steps range over $\pm 3\sigma$, where σ^2 is the variance of the input.

8.4.6 A signal is Gaussian distributed with zero mean and variance of σ^2. The signal is to be transmitted using four-bit PCM, where the quantization steps are uniformly distributed over the range $\pm 3\sigma$. Find the mean square quantization error.

8.4.7 (a) A signal is derived by passing a 1-kHz square wave (zero average value) through an ideal lowpass filter with cutoff at 5.1 kHz. The resulting signal is quantized using six-bit PCM. Find the signal-to-quantization noise ratio.

(b) Repeat the analysis, assuming that μ-255 companding is used.

8.4.8 A signal is given by

$$s(t) = 20 \cos 100\,\pi\,t + 17 \cos 500\,\pi\,t$$

How many bits of quantization are required so that the signal-to-quantization noise ratio is greater than 40 dB?

8.4.9 If μ-255 companding is used in a quantization system with the signal

$$s(t) = 20 \cos 100\,\pi\,t + 17 \cos 500\,\pi\,t$$

how many bits of quantization are required to achieve a signal-to-quantization noise ratio of 50 dB? Compare your answer to that of Problem 8.4.8.

8.4.10 Find the mean square quantization error when a signal $s(t)$ with probability density as shown in Fig. P8.4.10 is:

(a) Uniformly quantized using five-bit quantization.

(b) Compressed according to the formula

$$F(s) = \sqrt{|s|}\ \text{sgn}\,(s)$$

and then uniformly quantized using five-bit quantization.

(c) Compressed according to the μ-255 rule and then uniformly quantized using five-bit quantization.

Repeat parts (a), (b), and (c) for four-bit quantization.

8.4.11 A compressor operates along a sinusoidal curve given by

$$F(s) = V_{\max} \sin \frac{\pi s}{2V_{\max}}$$

This curve is shown in Fig. P8.4.11. Find the improvement factor if:

(a) The signal is a sinusoid.

(b) The signal is triangularly distributed as shown in part (b) of the figure.

(c) The signal is uniformly distributed as shown in part (b) of the figure.

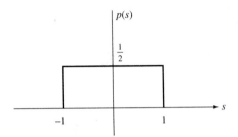

Figure P8.4.10

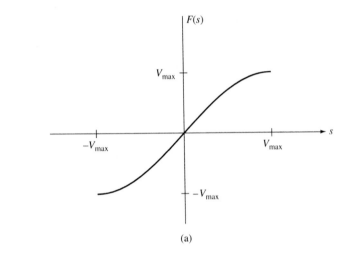

(a)

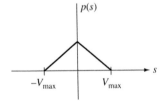
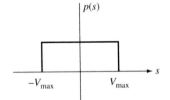

Figure P8.4.11

8.4.12 Repeat the analysis of Example 8.6, but using quantization regions that are selected according to Eq. (8.30). Compare your results with those obtained in the example.

8.4.13 A signal is derived by passing a 1-kHz square wave through a lowpass filter with cutoff at 5.1 kHz. Choose appropriate parameters for quantization by delta modulation, and find the signal-to-quantization noise ratio.

8.4.14 A signal to be transmitted is of the form

$$s(t) = 10\cos 1{,}000\,\pi\,t + 5\cos 1{,}500\,\pi\,t$$

The signal is quantized using delta modulation.

(a) Choose an appropriate sampling rate and step size.

(b) Find the signal-to-quantization noise ratio for your design of part (a).

8.4.15 You are given a signal $s(t)$ with probability density $p(s)$, as shown in Fig. P8.4.15. The signal is to be converted from analog to digital form using five bits of uniform quantization.

(a) Design a serial analog-to-digital converter.

(b) Find the signal-to-quantization noise ratio for the converter of part (a).

(c) Suppose that the signal is now *uniformly* distributed between 0 and 2 (instead of having the ramp distribution shown in the figure). How many decibels of improvement in signal-to-quantization noise ratio would result relative to the signal of part (a)?

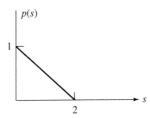

Figure P8.4.15

8.4.16 A signal has a probability density

$$p(x) = \begin{cases} Ke^{-2|x|} & -3 < x < 3 \\ 0 & \text{otherwise} \end{cases}$$

(a) Find the signal-to-quantization noise ratio if three bits of uniform quantization are used.

(b) Find the signal-to-quantization noise ratio if three bits of quantization are used for μ-255 companding.

8.4.17 A PCM system is to be used to transmit a monochrome video signal. Assume that each frame contains 100,000 picture elements and that 32 levels of gray are desired. At least 50 frames per second are required to prevent flicker. Additive white Gaussian noise of power $N_0 = 4 \times 10^{-6}$ watt/Hz is present in the channel, which can support voltages between -3 V and $+3$ V. The required signal to noise ratio at the receiver is 30 dB. Design the system, and specify the required bandwidth for the channel.

9

Channel Encoding

9.0 PREVIEW

What You Will Learn and Why You Should Care

Chapter 8 presented various techniques for encoding a source signal to produce a sequence of bits. We saw that certain errors were made in the process of digitizing the information. We called these errors *quantization noise*. We also saw that the faster you want to transmit bits, the more bandwidth you need.

Chapter 10 explores methods of transmitting these bits. When you get to that chapter, you will find that additional errors are introduced during transmission. Ones sometimes change to zeros and vice versa. The present chapter discusses a number of techniques used to modify the digital signal. The modifications are performed in a manner that decreases the effects of transmission errors.

Since the bandwidth of the channel is dependent on the number of bits per second to be transmitted, we can save bandwidth by decreasing the bit rate. This can be done either by slowing up the transmission speed or by reducing the number of bits in the message. The latter technique comes under the broad heading of *data compression* and is the subject of the first section of this chapter.

Necessary Background

To comprehend the channel-encoding techniques presented in this chapter, you need an understanding of basic probability. A working knowledge of linear algebra is also helpful, although we review the necessary concepts within the chapter.

9.1 SIGNAL COMPRESSION AND ENTROPY CODING

Signal compression is the process of shortening a signal without losing any essential information. It is applied to both analog and digital signals. For example, analog techniques have been developed that allow compressed reading of literature onto audiotape by eliminating or reducing pauses between words and sentences. These techniques are used to compress recordings for those with impaired vision; they are analogous to the various forms of shorthand that secretaries use to compress written messages. An audio recording could, of course, be shortened by increasing the playback speed, but this also increases the frequencies, leading to unnatural sounds and to increased demands on both the channel and the receiver. Analog compression techniques expand on some of the approaches we used in A/D conversion in Chapter 8. In particular, when we predicted the value of a waveform and transmitted only the correction factor, we were using a form of signal compression.

Before getting to the details, we need to clarify the terminology. The word *coding* is used in several different contexts. *Source encoding* is the technique for changing an analog signal into a digital sequence. We discussed PCM and related forms of source encoding in the previous chapter. There are at least three other common interpretations of the word *coding* that are broadly classified under the term *channel encoding*.

Entropy coding, the topic of this section, is the method of associating a digital word with each message to be sent. The association is done in such a manner as to minimize the length of the transmitted message.

Later in this chapter we explore *error control coding*, which is quite different from entropy coding. By way of contrast, we will see that we usually intentionally *lengthen* a word to accomplish error control.

The word *coding* is also sometimes used to describe techniques for disguising a message so that an unauthorized eavesdropper cannot hear or alter the message. These techniques fall under the broad topic of *encryption*.

Even in the absence of additive noise, entropy codes must be carefully designed to avoid decoding errors. The first potential problem relates to the concept of *unique decipherability*. As an example, suppose there are four possible messages, M_1 to M_4, to be transmitted, which are coded into binary numbers as follows:

$$M_1 = 1 \quad M_2 = 10 \quad M_3 = 01 \quad M_4 = 101$$

Suppose now that you are sitting at the receiver and receive the sequence 101. You would not know whether this constituted M_4 or either of the paired message sets M_2M_1 and M_1M_3. Therefore, this choice of code words yields a code that is *not uniquely decipherable*.

A code is *uniquely decipherable* if no code word forms the starting sequence (known as the *prefix*) of any other code word. Thus, the following four-message code is uniquely decipherable:

$$M_1 = 1 \quad M_2 = 01 \quad M_3 = 001 \quad M_4 = 0001$$

The restriction on the prefix is sufficient, but not necessary, for unique decipherability. For example, the code

$$M_1 = 1 \quad M_2 = 10 \quad M_3 = 100 \quad M_4 = 1000$$

is uniquely decipherable even though each code word is the prefix of every other code word to its right. The major difference between this and the earlier example that was not uniquely decipherable is that in the current example no code word can be formed as a combination of other code words. There is, however, a disadvantage to this code: It is uniquely decipherable, but not *instantaneous*. To see this, suppose again that you are sitting at the receiver, and you now receive 10. Until you look at the next two bits received, you do not know whether you are receiving the message M_2, M_3, or M_4.

Example 9.1

Which of the following codes are uniquely decipherable? For those that are uniquely decipherable, determine whether they are instantaneous.

(a) 0, 01, 001, 0011, 101

(b) 110, 111, 101, 01

(c) 0, 01, 011, 0110111

Solution:

(a) This is not uniquely decipherable, since the sequence consisting of the first and last words, namely, 0101, could be interpreted as two transmissions of the second word.

(b) This is uniquely decipherable, since all words but one start with a 1 and have length 3. If the start of a three-bit sequence is not 1, we know that we are receiving the only two-bit word. This code is also instantaneous, since no code word is the prefix of another word.

(c) This is uniquely decipherable, since all words start with a single zero, there is no repeated zero in any word, and no word is a combination of other words. It is not instantaneous, because each of the first three words is a prefix of at least one other word.

A helpful visual construct is the *code tree*. The restriction on the prefix and the instantaneousness of the code can easily be verified using code trees. The tree is formed of branches, each representing one bit. The nodes at the ends of branches represent binary numbers. By convention, we branch *up* for a binary zero and *down* for a binary one. (We could just as well have chosen the opposite convention.) Figure 9.1 shows the code tree for codes (b) and (c) of Example 9.1. The code words are indicated by heavy dots. Note that for code (b), the path to any code word *does not* pass through another code word, so the code is instantaneous. By contrast, in code (c) we must go through code words to arrive at other code words.

9.1.1 Information and Entropy

The more *information* a message contains, the more work we have to do to transmit it from one point to another. Some nontechnical examples should help to illustrate this intuitive concept.

Suppose you enjoy Mexican cuisine. You enter your favorite restaurant to order one cheese enchilada, one beef taco, one order of refried beans, one order of rice, two corn tortillas, butter, a sprig of parsley, chopped green onions, and a glass of water. Instead of writing out this entire list in longhand, the person taking the order probably communicates something of the form "*one number-two combination.*" Stop and think about this; rather than attempting to decide whether to transmit the specific words via English spelling or by some form of shorthand, a much more basic approach has been taken.

Let us look at another example. A popular traditional form of written communication is the telegram. Have you ever called Western Union to send a telegram congratulating someone on his or her wedding? If so, you probably found that the operator tried to sell you a standard wording, such as "Heartiest congratulations upon the joining of two wonderful people. May your lives together be happy and productive. A gift is being mailed under separate cover." In the early days of telegraphy, the company noticed that many wedding telegrams sounded almost identical to this. A primitive approach to transmitting such a message might be to send it letter by letter. However, since the wording is so pre-

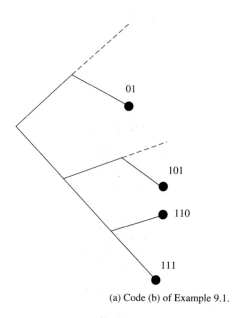

(a) Code (b) of Example 9.1.

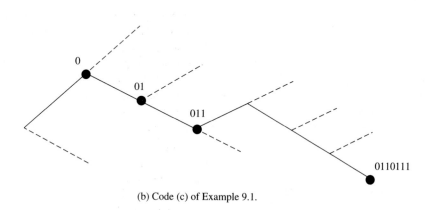

(b) Code (c) of Example 9.1.

Figure 9.1   Code tree for Example 9.1.

dictable, why not have a short code for this message? For example, suppose we call this particular message "#79." The operator would then simply type the names of the sender and of the addressee and then #79. At the receiver, form #79 is pulled from the file, and the address and signature are filled in. Think of the amount of transmission time saved!

As yet a third example, consider the (atypical, I hope) university lecture in which a professor stands before the class and proceeds to read directly from the textbook for two hours. A great deal of time and effort could be saved if the professor instead read the messages, "pages 103 to 120." In fact, if the class expected this and knew they were past page 100, the professor could shorten the message to "0320." The class could then read the 18 pages on their own and get the same information. Also, the professor could dismiss the class after 20 seconds and return to more scholarly endeavors.

By now, the point should be clear: A basic examination of the *information content* of a message has the potential for saving considerable effort in transmitting the message from one point to another.

Information

Some of the examples given in this section indicate that, in an intuitive sense, some relatively long messages do not contain a great deal of information. For instance, every detail of the restaurant order or the telegram is so commonplace that the person at the receiving end can almost guess what comes next.

The concept of the information content of a particular message must now be formalized. After doing this, we shall find that the less information in a particular message, the quicker we can communicate the message.

Information content is related to predictability. That is, the more probable a particular message, the less information is given by transmitting that message. For example, if today is Tuesday, and you phoned someone to tell her or him that tomorrow will be Wednesday, you would certainly not be communicating any information. This is true because the probability of that particular message is 1.

The definition of information content should be such that it monotonically decreases with increasing probability and goes to zero for a probability of unity. Another property we would like this measure of information to have is *additivity*. If one were to communicate two (independent) messages in sequence, the total information content should be equal to the sum of the individual information contents of the two messages. Now, if the two messages are independent, we know that the total probability of the composite message is the product of the two individual probabilities. Therefore, the definition of information must be such that when probabilities are multiplied together, the corresponding information contents are added.

The logarithm function satisfies these requirements. We thus *define* the information content of a message x as

$$I_x = \log \frac{1}{P_x} \tag{9.1}$$

where P_x is the probability of occurrence of x. This definition satisfies the requirements that information content be additive and monotonic with increasing probability and also that, for $P_x = 1$, $I_x = 0$. Note that this is true regardless of the base chosen for the logarithm.

We usually use base 2 for the logarithm. To understand the reason for this choice, let us return to the restaurant example. Suppose that there are only two selections on the menu. Further, suppose that past observation has shown that these two are equally probable (i.e., each is ordered half of the time). The probability of each message is therefore $\frac{1}{2}$, and if base-2 logarithms are used in Eq. (9.1), the information content of each message is

$$I = \log_2(2) = 1 \tag{9.2}$$

Thus, one unit of information is transmitted each time an order is placed.

Now let us think back to digital communication and decide on an efficient way to transmit this order. Since there are only two possibilities, one binary digit would be used to send the order to the kitchen. A "0" could represent the first dinner and a "1" the second dinner on the menu.

Suppose that we now increase the number of items on the menu to four, each with a probability of $\frac{1}{4}$. The information content of each message is now $\log_2 4$, or 2 units. If binary digits are used to transmit the order, 2 bits are required for each message. The various dinners could then be coded as 00, 01, 10, and 11. We can therefore conclude that if the various messages are equally probable, the information content of each message is exactly equal to the minimum number of bits required to send the message (provided that this is an integer). This is the reason for commonly using base-2 logarithms, and in fact, the unit of information is called the *bit of information*. Thus, one would say that in the current example, each menu order contains two bits of information.

When all the possible messages are equally likely, the information content of any single message is the same as that of any other message. In cases where the probabilities are not equal, the information content depends on which particular message is being transmitted.

Entropy

Entropy is defined as the average amount of information per message. To calculate the entropy, we take the various information contents associated with the messages and weight each by the fraction of time we can expect that particular message to occur. This fraction is the probability of the message. Thus, given n messages, x_1 through x_n, the entropy is defined by

$$H = \sum_{i=1}^{n} P_{xi} I_{xi} = \sum_{i=1}^{n} P_{xi} \log\left(\frac{1}{P_{xi}}\right) \tag{9.3}$$

By convention, the letter H is used for entropy.

Example 9.2

A communication system consists of six possible messages with probability $\frac{1}{4}, \frac{1}{4}, \frac{1}{8}, \frac{1}{8}, \frac{1}{8}$, and $\frac{1}{8}$, respectively. Find the entropy of the system.

Solution: The information content of the six messages is 2 bits, 2 bits, 3 bits, 3 bits, 3 bits, and 3 bits, respectively. The entropy is therefore

$$H = \frac{1}{4} \times 2 + \frac{1}{4} \times 2 + \frac{1}{8} \times 3 + \frac{1}{8} \times 3 + \frac{1}{8} \times 3 + \frac{1}{8} \times 3$$

$$= 2.5 \text{ bits/message}$$

It is instructive to examine Eq. (9.3) for the case of a binary message. That is, we consider a communication scheme made up of two possible messages, x_1 and x_2. For this case,

$$P_{x2} = 1 - P_{x1}$$

and the entropy is

$$H = P_{x1}\log\left(\frac{1}{P_{x1}}\right) + (1 - P_{x1})\log\left(\frac{1}{1 - P_{x1}}\right) \tag{9.4}$$

This result is sketched in Fig. 9.2.

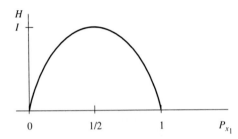

Figure 9.2 Entropy of binary communication system.

Note in the preceding example that, as either of the two messages becomes more likely, the entropy decreases. When either message has probability 1, the entropy goes to zero. This is reasonable because, in either case, the outcome of the transmission is certain without having to go through the trouble of sending any message. Thus, if $P_{x1} = 1$, we know that message x_1 will be sent all the time, and alternatively, if $P_{x1} = 0$, we know that x_2 will be sent all the time. In these two cases, no information is transmitted by sending the message.

If we were able to show a four-dimensional plot for the four-message case, we would find a similar result, with a peak in entropy for message probabilities of $\frac{1}{4}$.

The entropy function is symmetrical with a maximum for the case of equally likely messages. This is an important property that is used in later studies of coding.

9.1.2 Channel Capacity

In this section, we investigate the rate at which information can be sent through a channel and the relationship of this rate to errors. Before we can do this, we must define the rate of information flow.

Assume that a source can send any one of a number of messages at a rate of r messages per second. For example, if the source is a keyboard on a computer, the rate would be the number of symbols per second. Once we know the probability of each individual message, we can compute the entropy H in bits per message. If we now take the product of the entropy with the message rate r, we get the information rate in bits per second, denoted R. Thus,

$$R = rH \text{ bits/sec} \tag{9.5}$$

For example, if the messages given in Example 9.2 were sent at a rate of 2 messages per second, the information rate would be 5 bps.

One can probably intuitively reason that, for a given communication system, as the information rate increases, the number of errors per second will also increase.

C. E. Shannon has shown that a given communication channel has a maximum rate of information, C, known as the *channel capacity*. If the information rate R is less than C, one can approach arbitrarily small error probabilities by intelligent coding techniques. This is true even in the presence of noise, a fact that probably contradicts your intuition.

The converse of Shannon's theorem is also true. That is, if the information rate R is greater than the channel capacity C, errors cannot be avoided regardless of the coding technique employed.

We shall not prove Shannon's theorem here. We shall, however, discuss its application to several commonly encountered cases.

Consider the bandlimited channel operating in the presence of additive white Gaussian noise. In this case, the channel capacity, in bits per second, is

$$C = B\log_2\left(1 + \frac{S}{N}\right) \tag{9.6}$$

where B is the bandwidth of the channel in Hz and S/N is the signal to noise ratio. Equation (9.6) makes intuitive sense. As the bandwidth of the channel increases, it should be possible to make faster changes in the information signal, thereby increasing the information rate. As S/N increases, one would expect to be able to increase the information rate while still preventing errors due to the noise. Note that with no noise at all, the signal to noise ratio is infinite, and an infinite information rate is possible regardless of the bandwidth.

Equation (9.6) might lead one to conclude that if the bandwidth approaches infinity, the capacity also approaches infinity. This observation is not correct, however. Since the noise is assumed to be white, the wider the bandwidth, the more noise is admitted to the system (unless the noise is identically equal to zero). Thus, as B increases in Eq. (9.6), S/N decreases.

It is instructive to expand upon this last point. Suppose that the total signal and noise power per Hz is fixed and we are trying to design the best possible channel. There is an expense associated with increasing the system bandwidth, so we should attempt to find the maximum channel capacity. Suppose that the noise is white with power spectral density $N_0/2$. Suppose further that the signal power is fixed at a value S. The channel capacity is given from Eq. (9.6) as

$$C = B\log_2\left(1 + \frac{S}{N_0 B}\right) \tag{9.7}$$

Note that B is in Hz and that N_0 is the noise power in watts per Hz. We now find the value that the channel capacity approaches as B goes to infinity:

$$C = \lim_{B \to \infty} B\log_2\left(1 + \frac{S}{N_0 B}\right)$$

$$= \lim_{B \to \infty} \frac{S}{N_0} \log_2\left(1 + \frac{S}{N_0 B}\right)^{N_0 B/S} \tag{9.8}$$

$$= \frac{S}{N_0} \log_2 e = 1.44 \frac{S}{N_0}$$

Equation (9.8) shows the maximum possible channel capacity as a function of signal power and noise spectral density. In designing an actual system, the channel capacity will be compared to this figure, and a decision will be made as to whether a further increase in bandwidth is worth the expense.

Suppose now that you were asked to design a binary communication system, and you "cranked" the bandwidth of your communication channel, the maximum signal power, and the noise spectral density into Eq. (9.8). You come up with a maximum information rate C. You "plug in" the sampling rate and the number of bits of quantization, and you arrive at a certain information transmission rate, R bps. Aha, you observe, R is less than C, so Shannon tells you that if you do enough work encoding this binary data train, you can achieve an arbitrarily low probability of error. But you have already coded the original signal into a train of binary digits. What do you do next?

The next section discusses answers to this question. As applied to the foregoing example, most procedures would entail grouping combinations of bits and calling the result a new word. For instance, we can group in pairs of bits to yield four possible messages: 00, 01, 10, and 11. Each of these can be transmitted using some code word. By so doing, the probability of error can be reduced.

As an example, if you wished to transmit this textbook to a class of students, you could read each letter aloud as in the following:

r–e–a–d–e–a–c–h–l–e–t–t–e–r–a–l–o–u–d

Alternatively, you could group letters together and send the groups as a code word from an acceptable dictionary. Thus, in place of the preceding, you send

read–each–letter–aloud

Suppose that in reading the individual letters, you slur the d sound, and one student receives it as v. That student then receives the erroneous message,

r–e–a–v–e–a–c–h–l–e–t–t–e–r–a–l–o–u–d

However, with the code word technique, the student would receive

reav–each–letter–aloud

Since *reav* is not an acceptable code word, the student could correct the error and interpret the received message as

read–each–letter–aloud

thus making no error. Accordingly, the essence of coding to achieve arbitrarily small errors amounts to grouping into longer and longer code words. The longer the code words, the more different the dictionary entries will be and the less likely that individual errors cannot be corrected. (We are getting ahead of the game, so don't worry about details at this time.)

9.1.3 Entropy Coding

Our challenge is to find uniquely decipherable codes of *minimum length*. This will allow maximum transmission rates through the channel. Examination of the codes presented earlier shows that different messages are coded into words of different length. When talking about the length of a code, we therefore refer to the *average* length of the code words. This average is computed by taking the probability of each message (i.e., the fraction of time that the message occurs) into account. It is given by

$$\overline{L} = \sum_i L_i P_i \tag{9.9}$$

where $\overline{L}$ is the average length, L_i is the length of the *i*th code word, and P_i is the probability of the *i*th message. Clearly, it is advantageous to assign the shorter code words to the most probable messages. The historically important *Morse code* follows this rule by assigning the shortest code to the letter *E*. The fundamental theorem in noiseless coding theory states that *for binary-coding alphabets, the average code word length is greater than or equal to the entropy.*

With the average word length denoted by $\overline{L}$, the theorem then states,

$$\overline{L} \geq H \tag{9.10}$$

Equation (9.10) is not difficult to prove. We begin with a property of the logarithm given by the following equation, where ln is the natural logarithm:

$$\log_2(x) \leq \frac{x - 1}{\ln 2} \tag{9.11}$$

This is true because the logarithm to the base 2 is convex, and its slope at $x = 1$ is $1/(\ln 2)$. The two sides of the inequality are sketched in Figure 9.3.

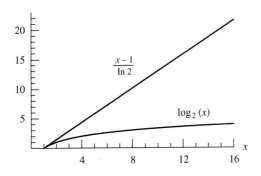

Figure 9.3   Comparison of two sides of inequality.

Let us define

$$q_i = \frac{2^{-L_i}}{\sum\limits_{i=1}^{M} 2^{-L_i}} \tag{9.12}$$

and let the *x* of Eq. (9.11) be given by

$$x = \frac{q_i}{P_i} \tag{9.13}$$

In Eq. (9.12), L_i is the length of code word i and M is the total number of messages. From Eq. (9.11), we have

$$\log \frac{q_i}{p_i} \leq \frac{q_i - p_i}{p_i \ln 2}$$

$$\sum_{i=1}^{M} p_i \log \frac{q_i}{p_i} \leq \sum_{i=1}^{M} \frac{q_i - p_i}{\ln 2} = 0 \tag{9.14}$$

The reason the last summation is zero is that both the p_i and the q_i sum to unity. Finally,

$$\sum_{i=1}^{M} p_i \log q_i \leq \sum_{i=1}^{M} p_i \log p_i \tag{9.15}$$

Substituting for q_i yields

$$\sum_{i=1}^{M} p_i L_i \geq \sum_{i=1}^{M} p_i \log \frac{1}{p_i} - \log \left(\sum_{i=1}^{M} 2^{-L_i} \right) \geq \sum_{i=1}^{M} p_i \log \frac{1}{p_i} \tag{9.16}$$

The rightmost inequality is true because, for instantaneous codes,

$$\sum_{i=1}^{M} 2^{-L_i} \leq 1 \tag{9.17}$$

The left side of Eq. (9.16) is the average code word length, and the right side is the entropy. This, therefore, proves the theorem.

Example 9.3

Find the minimum average length of a code with four messages with probabilities $\frac{1}{8}, \frac{1}{8}, \frac{1}{4}$, and $\frac{1}{2}$, respectively.

Solution: The entropy is

$$H = \frac{1}{8} \times 3 + \frac{1}{8} \times 3 + \frac{1}{4} \times 2 + \frac{1}{2} \times 1 = 1.75 \text{ bits}$$

The fundamental theorem of Eq. (9.10) tells us that this entropy is also the minimum average length for this code. We note that one possible code is:

$$M_1 = 000 \quad M_2 = 001 \quad M_3 = 01 \quad M_4 = 1$$

This code is uniquely decipherable, and the average length is 1.75 bits.

Variable-Length Codes

If the various messages to be transmitted do not have equal probabilities, the shortest average code length is achieved if the code words have unequal length. For example, suppose that there are four messages with probabilities $\frac{1}{8}, \frac{1}{8}, \frac{1}{4}$, and $\frac{1}{2}$, respectively (as in Example 9.3). One way to code these into binary words is to use 00, 01, 10, and 11. The resulting average length is 2 bits/message. If we instead use 111, 110, 10, and 0 (which is uniquely decodable and instantaneous—convince yourself of this!), the average length is

$$\bar{L} = \frac{1}{8} \times 3 + \frac{1}{8} \times 3 + \frac{1}{4} \times 2 + \frac{1}{2} \times 1 = 1.75 \text{ bits} \tag{9.18}$$

We are coding the more probable messages into shorter code words. In this particular case, the average word length matches the entropy, so we cannot find a code with a smaller average length.

One way to design variable-length codes is to start with constant-length codes and expand subgroups. For example, starting with a one-bit code, we have two code words, 0 and 1. We can expand this to five code words by taking the 1 and expanding it to 100, 101, 110, and 111, yielding the five code words,

$$
\begin{array}{c}
0 \\
100 \\
101 \\
110 \\
111
\end{array}
$$

Another way to achieve five code words is to start with the two-bit code 00, 01, 10, 11 and expand any one of these four words into two words. If 01 is chosen for expansion, we get the five-word code,

$$
\begin{array}{c}
00 \\
010 \\
011 \\
10 \\
11
\end{array}
$$

The question now is, Of the many ways to perform this expansion, which results in the code of minimum average length?

We present two techniques for finding efficient variable-length codes.

The *Huffman code* provides an organized technique for finding the best possible variable-length code for a given set of messages. We present the procedure using a specific example.

Suppose that we wish to code five messages, s_1, s_2, s_3, s_4, and s_5 with probabilities $\frac{1}{16}$, $\frac{1}{8}$, $\frac{1}{4}$, $\frac{1}{16}$, and $\frac{1}{2}$, respectively. The Huffman procedure is accomplished in four steps, as follows:

Step 1. Arrange the messages in order of decreasing probability. If there are messages with equal probabilities, choose any of the various orderings.

Word	Probability
s_5	$\frac{1}{2}$
s_3	$\frac{1}{4}$
s_2	$\frac{1}{8}$
s_1	$\frac{1}{16}$
s_4	$\frac{1}{16}$

Step 2. Combine the bottom two entries to form a new entry whose probability is the sum of the probabilities of the original entries. If necessary, reorder the list so that the new probabilities are in descending order.

Word	Probability	Probability
s_5	$\frac{1}{2}$	$\frac{1}{2}$
s_3	$\frac{1}{4}$	$\frac{1}{4}$
s_2	$\frac{1}{8}$	$\frac{1}{8}$
s_1	$\frac{1}{16}$	$\frac{1}{8}$
s_4	$\frac{1}{16}$	

Note that the bottom entry in the right-hand column is a combination of s_1 and s_4.

Step 3. Continue combining in pairs (i.e., repeat Step 2) until only two entries remain.

Word	Prob	Prob	Prob	Prob
s_5	$\frac{1}{2}$	$\frac{1}{2}$	$\frac{1}{2}$	$\frac{1}{2}$
s_3	$\frac{1}{4}$	$\frac{1}{4}$	$\frac{1}{4}$	$\frac{1}{2}$
s_2	$\frac{1}{8}$	$\frac{1}{8}$	$\frac{1}{4}$	
s_1	$\frac{1}{16}$	$\frac{1}{8}$		
s_4	$\frac{1}{16}$			

Step 4: Assign code words by starting at the right with the most significant bit. Move to the left, and assign another bit if a split has occurred. The assigned bits are underlined in the following table:

Word	Prob		Prob		Prob		Prob	
s_5	$\frac{1}{2}$		$\frac{1}{2}$		$\frac{1}{2}$		$\frac{1}{2}$	<u>0</u>
s_3	$\frac{1}{4}$		$\frac{1}{4}$		$\frac{1}{4}$	<u>10</u>	$\frac{1}{2}$	<u>1</u>
s_2	$\frac{1}{8}$		$\frac{1}{8}$	<u>110</u>	$\frac{1}{4}$	<u>11</u>		
s_1	$\frac{1}{16}$	<u>1110</u>	$\frac{1}{8}$	111				
s_4	$\frac{1}{16}$	<u>1111</u>						

Finally, the code words are as follows:

s_1	1110
s_2	110
s_3	10
s_4	1111
s_5	0

Note that at each assignment point, two assignments are possible. In addition, when there are three or more equal lowest probabilities, the choice for combination is arbitrary.

The average length is

$$\bar{L} = 4 \times \frac{1}{16} + 3 \times \frac{1}{8} + 2 \times \frac{1}{4} + 4 \times \frac{1}{16} + 1 \times \frac{1}{2} = \frac{15}{8}$$

If a block code of equal word length had been used, we would have needed three bits per message, and the average length would have been 3. The entropy of the code is

$$H = \frac{2}{16} \log(16) + \frac{1}{8} \log(8) + \frac{1}{4} \log(4) + \frac{1}{2} \log(2) = \frac{15}{8}$$

This is the same as the average length; thus, the Huffman procedure yields a code with maximum efficiency. In both of the examples in this subsection, the average word length was equal to the entropy. This is *not* common in real life. It results when all message probabilities are powers of $\frac{1}{2}$.

One disadvantage of the Huffman code is that we cannot start assigning code words until the entire combination process is completed. That is, every one of the columns must be developed before the first code word can be assigned. The coding process is often performed by a special-purpose microcomputer. This indicates that considerable storage may be required.

The *Shannon-Fano* code is similar to the Huffman; a major difference is that in the former the operations are performed in a forward, rather than backward, direction. Thus, the storage requirements are considerably relaxed, and the code is easier to implement. While the Shannon-Fano code can lead to average lengths that are the same as those of the Huffman code, its results are not always as good as those of Huffman.

Again, we illustrate the technique with an example, the same one we used to present the Huffman code.

Step 1.   Arrange the messages in order of decreasing probability. If there are messages with equal probabilities, choose any of the various orderings.

Word	Probability
s_5	$\frac{1}{2}$
s_3	$\frac{1}{4}$
s_2	$\frac{1}{8}$
s_1	$\frac{1}{16}$
s_4	$\frac{1}{16}$

Step 2. Partition the messages into the most equiprobable subsets. That is, start at the top or bottom, and divide the group into two sets. Find the total probability of the upper set and the total probability of the lower set. Choose a location for the dividing line that results in the closest two probabilities between the upper and lower sets. In this case, the dividing line falls below the first entry, resulting in exactly $\frac{1}{2}$ for the probability of the entries above and below the line. This is illustrated in the following table:

Word	Probability	
s_5	$\frac{1}{2}$	0
s_3	$\frac{1}{4}$	
s_2	$\frac{1}{8}$	
s_1	$\frac{1}{16}$	
s_4	$\frac{1}{16}$	

We now assign a 0 to all members of one of the two sets, and a 1 to all members of the other. (The choice is arbitrary.) Suppose we choose 0 for the top set and 1 for the bottom. If a set contains only one entry, the process for that set is concluded. Thus, the code word used to send s_5 is 0, and we need no longer look at that message. Hence, we concentrate on the other set and repeat the subdivision process. After one more subdivision, we have the following:

Word	Probability	
s_3	$\frac{1}{4}$	Code word 10
s_2	$\frac{1}{8}$	
s_1	$\frac{1}{16}$	Code word 11
s_4	$\frac{1}{16}$	

Note that once again, the subdivision works out perfectly, as the probability both above and below the line is exactly $\frac{1}{4}$. We have added a second bit to the code words, using a 0 for that bit above the line and a 1 below. Since there is only one entry in the top set, we are finished, and the code for s_3 is 10. Continuing the subdivision with the bottom set, we have:

Word	Probability	
s_2	$\frac{1}{8}$	Code word 110
s_1	$\frac{1}{16}$	Code word 111
s_4	$\frac{1}{16}$	

Finally, we subdivide the bottom set to get the following:

Word	Probability	
s_1	$\frac{1}{16}$	Code word 1110
s_4	$\frac{1}{16}$	Code word 1111

The resulting code words are thus:

s_1	1110
s_2	110
s_3	10
s_4	1111
s_5	0

In this example, the result turned out to be exactly the same as that using Huffman coding.

The foregoing two techniques for reducing a set of messages to a binary code do so with excellent efficiency. In both of them, however, we assume that the messages are given and that they cannot be combined prior to coding. In fact, greater efficiencies are often possible with combinations of messages. We illustrate this for the case of two messages.

Suppose the two messages have the following probabilities:

Message	Probability
s_1	0.9
s_2	0.1

The entropy is

$$H = -0.9 \log 0.9 - 0.1 \log 0.1 = 0.47 \text{ bit}$$

We would therefore hope to arrive at a code with average length close to this value. However, using either the Huffman or Shannon-Fano technique results in assigning a 0 to one of the words and a 1 to the other word. The average length is then 1 bit per message, more than twice the entropy.

Suppose we combine the messages in pairs. We then have four possible two-message sets. Assuming that the messages are independent, the possible sets and resulting probabilities are:

Message Set	Probability
$s_1 s_1$	0.81
$s_1 s_2$	0.09
$s_2 s_1$	0.09
$s_2 s_2$	0.01

Using the Shannon-Fano method, we assign the code words as follows:

Message Set	Probability	Code Word
$s_1 s_1$	0.81	0
$s_1 s_2$	0.09	10
$s_2 s_1$	0.09	110
$s_2 s_2$	0.01	111

The average word length is then

$$\overline{L} = 1 \times 0.81 + 2 \times 0.09 + 3 \times 0.10 = 1.29 \text{ bits}$$

Now, since each combined message represents two of the original messages, we divide 1.29 by 2 to find that 0.645 bit is being used to send each of the original messages. This is better than the one bit per message we obtained earlier, but it is still considerably above the entropy.

Suppose, then, that we combine *three* messages at a time to get the following message probabilities and code words (using Huffman coding):

Message Set	Probability	Code Word
$s_1 s_1 s_1$	0.729	0
$s_1 s_1 s_2$	0.081	100
$s_1 s_2 s_1$	0.081	101
$s_1 s_2 s_2$	0.009	11100
$s_2 s_1 s_1$	0.081	110
$s_2 s_1 s_2$	0.009	11101
$s_2 s_2 s_1$	0.009	11110
$s_2 s_2 s_2$	0.001	11111

The average length of the codes is now 1.598 bits, so the average length per original message is

$$\overline{L} = \frac{1.598}{3} = 0.533 \text{ bit}$$

Note that as we combine more and more messages, the average length approaches the entropy. The average length will equal the entropy if the probabilities are all inverse powers of 2. As more and more messages are combined, the probabilities approach such powers more and more closely.

9.1.4 Data Compression

Data compression is a term applied to a wide variety of techniques for reducing the number of bits required to send a given message. Entropy coding is one form of data compression.

The success of data compression techniques depends on the properties of the message. For example, entropy coding becomes most effective when the probabilities of the various messages are far from being equal. The other techniques we shall describe depend on sequential properties of the messages. That is, they depend on symbols occurring in a predictable order.

As an example, consider the encoding of a television picture. In the United States, a (non-high-definition) TV picture contains 426 visible picture elements (pixels) in each horizontal line. If we talk about black-and-white TV, we need to send the brightness (luminance) of each pixel. Suppose we decide to transmit 7 bits of information. That is, we quantize the luminance into 2^7, or 128, different levels. This represents a high quality of resolution. We need 7×426, or 2,982, bits to transmit the information in each line using PCM. A standard TV picture often contains a sequence of adjacent pixels with the same luminance. That is, as you trace across a horizontal line, you may encounter as many as several hundred pixels with the same brightness. (Suppose there is one figure in the middle of the screen, and the background is uniform; alternatively, suppose you are sending text on a uniform background.) In such cases, we can use a data compression technique known as *run-length coding* to reduce the number of bits required to send the signal. Instead of sending the luminance of each pixel, we send the starting position and luminance of the first of a number of pixels with the same brightness. In order to send the position, we need 9 bits of information (since $2^9 = 512$, and there are 426 different positions). Thus, we need 9 bits for the location and 7 bits for the luminance, for a total of 16 bits. For runs of three or more pixels of identical brightness, we save bits by using this technique. For example, if 10 adjacent pixels have the same luminance, we would need $10 \times 7 = 70$ bits to send these individually, but only 16 bits to send them using run-length coding. This concept can lead to even greater savings when extended to two dimensions. Thus, we would have to specify only the luminance and corner coordinates of any rectangle of constant luminance.

One disadvantage of run-length coding is that the data signal is generated at a nonuniform rate. That is, without such coding, bits are being sent at a constant rate, whereas with the coding, large areas of uniform brightness result in lower data transmission rates. The system therefore requires a buffer. An additional shortcoming is that errors propagate, since the system has memory. A bit error in a system that uses PCM to send information on each individual pixel results in a luminance error for that pixel only. However, if run-length coding is used, a bit error can affect the brightness or location of the entire line segment (or rectangle).

Prediction can be used in other forms of data compression. If the future values of data can be predicted from the present and past values, there is no need to send all the data. It may be sufficient merely to send the current data values, plus some key parameters to aid in the prediction. This can be likened to sending a particular curve. We can send the value of the ordinate at every value of the abscissa, or we can send key parameters. For example, if the curve is a straight line, we need only send the slope and intercept. If it is a parabola, other parameters are sufficient. We can even send the various coefficients in a series expansion if this results in a saving of transmission time.

9.2 ERROR CONTROL CODING

Normally, we attempt to design a system so as to minimize the probability of bit error, subject to certain constraints. (We explore the design of transmission systems in the next chapter.) However, in a noisy environment, it is often not possible to reduce the bit error rate to acceptable levels. Doing so may require raising the signal power beyond practical limits. Alternatively, lowering the error rate might require communicating at an unacceptably slow rate.

There is another available option to improve the performance of a digital communication system: *Error control coding* can be used to provide a structure for the signal. The structure is such that errors can be recognized at the receiver.

We divide error control into two categories: detection and correction. *Error detection* is the process of providing enough structure so that the receiver knows when errors occur. In effect, we create a situation where errors result in the reception of an invalid message. If the added structure is sufficiently detailed to allow pinpointing of the location of these errors, the code is an *error-correcting* code, and it is possible to correct errors at the receiver without requesting additional information from the transmitter (e.g., a *retransmit* instruction). The process is known as *forward error correction* (FEC), since there is no need for any transmission in the reverse direction (i.e., a request to retransmit). Forward error correction normally requires adding redundancy to the signal; we intentionally send more bits than are required.

We consider two broad categories of error control coding: block encoding and convolutional encoding. Our treatment is intentionally abbreviated; entire texts are devoted to coding, and you are encouraged to seek out additional references.

9.2.1 Linear Block Encoding

In linear block encoding, constant-length groups of message symbols are coded into constant-length groups of code symbols. Although this theory applies to any symbols being transmitted, we restrict our attention to binary communications. Thus, the symbols are bits: ones and zeros. As an example, if we wish to transmit a continuing sequence of bits, we could combine them into groups (blocks) of three. This results in eight possible words: 000, 001, 010, 011, 100, 101, 110, and 111. Each of these can be coded into one of eight different code words, which would normally be longer than three bits in order to achieve *redundancy* and create invalid bit combinations.

We begin by assuming that the bit errors are randomly distributed. That is, we assume that the probability of a particular bit being in error is independent of the probability of any other bit being in error. In other words, the mechanism that causes errors operates on individual bits. Alternatively, we will later consider the case of *burst errors*, that is, errors which normally occur over a sequence of adjacent bits. The types of errors that are present depend on the nature of the disturbance (e.g., additive noise).

The *distance* between two binary words of equal length is defined as the number of bit positions in which the two words differ. For example, the distance between 000 and 111 is 3, while the distance between 010 and 011 is 1. There is a distance of 1 between any word and the word formed by changing one bit in the original word.

Suppose now that we transmit one of the eight possible three-bit words. Suppose that the channel is noisy and that one bit position is incorrectly received. Since every possible three-bit combination is used as a message, the three-bit combination that is received is identical to one of the code words, and an error is made. For example, if 101 is transmitted, and an error occurs in the third bit, 100 is received. Since 100 is a valid word, the receiver has no way of knowing that an error has occurred.

Now suppose that the *dictionary* of code words is such that the distance between any two words is at least 2. The following eight code words have this property (prove it!):

$$0000, 0011, 0101, 0110, 1001, 1010, 1100, 1111$$

Now suppose we send one of these eight words, and a single bit error occurs during transmission. Then, since the distance between the received word and the transmitted word is 1, the received word cannot match any of the dictionary words. For example, if 0101 is transmitted, and an error occurs in the third bit, 0111 is received. This is *not* one of the eight acceptable words, so the system *detects* an error. It cannot reliably *correct* the error, since it does not "know" whether it resulted from a single bit error in the transmission of 0011, 0101, 0110, or 1111. That is, we do not know which bit is in error.

To see this graphically, let us plot the code words in an n-dimensional space. In three dimensions, the eight possible words are corners of a unit cube, as shown in Fig. 9.4. Starting at each corner of the cube, if a single bit error is made, we move along one of the edges to an adjacent corner a distance one unit away. Thus, the distance between two words is the minimum number of edges that must be traversed to move from one word to the other.

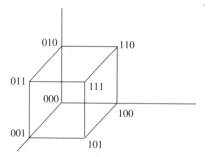

Figure 9.4   Three-dimensional space.

In the four-bit code word example given previously, we need a drawing with eight points representing the four-dimensional code words. This is a cube in four-dimensional space. We find that at least two edges must be traversed to move from one word to another.

In the general case, if the minimum distance between code words is 2, the code words are separated by at least two edges in an n-dimensional space. We illustrate this in

Fig. 9.5, where we indicate three of the code words from the example. The *n*-dimensional sphere of radius 1 includes all possible words within a distance 1 from the center.

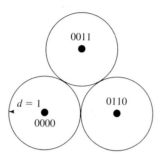

Figure 9.5   *n*-dimensional space.

Now suppose that the minimum distance between code words is increased to 3. Rather than trying to draw this in *n*-dimensional space, we shall use a planar figure in which each edge of the cube is represented by a straight line and code words are indicated by small circles. This is shown in Fig. 9.6. The figure is conceptual, and the actual directions of the lines are of no consequence—of course, two adjacent lines cannot be drawn colinear, or we would lose the concept. We see from the figure that if a single error is made, the received word is a distance 1 from the correct code word and at least 2 from every other code word. We decode as the nearest acceptable word, so this code is capable of *correcting* single bit errors.

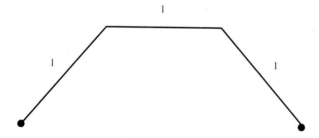

Figure 9.6   Distance of three.

Note that the error could have occurred from sending a word and making *two* bit errors, in which case our decoder would give the wrong answer. However, the probability of two bit errors occurring is usually smaller than the probability of a single bit error occurring. For example, if we transmit five-bit words, and the probability of a bit error is 10^{-4}, the probability of a single bit error in a five-bit word is

$$\text{PR[1 bit error]} = 5 \times 10^{-4} \times (1 - 10^{-4})^4 \approx 5 \times 10^{-4} \qquad (9.19)$$

The probability of two bit errors is

$$\text{PR[2 bit errors]} = 10 \times (10^{-4})^2 \times (1 - 10^{-4})^3 \approx 10 \times 10^{-8} = 10^{-7} \qquad (9.20)$$

Thus, one bit error is about 500 times more likely than two bit errors, so our strategy can be expected to result in an average of 500 correct error cancellations for each incorrect error cancellation.

If we are interested in error *detection* instead of correction, we see that up to two bit errors can be detected with a distance of 3. That is, we need to make three bit errors before one acceptable code word is turned into another. With any number of errors less than three, we arrive at an invalid word and know that an error has been made.

In general, if the minimum distance between code words is D_{min}, we can *detect* up to $D_{min} - 1$ errors. We can *correct* $(D_{min} - 2)/2$ errors for D_{min} even and $(D_{min} - 1)/2$ errors for D_{min} odd. You should convince yourself that these statements are the same as saying that we can correct errors as long as the received word is closer to the correct code word than to any incorrect code word.

We have been referring to the *minimum* distance between code words. The distance between pairs of code words is not uniform throughout a code. For example, for the words

$$0000, 0011, 0101, 0110, 1001, 1010, 1100, 1111$$

the minimum distance is 2. Note that the distance between 0000 and 1111 is four, as is the distance between 0011 and 1100. The effectiveness of the code depends on the "weakest link." You would not want to tell someone, "I have designed a code that corrects three bit errors as long as you don't send the sequence xxxxx." Thus, a code capable of correcting all possible combinations of two bit errors may be capable of correcting certain combinations of three bit errors. However, we would not say that the code corrects three errors unless it does so for every possible transmission and combination of errors.

In the preceding discussion, we made it sound as if you must decide in advance whether you want to design a code to detect or to correct errors. However, codes can be designed to detect and correct errors *simultaneously*. As an example, consider a code with a minimum distance that is even. Figure 9.7 shows an example of a minimum distance of 4. We have labeled the two code words v_a and v_b. This notation is consistent with what we will introduce in the next section. Now suppose you receive either word A or word B. Each of these is closer to the correct word than to the nearest incorrect word. You can therefore correct single errors by having your decoder move to the nearest valid code word. Suppose

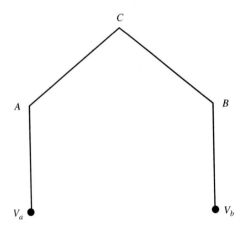

Figure 9.7   Minimum distance of 4.

next that you receive word C. This word is equidistant from each of the two valid words, so we cannot correct the errors. However, since it is not itself a valid word, you know that errors have occurred. We would therefore say that this code is capable of simultaneously *correcting* single bit errors and *detecting* double bit errors. Our earlier discussion indicated that such a code could detect up to three errors, and now we are saying that it can detect only two errors. Of course, in the three-error detection mode, you are not doing any correction. Which approach is better depends on the application. The code that does simultaneous detection and correction would incorrectly receive a word with three bit errors. That is, it would assume that only one bit error occurred and decode the message as the nearest valid word. The probability of three errors occurring is normally much smaller than the probability of one or two errors occurring. But suppose you had a very sensitive system in which you could not tolerate three errors. You might then prefer to detect up to three errors without doing any correction. Even though more messages will be "thrown out," you would not misinterpret the message received.

9.2.2 Linear Algebra

Clearly, the distance between code words is an important concept. To find the minimum distance, we need to check every possible pair of words. Combinatorial logic indicates that the number of pairs in a k-word code is $k!/2!(k-2)!$, which is $k(k-1)/2$. If, for example, there are 16 words, you would have to check 120 pairs of code words to find the minimum distance. It would be helpful if we had an easier way to find that distance. Fortunately, some concepts of linear algebra provide that way. In addition, we need such concepts to describe the block codes we will study.

A *group* is a set of elements with an associated binary operation that satisfies certain conditions. If we denote the group as G and the operation as $*$, the conditions are as follows:

(a) The binary operation (and hence the group) is *associative*, which means that, for any x, y, and z in the group,

$$x*(y*z) = (x*y)*z$$

(b) There is an identity element i in G such that, for any x in G,

$$x*i = i*x = x$$

(c) For any element x, there is an inverse x', such that

$$x*x' = i$$

(d) The binary operation (and hence the group) is said to be *commutative* if the order of operation does not matter—that is, if

$$x*y = y*x$$

Example 9.4

Show that the set of binary integers, $G = \{0, 1\}$, is a commutative group under modulo-2 addition. Modulo-2 addition is defined by

$$0 \oplus 0 = 0; \quad 0 \oplus 1 = 1; \quad 1 \oplus 0 = 1; \quad 1 \oplus 1 = 0$$

This is the same as binary addition, except that we ignore the carry bit. You can think of it as the *exclusive-OR* operation in Boolean algebra. Note that the standard notation for binary addition is to use a "+" enclosed in a circle. However, since all of the addition in this chapter is modulo-2, we will omit the circle in all future equations.

Solution: We check the conditions for a group:

(a) We can see that G is associative through exhaustive testing of the eight possible combinations (i.e., form a truth table). For example, one combination is

$$1 + (1 + 0) = (1 + 1) + 0 = 0$$

(b) The identity element is $i = 0$. Thus

$$1 + 0 = 1 \quad \text{and} \quad 0 + 0 = 0$$

(c) The inverse element is given by

$$0' = 0; \text{ that is, } 0 + 0 = i = 0$$

$$1' = 1; \text{ that is, } 1 + 1 = i = 0$$

(d) The definition of modulo-2 addition makes G commutative. That is, $1 + 0 = 0 + 1$.

The concept of a *group* covers one operation among its elements. To manipulate the various equations in block encoding, we need two operations. We therefore require something broader than a group, and that leads to the concept of a *field*.

A *field F* is a set of elements with two defined operations, $\oplus$ (addition) and $\cdot$ (multiplication), that obey the following conditions:

(a) F is a commutative group under addition.

(b) F is a commutative group under multiplication, provided that we omit the zero element. The zero element is the identity element under addition.

(c) The following distributive law is followed for any three elements in the field:

$$x \cdot (y + z) = x \cdot y + x \cdot z$$

Example 9.5

Show that the set of binary integers, $F = \{0, 1\}$, forms a field.

Solution: We check the properties as follows:

(a) We already know that F forms a commutative group under addition. The identity element is 0.

(b) Multiplication is defined by the following:

$$0 \cdot 0 = 0$$
$$0 \cdot 1 = 0$$
$$1 \cdot 0 = 0$$
$$1 \cdot 1 = 1$$

Since we require F to be a commutative group under multiplication when we exclude the identity element, we are only talking about the single element, 1. Clearly, F obeys all of the rules for an identity element of 1.

(c) We can easily check the various distributive combinations. For example,

$$1 \cdot (0 + 1) = 1 \cdot 0 + 1 \cdot 1 = 1$$

F is usually called the *binary field* and is used throughout our studies. It is often referred to as *GF*(2), where the G stands for *Galois*.

This text emphasizes a practical treatment of communications. If we wanted to be more mathematical, we would preface our coding examples with words such as "a code is defined over $GF(2)$ as. . . ." This terminology becomes important as you reference the literature to determine newer approaches to optimum coding. It is also important when alphabets are larger than binary.

9.2.3 Binary Arithmetic

We have essentially covered binary arithmetic in the previous description of $GF(2)$. We need simply become comfortable with the application of this theory to matrix operations and to polynomials. In our treatment of block codes, we will often be multiplying two matrices together. We assume that you are already familiar with matrix operations, so we need simply make sure that you are comfortable in $GF(2)$. We illustrate with a simple example.

Example 9.6

Perform the following multiplication:

$$[1\ 0\ 1] \begin{bmatrix} 1 & 0 & 1 & 0 & 0 \\ 1 & 1 & 0 & 1 & 0 \\ 0 & 1 & 0 & 0 & 1 \end{bmatrix}$$

Solution: The dimension of the vector is 1×3, and the dimension of the matrix is 3×5, where the first entry is the number of rows and the second is the number of columns. This is a valid multiplication, since the number of columns in the first matrix matches the number of rows in the second. The result is a 1×5 matrix (i.e., a row vector). The entries are found by taking the dot product of the appropriate vectors. The first entry is

$$1 \cdot 1 + 0 \cdot 1 + 1 \cdot 0 = 1$$

where the operations are in $GF(2)$. We have multiplied the row vector by the first column of the matrix. The second entry is

$$1 \cdot 1 + 0 \cdot 1 + 1 \cdot 0 = 1$$

In a similar manner, we develop the other three entries to obtain the following result:

$$[1\ 1\ 1\ 0\ 1]$$

Since the original vector contains a 1 in the first and third columns, we could have gotten this result by adding the first and third rows of the matrix together. That is,

$$\begin{array}{r} 1\ 0\ 1\ 0\ 0 \\ +\ \ 0\ 1\ 0\ 0\ 1 \\ \hline =\ \ 1\ 1\ 1\ 0\ 1 \end{array}$$

In performing this addition, don't forget that you are in $GF(2)$: Don't ever "carry the one." You will have lots of practice in this type of multiplication in the context of devising actual codes, so we stop at this simple example.

We will need polynomials when we examine cyclic codes and convolutional codes later in the chapter. Polynomials in $GF(2)$ are the same as the polynomials you have dealt with throughout your algebra courses, except that all coefficients are binary numbers, and operations are in $GF(2)$. Polynomials can be added together, subtracted from each other, and multiplied and divided by each other. We show this with several examples.

Example 9.7

Add the two polynomials

$$1 + X + X^3 + X^5 \quad \text{and} \quad X + X^2 + X^4 + X^5$$

Solution: The result is

$$1 + X^2 + X^3 + X^4$$

The reason the X term and the X^5 terms have disappeared is that their coefficients are $1 + 1$, which is 0.

Example 9.8

Do the following:

$$1 + X + X^3 + X^5 \quad \text{minus} \quad X + X^2 + X^4 + X^5$$

Solution: The result is

$$1 + X^2 + X^3 + X^4$$

This is true because, in modulo-2 arithmetic, the decimal number -1 is the same as the modulo-2 number 1. That is,

$$0 - 0 = 0$$
$$0 - 1 = 1$$
$$1 - 0 = 1$$
$$1 - 1 = 0$$

The only equation of these four that may bother you is $0 - 1 = 1$. We can verify this by adding 1 to each side. Thus, we wish to find X in the following:

$$0 - 1 = X$$
$$0 = X + 1$$

The solution is clearly $X = 1$.

Example 9.9

Do the following:

$$(1 + X^2 + X^4) \cdot (1 + X + X^2)$$

Solution: Doing long multiplication, we obtain

$$(1 + X^2 + X^4) \cdot (1 + X + X^2)$$
$$= 1 + X^2 + X^4 + X + X^3 + X^5 + X^2 + X^4 + X^6$$
$$= 1 + X + (1 + 1)X^2 + X^3 + (1 + 1)X^4 + X^5 + X^6$$
$$= 1 + X + X^3 + X^5 + X^6$$

Example 9.10

Form the quotient and remainder of

$$(1 + X^2 + X^4) \div (1 + X + X^2)$$

Solution: Doing long division, we obtain

$$
\begin{array}{r}
X^2 + X + 1 \\
X^2 + X + 1 \overline{)\ X^4 + X^2 + 1} \\
X^4 + X^3 + X^2 \\
\hline
X^3 + 1 \\
X^3 + X^2 + X \\
\hline
X^2 + X + 1 \\
X^2 + X + 1 \\
\hline
0
\end{array}
$$

We perform the long division just as is done in decimal arithmetic, but remembering that the operations are in $GF(2)$. In this case, the remainder turned out to be zero. Therefore, we have shown that

$$(X^2 + X + 1)(X^2 + X + 1) = (X^2 + X + 1)^2 = X^4 + X^2 + 1$$

9.2.4 Algebraic Codes

We now formalize the foregoing concepts using linear algebra. Suppose that our message word consists of k bits, and we add redundancy in the form of m additional bits. The length of each code word is then $k + m$ bits. Thus, every k-bit information word is associated with an n-bit code word. If the information word appears explicitly as part of the code word, we refer to the scheme as a *systematic code.*

If we denote the information bits as u_i *and the added redundancy bits as* c_i (the c stands for *check*), the code word can be written as

$$c_1 c_2 \ldots c_m u_1 u_2 \ldots u_k$$

We have placed the information bits at the right end of the code word. This is not necessary, and they can appear anywhere in the word.

An *algebraic code* is a code for which the code words and the information words are related by a matrix equation of the form

$$v = u[G] \tag{9.21}$$

where

> u is a $[1 \times k]$ vector representing the information
> v is a $[1 \times n]$ vector representing the code word
> $[G]$ is a $[k \times n]$ *generating matrix*

This is known as an (n,k) linear code, where n is the length of the code words and k is the number of bits in each uncoded message.

Example 9.11

A (4,3) linear code is generated by the matrix

$$[G] = \begin{bmatrix} 1 & 1 & 0 & 0 \\ 1 & 0 & 1 & 0 \\ 1 & 0 & 0 & 1 \end{bmatrix}$$

Find the code words associated with each possible information word.

Solution: A (4,3) code has information words that are three bits in length and code words that are four bits in length. There are eight possible three-bit information words. We multiply each by the generating matrix to find the code words, as follows:

Information	Code Word
000	0000
001	1001
010	1010
011	0011
100	1100
101	0101
110	0110
111	1111

Before leaving this example, we make several observations. First, note that the last three code bits match the information word. The code is therefore *systematic*. This match occurs because the right portion of $[G]$ is a three-dimensional *identity matrix*. Observe also that the added redundancy bit is a *parity check bit* that is chosen to provide overall even parity. The added bits in an algebraic code will always be parity check bits. This is why we choose the symbol c_i for these redundancy bits.

Example 9.12

A (7,4) linear code is generated by the matrix

$$[G] = \begin{bmatrix} 1 & 1 & 0 & 1 & 0 & 0 & 0 \\ 0 & 1 & 1 & 0 & 1 & 0 & 0 \\ 1 & 1 & 1 & 0 & 0 & 1 & 0 \\ 1 & 0 & 1 & 0 & 0 & 0 & 1 \end{bmatrix}$$

Find the code words associated with each information word, and find the minimum distance between code words.

Solution: For a (7,4) code, there are four information bits and three parity bits. The information words and associated code words are given in the following table:

Information	Code
0000	0000000
0001	1010001
0010	1110010
0011	0100011
0100	0110100
0101	1100101
0110	1000110
0111	0010111
1000	1101000
1001	0111001
1010	0011010
1011	1001011
1100	1011100
1101	0001101
1110	0101110
1111	1111111

Examination of the [G] matrix indicates that:

The first parity bit provides even parity when combined with the first, third, and fourth information bits.

The second parity bit provides even parity when combined with the first, second, and third information bits.

The third parity bit provides even parity when combined with the second, third, and fourth information bits.

We can check the distance between every pair of code words. (There are 120 pairs to check!) If we do so, we find a minimum distance of 3. This code can therefore correct single bit errors or detect double bit errors. In a sense, checking the three parities of the received word allows us to locate errors by "triangulation" (as used in navigation).

Checking the distance in the previous example is an exhaustive process. Some simple algebra makes the process almost trivial.

We begin by defining the *weight* of a code word as the number of 1's contained in the word. If we add two words together (modulo-2 bit-by-bit arithmetic—the *exclusive-OR* operation), the sum contains a 1 in each bit position where the two words differ. Thus, the distance between two words is the weight of their sum. Denoting the distance operator by D and the weight operator by W, we have

$$D(x,y) \; = \; W(x \; + \; y) \tag{9.22}$$

where the addition of the two vectors, x and y, is as described before.

We can see from Eq. (9.21) that the sum of two code words is itself an acceptable code word. That is, if we add together two information words, the resulting code word is the sum of the two original code words. This is a basic property of an algebraic code and is a simple result of linearity. That is,

$$[u_1 + u_2][G] = u_3[G]$$

where   (9.23)

$$u_3 = u_1 + u_2$$

Since the code contains all possible k-bit information words, u_3 must be an acceptable message. Looking at the previous example, we see that the sum of any two of the 16 code vectors must be equal to another one of the code vectors. Therefore, each of the nonzero code vectors represents a sum of two other vectors. (The zero vector is a sum of a code vector with itself.) *The minimum distance between code words is therefore the minimum weight of the nonzero code words.* This is clearly seen to be 3 for the code in the previous example, where we need now only check 15 weights instead of 120 distances.

For every $[k \times n]$ generator matrix, there is an associated $[(n - k) \times n]$ *parity check matrix* denoted $[H]$. Let us derive this matrix for a [6,3] code and then generalize the results. The (6,3) code is formed by taking the product of the message with $[G]$, as follows:

$$[c_1\ c_2\ c_3\ u_1\ u_2\ u_3] = [u_1\ u_2\ u_3] \begin{bmatrix} g_{11} & g_{12} & g_{13} & 1 & 0 & 0 \\ g_{21} & g_{22} & g_{23} & 0 & 1 & 0 \\ g_{31} & g_{32} & g_{33} & 0 & 0 & 1 \end{bmatrix} \quad (9.24)$$

Equating the first three entries on each side of Eq. (9.24), we have

$$\begin{aligned} c_1 &= u_1 g_{11} + u_2 g_{21} + u_3 g_{31} \\ c_2 &= u_1 g_{12} + u_2 g_{22} + u_3 g_{32} \\ c_3 &= u_1 g_{13} + u_2 g_{23} + u_3 g_{33} \end{aligned} \quad (9.25)$$

Adding the appropriate check bit to each side of the equations, and noting that the sum of anything with itself is zero, we can rewrite these three equations as

$$\begin{aligned} c_1 + u_1 g_{11} + u_2 g_{21} + u_3 g_{31} &= 0 \\ c_2 + u_1 g_{12} + u_2 g_{22} + u_3 g_{32} &= 0 \\ c_3 + u_1 g_{13} + u_2 g_{23} + u_3 g_{33} &= 0 \end{aligned} \quad (9.26)$$

or, in matrix form,

$$[c_1\ c_2\ c_3\ u_1\ u_2\ u_3] \begin{bmatrix} 1 & 0 & 0 \\ 0 & 1 & 0 \\ 0 & 0 & 1 \\ g_{11} & g_{12} & g_{13} \\ g_{21} & g_{22} & g_{23} \\ g_{31} & g_{32} & g_{33} \end{bmatrix} = [0\ 0\ 0] \quad (9.27)$$

The matrix in Eq. (9.27) is the transpose of the *parity check matrix*, $[H]^T$. Note that the number of columns in this matrix is the number of parity check bits $(n - k)$ and the number of rows is n, the number of bits in the code word. The parity check matrix for this code is therefore

$$[H] = \begin{bmatrix} 1 & 0 & 0 & g_{11} & g_{21} & g_{31} \\ 0 & 1 & 0 & g_{12} & g_{22} & g_{32} \\ 0 & 0 & 1 & g_{13} & g_{23} & g_{33} \end{bmatrix} \tag{9.28}$$

Note that the parity check matrix is formed by starting with an identity matrix and appending the transpose of the "nonidentity" portion of $[G]$.

We see from Eq. (9.26) that the parity check matrix has the property

$$v[H]^T = 0 \tag{9.29}$$

That is, any code word multiplied by the transpose of $[H]$ yields a zero vector. As an illustration, consider again the code of Example 9.12. The parity check matrix for this code is

$$[H] = \begin{bmatrix} 1 & 0 & 0 & 1 & 0 & 1 & 1 \\ 0 & 1 & 0 & 1 & 1 & 1 & 0 \\ 0 & 0 & 1 & 0 & 1 & 1 & 1 \end{bmatrix} \tag{9.30}$$

Let us check Eq. (9.29) for the third code word, 1001011. We find that

$$[1 \ 0 \ 0 \ 1 \ 0 \ 1 \ 1]\begin{bmatrix} 1 & 0 & 0 \\ 0 & 1 & 0 \\ 0 & 0 & 1 \\ 1 & 1 & 0 \\ 0 & 1 & 1 \\ 1 & 1 & 1 \\ 1 & 0 & 1 \end{bmatrix} = [0 \ 0 \ 0] \tag{9.31}$$

The parity check matrix is extremely useful in correcting errors. Suppose that we transmit the code vector v and that an error occurs in the fourth bit position. This is equivalent to adding the error vector

$$[e] = [0 \ 0 \ 0 \ 1 \ 0 \ 0 \ 0] \tag{9.32}$$

to the transmitted vector v. We then take the received vector r and multiply it by $[H]^T$. The result is

$$[r][H]^T = [v + e][H]^T = v[H]^T + e[H]^T = e[H]^T \tag{9.33}$$

The final equality results because $v[H]^T$ is zero. Now, if $[e]$ contains a single 1, $e[H]^T$ matches the row of $[H]^T$ corresponding to the error position. For example, if we change the fourth bit in our example, we receive 1000011. Multiplying this by $[H]^T$ yields $[1 \ 1 \ 0]$, which is the fourth row of $[H]^T$. Once we identify this as matching the fourth row, we know where the error occurred and can correct it. For that reason, the product of the received vector with $[H]^T$ is known as the *syndrome*.

The previous discussion shows that the parity check matrix is very important in error correction. If a single error occurs, the syndrome (the result of multiplying the received vector by the transpose of $[H]$) is the row of $[H]^T$ corresponding to the location of the error. Thus, to correct single errors, $[H]$ must have the following properties (note that rows of $[H]^T$ are columns of $[H]$):

(a) All columns of $[H]$ must be distinct. (There must be no repetitions.)

(b) No column of $[H]$ can consist of all zeros.

The first property is necessary to pinpoint the location of the error uniquely. The second property is needed because, if no errors occur, the syndrome is zero.

Syndrome generation is a simple operation. Figure 9.8 shows a system that generates the syndrome for the code of Example 9.12.

Hamming codes are linear block codes with a distance of 3. The parity check matrix $[H]$ consists of all possible nonzero bit combinations. For example, with three parity check bits, the columns of $[H]$ would each contain three bits. Since there are seven nonzero three-bit combinations, the matrix would contain seven columns. We are therefore describing a (7,4) code. If the number of check bits is raised to four, there are now 15 unique nonzero combinations, and the code becomes a (15,11) code. The parity check matrix is then

$$[H] = \begin{bmatrix} 0 & 0 & 0 & 0 & 1 & 1 & 1 & 1 & 1 & 1 & 1 & 1 & 0 & 0 & 0 \\ 0 & 1 & 1 & 1 & 0 & 0 & 0 & 1 & 1 & 1 & 1 & 0 & 1 & 0 & 0 \\ 1 & 0 & 1 & 1 & 0 & 1 & 1 & 0 & 0 & 1 & 1 & 0 & 0 & 1 & 0 \\ 1 & 1 & 0 & 1 & 1 & 0 & 1 & 0 & 1 & 0 & 1 & 0 & 0 & 0 & 1 \end{bmatrix} \qquad (9.34)$$

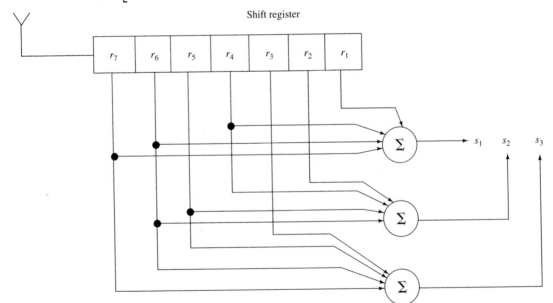

Figure 9.8   Syndrome generator for Example 9.12.

In general, with m equal to the number of parity check bits, the dimension of the (n,k) Hamming code is

$$(2^m - 1, \quad 2^m - 1 - m)$$

In an intuitive sense, Hamming codes have high efficiencies. For example, the (15,11) Hamming code has a ratio of information bits to transmitted bits of 11/15, or 73 percent. If one column of the $[H]$ matrix were removed, we would have a (14,10) code capable of correcting single errors. Its ratio of information bits to transmitted bits would be 10/14, or 71 percent.

Hamming codes are very simple; far more powerful codes have been devised, and are presented in the literature.

Suppose now that two errors are made during transmission. The error vector $[e]$ will accordingly contain two 1's. Equation (9.26) would then result in a vector that is a sum of the two corresponding rows of $[H]^T$. As an example, suppose in the previous example that two bit errors occurred, one in the second and one in the fourth position. The error vector is then

$$[e] = [0\ 1\ 0\ 1\ 0\ 0\ 0] \tag{9.35}$$

The syndrome is

$$[s] = [r][H]^T = [e][H]^T = [0\ 1\ 0\ 1\ 0\ 0\ 0] \begin{bmatrix} 1 & 0 & 0 \\ 0 & 1 & 0 \\ 0 & 0 & 1 \\ 1 & 1 & 0 \\ 0 & 1 & 1 \\ 1 & 1 & 1 \\ 1 & 0 & 1 \end{bmatrix} = [100] \tag{9.36}$$

Although this syndrome is the sum of the second and fourth rows of $[H]^T$, it is also the same as the first row. The error correction circuitry at the receiver would then "correct" the first bit received and end up with a word that is a distance of 3 from the correct word. Some thought should convince you that this is a case of a distance-3 code. With two transmission errors, we end up a distance of 1 from an incorrect word.

In order for the code to be capable of correcting two errors, not only must the columns of $[H]$ be distinct and non-zero, but also, sums of two columns must be distinct. That is, given the sum, there must be only one way to add two columns to arrive at that sum. This approach gets quite complex in designing codes capable of correcting multiple bit errors. In fact, such codes are approached in a different manner, using polynomials. (We will use polynomials in the next section.)

Cyclic Codes

The implementation of algebraic codes requires circuits capable of performing matrix multiplication and of comparing the result to various binary numbers. An alternative method is a look-up table. The most popular codes have been implemented as integrated circuits. However, with very long blocks, the circuitry becomes quite complex.

Cyclic codes, a special case of block codes, can be implemented in a very simple manner. A cyclic code has the property that any cyclic shift of a code word results in another code word. A *cyclic shift* is a shift by one position (to the right or left), with the end bit cycling around to the other end. For example, if 0100111 is shifted to the right, the result is 1010011. In general, if the code word is designated as

$$(v_1, v_2, v_3, v_4, \ldots v_{n-1}, v_n) \tag{9.37}$$

then a cyclic shift to the right is given by

$$(v_n, v_1, v_2, v_3, \ldots v_{n-1}) \tag{9.38}$$

While cyclic codes can be approached using a generator matrix, we begin our study of these important codes with a polynomial approach in $GF(2)$. A polynomial can be used to represent a binary number by setting the coefficients equal to the bits of the number. For example, the binary number 1011101 can be viewed as the polynomial

$$1 \times X^0 + 0 \times X^1 + 1 \times X^2 + 1 \times X^3 + 1 \times X^4 + 0 \times X^5 + 1 \times X^6$$

$$= 1 + X^2 + X^3 + X^4 + X^6 \tag{9.39}$$

We define a *generating polynomial* $g(X)$. After presenting an example, we will explore the manner in which the generating polynomial is chosen. For now, we select

$$g(X) = 1 + X + X^3 \tag{9.40}$$

This corresponds to the binary number 1101. The polynomial can be used to generate a (7,4) code. The code word polynomials are given by

$$v(X) = u(X) \, g(X) \tag{9.41}$$

For this example, we have the 16 words shown in Table 9.1.

TABLE 9.1 CYCLIC CODE FROM GENERATING POLYNOMIAL

u	$u(X)$	$v(X) = u(X)g(X)$	v
0000	0	0	0000000
0001	X^3	$X^3 + X^4 + X^6$	0001101
0010	X^2	$X^2 + X^3 + X^5$	0011010
0011	$X^2 + X^3$	$X^2 + X^4 + X^5 + X^6$	0010111
0100	X	$X + X^2 + X^4$	0110100
0101	$X + X^3$	$X + X^2 + X^3 + X^6$	0111001
0110	$X + X^2$	$X + X^3 + X^4 + X^5$	0101110
0111	$X + X^2 + X^3$	$X + X^5 + X^6$	0100011
1000	1	$1 + X + X^3$	1101000
1001	$1 + X^3$	$1 + X + X^4 + X^6$	1100101
1010	$1 + X^2$	$1 + X + X^2 + X^5$	1110010
1011	$1 + X^2 + X^3$	$1 + X + X^2 + X^3 + X^4 + X^5 + X^6$	1111111
1100	$1 + X$	$1 + X^2 + X^3 + X^4$	1011100
1101	$1 + X + X^3$	$1 + X^2 + X^6$	1010001
1110	$1 + X + X^2$	$1 + X^4 + X^5$	1000110
1111	$1 + X + X^2 + X^3$	$1 + X^3 + X^5 + X^6$	1001011

The 16 code words can be divided into three sets.

The first set consists of 0000000 and 1111111. Any cyclic shift of either word leaves that word unchanged.

The second set consists of 0001101 and its six distinct cyclic shifts, that is,

0001101, 1000110, 0100011, 1010001, 1101000, 0110100,

001101

The third set consists of 0010111 and its six distinct cyclic shifts, that is,

0010111, 1001011, 1100101, 1110010, 0111001, 1011100,

0101110

It should be noted that this code is *not* systematic; that is, the message does not appear explicitly as part of the code word. Before discussing techniques to make the code systematic, let us look at the considerations involved in choosing the generating polynomial $g(X)$. We present two theorems that can be proved using concepts of linear algebra.

Theorem 1.   If $g(X)$ is a polynomial of degree $n - k$ and is a factor of $X^n + 1$, then $g(X)$ generates an (n,k) cyclic code. You should take a moment to show that $1 + X + X^3$ is a factor of $X^7 + 1$. That is, if you divide $X^7 + 1$ by $1 + X + X^3$, the remainder is zero.

If we were to prove the theorem (as we shall not do here), we would find that the condition is both necessary and sufficient. That is, any factor of $X^n + 1$ with degree $n - k$ generates an (n,k) cyclic code. This is very useful. For example, we can factor $X^7 + 1$ as follows:

$$X^7 + 1 = (X + 1)(1 + X + X^3)(1 + X^2 + X^3) \tag{9.42}$$

The theorem then tells us that $X + 1$ can be used to generate a (7,6) cyclic code; either $1 + X + X^3$ or $1 + X^2 + X^3$ can be used to generate a (7,4) cyclic code.

A second theorem is useful for deriving generating polynomials without having to factor $X^n + 1$. This theorem is as follows:

Theorem 2.   Any *irreducible* polynomial of degree i is a factor of $X^{2^i - 1} + 1$. The irreducible polynomial of degree i can therefore be used to generate a $(2i + 1, i + 1)$ cyclic code.

An irreducible polynomial is a polynomial that has no factors other than itself or 1. For example, we can show that $1 + X + X^3$ is irreducible. Thus, without having to do any factoring, we know that it is a root of $X^{2^3 - 1} + 1$. We therefore know that it generates a (7,4) cyclic code. Note that since $X + 1$ is irreducible, it is a factor of $X^3 + 1$. It can therefore be used to generate a (3,2) cyclic code *in addition to* generating a (7,6) cyclic code.

Generating Matrix

We have seen how to generate a cyclic code from the product of two polynomials, $u(X)$ and $g(X)$. Let us examine this product for a (7,4) code generated by a third-order generating polynomial. We have

$$u(X)g(X) = (u_1 + u_2X + u_3X^2 + u_4X^3)(g_1 + g_2X + g_3X^2 + g_4X^3)$$

$$= u_1g_1 + X(u_1g_2 + u_2g_1) + X^2(u_1g_3 + u_2g_2 + u_3g_1)$$

(9.43)

$$+ X^3(u_1g_4 + u_2g_3 + u_3g_2 + u_4g_1) + X^4(u_2g_4 + u_3g_3 + u_4g_2)$$

$$+ X^5(u_3g_4 + u_4g_3) + X^6(u_4g_4)$$

Examination of Eq. (9.43) indicates that we could have solved for the code word using matrix multiplication. That is,

$$v = u[G]$$

$$(v_1 \ v_2 \ v_3 \ v_4 \ v_5 \ v_6 \ v_7) = (u_1 \ u_2 \ u_3 \ u_4) \begin{bmatrix} g_1 & g_2 & g_3 & g_4 & 0 & 0 & 0 \\ 0 & g_1 & g_2 & g_3 & g_4 & 0 & 0 \\ 0 & 0 & g_1 & g_2 & g_3 & g_4 & 0 \\ 0 & 0 & 0 & g_1 & g_2 & g_3 & g_4 \end{bmatrix}$$

(9.44)

The generating matrix can therefore be derived from the coefficients of the generating polynomial by listing the coefficients in the first row and then shifting them one position to the right in order to generate each of the remaining rows.

For $g(X) = 1 + X + X^3$, the generating matrix is

$$[G] = \begin{bmatrix} 1 & 1 & 0 & 1 & 0 & 0 & 0 \\ 0 & 1 & 1 & 0 & 1 & 0 & 0 \\ 0 & 0 & 1 & 1 & 0 & 1 & 0 \\ 0 & 0 & 0 & 1 & 1 & 0 & 1 \end{bmatrix}$$

(9.45)

Note that this matrix does not lead to a systematic code: There is no identity matrix existing as part of the generating matrix. We already knew from Table 9.1 that this code was not systematic.

Systematic Cyclic Codes

Table 9.1 shows the nonsystematic cyclic code we generated. It would still be a cyclic linear code if we reordered the code words. That is, suppose we assign the code words to different messages. For example, in the table, the code word 0001101 is associated with the message 0001. Suppose that we instead associate this code word with the message 1101. Continuing in this manner, we form the systematic cyclic codes shown in Table 9.2.

TABLE 9.2
SYSTEMATIC CODE
DERIVED BY
REORDERING
NONSYSTEMATIC CODE

u	v
0000	0000000
0001	1010001
0010	1110010
0011	0100011
0100	0110100
0101	1100101
0110	1000110
0111	0010111
1000	1101000
1001	0111001
1010	0011010
1011	1001011
1100	1011100
1101	0001101
1110	0101110
1111	1111111

This example shows that one way to derive a systematic cyclic code is to start with a nonsystematic code (using generating matrix or polynomial approach) and then reorder the code words so as to associate them with the correct message. While this technique may seem reasonable for relatively small codes, it is unacceptable for large codes. We prefer to find an analogy to the matrix and polynomial approaches.

Let us begin with the nonsystematic generating matrix, that is,

$$[G] = \begin{bmatrix} g_{11} & g_{12} & g_{13} & \cdots & g_{1n} \\ g_{21} & g_{22} & g_{23} & \cdots & g_{2n} \\ g_{31} & g_{32} & g_{33} & \cdots & g_{3n} \\ \cdot & \cdot & \cdot & \cdot & \cdot \\ g_{k1} & g_{k2} & g_{k3} & \cdots & g_{kn} \end{bmatrix} \tag{9.46}$$

Suppose we were to add two rows together—for example, modify the third row by adding the second row to it. This yields

$$[G'] = \begin{bmatrix} g_{11} & g_{12} & g_{13} & \cdots & g_{1n} \\ g_{21} & g_{22} & g_{23} & \cdots & g_{2n} \\ g_{31} + g_{21} & g_{32} + g_{22} & g_{33} + g_{23} & \cdots & g_{3n} + g_{2n} \\ \cdot & \cdot & \cdot & \cdot & \cdot \\ g_{k1} & g_{k2} & g_{k3} & \cdots & g_{kn} \end{bmatrix} \tag{9.47}$$

The effect is the same as would occur if we modified the original messages by adding u_3 to u_2. For example, in a (7,4) code, the code word corresponding to 1111 using $[G]$ is the same as that corresponding to 1011 using $[G']$. We have added the third bit to the second.

If we perform this operation on all messages, the result is simply to reorder the messages. Therefore, performing the linear operation on the generating matrix simply reorders the resulting code words.

We now know how to derive a systematic code from a nonsystematic code: We simply perform linear operations on the rows of the generating matrix, resulting in a matrix that contains an identity. Let us illustrate this for the (7,4) code presented earlier in this section. We derived the following generating matrix for that code:

$$[G] = \begin{bmatrix} 1 & 1 & 0 & 1 & 0 & 0 & 0 \\ 0 & 1 & 1 & 0 & 1 & 0 & 0 \\ 0 & 0 & 1 & 1 & 0 & 1 & 0 \\ 0 & 0 & 0 & 1 & 1 & 0 & 1 \end{bmatrix} \tag{9.48}$$

We wish to perform linear operations in order to place an identity matrix as the right-hand partition of this matrix. We note that the diagonal of the partition formed by the rightmost four columns of $[G]$ is already correct, as is the upper half of the matrix (above the diagonal). This will always be true because of the way this matrix was formed. We need simply perform operations to "clear" the portion below the diagonal. In the third row, we must change the last four entries from 1010 to 0010. We can do this by adding the first row to the third row, which yields

$$[G'] = \begin{bmatrix} 1 & 1 & 0 & 1 & 0 & 0 & 0 \\ 0 & 1 & 1 & 0 & 1 & 0 & 0 \\ 1 & 1 & 1 & 0 & 0 & 1 & 0 \\ 0 & 0 & 0 & 1 & 1 & 0 & 1 \end{bmatrix} \tag{9.49}$$

Finally, we need to modify the last row so that the rightmost entries read 0001 instead of 1101. We do this by adding the first and second rows to the fourth row to form

$$[G_{sys}] = \begin{bmatrix} 1 & 1 & 0 & 1 & 0 & 0 & 0 \\ 0 & 1 & 1 & 0 & 1 & 0 & 0 \\ 1 & 1 & 1 & 0 & 0 & 1 & 0 \\ 1 & 0 & 1 & 0 & 0 & 0 & 1 \end{bmatrix} \tag{9.50}$$

This matrix generates the systematic cyclic code. You should take the time to check that this is the code given in Table 9.2.

Now that we know how to generate a systematic cyclic code using the generating matrix, we turn our attention to the polynomial approach. We begin by making some additional observations about the polynomial form of a code word.

Suppose you start with a polynomial corresponding to a particular code word. If you multiply that polynomial by X, the corresponding code word is a (right) shifted version of the original word, with a 0 inserted in the left-hand position. If you multiply by X^j, you shift the word by j positions to the right. However, if you start with a polynomial corresponding to an n-bit code word, that polynomial must have degree less than n. If the multiplication by X^j yields a polynomial of degree n or higher, the product can no longer represent an n-bit code word. The problem arises if there are any 1's among the rightmost j bits of the original word. In a cyclic shift, these 1's must move from the right end to the left end of the word. This requires an additional polynomial operation.

If a polynomial $v(X)$ represents a code word in an (n,k) code, a cyclic shift to the right by j positions is represented by the polynomial

$$X^j v(X) \text{ modulo } (X^n + 1)$$

where "modulo $(X^n + 1)$" means that you divide $X^j v(X)$ by $(X^n + 1)$ and then take the remainder. Note that the modulo operation is necessary only if $X^j v(X)$ is of order n or higher. If its order is less than n, taking the modulo does not change the polynomial. This agrees with the previous intuitive argument.

Example 9.13

The bit string 0001101 is a word in a (7,4) code. Use the polynomial approach to perform a cyclic shift of two positions to the right.
Solution: The product is

$$X^j v(X) = X^2(X^3 + X^4 + X^6) = X^5 + X^6 + X^8$$

We perform the division to get

$$
\begin{array}{r}
X \\
X^7 + 1 \overline{) X^8 + X^6 + X^5} \\
X^8 + X \\
\hline
X^6 + X^5 + X
\end{array}
$$

The remainder corresponds to the code word 0100011, which is the required shifted word.

If the code is now systematic, the right-hand bits of the code word are the information u. This is represented by the polynomial $X^{n-k}u(X)$, since it signifies a shift of $n - k$ positions to the right. Suppose that we now take this polynomial modulo $g(X)$, the generating polynomial. Let us call the result $r(X)$. We then have

$$r(X) = X^{n-k} u(X) \text{ modulo } g(X) \qquad (9.51)$$

The modulo operation means that

$$X^{n-k} u(X) = q(X)g(X) + r(X) \qquad (9.52)$$

where $q(X)$ is the quotient and $r(X)$ is the remainder. We can now add $r(X)$ to both sides (recall that in modulo-2 arithmetic, adding anything to itself yields zero) to get

$$r(X) + X^{n-k} u(X) = q(X)g(X) \qquad (9.53)$$

We note that $q(X)g(X)$ is a valid code word because $g(X)$ is a polynomial of degree less than $n - k$. We can think of $q(X)g(X)$ as one of the valid $u(X)$ polynomials. But $u(X)g(X)$ is also a code word. Therefore, $r(X) + X^{n-k}u(X)$ is a code word that is part of the cyclic code. The right-hand bits of this code word correspond to the information message, $u(X)$. We thus have derived a method of generating a systematic cyclic code using polynomials.

Example 9.14

A (7,4) systematic cyclic code is specified by the generating polynomial $g(X) = 1 + X + X^3$. Find the code word corresponding to $u = 0111$.

Solution: The code word polynomial is given by

$$v(X) = r(X) + X^{n-k}u(X)$$

where $r(X)$ is the remainder after dividing $X^{n-k}u(X)$ by $g(X)$. Thus,

$$X^{n-k}u(X) = X^3(X + X^2 + X^3) = X^4 + X^5 + X^6$$

Performing the division, we obtain

$$
\begin{array}{r}
X^3 + X^2 \\
X^3 + X + 1 \overline{\smash{)}X^6 + X^5 + X^4} \\
\end{array}
$$

$$
\begin{array}{ll}
X^6 & + X^4 + X^3 \\
X^5 & + X^3 \\
X^5 & + X^3 + X^2 \\
\hline
& X^2 = r(X)
\end{array}
$$

Finally, the code word polynomial is

$$v(X) = r(X) + X^{n-k}u(X)$$

$$= X^2 + X^4 + X^5 + X^6$$

and the code word is $v = 0010111$.

Practical Implementation

At this point, you are probably wondering why we have gone through so much trouble to approach cyclic codes using polynomial theory. Indeed, the generating matrix approach appears to be a much easier way to generate the code. However, while that might be true when generating a code using pencil and paper, it is not true electronically. Matrix multiplication requires circuitry that performs computing. Alternatively, polynomial arithmetic lends itself to simple circuit implementations.

A feedback shift register can be used to perform polynomial division. For example, suppose we wish to divide $a(X)$ by $b(X)$, where the order of $a(X)$ is m and the order of $g(X)$ is n. Then

$$a(X) = a_0 + a_1X + a_2X^2 + ... + a_mX^m$$

$$b(X) = b_0 + b_1X + b_2X^2 + ... + b_nX^n$$

The circuit of Figure 9.9 performs this division. Each of the boxes marked z^{-1} represents a delay of one bit. The circuit therefore represents a shift register. The b_i represents switches or connections (taps). If the appropriate coefficient is 1, the switch is closed, and that feedback path is closed. If the coefficient is 0, the switch is open, and the feedback path is not connected.

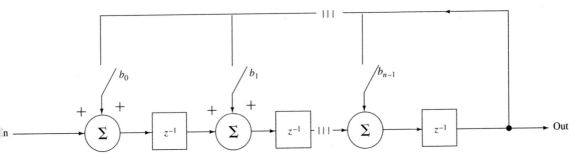

Figure 9.9   Shift register divider.

Pseudonoise

There is an important class of binary sequences related to the cyclic code. This class occurs when the generator polynomial has certain properties (the terminology relates to *max-
imal irreducible*—you are referred to the references for details). As an example, suppose
the shift register taps are as shown in Figure 9.10. Then the resulting output sequence is

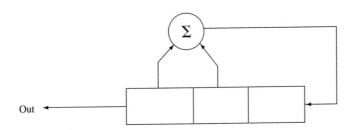

Figure 9.10   Shift register implementation of pseudonoise.

known as *pseudonoise* (PN).

Suppose the initial state of the shift register is 100. The resulting output sequence
will then be

   1001011100101 . . .

Note that once the input three bits repeat, all remaining bits also repeat. Thus, the output is
periodic with a period of 7. The seven code words generated by this cyclic code are:

                    1001011
                    0010111
                    0101110
                    1011100
                    0111001
                    1110010
                    1100101

The code has uniform properties: The distance between any pair of code words is 4.

If we add one cell to the shift register, we would generate code words of length 15. If the generator polynomial is selected to generate a pseudonoise sequence, the distance between code words will be 8.

In general, with n cells in the shift register generator, the code length will be $2^n - 1$, the distance between code words will be 2^{n-1}, and each code word will have 2^{n-1} 1's in it.

The periodic output sequence is known as pseudonoise because it has properties that approximate those of random white noise. The autocorrelation of the noise approaches an impulse (i.e., for any shift, the number of bit agreements is approximately equal to the number of disagreements), and the sequence has an approximately equal number of 1's and 0's. The sequence differs from random noise, however, in one very important respect: It is perfectly predictable and reproducible. This becomes very important in *spread spectrum* communications and in *code division multiple access*. (See Chapter 11 and the references.)

9.3 CONVOLUTIONAL CODING

Convolutional coding is fundamentally different from block coding. In block coding, the data stream is divided into blocks of bits, and each block is transformed into a sequence of bits called the code word. The code word is usually longer than the data word. The system *has no memory* between one block and the next. That is, the encoding of a specific block of data bits does not depend on what happens before or after that block of bits occurs.

Convolutional coding also divides the data stream into blocks, but these blocks are typically much shorter than those used in block coding. Indeed, the convolutional blocks can be as short as one bit in length. The information blocks are coded sequentially, and the system *does contain memory*. Therefore, the particular code corresponding to a sequence of data bits depends on what happened before the sequence arrived.

We shall illustrate a convolutional code using an example. Unlike the (n,k) block codes we studied, a convolutional code is specified by three parameters. We refer to it as an (n,k,m) convolutional code. The first parameter, n, describes the length of the output or code block, while k specifies the length of the input block. Thus, bits are fed into the code in groups of k bits, and after each block is input, an output of n bits is produced. The third parameter, m, describes the memory of the code. For example, a $(2,1,3)$ convolutional code spits out two code bits each time a single bit forms the input, and the output depends not only on the current input, but on the three previous input blocks.

Shift registers are well suited to modeling systems with memory. Figure 9.11 shows an example of implementing a $(2,1,3)$ convolutional code using a shift register. The figure shows two different ways of drawing the register. In Fig. 9.11(a), each of the three blocks is a delay (sometimes shown as D or as z^{-1}, the latter notation coming from z-transform theory). Thus, if a particular bit appears at the input at the left, that bit will transfer to the output of the first delay block at the second sampling interval. At each clock pulse, the bit at D is erased from the system, while the bit at C transfers to D, the bit at B transfers to C, and the bit at A transfers to B. In analyzing this diagram, it is useful to start at the right and work toward the left. Otherwise, one can lose track of the timing.

In Fig. 9.11(b), we illustrate the shift register as a line of four storage cells. At each sampling point, a new input bit enters the left-hand cell, and the bit previously in each cell

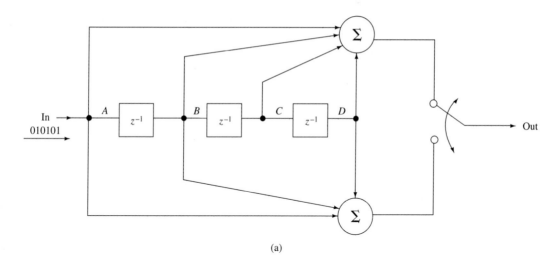

(a)

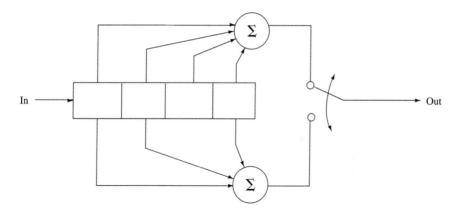

Figure 9.11   Shift register implementation of (2,1,3) code.

moves to the right by one position. In either case, the memory of the system is three previous bits. Thus, the current output depends on the contents of the four cells, but the left-hand cell contains the current input.

Now that we have described the two forms of notation, let us examine the actual convolutional code represented by the figure. Each time a bit enters the system, two output bits are produced. The output switch starts in the top position, so the first output bit is the sum of the current input and the three previous input bits. Then the output switch moves to the lower position, and the second output bit is the sum of the current input, the previous input, and the input that occurred three slots before. As an example, suppose the input to this system is 101010. Since the system has memory, we need to specify initial conditions. In most cases, we assume that the system starts at rest (i.e., zero is stored in each cell). We must also be clear on the order of input. When we say that the input is 101010, we mean that the first bit into the system is 1. Note that in drawing this in Fig. 9.11(a), we have reversed the sequence so that the nearest bit is the first to enter the system.

When the first bit arrives, the current bits at A, B, and C shift to the right, and the new input becomes A. The top summation is 1, as is the bottom. The following table summarizes the sequence of inputs and corresponding outputs:

IN	A	B	C	D	$A + B + C + D$	$A + B + D$
1	1	0	0	0	1	1
0	0	1	0	0	1	1
1	1	0	1	0	0	1
0	0	1	0	1	0	0
1	1	0	1	0	0	1
0	0	1	0	1	0	0
0	0	0	1	0	1	0
0	0	0	0	1	1	1
0	0	0	0	0	0	0

Since the system has memory, we have assumed that the six specified input bits are followed by zeros. The total output is then given by combining the two outputs to yield 111101000100101100.

If you followed the preceding example carefully, you are probably hoping that there is a simpler way to generate convolutional codes. In fact, we will see that considerable simplification is possible as we further analyze the code.

The reason that this code is called a convolutional code is that the feedback shift register is performing a convolution operation. For example, each of the two outputs of the encoder of Fig. 9.11 can be described by the convolution operation

$$v_n = \sum_{k_{n-\infty}}^{n} u_k h_{n-k} \tag{9.54}$$

For the top summer, the impulse response would be $h = 1111$. In fact, we can easily analyze the system to find the impulse response. If we divide the response into two parts corresponding to the top and bottom summers, we find that

$$g^{(1)} = (1\ 1\ 1\ 1)$$

$$g^{(2)} = (1\ 1\ 0\ 1) \tag{9.55}$$

You can generate these two impulse responses by assuming that a single 1 (a discrete impulse) enters the left side of the system and propagates through the storage cells of the shift register. We use the notation g rather than h because we will be employing this technique in a manner similar to that of the generating matrices of block encoding. The two output sequences are then the convolutions of the input with the generating (impulse response) sequences. That is,

$$v^{(1)} = u * g^{(1)}$$

$$v^{(2)} = u * g^{(2)} \tag{9.56}$$

where the asterisk denotes convolution. Thus,

$$v_i^{(1)} = \sum_{j=0}^{m} u_{i-j} g_j^{(1)}$$

(9.57)

$$v_i^{(2)} = \sum_{j=0}^{m} u_{i-j} g_j^{(2)}$$

The two output sequences are interspersed to form the total output in the form,

$$v_0^{(1)} v_0^{(2)} v_1^{(1)} v_1^{(2)} \; v_2^{(1)} v_2^{(2)} \; v_3^{(1)} v_3^{(2)} \; \ldots$$

(9.58)

The convolution can be performed by a matrix multiplication. The input vector is multiplied by a generating matrix of the form

$$[G] = \begin{bmatrix} g_0^{(1)} g_0^{(2)} & g_1^{(1)} g_1^{(2)} & g_2^{(1)} g_2^{(2)} & \cdots & g_m^{(1)} g_m^{(2)} & 00 & 00 & 00 \\ 00 & g_0^{(1)} g_0^{(2)} & g_1^{(1)} g_1^{(2)} & \cdots & g_{m-1}^{(1)} g_{m-1}^{(2)} & g_m^{(1)} g_m^{(2)} & 00 & 00 \\ 00 & 00 & g_0^{(1)} g_0^{(2)} & \cdots & g_{m-2}^{(1)} g_{m-2}^{(2)} & g_{m-1}^{(1)} g_{m-1}^{(2)} & g_m^{(1)} g_m^{(2)} & 00 \\ 00 & 00 & 00 & \cdots & \cdots & \cdots & \cdots & \cdots \\ 00 & 00 & 00 & \cdots & g_{m-3}^{(1)} g_{m-3}^{(2)} & g_{m-2}^{(1)} g_{m-2}^{(2)} & g_{m-1}^{(1)} g_{m-1}^{(2)} & g_m^{(1)} g_m^{(2)} \end{bmatrix}$$

(9.59)

The number of rows in this matrix is equal to the length of the input sequence.

Example 9.15

Find the generating matrix and output for the (2,1,3) convolutional encoder given at the beginning of this section. The two generator sequences are

$$g^{(1)} = (1 \; 1 \; 1 \; 1)$$

$$g^{(2)} = (1 \; 1 \; 0 \; 1)$$

The input sequence is 101010.

Solution: The output is the product of the input vector and the generating matrix:

$$v = u[G] = [1 \; 0 \; 1 \; 0 \; 1 \; 0] \begin{bmatrix} 11 & 11 & 10 & 11 & 00 & 00 & 00 & 00 & 00 \\ 00 & 11 & 11 & 10 & 11 & 00 & 00 & 00 & 00 \\ 00 & 00 & 11 & 11 & 10 & 11 & 00 & 00 & 00 \\ 00 & 00 & 00 & 11 & 11 & 10 & 11 & 00 & 00 \\ 00 & 00 & 00 & 00 & 11 & 11 & 10 & 11 & 00 \\ 00 & 00 & 00 & 00 & 00 & 11 & 11 & 10 & 11 \end{bmatrix}$$

$$= [11 \; 11 \; 01 \; 00 \; 01 \; 00 \; 10 \; 11 \; 00]$$

This matches what we found earlier.

Since convolutional encoders normally operate on continuous strings of data bits, the generating matrix approach is a very cumbersome method of evaluating the output of the system (on paper—the circuit does what it does without any calculations).

The output of the encoder depends on the input and on a particular number of previous inputs (i.e., stored bits). Hence, it lends itself to a state-space approach. The memory of the system establishes a state, and the input combined with this state yields the output and the next state. Suppose we identify the state as the contents of the three memory storage cells in the shift register. This corresponds to points B, C, and D in Fig. 9.11(a). The order is important, and we will write the state as a binary number where the most significant bit is the most recent input and the least significant bit is the input furthest in the past. Thus, if the bits stored in B, C, and D are 011, respectively, we will call this state 011, or simply, S_3. If we are in S_3 and the next input is a 0, the state we move to is 001, or S_1. The input zero has replaced the most significant bit, and each bit moves one position to the right (with the least significant bit dropping from the memory). In this manner, we can establish the state diagram shown in Fig. 9.12. Since the memory of our code stretches among the three most recent bits, the code has eight states. Note that the transitions shown in the figure are independent of the tap connections. This will apply for all eight-state codes (three-bit memory). We now need to add outputs to our state diagram.

Suppose we are in state S_0 and the input is 1. The bits in A, B, C, and D are now 1000, and the output is 11. We develop the 16 possible output pairs in a similar manner. These are shown in Fig. 9.13, where the input is the first notation on the directed arrow and the output follows the slash (/).

The state diagram can be used to find the output due to a particular input. Once again using the same example as before, let us assume that the input is 101010. We start in the rest position, S_0, and trace the path shown in Fig. 9.13. Notice that we end up in state S_2. If we assume that the specified input is followed by zeros, we can continue tracing a path that goes through S_1 and then to S_0. In fact, from any state in the system, it takes at most three transitions resulting from inputs of 0 to return us to state S_0. This becomes obvious when you consider that the system has a memory of three bits, and after three bits, the system is at rest. The outputs resulting from this return to the zero state are the transient response of the system.

Note that the outputs on the two lines leaving each state (for inputs of 0 and 1) are complements of each other. For example, leaving State 2, the output corresponding to an

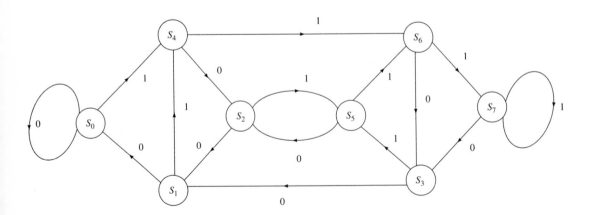

Figure 9.12   State transition diagram for example of (2,1,3) code.

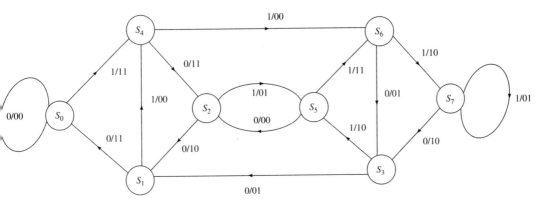

Figure 9.13   State diagram showing outputs.

input of 1 is 01, while that corresponding to an input of 0 is 10. This could have been anticipated from the structure of the code. Note that in the shift registers of Fig. 9.11, the input node is connected to both output summers. Therefore, if all inputs to the summer stay the same except in one connected to the input node, the output of the summer must change. Consequently, both outputs change when the input changes.

The convolutional encoder has added structure to the bit sequence. Even if the input is totally random (i.e., if a coin is flipped to determine each input bit), the output is constrained to follow certain sequences. It is this constraint that leads to the error correction capability of convolutional codes.

The state diagram can be used to generate an output for a given input. We simply trace the appropriate path through the diagram. Alternatively, we can conceptually invent a decoding technique based on the state diagram. Suppose you receive a long string of bits that (probably) contains errors. The job of the decoder is to find a route through the state diagram which results in a code word that is as close as possible to that received. For example, if you received the sequence 111101, you might begin decoding it in the following manner. If we assume the system starts at rest, we begin in State 0. The first two bits received exactly match the path between State 0 and State 4, so we assume that a 1 was sent. The next two bits exactly match the path to State 2, so we assume that a 0 was sent. The fifth and sixth bits received match the path from State 2 to State 5, so we assume that a 1 was sent. So far this seems easy, but be patient. The seventh and eighth bits, 10, do not match either path leaving State 5. Since the distance between the two received bits and each path code output is 1, you might think you can "flip a coin" to decide. However, in contrast to block coding, convolutional codes have memory. An incorrect choice at this point might position us for worse decisions at later points. Therefore, just as in a chess game, it is important to look ahead.

You might suggest trying all possible paths through the state diagram, but this rapidly becomes impractical. Suppose you are trying to decode 500 received bits (a relatively small number). Since this corresponds to 250 transmitted bits, and there are two paths to choose from at each input bit, the number of possible paths through the state diagram is $2^{250} = 1.8 \times 10^{75}$. This number increases geometrically as the number of bits increases. Clearly, it is not practical to look at all paths. Fortunately, this is not necessary, since the system has limited memory. With three bits of memory, it should be necessary to check only eight paths.

To point the way to a simple decoding technique, we need a method of indicating long paths through the state diagram. One way of doing this is to "stretch out" and repeat the state diagram. A linear form of such a diagram is known as a *trellis*. We illustrate the trellis for the eight-state code in Fig. 9.14. Note that we have not marked inputs on the lines of the trellis. Doing so is not necessary, since the upper of the two paths leaving each node corresponds to an input of 0, while the lower corresponds to 1—but only if the states are selected in the proper order. The trellis has a structure that applies regardless of the number of states. The lines leaving the top state go to the top (first) and the second state. Those leaving the second state go to the third and fourth states. The lines leaving the third state go to the fifth and sixth states. The process continues until you run out of states. At that point, modular arithmetic is used, and we cycle back to the first state. Thus, in the eight-state code, instead of the fifth state going to the ninth and tenth states (which do not exist), it goes back to the first and second states.

Note that because we assume that the system starts at rest, the full trellis does not develop until the third input bit. Note also that we mark the resulting output on the lines, just as we did with the state transition diagram.

Viterbi decoding uses a trellis to find the code that is a minimum distance from the received word. It is a bookkeeping technique that simplifies the decoding process considerably.

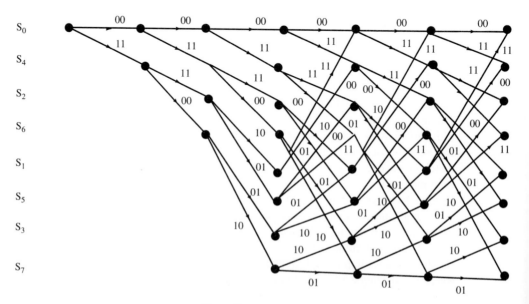

Figure 9.14   Trellis for eight-state code.

9.4 CRITERIA FOR CODE SELECTION

We shall briefly examine several criteria for selecting the best code, including coding gain, error detection and correction capability, and throughput.

When we study techniques for sending binary information, we will see that the probability of a bit error occurring is dependent upon the signal to noise ratio. Specifically, as the signal to noise ratio is raised, the probability of a bit error, P_e, will decrease. Error control coding is an alternative to increasing the signal to noise ratio. That is, by providing for correction of a number of bit errors, we can achieve the same overall performance as we would obtain with increases in the signal to noise ratio (without error control coding). The *coding gain* is defined as the reduction of signal to noise ratio permitted by the error control coding. For example, if the same overall correct message transmission can occur using coding and half of the signal to noise ratio, the coding gain for that particular code is 3 dB. Thus, coding can be thought of as either providing a lower probability of error for a given signal to noise ratio, or permitting a lower signal to noise ratio yet achieving the same probability of error.

If the important criterion for code selection is *throughput*, then it is important that the probability of rejection be made small. We therefore wish to maximize the probability of code acceptance, namely,

$$P_{CA} = (1 - P_e)^n \tag{9.60}$$

If the number of parity bits (indicating the redundancy of the code) is increased, the throughput will decrease. This occurs because, in a practical system, increasing redundancy implies less time to send each bit. For example, if we must send 10^4 words per second, and each word is 10 bits long, we have 10 microseconds to send each bit. On the other hand, if each word is increased to 20 bits, we have only 5 microseconds to send each bit. This decreases the energy per bit (nominally by a factor of 2), which decreases the signal to noise ratio, thereby increasing the probability of bit error and decreasing the probability of code acceptance. Obviously, code design requires some significant trade-off decisions.

On the other hand, if the criterion is to decrease the probability of false alarm, we need to increase the distance between code words. This is accomplished by having high redundancy (i.e., many check bits).

Code performance can also be evaluated against the Shannon limit. Recall that, assuming that noise is white, the channel capacity is given by

$$C = B \log_2(1 + S/N_0B) \tag{9.61}$$

with an upper limit of

$$C_{max} = 1.44S/N_0 \tag{9.62}$$

The signal power per bit is given by the energy per bit, E_b, divided by the time per bit, T_b, or

$$S = E_b/T_b \tag{9.63}$$

But if we communicate at the channel capacity C, then

$$T_b = 1/C. \tag{9.64}$$

Thus,

$$S = E_b C \tag{9.65}$$

and plugging this into equation (9.62), we obtain

$$C = 1.44 E_b C / N_0 \tag{9.66}$$

or

$$E_b / N_0 = 1/1.44 = -1.6 \text{ dB} \tag{9.67}$$

Thus, if the signal to noise ratio (bit energy divided by N_0) is above -1.6 dB, we can theoretically communicate up to the channel capacity. This provides an additional criterion for selecting a code.

Block codes are excellent for error detection, but typically provide relatively low coding gains when used for error correction. The reason is that too high a percentage of bits is used for parity.

On the other hand, convolutional codes provide relatively high coding gains. Accordingly, they are very useful for error correction and are useless for error detection. This is so because *something* comes out of the decoder regardless of the number of errors in the received sequence.

The fact that block codes excel for error detection while convolutional codes excel for error correction has led to a hybrid application known as *concatenated codes*. This is a combination of two coding techniques into an "inner" and an "outer" code. A message can be encoded using block codes, which leads to excellent error detection. Then the resulting block code can be encoded using convolutional coding, thereby providing for excellent error correction.

PROBLEMS

9.1.1 Which of the following are uniquely decipherable codes? Of those that are uniquely decipherable, which are instantaneous?
   (a) 010, 0110, 1100, 00011, 00110
   (b) 0, 010, 01, 10
   (c) 0, 100, 101, 11

9.1.2 A communication system consists of three possible messages. The probability of message A is p, and the probability of message B is also p. Plot the entropy as a function of p.

9.1.3 A communication system consists of four possible messages. $\Pr(A) = \Pr(B)$ and $\Pr(C) = \Pr(D)$. Plot the entropy as a function of $\Pr(A)$.

9.1.4 The probability of rain on any particular day in Las Vegas, Nevada, is 0.01. Suppose that a weather forecaster in that city decides to save effort by predicting no rain every day of the year. What is the average information transmitted in each forecast (in bits per day)? Make the (simplistic) assumption that the information content of an incorrect prediction is zero. Suppose the forecaster now decides to predict rain every day of the year. Repeat your calculations for this case.

9.1.5 A TV picture contains 211,000 picture elements. Suppose that this can be considered a digital signal with eight brightness levels for each element. Assume that each brightness level is equally probable.

(a) What is the information content of a picture?

(b) What is the information transmission rate (bits per second) if 30 separate pictures (frames) are transmitted per second?

9.1.6 The following list shows the probability of occurrence of each letter of the alphabet in a standard English text. (Note that these would be different in a technical text.)

A	0.081	B	0.016	C	0.032
D	0.037	E	0.124	F	0.023
G	0.016	H	0.051	I	0.072
J	0.001	K	0.005	L	0.040
M	0.022	N	0.072	O	0.079
P	0.023	Q	0.002	R	0.060
S	0.066	T	0.096	U	0.031
V	0.009	W	0.020	X	0.002
Y	0.019	Z	0.001		

(a) Find the entropy associated with sending a single letter.

(b) What is the information content of the three-letter word *THE*? Assume that the letters are independent of each other (a highly unrealistic assumption).

(c) You are told that the probability of occurence of the word *THE* is 0.027. That is, out of every 1,000 words, there will be 27 occurrences of the word *THE*, on average. Use this result to find the information content of the word *THE*. What does this imply about the independence assumption of part (b)?

9.1.7 Find the minimum average length of a code with five messages having probabilities:

(a) $\frac{1}{8}, \frac{1}{8}, \frac{1}{8}, \frac{1}{8}, \frac{1}{2}$

(b) 0.2, 0.2, 0.3, 0.15, 0.15

9.1.8 In a particular binary communication system, the probability of a 0 is 0.95. Bits are independent of each other. Since a 0 is much more likely than a 1, you decide to design a run-length coding scheme to combine multiple consecutive 0's into a single code word. Propose a scheme that would compress a run of 0's into a code indicating the length of the run. Be sure that your system is uniquely decipherable. Find the average bit transmission rate for your system, and find the percentage of improvement when this rate is compared to the original transmission rate.

9.1.9 Derive the Huffman code for five messages with probabilities as follows:

(a) $\frac{1}{8}, \frac{1}{8}, \frac{1}{8}, \frac{1}{8}, \frac{1}{2}$

(b) 0.2, 0.2, 0.3, 0.15, 0.15

(c) 0.1, 0.2, 0.3, 0.2, 0.2

Find the average code word length and the entropy for each code.

9.1.10 Find the Huffman code for messages with the following probabilities:

(a) 0.2, 0.2, 0.2, 0.2, 0.2

(b) 0.1, 0.15, 0.2, 0.2, 0.25, 0.05, 0.025, 0.025

(c) 0.4, 0.2, 0.2, 0.1, 0.1

Find the entropy and the average code word length.

9.1.11 Repeat Problems 9.1.9 and 9.1.10 using the Shannon-Fano code, and compare the resulting average code word lengths.

9.1.12 You are given six messages with the following probabilities:

M_1	M_2	M_3	M_4	M_5	M_6
0.1	0.25	0.2	0.25	0.1	0.1

(a) Find the entropy of this code.
(b) Use either Huffman or Shannon-Fano coding to design a variable-length code that is efficient.
(c) Find the average length per code word for your code.

9.1.13 Find the channel capacity C if the SNR is 6 dB and a standard broadcast AM channel is used (5-kHz bandwidth). Repeat for a broadcast TV channel (6-MHz bandwidth), and compare your answers.

9.1.14 A voice-grade telephone line has a bandwidth of 3 kHz. If you wished to transmit at a rate of 5 kbps, what SNR is required?

9.2.1 Find the minimum distance for the following code consisting of four code words:

$$0111001, \quad 1100101, \quad 0010111, \quad 0101001$$

How many bit errors can be detected? How many bit errors can be corrected?

9.2.2 Four-bit message words are encoded by adding a fifth parity bit. Find the probability of a word error if the bit error rate is 10^{-4}.

9.2.3 A 10-bit word has an 11th parity bit added. What fraction of all possible 11-bit words is not used by this code?

9.2.4 The parity check matrix for a particular (7,4) code is

$$[H] = \begin{bmatrix} 1 & 0 & 0 & 1 & 0 & 1 & 1 \\ 0 & 1 & 0 & 1 & 1 & 1 & 0 \\ 0 & 0 & 1 & 0 & 1 & 1 & 1 \end{bmatrix}$$

(a) Find the generating matrix $[G]$.
(b) List all code words.
(c) What is the minimum distance between code words?
(d) How many errors can be detected? How many errors can be corrected?

9.2.5 You are given a (9,3) code with the following generator matrix:

$$[G] = \begin{bmatrix} 1 & 0 & 0 & 1 & 0 & 0 & 1 & 0 & 0 \\ 0 & 1 & 0 & 0 & 1 & 0 & 0 & 1 & 0 \\ 0 & 0 & 1 & 0 & 0 & 1 & 0 & 0 & 1 \end{bmatrix}$$

(a) Find the minimum distance between code words.
(b) How many bit errors can be corrected?
(c) How many bit errors can be detected?
(d) Find the parity check matrix $[H]$ for this code.

9.2.6 You are given an (n,k) code with the following parity check matrix:

$$[H] = \begin{bmatrix} 1 & 0 & 0 & 1 & 1 & 1 & 0 \\ 0 & 1 & 0 & 1 & 1 & 0 & 1 \\ 0 & 0 & 1 & 1 & 0 & 1 & 1 \end{bmatrix}$$

(a) What are the values of n and k?
(b) Is this a Hamming code?

(c) What is the minimum distance between code words?

(d) You receive the word (0000011). What is the correct message (data, information) word?

9.2.7 An eight-bit word is formed from four information bits and four parity bits, with the parity bits given by the following equations:

$$c_1 = u_2 + u_3 + u_4$$
$$c_2 = u_1 + u_2 + u_3$$
$$c_3 = u_1 + u_2 + u_4$$
$$c_4 = u_1 + u_3 + u_4$$

(a) Find the generating matrix $[G]$.

(b) How many errors can be detected?

(c) How many errors can be corrected?

(d) Demonstrate the decoding process for any single message.

9.2.8 You are asked to design a (5,2) linear systematic block code. The code must be capable of correcting single bit errors. That is, you wish the minimum distance to be 3.

(a) Find $[G]$ for your code.

(b) List all code words.

(c) Determine the parity check matrix $[H]$.

(d) Now redesign the code so that it can only *detect* single errors, but not correct them. (I'm asking you to do a poor design.) That is, find one possible $[G]$ for which the minimum distance is only 2.

(e) List all code words for the $[G]$ of part (d).

(f) Determine the parity check matrix $[H]$ for the $[G]$ of part (d).

9.2.9 You are asked to design a (7,3) code that can correct single bit errors.

(a) Derive the parity check matrix $[H]$ for your code.

(b) Derive the generator matrix $[G]$ for your code.

(c) Find the code word for the message 111.

(d) Demonstrate how an error in the seventh bit transmitted is corrected.

9.2.10 Telephone companies sometimes use a *"best-of-five"* encoder for digital transmission. Each bit is repeated five times. The receiver uses a "majority rule" to decode; that is, if there are more 1's than 0's in a sequence of five received bits, the receiver calls the sequence a 1.

You can think of this as a (5,1) block code with generator matrix

$$[G] = [1\ 1\ 1\ 1\ 1]$$

(a) What is the minimum distance for this code?

(b) How many errors can be corrected?

(c) If the probability of a single bit error is 10^{-3}, what is the probability of a coded bit being decoded as the wrong message?

9.2.11 You are asked to design an (8,3) linear systematic algebraic block code that is capable of correcting single bit errors.

(a) What is the parity check matrix $[H]$ for your code?

(b) List all code words.

(c) What is the minimum distance for your code?

(d) What is the syndrome if an error occurs in the fifth bit of a transmitted code word?

9.2.12 (a) Is the polynomial

$$f(X) = 1 + X^3$$

irreducible?

(b) Factor the polynomial

$$f(X) = X^4 + X^3 + X + 1$$

9.2.13 The generator polynomial of a cyclic code is

$$g(X) = 1 + X + X^2 + X^4$$

This polynomial is a factor of $X^7 + 1$, so it generates a (7,3) cyclic code.

(a) Assuming that this is a nonsystematic code, find the code word for each possible message.
(b) Now change this to a systematic code. List all possible messages and the associated code words.
(c) What is the minimum distance for this code?

9.2.14 (a) Which of the following fourth-order polynomials are factors of $X^7 + 1$?

$$1 + X^2 + X^4$$
$$1 + X + X^2 + X^4$$
$$1 + X^3 + X^4$$

(b) Choose one of the factors you found in part (a), and use it to generate a (7,3) cyclic code. List all of the message words and the associated code words.
(c) Now make this into a systematic code. List all of the messages and associated code words.
(d) What is the generator matrix for the systematic code of part (c)?
(e) Derive the systematic code word for the message [111] using the polynomial approach (remainder property), and show that the result agrees with that found in part (c).
(f) What is the minimum distance between code words?
(g) How many errors can this code correct? Can it simultaneously correct and detect? If so, how many errors can it simultaneously correct and detect?

9.3.1 Three different convolutional codes are derived from the shift register circuits shown in Fig. P9.3.1. Derive the state diagram for each code.

9.3.2 A (2,1,3) convolutional coder is shown in Fig. P9.3.2.

(a) Find the state diagram for this coder.
(b) What is the output when the input is

$$1\,1\,1\,0\,1\,0\,1\,1\,1\,1$$

(You do not need to include the transient response.)
(c) Find the trellis for the coder.

9.3.3 A trellis for a (3,1,3) convolutional code is shown in Fig. P9.3.3 (for only four input bits). Note that this is unrealistically short. The outputs are shown on each branch of the trellis. (They do not necessarily correspond with any real convolutional code.)

Suppose you receive the following bit sequence (the bit on the left is the first one received):

$$1\,1\,1\,1\,1\,1\,0\,1\,1\,1\,0\,1$$

Use the Viterbi approach to find the most likely message. Note that the message will have only four bits.

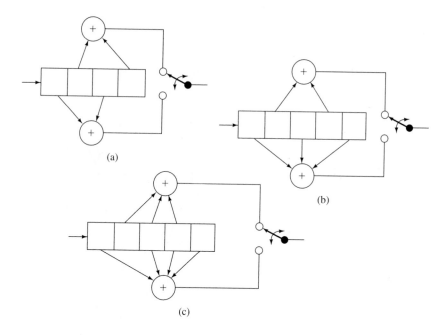

(a)

(b)

(c)

Figure P9.3.1

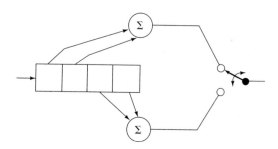

Figure P9.3.2

9.3.4 Draw the trellis and state diagram for a convolutional code with a rate of $\frac{1}{3}$, and the following generator polynomials:

$$g_1(X) = 1 + X^2$$

$$g_2(X) = 1 + X$$

$$g_3(X) = 1 + X + X^2$$

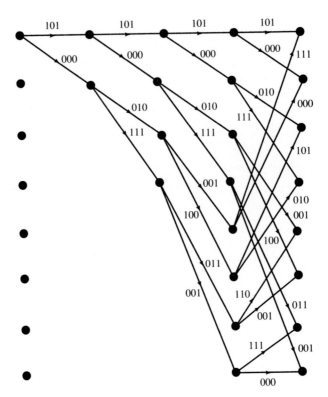

Figure P9.3.3

10

Baseband Digital Communication

10.0 PREVIEW

What We Will Cover and Why You Should Care

After reading the previous two chapters, you should have a good idea of how to convert a signal $s(t)$ into a train of bits. You also know how to modify that train of bits to allow for error control.

Now it is time to explore various techniques for transmitting the digital signals. We will find parallels to the techniques used in analog communication. After reading this material, you will know the basics of designing digital communication systems, including designing the transmitter and the receiver. In addition, you will know how to calculate the performance of the various systems, and will have a feel for the trade-offs that make our profession so interesting. You should also be in a position to understand the advanced technical literature that relates to this material. With all this background, you can follow the newest developments in what is one of the fastest moving fields within electrical engineering.

Necessary Background

You will have to recall a lot of material from the recesses of your memory, including that covered earlier in the book. You should now be thoroughly familiar with the digital baseband material in Chapter 8. While it is helpful to know how a particular sequence of digital signals arises (e.g., through A/D conversion and coding), that knowledge is not absolutely necessary; we can concentrate on the techniques for transmitting 1's and 0's without regard to where they came from.

To understand some of the timing systems presented in Section 10.1, you need some elementary control systems theory relating to feedback loops. Also, the discussion of performance analysis in Section 10.4 requires a knowledge of Gaussian probability densities. The systems of Section 10.4 will have far more meaning if you already understand analog AM, FM, and PM.

10.1 TIMING

Timing is the operation that significantly distinguishes a digital communication system from an analog system. Before a receiver can even begin to determine which of the various symbols it is receiving (e.g., 0's or 1's in a binary system), it must establish symbol timing. That is, it must synchronize a local clock with the timing of the received symbols.

Once timing is established and the receiver decoder is determining which symbols are being received, it is necessary to establish other timing relationships. These include character/word and block/message synchronization. It is not sufficient to decode a long string of symbols, unless the symbols can be associated with the correct message. This is particularly important in time division multiplexed transmissions where the symbols must be *decommutated*.

10.1.1 Symbol Synchronization

Since the received waveform is usually in the form of an electrical signal extending over all time, it is important to be able to chop the time axis into pieces corresponding to the segments of the signal for each symbol.

Data transmission can be either *synchronous*, in which symbols are transmitted at a regular, periodic rate, or *asynchronous*, where the spacing between words or message segments is irregular. Asynchronous transmission is often given the descriptive name, *start/stop*. Asynchronous communication requires that symbol synchronization be established at the start of each message segment or code word. This entails a great deal of *overhead*. In the synchronous mode, symbol timing can be established at the very beginning of the transmission, and only minor adjustments are needed thereafter.

The problem of symbol (or bit) synchronization is greatly simplified if a periodic component exists in the incoming symbol sequence. This is the case with some forms of signal.

Now is an appropriate time to review the baseband signal formats of Section 8.3.1. In the return-to-zero (RZ) signaling system, a transition occurs at the midpoint of every interval in which a 1 is being sent. We can therefore decompose the waveform into a sum of a periodic square wave and a random pulse signal. This is shown in Fig. 10.1 for a representative binary sequence. Because the RZ signal contains a periodic component, its power spectrum consists of a continuous portion (due to the random signal component) and a series of discrete spectral lines at the bit rate and its harmonics. A very narrow bandpass filter tuned to the vicinity of the bit rate can extract the clock information. Alternatively, a phase lock loop can be utilized to lock on to the periodic component.

The RZ system can be used in those cases where simplicity is important and signal to noise ratio is high. The system is not very power efficient, since no signal is being sent half of the time. We therefore suffer a factor-of-two, or 3-dB, degradation in performance, compared to NRZ or biphase signaling. However, with NRZ or biphase transmission and random data, the received signal does *not* contain a periodic component at the bit rate. We must therefore resort to more sophisticated bit synchronization approaches.

We shall explore three of these techniques: the maximum a posteriori, the early-late gate, and the digital data transition tracking loop.

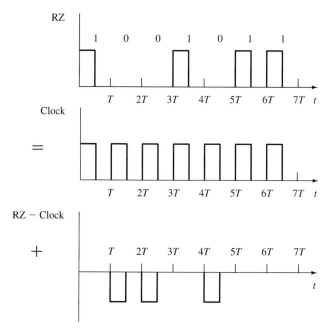

Figure 10.1 Periodic component in RZ bipolar signal.

In the *maximum a posteriori* (*MAP*) approach to bit synchronization, we observe the received signal over an interval of finite length and decode the signal using an assumed locally generated clock. The decoded signal is then correlated with a stored replica of the known symbol sequence, and the clock is adjusted to achieve maximum correlation. Perfect correlation is usually not obtainable, due to bit errors that occur in transmission. Note that this scheme requires that part of the transmitted signal be dedicated to the bit synchronization function; that is, the receiver must "know" what it is looking for, so information cannot be transmitted during this synchronization period. MAP synchronization can be performed in serial or parallel, but problems are associated with either technique. In the serial mode, adjustments are made in the local clock, and the next group of received symbols is correlated with the stored sequence. This requires a long acquisition time, as the "known" sequence must be transmitted more than once. Alternatively, the receiver could try different clock adjustments, simultaneously using parallel processing. However, this is expensive in terms of hardware.

Because of these shortcomings, practical approaches have been devised that approximate the MAP synchronizer. All of these approaches include adaptive control loops that lock onto the desired timing by minimizing an error term.

The *early-late gate* is one such approach. A block diagram of the system is shown in Fig. 10.2. The error signal is defined as the difference between the square of the area under the early gate and the square of the area under the late gate. The error goes to zero when the early and late gates coincide. In this case, the correct timing is at the midpoint of the gate periods. When the error is nonzero, the timing changes in a direction such as to cause the error to go to zero. Note that when the signal is zero, both the early and late gates have zero output, and no adjustments occur.

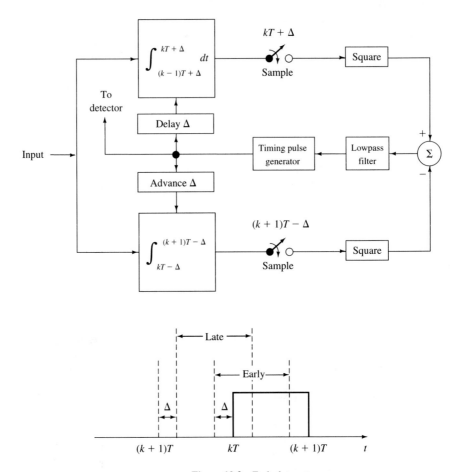

Figure 10.2 Early-late gate.

The *data transition tracking loop* (DTTL) is shown in Fig. 10.3. When the loop is in lock, the upper portion integrates the input pulse over periods corresponding to the bit period, while the lower portion integrates over intervals spanning transitions. This is best understood by examining typical waveforms when the loop is in lock. (See Fig. 10.4.)

Figure 10.4(a) shows a representative NRZ-L baseband waveform, and Figs. 10.4(b) and (c) show the output waveforms for the in-phase and quadrature integrators, respectively. Note that the quadrature integrator dumps (sets to zero) at the midpoints of the intervals.

The remainder of the system compares integrator outputs from one interval to the next. Figure 10.4(d) shows the output of the decision device, which is simply a sampler and hard limiter. This device tests the output of the in-phase integrator to see whether it is positive or negative. We indicate a positive output with a $+1$ pulse and a negative output with a -1 pulse. If two adjacent pulses are the same, the device indicates that no transition occurred in the original waveform. On the other hand, if two pulses are different, a transition occurs. The transition detector compares adjacent outputs and creates the I_k waveform

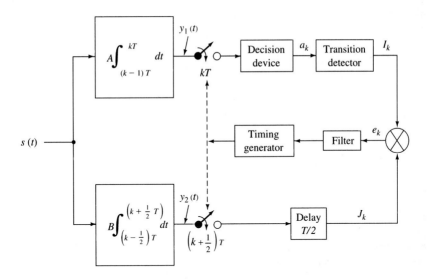

Figure 10.3   Data transition tracking loop.

of Fig. 10.4(e). In the case where no transition occurs, we view the quadrature channel. Since the integrator for this channel is offset by one-half of the symbol period, its output should be zero when a transition occurs and nonzero when no transition occurs. This output is sampled and delayed by one-half of the bit period to form the waveform of Fig. 10.4(f). Thus, the output I_k contains a pulse at each transition point, and J_k contains a pulse at those sample points where no transition occurs. These two waveforms combine to form the reconstructed clock.

We have analyzed the loop in the lock condition. In this situation, the error signal, e_k, is zero. This is found by multiplying I_k by J_k. If the product is not zero, the timing pulse generator is adjusted in a direction that forces the error toward zero.

In analyzing the control loop approaches to symbol synchronization, it is important to be aware of acquisition times (i.e., the transient response) and of the chance of falsely locking at the wrong point. The approaches presented in this section effectively recover timing in the absence of noise. When noise is present or when data transmission is at high speeds relative to the loop acquisition time, the loop performance must be analyzed to assure that system specifications are met. The literature presents curves of synchronization error as a function of signal to noise ratio, mean time to acquisition, and probability of false lock.[1]

10.1.2 Nonlinear Clock Recovery

Nonlinear filtering approaches attempt to operate on the incoming signal in a manner that produces a periodic component at the clock rate. For example, with biphase-L coding, there is always a transition at the midpoint of each interval. The tracking loop shown in

[1]For details, see William C. Lindsey and Marvin K. Simon, *Telecommunication Systems Engineering,* Englewood Cliffs, NJ, Prentice-Hall, Inc., 1973.

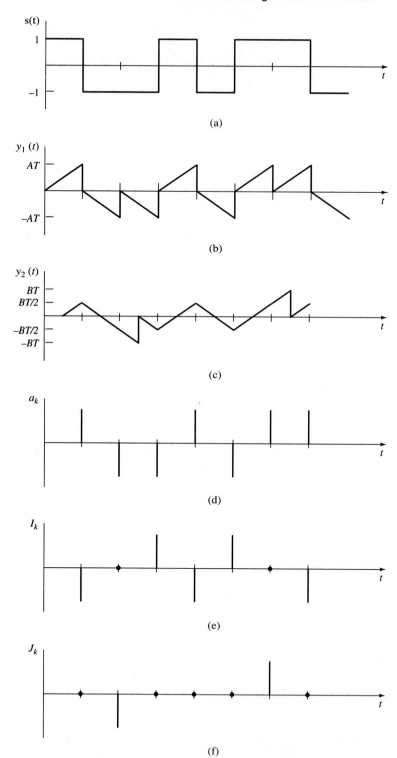

Figure 10.4   Typical DTTL waveforms.

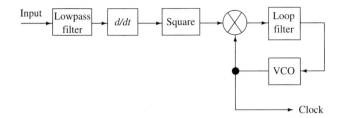

Clock   **Figure 10.5**   Nonlinear tracking loop.

Fig. 10.5 can be used to extract this component. The system first differentiates the incoming waveform, producing a sharp pulse (ideally, an impulse) at the center of each interval. Some of these pulses will be positive and others negative, depending on the direction of the original transition. A squaring device is used to make each of the pulses positive and therefore generate a periodic element within the signal waveform. This periodic element is extracted either with a narrowband filter tuned to a frequency close to the bit rate or by a phase lock loop.

10.1.3 Frame Synchronization

With time division multiplexing it is important to establish the timing of the overall frame. The receiver must be able to associate individual bits with the proper transmitting source. The decommutator must recognize when a new frame begins.

Frame synchronization can be accomplished by transmitting a very distinctive voltage waveform at the start of each frame, much as a comma or period is used to separate phrases or sentences. However, because of implementation considerations, it is desirable to require the receiver to recognize only the various symbol waveforms. Frame synchronization should therefore be accomplished as an integral part of transmitting the symbols. This is usually done by sending a specified sequence, or *prefix*, at the start of each frame.

As a first approach, let us assume that we send a binary 1 at the beginning of each frame. Thus, if the words are k bits long, a 1 would be inserted as every $(k + 1)$th bit. The receiver would add together bits separated by k positions. The addition could be performed by an integrator. A shift-register arrangement, as shown in Fig. 10.6, performs this addition for five words.

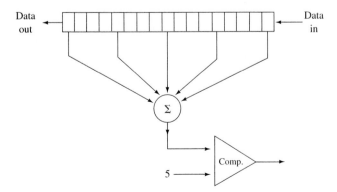

Figure 10.6   Addition of five framing bits.

Now, when the system is locked on to the one bit marking the beginning of each frame, the output of the summing device is 5. The probability of achieving a sum of 5 for any other position would be $(0.5)^5$, or 1/32, assuming that the bits within words are equally likely to be 0 or 1. This probability of false synchronization could be reduced by including more frames in the sum. In any case, the system would continually check the sum and adjust if necessary. Therefore, if the system goes out of synchronization, it will usually reachieve synchronization within $k + 1$ input bits.

The reliability of this technique can be improved by enlarging the prefix code to be longer than one bit.

We now formalize the strategies used to establish and maintain frame synchronization. Synchronization occurs in three modes: search, check, and lock. In some systems, the search and check modes are combined into a single *acquisition* mode.

Suppose that the prefix is N bits long. In the *search* mode, the frame synchronizer correlates each successive N data bits with a locally stored reference word. A threshold is set, depending upon system noise (bit error rate), and if the correlation exceeds this threshold, the search mode is exited, and the check mode is entered. The selected length of the prefix and the number of errors tolerated in the search mode depend upon the noise level and critical nature of the signal. For example, in the *T1* voice communication system, the frame consists of 192 bits and only 1 additional prefix bit. In other systems (particularly military and low-SNR applications), the frame might be several thousand bits long, and the prefix might be on the order of 20 bits in length. In such cases, the tolerable errors during search may range from zero (in a high-SNR case) to several bits (in a low-SNR case). There is a trade-off between the acquisition time and probability of false lock: If the threshold is set too low, the loop spends very little time in the search mode, but the probability of locking onto the wrong point in the frame increases.

In the *check* mode, the synchronizer looks at the next frame and correlates the prefix with the stored word. In some cases, more than one additional frame is examined. A threshold is set (not necessarily the same as that used in the search mode), and if it is exceeded, the lock mode is entered. If the threshold is not reached, we assume that the proper synchronization point has not been attained, and the search mode is reentered.

The *lock* mode is the normal mode in which we hope operation occurs. The prefix is continually correlated against the stored word, and the extent of the correlation between them is compared to a threshold value. This value is usually lower than that used in the search or check mode. If the threshold is not broken in a specified number of successive frames, the synchronizer is considered out of lock, and the search mode is reentered. In some more complex synchronizers, a modified search mode is entered in which slight perturbations of the synchronization position are tried before the full search is reinitiated.

The search, check, and lock process is illustrated in the block diagram of Fig. 10.7.

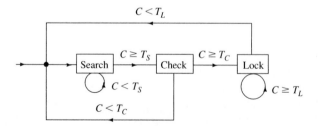

Figure 10.7 Strategy for frame synchronization.

10.1.4 Codes for Synchronization

The distinctive sequence of symbols marking the beginning of each frame is known as a *marker*, or *prefix*. Let us assume that this sequence is m bits long. The receiver then correlates each m-bit received sequence against a stored version of the known prefix. If the correlation exceeds a specified threshold, the search mode is exited.

We desire a prefix sequence for which there is a minimum probability of falsely establishing frame synchronization at the wrong location.

For example, suppose we choose the seven-bit prefix sequence, 0001101. The received sequence will then be of the form

$$....x\,x\,x\,x\,x\,x\,0\,0\,0\,1\,1\,0\,1\,x\,x\,x\,x\,x....$$

where x is used to denote a portion of the information message. We consider the x's to be randomly distributed, with equal probability of each being 0 or 1. Let us define the *correlation* between the stored sequence and the portion of the received sequence as the difference between the number of bits in which there is agreement and the number of bits in which there is disagreement. Thus, the correlation ranges from $+7$, when all bits are identical in the two sequences, to -7, when all bits are different. Accordingly, when synchronization is achieved, in the absence of bit errors, the correlation is 7. Suppose we compare the two sequences, and suppose the timing is incorrect by one bit period. We then would compare

$$x\,0\,0\,0\,1\,1\,0$$

with

$$0\,0\,0\,1\,1\,0\,1$$

The correlation is either 1 or -1, depending upon whether x is 0 or 1, respectively. In the other direction, we compare

$$0\,0\,1\,1\,0\,1\,x$$

with

$$0\,0\,0\,1\,1\,0\,1$$

Again, the correlation is either 1 or -1, depending upon whether x is 0 or 1, respectively.

Clearly, in designing prefix sequences, we wish to obtain minimum correlation of the sequence with truncated shifted versions of itself.

The truncated correlation, for a shift of k positions, is given by

$$C_k = \sum_{i=1}^{m-k} x_i x_{i+k} \tag{10.1}$$

where x_i is either $+1$ or -1, and m is the length of the prefix.

One of the most promising codes for use in synchronization is the *Barker code*, for which the correlation is limited to a magnitude of 1 for $k \neq 0$. Barker codes have been discovered for lengths of 3, 7, 11, and 13. For example, for a length of 13, the correlation for a zero shift is 13, while for any nonzero shift, the correlation is bounded in magnitude by

1. Unfortunately, Barker codes do not exist for lengths greater than 13. The known Barker codes are as follows:

$$1\ 1\ 0$$

$$1\ 1\ 1\ 0\ 0\ 1\ 0$$

$$1\ 1\ 1\ 0\ 0\ 0\ 1\ 0\ 0\ 1\ 0$$

$$1\ 1\ 1\ 1\ 1\ 0\ 0\ 1\ 1\ 0\ 1\ 0\ 1$$

In a very noisy environment, we may require prefixes longer than 13 bits. *Neuman-Hofman* codes are designed to minimize the maximum value of cross-correlation between the prefix pattern and the incoming data stream. As an example, the Neuman-Hofman code of length 24 is

$$0\ 0\ 0\ 0\ 0\ 1\ 1\ 1\ 0\ 0\ 1\ 1\ 1\ 0\ 1\ 0\ 1\ 0\ 1\ 1\ 0\ 1\ 1\ 0$$

The correlation, with zero shift, is 24. The maximum-magnitude correlation for other shifts occurs at a shift of 10 and results in a correlation of -4. The reason we are concerned with negative correlations is that Neuman-Hofman codes are often used in differential systems where bit reversals may occur. Thus, rather than look for the maximum positive correlation in order to declare frame synchronization, we look at the *magnitude* of the correlation.

Example 10.1

A frame synchronization scheme is to be designed with a search mode, a check mode, and a lock mode. Each frame contains 250 bits of information and a preamble of length N bits. The frame must be transmitted within one sec. The bit error rate is

$$P_e = 0.5 - 0.5e^{-\frac{f}{2,500}}$$

where f is the number of bits per second being transmitted.

(a) If $m = 10$ and frame synchronization must be acquired within 10 frames (with 99-percent probability), find the threshold to be used in the search mode.

(b) Once the system is in lock, we wish it to remain in this condition with probability 99.9 percent. Find the lock threshold value.

Solution: Before solving this problem, we note that the given bit error rate is a function of the bit transmission rate. As the bit rate increases, the time allocated to each bit decreases. This decreases the transmitted energy per bit. We shall see in the remainder of this chapter that the decrease in energy per bit has the effect of increasing the bit error rate. Therefore, adding additional framing bits is not without cost.

(a) With a preamble length of 10 bits, the bit error rate is

$$P_e = 0.5\left(1 - e^{-\frac{260}{2,500}}\right) = 0.0493$$

Let us denote the search mode threshold as T_S. That is, if there are at least T_S matches out of the 10 prefix bits, the search mode is exited. The probability of exactly T_S matches out of 10 bits is

$$\binom{10}{T_S}(1 - 0.0493)^{T_S}(0.0493)^{10 - T_S}$$

where the first term is the number of ways that T_S matches can be distributed among 10 bit positions. The probability of 10 matches is $(0.9507)^{10} = 0.603$. The probability of nine matches is 0.916. The specifications in the problem are loose enough that we need look no further. We are asked to acquire synchronization within 10 frames with a probability of 99 percent. If we require the full 10 bits to agree, the probability of this happening at least once during 10 frames is

$$1 - (1 - 0.603)^{10} = 0.9999$$

We have calculated this quantity by taking the difference between 1 and the probability of fewer than 10 matches occuring 10 times in a row. Thus, the probability of acquiring frame synchronization within 10 frames is 99.99 percent, assuming that we require 10 matches. We therefore set T_S equal to 10.

(b) Once the search mode is terminated, we desire to stay in synchronization with a probability of 99.9 percent. Let us assume that the check is made over only one frame at a time. (This is a high-SNR environment.) We then need to check the earlier formula to find out how many agreements are required to meet the specification. The probability of at least seven agreements is

$$\sum_{n=7}^{10}\binom{10}{n}(1 - 0.0493)^n(0.0493)^{10 - n} = 0.999$$

Therefore, the threshold during lock should be set at 7 matches.

10.1.5 Design Example

Suppose you are asked to design a system for military applications that must transmit 1 million information bits per second. The information is arranged in frames of 1,000 bits each. The environment is extremely noisy, and the application is highly sensitive. You must design the frame synchronization system to have a probability of less than 10^{-6} of falsely locking at the wrong point. Frame synchronization must be acquired within five frames with a probability greater than 99.9 percent.

We are at quite an early point in our study of digital communications to tackle a design problem of this magnitude. However, we shall use it to establish parameters and motivate the further studies.

A very important quantity that the problem has not specified is the bit error rate. The problem states that the environment is very noisy, so we could expect bit error rates on the order of 10^{-2} or more. As the bit transmission rate increases, less time is available to send each bit, so the energy per bit decreases, and the pulse spreading (intersymbol interference) increases. Thus, as the transmission rate increases, the bit error rate also increases. This is an important observation, since lengthening of the frame synchronization prefix must increase the transmission rate. The problem specifies that each frame contains 1,000 information bits, and the information must be transmitted at 1 Mbps. Thus, the bit transmission rate will be $(1,000 + N)/1,000$ times 1 Mbps.

The actual relationship between bit error rate and transmission speed depends upon the noise power, the bandwidth of the channel, channel factors such as whether there is a multipath, the transmission scheme, and the detector design. Since we need this information to solve the problem, and many of these topics are not covered until later in the text, let us assume that the bit error rate is related to the transmission rate by the graph of Fig. 10.8. The key values are summarized in the following table:

Transmission rate (Mbps)	Bit error rate ($\times 10^{-2}$)
1	1
1.001	1.0005
1.002	1.0011
1.003	1.0018
1.004	1.0026
1.005	1.0035
1.006	1.0045
1.007	1.0056
1.008	1.0068
1.009	1.0081
1.01	1.0095

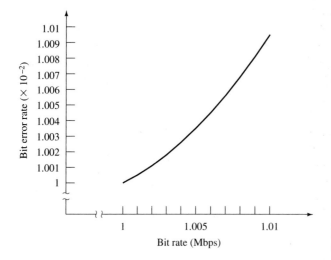

Figure 10.8 Bit error rate vs. transmission rate for design example.

We are now ready to try various framing techniques. To start the analysis, let us assume that a single framing bit is added to each frame and that the receiver looks for that bit over a single frame. This is clearly not acceptable, since the probability of false lock would be 50 percent—that is, the probability that any information bit would be the same as the transmitted frame synchronization bit. Even if the receiver scans over five adjacent frames (as permitted in the specifications), the probability of the information's having any specified five-bit pattern is 2^{-5}, or 3.1×10^{-2}. The specifications call for a false-lock probability of less than 10^{-6}. Clearly, we must go to longer frame synchronization sequences.

We will start by concentrating on the probability of acquiring synchronization in order to gain facility with the probability calculations.

Suppose we decide to try a Barker code of length 3—that is, 110. Suppose further that we set the threshold at the correlation value 3. That is, we look for perfect agreement between the received prefix and the locally generated replica. The probability of locking in a single frame will then be the probability that no bit errors occurred in the prefix, or

$$Pr(lock) = (1 - BER)^3 = (1 - 0.010018)^3 = 97\%$$

BER is the bit error rate. This does not meet the specifications, so there is no need even to examine the probability of falsely locking onto the wrong point. We must investigate either lowering the threshold or viewing more than a single frame at a time.

Suppose we decide to lower the correlation threshold to 2. Then the probability of locking in a single frame is still the probability of there being no bit errors. This is true because a single bit error will reduce the correlation by 2. Thus, the probability of locking in a single frame still does not meet the specifications. We will not even waste time trying to lower the required correlation threshold to 1, since it should be clear that the probability of false locking under this condition would also be unacceptable.

The alternative is to increase the number of frames over which we look for agreement. Suppose we examine the received sequence over two frames and look for a total correlation of at least 4. We can then permit a single bit error over the six bits, so the probability of acquisition becomes

$$(1 - BER)^6 + 6 \times BER \times (1 - BER^5) = 99.85\%$$

This still does not meet the specifications. If we wish to permit two bit errors over the two frames (i.e., look for a total correlation of at least 2), we will meet this specification, but not be close to the false-lock specification. Hence, we must look over at least three frames. Suppose we do this and require a total correlation of at least 5. That is, we are examining nine bits and are permitting up to two bit errors. The probability of acquisition is then

$$(1 - BER)^9 + 9 \times BER \times (1 - BER)^8 + 36 \times BER^2 \times (1 - BER)^7 = 99.99\%$$

We have therefore met the first specification and must now examine the probability of false lock. Unfortunately, this is where we get into trouble. The probability of falsely locking is the probability of getting at least seven agreements between the synchronization sequence (three 3-bit transmissions over three frames) and the data. If the data are considered to be random, the probability of false lock is

$$(0.5)^9 + 9 \times (0.5)^8(0.5)^1 + 36 \times (0.5)^7(0.5)^2 = 46 \times (0.5)^9 = 8.98 \times 10^{-2}$$

which clearly does not meet the specification. Even if we were to look over the full five frames permitted by the specifications and require perfect agreement (doing so could not achieve the required probability of acquisition), the probability of false alarm would be

$$(0.5)^{15} = 3 \times 10^{-5}$$

This does not meet the specification of 10^{-6} maximum probability of false lock. Obviously, we need a longer synchronization sequence.

Suppose we go to the 7-bit Barker sequence. Let us approach the analysis by examining the acquisition probability as a function of the required correlation. Note that the bit error rate increases to 1.0056×10^2, since we must transmit at a faster rate. If we use the full five frames, we are transmitting 35 synchronization bits. The probability of errors among these 35 bits is found from the binomial distribution and is as follows:

$$\Pr(0 \text{ errors in 35 bits}) = (1 - \text{BER})^{35} = 0.702$$

$$\Pr(1 \text{ error in 35 bits}) = 35 \times (\text{BER})^1 (1 - \text{BER})^{34} = 0.2496$$

$$\Pr(2 \text{ errors in 35 bits}) = \binom{35}{2}(\text{BER})^2 (1 - \text{BER})^{33} = 8.62 \times 10^{-2}$$

$$\Pr(3 \text{ errors in 35 bits}) = \binom{35}{3}(\text{BER})^3 (1 - \text{BER})^{32} = 4.82 \times 10^{-3}$$

$$\Pr(4 \text{ errors in 35 bits}) = \binom{35}{4}(\text{BER})^4 (1 - \text{BER})^{32} = 3.914 \times 10^{-4}$$

$$\Pr(5 \text{ errors in 35 bits}) = \binom{35}{5}(\text{BER})^5 (1 - \text{BER})^{31} = 2.465 \times 10^{-5}$$

Since the specifications require a 99.9-percent probability of locking within five frames, we see that permitting up to three errors among the 35 synchronization bits would meet the specifications. That is, if three errors are permitted, the probability of *not* locking is the probability of four or more errors, which is approximately 4×10^{-4}. Thus, the probability of acquisition is approximately 99.96 percent, which meets the specifications.

We require a correlation of at least 29 [35 $-$ 2 $\times$ 3)] to declare that frame synchronization is achieved. It is now necessary to check that the probability of false lock is within the specification. The probability of achieving agreement between 32 of the 35 bits, assuming random data, is

$$\Pr(\text{false lock}) = 1 \times (0.5)^{35} + 35 \times (0.5)^{34}(0.5)^1$$

$$+ \binom{35}{2}(0.5)^{33}(0.5)^2 + \binom{35}{3}(0.5)^{32}(0.5)^3$$

$$= 2.09 \times 10^{-7}$$

This meets the specification, so we have been successful at configuring one possible frame synchronization scheme.

We have ignored other design trade-off considerations, including the entire issue of hardware and implementation. We will be ready to tackle these issues after learning about specified transmission techniques.

10.2 INTERSYMBOL INTERFERENCE

Before performing a mathematical analysis of intersymbol interference, we shall consider a pictorial approach in order to introduce the concept of *eye patterns*. Suppose that binary information is transmitted by means of a baseband digital waveform. Let us assume that a 1-V pulse is used to send a binary 1 and a 0 -V pulse is used to send a binary 0 (i.e., we have a unipolar baseband). When the waveform goes through the system, it is distorted. Any sharp corners of the wave are rounded, since practical systems cannot pass infinite frequencies. Therefore, the values in previous sampling intervals affect the value within the present interval. If, for example, we send a long string of 1's, we would expect the channel output to eventually settle to being a constant, 1. Similarly, if we send a long string of 0's, the output should eventually settle toward 0. If we alternate 1's and 0's, the output might resemble a sine wave, depending on the frequency cutoff of the channel.

If, assuming all possible past transmissions, we now plot all possible waveforms within the interval, including those for a 1 and those for a 0, we get a pattern that resembles a picture of an eye. Figure 10.9 shows some representative transmitted waveforms

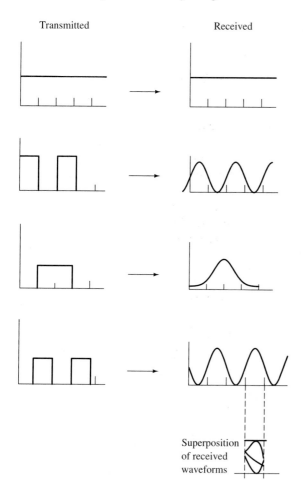

Transmitted   Received

Superposition
of received
waveforms

Figure 10.9   Generation of eye pattern.

and the resulting received waveform. The partial eye pattern is sketched in the figure. The number of individual waveforms contributing to this pattern depends on the memory of the system.

Eye patterns are easy to generate in the laboratory. Suppose that a random NRZ waveform forms the input to a system and that the output is observed on an oscilloscope. Suppose also that the oscilloscope is synchronized to the sampling interval. The eye pattern is then observed on the screen. The eye opening (a blank space in the center) is significant. If the eye is open, it indicates that all waveforms for a transmitted 1 are separated from those for a transmitted 0, and accurate decisions can be made. If the eye is closed, errors will be made in receiving certain sequences. Note that for the ideal distortionless channel, the eye consists of two horizontal lines, one at the top and one at the bottom of the interval.

Now suppose that we transmit a particular bipolar signal and the square pulses that are transmitted are distorted by the channel. Suppose the received waveform due to one transmitted pulse is $s(t)$. The signal at the receiver is then

$$r(t) = \sum_{n=-\infty}^{K} a_n s(t - nT_b) + n(t) \qquad (10.2)$$

where $n(t)$ is the additive channel noise, a_n is either $+1$ or -1, depending on the bit transmitted, and K is the largest integer less than t/T_b. We are assuming the channel is causal, so the received signal does not depend on future values of the data.

We now look at the received signal at the middle of a particular bit interval:

$$r\left(mT_b + \frac{T_b}{2}\right) = \sum_{n=-\infty}^{m} a_n s\left(\frac{T_b}{2} + (m - n)T_b\right) + n\left(mT + \frac{T_b}{2}\right) \qquad (10.3)$$

For simplicity, we will adopt the following notation to change these time samples into a discrete numerical sequence:

$$t_i = iT_b + \frac{T_b}{2}$$

$$s_i = s\left(iT_b + \frac{T_b}{2}\right)$$

$$r_i = r\left(iT_b + \frac{T_b}{2}\right) \qquad (10.4)$$

$$n_i = n\left(iT_b + \frac{T_b}{2}\right)$$

With this notation, Eq. (10.3) becomes

$$r_m = \sum_{n=-\infty}^{m} a_n s_{m-n} + n_m \qquad (10.5)$$

The intersymbol interference is represented by the summation in Eq. (10.5), where $n = m$ represents the desired signal component and the other terms represent the interference from past binary information. Separating out the correct term, we have

$$r_m = a_m s_0 + \sum_{n=-\infty}^{m-1} a_n s_{m-n}$$
$$= a_m s_0 + I_m \tag{10.6}$$

I_m is the intersymbol interference, and it depends on the specific shape of the signal $s(t)$ and on the particular sequence of binary digits transmitted. The peak value of the interference can be found by assuming that the particular sequence is such that all terms in the sum are positive. Therefore,

$$I_m(\text{peak}) = \sum_{n=-\infty}^{m-1} |s_{m-n}| \tag{10.7}$$

The mean square value of the interference may now be calculated:

$$\overline{I^2} = E\left\{ \left[\sum_{n=-\infty}^{m-1} a_n s_{m-n} \right]^2 \right\}$$
$$= \sum_{j=-\infty}^{m-1} \sum_{i=-\infty}^{m-1} E\{a_i a_j\} s_{m-i} s_{m-j} \tag{10.8}$$

We assume that the a_i are independent of each other and that 1's and 0's are equally likely. The cross products then go to zero, and the double summation reduces to

$$\overline{I^2} = \sum_{n=-\infty}^{m-1} s^2_{m-n} \tag{10.9}$$

Example 10.2

Find the peak and mean square intersymbol interference for a signal resulting from sending ideal impulse samples through a channel with triangular passband characteristic, as shown in Fig. 10.10.

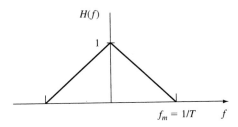

Figure 10.10   Channel characteristics for Example 10.2.

Solution: We first find the impulse response of the channel:

$$h(t) = \frac{T}{n^2 t^2} \sin^2\left(\frac{\pi t}{T} \right)$$

From this result, we find the sample values:

$$s_0 = \frac{4}{n^2 T}$$

$$s_1 = \frac{4}{\pi^2 3^2 T}$$

$$\vdots$$

$$s_n = \frac{4}{\pi^2 (2n + 1)^2 T}$$

The peak and mean square intersymbol interference are found as follows:

$$I_m(\text{peak}) = \frac{4}{\pi^2 T} \sum_{n=-\infty}^{m-1} \frac{1}{|2(m - n) + 1|^2}$$

$$\overline{I_m^2} = \left(\frac{4}{\pi^2 T}\right)^2 \sum_{n=-\infty}^{m-1} \frac{1}{|2(n - m)|^4}$$

What Is the Best Pulse Shape?

Suppose that a waveform $s(t)$ is used to transmit a binary 1 and that $-s(t)$ is used to send 0. The question then arises as to the best choice of $s(t)$. We start by assuming that the transmitted pulses are generated using an impulse fed through an ideal lowpass filter. This is shown in Fig. 10.11. The assumed pulse is therefore of the form

$$\frac{\sin (\pi t / T)}{\pi t / T}$$

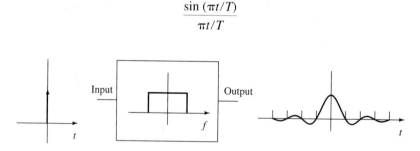

Figure 10.11   Pulses formed by ideal lowpass filter.

This pulse has the property that it goes through zero at every sampling point except for the point at time zero. Therefore, in the intersymbol interference analysis, s_m is zero for all non-zero m, and we have no intersymbol interference. Figure 10.12 illustrates this for three pulses.

The preceding observation is actually a restatement of the sampling theorem. If the channel passes frequencies up to $T_b/2$, then the individual samples values are independent.

The ideal bandlimited pulse is difficult to achieve because of the sharp corners in the frequency spectrum. Even if we could achieve this channel characteristic, the relatively significant side lobes of the impulse response place demands on the detection system. If

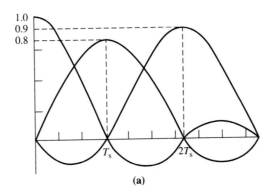

(a)

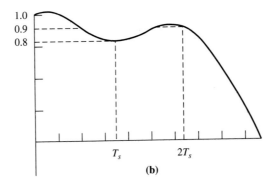

(b)

Figure 10.12   Sequence of pulses formed by ideal lowpass filter.

sampling is not done at the precise multiples of T_b, intersymbol interference becomes a significant factor. Thus, synchronization must be precise, and timing jitter must be kept to a minimum.

A desirable compromise is the *raised cosine* characteristic. The Fourier transform of this pulse is similar to the square transform presented in Fig. 10.11, but the transition from maximum to minimum follows a sinusoidal curve. The raised cosine Fourier transform is shown in Fig. 10.13. The value of K determines the width of the constant portion of the transform. If $K = 0$, the transform becomes that of the ideal bandlimited pulse. If $K = 1$, the flat portion is reduced to a point at the origin, and the bandwidth is $2f_0$. Because of the rounded corners, we can approximate this characteristic to a far greater degree of accuracy than is possible for the square transform.

The function of time corresponding to the raised cosine transform is

$$h(t) = A \frac{\sin 2\pi f_0 t}{2\pi f_0 t} \frac{\cos KT2\pi f_0}{1 - \frac{8K^2 f_0^2 t^2}{\pi}} \tag{10.10}$$

This function is sketched for several representative values of K in Fig. 10.14. Note that at $K = 1$, the response goes to zero not only at the zeros of the $\sin(t)/(t)$ function, but also at points midway between these sample values. It is therefore possible to sample at the same

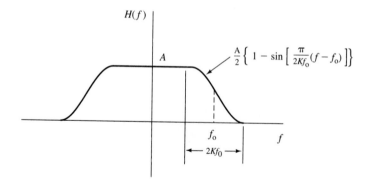

Figure 10.13   Raised cosine Fourier transform.

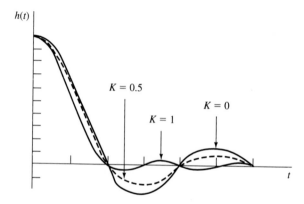

Figure 10.14   Raised cosine function of time.

rate as for the ideal channel, with no resulting intersymbol interference. The price we pay is increased bandwidth.

10.2.1 Equalization

The communication receiver must do two separate filtering jobs. The first is filtering to decrease the effects of additive noise. The second filter decreases the effects of intersymbol interference. This filter attempts to undo the channel distortion and is called an *equalizer.*

Ideal equalization requires the construction of a filter having a transfer function inverse to that of the channel. The product of the two transfer functions is then a constant. Synthesis of the equalizing filter is not simple in the general case. Equalizers are often constructed using a tapped delay line. This results in the *transversal equalizer,* as shown in Fig. 10.15. The impulse response and system function of the transversal equalizer are given by

$$h(t) = \sum_{n=0}^{\infty} a_n \delta(t - nT)$$

$$H(f) = \sum_{n=0}^{\infty} a_n e^{-jn2\pi fT}$$
(10.11)

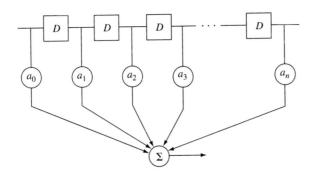

Figure 10.15   Transversal equalizer.

The design of equalizers depends on how much we know about the channel. If the channel characteristics are completely known, we can find the tap weights to achieve the desired filter characteristic. If the channel characteristics are not known in advance, a *training sequence* can be used to adapt the filter weights. The output of the filter is compared to the (known) transmitted sequence, and the weights are adjusted to force the difference to zero. The adjustment is often performed by a feedback control loop. In the worst case, neither the channel characteristics nor the transmitted signal are known. This leads to *blind equalization,* where the tap weights are chosen to minimize a parameter related to the error.

10.3 BASEBAND DETECTION

Baseband detectors, or *decoders*, have the challenge of receiving the baseband signal and converting it to a train of bits. A digital-to-analog converter is one component of the baseband decoder. However, as we use different waveforms to transmit 1's and 0's, we will need more sophisticated devices. In this section, we present the single-sample detector and the matched filter detector. We analyze the performance of these detectors in Section 10.4.

10.3.1 Single-sample Detector

Viewing an NRZ-L baseband signal (i.e., a signal in which $+V$ used to send a 1 and $-V$ a zero), we might start by simply sampling the signal at the midpoint of each interval. These sample values could then be compared to zero, with the signal decoded as a 1 if the sample is positive and as a 0 if the sample is negative. A receiver operating on this principle is illustrated in Fig. 10.16. The sample value is a random variable because of the presence of noise. The additive noise can be modeled as filtered white noise, where the filtering is done by the channel characteristic. The noise has zero mean and a variance σ^2. The sample would therefore be a random variable with mean value $\pm V$, depending upon which binary digit is being sent in the given interval. The variance of the sample would be the variance

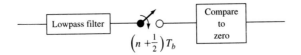

Figure 10.16   Single-sample detector.

of the noise, which is approximately equal to the bandwidth of the channel multiplied by the power per hertz of the white noise.

10.3.2 Binary Matched Filter Detector

Matched filters were presented in Section 4.7. Since the matched filter essentially "pulls" the signal out of the background noise, it makes intuitive sense to use it in the decision process. If we use two filters, one matched to each of the two possible transmitted signals, we can make decisions based on the filter outputs. The use of two matched filters leads to the *binary matched filter detector.*

We begin by generalizing the concept of the matched filter detector to apply to any signal waveforms. The results we obtain will be applicable to baseband, as well as to the various forms of modulation introduced in subsequent chapters. Figure 10.17 shows the *binary matched filter detector.* $s_0(t)$ and $s_1(t)$ are the signals used to transmit a 0 and a 1, respectively ($+V$ and $-V$ square pulses in NRZ-L). We assume that they are known at the receiver; that is, the functions of time that we use are the result of passing the transmitted signals through the channel. We illustrate the receiver operating on the first bit interval, 0 to T_b. Following this, the receiver would operate on the second bit interval, T_b to $2T_b$. It continues this operation for all future intervals $[nT_b$ to $(n + 1)T_b]$.

The illustration in Fig. 10.17 shows the receiver using correlators, but an exact equivalent receiver uses matched filters to replace the multipliers and integrators. The receiver compares the output of two filters, one matched to $s_0(t)$ and the other matched to $s_1(t)$. It takes the difference of these two outputs and compares it to a threshold value. We shall see that this threshold is often zero, in which case the detector is testing which matched filter output is largest. If the top filter output is larger than the bottom, y is positive; if the bottom output is larger, y is negative.

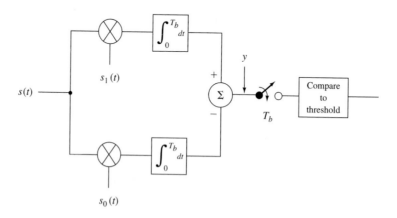

Figure 10.17   Binary matched filter detector.

The two signals, $s_0(t)$ and $s_1(t)$, could be the baseband pulse waveshapes (e.g., NRZ, biphase), or they could be more complex waveforms of the type we examine later in the text.

Our study of the matched filter detector is now reduced to two challenges. First, we must decide how to choose the threshold against which the output y is compared. Second, we must evaluate the performance of the detector.

The input to the comparator is a random variable. This is because the output of each integrator is a *number* (*not* a function of time) composed of a deterministic part (due to the signal) and a random part (due to the additive noise). The additive noise is assumed to be zero-mean Gaussian and white. Therefore, the input to the comparator is Gaussian. A Gaussian density is specified by only two parameters: the mean and the variance. We need only find their values.

Suppose the input to the detector is $s_i(t) + n(t)$, where i is either zero or unity, depending upon which signal is being transmitted. The input to the comparator is then given by

$$y = \int_0^{T_b} s_i(t)[s_1(t) - s_0(t)]dt + \int_0^{T_b} n(t)[s_1(t) - s_0(t)]dt \qquad (10.12)$$

We find the average value of y by adding together the average values of the two integrals. The first integral is nonrandom, so its average is equal to itself. The average of the second integral is zero, since we assume that the noise is zero mean. Therefore, the mean value of y is

$$m_y = \int_0^{T_b} s_i(t)[s_1(t) - s_0(t)]dt \qquad (10.13)$$

This mean value depends upon which signal is being sent.

The variance of y is the expected value of the square of the difference between y and its mean. That is,

$$\sigma_y^2 = E\{[y - m_y]^2\}$$

$$= E\left\{\left[\int_0^{T_b} n(t)[s_1(t) - s_0(t)]dt\right]^2\right\} \qquad (10.14)$$

When an integral is squared, it is not sufficient simply to square the integrand; doing so would ignore all of the cross products. The proper way to square an integral is to write it as the product of two integrals. Therefore,

$$\sigma_y^2 = E\left\{\int_0^{T_b}\int_0^{T_b} n(t)n(\tau)[s_1(t) - s_0(t)][s_1(\tau) - s_0(\tau)]dt\, d\tau\right\} \qquad (10.15)$$

The expected value of a sum is the sum of the expected values, so we can move the expected value symbol within the range of the integral signs. The only random part of the integral is that containing the noise n. We therefore need to find the expected value of the noise product,

$$E\{n(t)n(\tau)\} = R_n(t - \tau) \qquad (10.16)$$

This is the *autocorrelation* of the noise. The noise is assumed to be white with power spectral density $G_n(f) = N_0/2$. Therefore, its autocorrelation is the inverse Fourier transform of the power spectrum, or $R_n(t) = N_0\delta(t)/2$. Taking this into account, we find that Eq. (10.15) becomes

$$\sigma_y^2 = \left\{ \int_0^{T_b} \int_0^{T_b} \frac{N_0}{2} \delta(t - \tau)[s_1(t) - s_0(t)][s_1(\tau) - s_0(\tau)]dt\, d\tau \right\} \tag{10.17}$$

which, by the sampling property of the impulse, is equal to

$$\sigma_y^2 = \frac{N_0}{2} \int_0^{T_b} [s_1(t) - s_0(t)]^2 dt \tag{10.18}$$

Note that this result is independent of which signal is sent.

We have found the mean and variance of y. We can now sketch the probability density of y under the assumption that a 0 or a 1 is sent. Figure 10.18 shows the two probability density functions, labeled $p_0(y)$ and $p_1(y)$. Note that the two densities have the same variance but difference mean values.

We are now ready to select the threshold. If y is greater than the selected threshold,

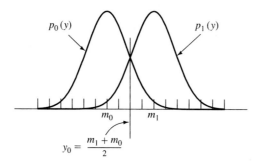

Figure 10.18   Probability densities of y

we assume that $s_1(t)$ is being sent; if y is less than the threshold, we assume that $s_0(t)$ is being sent. The correct choice of the threshold depends on the fraction of time a 1 is being sent and also on the cost of the errors. For example, if we know that 1's are being sent 99 percent of the time, we would bias our threshold to a lower value to assure that we are more likely to decide that a 1 is being received (i.e., the threshold is broken more often). Similarly, if the cost of one error is not equal to that of the other, we would bias the decision. Suppose, for example, you were told that accidentally calling a transmitted 0 a 1 costs you US$0.05, while erroneously calling a transmitted 1 a 0 costs you a passing grade in this class. You would probably set the threshold very low so that you hardly ever decide that a zero is being received. In fact, we would expect to hear you saying, "One, one, one, one, one, one, ..."

More practical examples of costs affecting the threshold exist in radar, where the cost of erroneously labelling a 0 as a 1 (a false alarm) might be a wasted antiballistic missile (only US$10 million), while the cost of calling a 1 a 0 (a miss) might mean the nation's capitol is wiped out. Under such a scenario, you would certainly bias the decision toward having false alarms rather than misses (assuming your country permits budget deficits).

This intuitive discussion can be formalized under the topic of decision theory. In the majority of digital communication applications, there is no reason to bias the decision, so we should make the decision corresponding to the highest conditional probability. For example, if we measure $y = 2$, and $p_1(2)$ is larger than $p_0(2)$, we would decide that a 1 is being transmitted. With regard to Fig. 10.18, this means that the threshold is chosen as the point at which the two probability density functions cross. This is labeled as y_0 on the diagram. Note that due to symmetry, y_0 is the midpoint between the two mean values. Thus,

$$y_0 = \frac{m_1 + m_0}{2} = \frac{1}{2}\int_0^{T_b} [s_1(t) + s_0(t)][s_1(t) - s_0(t)]dt$$

$$= \frac{1}{2}\int_0^{T_b} [s_1{}^2(t) - s_0{}^2(t)]dt \tag{10.19}$$

We recognize the integral of the square of the signal as the *signal energy* over the bit period. Denoting this energy as E_1 and E_0 for the two signals, we find that the threshold is

$$y_0 = \frac{E_1 - E_0}{2} \tag{10.20}$$

If the signals have equal energy, the threshold is at the origin. In such cases, we compare the difference between the two integrator outputs to zero, and we are effectively testing which of the outputs is larger.

10.4 PERFORMANCE OF DIGITAL BASEBAND

10.4.1 Single-sample Detector

We now evaluate the performance of the single-sample detector of Fig. 10.19(a). We defer analysis of the matched filter detector to the next section, where we derive a general result that applies to a broad class of detectors.

Each sample in the single-sample detector is composed of a signal ($\pm V$) plus a sample of additive noise. We assume that the additive noise is a sample of a Gaussian zero-mean noise process. We therefore find that the probability density of the sample follows one of two Gaussian curves, as shown in Fig. 10.19(b). The sample follows the curve labeled $p_0(y)$ if a 0 is being sent, and it follows $p_1(y)$ when a 1 is transmitted. We call the reception a 1 if the sample is positive and a 0 if it is negative. The error probabilities can therefore be found by integrating the tail of the curves. The probability of mistaking a transmitted 1 for a 0 is the cross-hatched area labeled P_M in the figure. The subscript M is a carryover from the radar problem, in which mistaking a target return (1) for no return (0) is considered a miss. The other transition probability is labeled P_{FA}, the probability of a false alarm. We can see from the symmetry of the diagram that these probabilities are equal. The model of this system is therefore a *binary symmetric channel*. This is illustrated in Fig. 10.19(c). The left side of this diagram represents the transmitted bit, while the right side shows the received bit. The horizontal paths therefore represent correct transmissions, and the diagonals represent errors. The labels on the various paths represent the (conditional) probabilities. The figure shows the *binary channel*. If P_{FA} is equal to P_M, the channel is known as a *binary symmetric channel* (BSC).

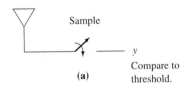

Sample

y

Compare to
threshold.

(a)

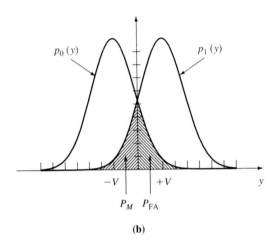

(b)

Figure 10.19 The single-sample detector.

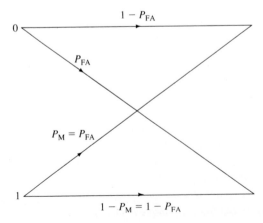

Figure 10.19(c)

The variance of the noise is proportional to the bandwidth f_m, where the proportionality constant is the noise power per Hz, N_0. Therefore, the probability of a bit error is given by either P_{FA} or P_M (they are the same) and is

$$P_e = \frac{1}{\sqrt{2\pi N_0 f_m}} \int_0^\infty \exp\left(\frac{-(y+V)^2}{2N_0 f_m}\right) dy \tag{10.21}$$

We make a change of variable to put this in the form of an error function and find that[2]

$$P_e = \frac{1}{2} \operatorname{erfc}\left(\frac{V}{\sqrt{2N_0 f_m}}\right)$$

$$= Q\left(\frac{V}{\sqrt{N_0 f_m}}\right) \tag{10.22}$$

We can identify $V^2/N_0 f_m$ as the signal to noise ratio. This is because V^2 is the signal power in a 1-Ω resistor and $N_0 f_m$ is the noise power in watts. Therefore,

$$P_e = \frac{1}{2} \operatorname{erfc}\left(\sqrt{\frac{S}{2N}}\right)$$

$$= Q\left(\sqrt{\frac{S}{N}}\right) \tag{10.23}$$

The probability of error is plotted as a function of signal to noise ratio in Fig. 10.20. Note that we express the signal to noise ratio in dB, which is 10 times the logarithm of the signal to noise ratio.

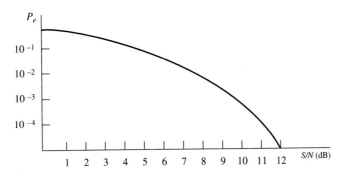

Figure 10.20   Bit error rate for single-sample detector.

[2]We give answers in terms of both the error function and the Q-function. The results are identical, and either approach can be used, depending on which particular table is available at the time.

10.4.2 Binary Matched Filter Detector

We now evaluate the probability of error for the binary matched filter detector of Fig. 10.21(a). As in our derivation for the single-sample detector, the error is given by the area under the tail of the Gaussian density. This is shown in Fig. 10.21(b).

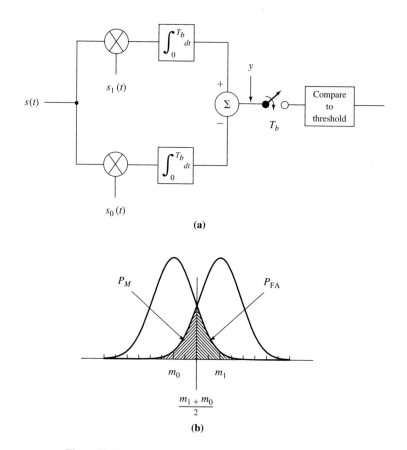

Figure 10.21   Probability densities for matched filter detector.

We shall derive the error for the special case where the two signals have equal energy. In the problems at the end of the chapter, you are requested to show that the result also applies to the case of unequal signal energies. The probability of error is

$$P_e = \frac{1}{\sqrt{2\pi}\sigma} \int_0^\infty \exp\left[\frac{(y - m_0)^2}{2\sigma^2}\right] dy \qquad (10.24)$$

To get this into the form of an error function, we change variables. Let

$$u = \frac{y - m_0}{\sqrt{2}\sigma} \qquad (10.25)$$

$$P_e = \frac{1}{2} \text{erfc}\left(\frac{-m_0}{\sqrt{2}\sigma}\right)$$

where

$$m_0 = \int_0^{T_b} s_0(t)[s_1(t) - s_0(t)]dt$$

$$(10.26)$$

$$\sigma_y^2 = \frac{N_0}{2}\int_0^{T_b} [s_1(t) - s_0(t)]^2 dt$$

T_b, the bit period, is the reciprocal of the number of bits per second. We need now simply plug Eqs. (10.25) and (10.24) into Eq. (10.23) in order to find the bit error rate. The result can be simplified considerably and also related to physical quantities by defining the *average energy*

$$E = \frac{E_0 + E_1}{2} \qquad (10.27)$$

and the *correlation coefficient*

$$\rho = \frac{\int_0^{T_b} s_0(t)s_1(t)dt}{E}$$

E_1 and E_0 are the energies of the two signals. That is,

$$E_i = \int_0^{T_b} s_i^2(t)dt \qquad (10.28)$$

For the equal-energy case, $E = E_0 = E_1$.

Substituting Eq. (10.27) into Eq. (10.26), we have

$$m_0 = \int_0^{T_b} s_0(t)[s_1(t) - s_0(t)]dt = \rho E - E$$

$$(10.29)$$

$$\sigma_y^2 = \frac{N_0}{2}\int_0^{T_b} [s_1(t) - s_0(t)]^2 dt = N_0 E(1 - \rho)$$

Equation (10.24) then becomes

$$P_e = \frac{1}{2} \text{erfc}\left(\sqrt{\frac{E(1 - \rho)}{2N_0}}\right)$$

$$(10.30)$$

$$= Q\left(\sqrt{\frac{E(1 - \rho)}{N_0}}\right)$$

Equation (10.30) is the desired result. It shows how to calculate the bit error rate from three parameters: the average energy per bit, the correlation of the two signals, and the noise power per hertz. We again emphasize that even though this result was derived assuming equal energy ($E_0 = E_1$), it applies to the more general case of unequal energies.

Both the complementary error function and the Q-function decrease as their argument increases. Therefore, the probability of error decreases as any of the following happen:

- The average energy increases
- The correlation decreases
- The noise power decreases

The observation about signal energy and noise power should be obvious, but that concerning correlation needs more investigation.

We shall see examples of calculating ρ in the coming sections. For now, we consider two extremes. First, assume that $s_0(t) = s_1(t)$. The correlation coefficient is then $\rho = 1$ [prove it], and Eq. (10.30) becomes

$$P_e = \frac{1}{2} \text{erfc}(0) = Q(0) = \frac{1}{2} \tag{10.31}$$

This is intuitively satisfying, since the transmission is supplying no information. The same signal is used to transmit both the 0 and the 1, so the receiver can only guess at the information. In such a situation, you could save your employer money by substituting a coin for the matched filter detector. Flip the coin to make a decision, and the probability of error is $\frac{1}{2}$.

At the other extreme, if $s_1(t) = -s_0(t)$, the correlation is $\rho = -1$, and the probability of error is a minimum at

$$P_e = \frac{1}{2} \text{erfc}\left(\sqrt{\frac{E}{N_0}}\right) = Q\left(\sqrt{\frac{2E}{N_0}}\right) \tag{10.32}$$

Figure 10.22 shows the bit error rate as a function of the signal to noise ratio, E/N_0, for three values of correlation: -1, 0, and 1. Note that the abscissa is in dB, which is 10 times the logarithm of E/N_0.

Example 10.3

Suppose binary information is transmitted using baseband signals of the form shown in Fig. 10.23. Design a matched filter detector, and find the probability of bit error, assuming that the additive noise has a power of 10^{-3} watt/Hz.

Solution: The matched filter detector is shown in Fig. 10.24(a). Note that because the signal energies are equal, the threshold against which the output is compared is zero. Although the detector is correctly designed as shown, for the first half of each bit interval $s_1(t) = s_0(t)$, and the signal portion of the output is zero. We can therefore simplify the detector to that shown in Fig. 10.24(b), where the bottom leg (multiplier and integrator) produces an output which is the negative of that of the top leg. Since the lower output is subtracted from the upper, the effect is the same as doubling the upper signal. We therefore further simplify the detector to that shown in Fig. 10.24(c). Finally, since the output is compared to zero, the multiplication by 2

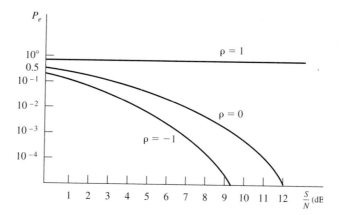

Figure 10.22   Bit error rate for matched filter detector.

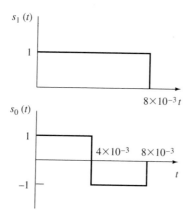

Figure 10.23   Baseband waveforms for Example 10.3.

is superfluous (and would cost extra money), so we get the simplified (*integrate-and-dump*) detector of Fig. 10.24(d).

The performance of all four detectors is identical, so we can calculate it based upon the original statement of the problem. We need only find E and ρ. The energy of each signal, and therefore the average energy, is 8×10^{-3} watt-sec. The correlation coefficient is

$$\rho = \frac{\int_0^{T_b} s_1(t)s_0(t)dt}{8 \times 10^{-3}} = 0$$

Therefore, the probability of error is

$$P_e = \frac{1}{2}\, \text{erfc}\!\left(\sqrt{\frac{E(1-\rho)}{2N_0}}\right) = \frac{1}{2}\, \text{erfc}\,(2) = 2.3 \times 10^{-3}$$

Using Q-functions, we find the probability of error to be

$$P_e = Q\!\left(\sqrt{\frac{E(1-\rho)}{N_0}}\right) = Q(2\sqrt{2}) = 2.3 \times 10^{-3}$$

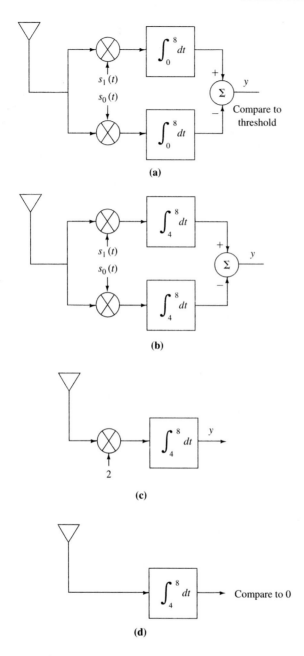

(a)

(b)

(c)

(d)

Figure 10.24 Matched filter detector for Example 10.3.

Example 10.4

Design a matched filter detector to choose between the two baseband signals shown in Fig. 10.25. Note that this is a biphase code. Assume that white Gaussian noise of power spectral density $N_0/2$ is added and that $E/N_0 = 3$.

Solution: The matched filter detector is shown in Fig. 10.26(a). Note that because the signals have equal energy, the output is compared to zero. Since the bottom leg produces an out-

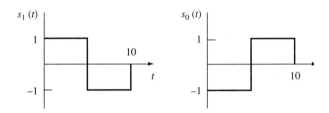

Figure 10.25   Baseband signals for Example 10.4.

put which is the negative of that of the top leg, we can remove it and double the upper leg signal. However, since the output is compared to zero, the doubling is not needed. This results in the simplified detector of Fig. 10.26(b).

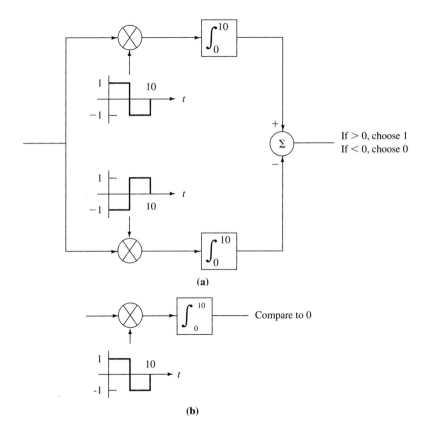

If > 0, choose 1
If < 0, choose 0

(a)

Compare to 0

(b)

Figure 10.26   Matched filter detector for Example 10.4.

Since the signals are the opposite of each other, the correlation coefficient is equal to −1. The performance is then

$$P_e = \frac{1}{2}\,\text{erfc}\left(\sqrt{\frac{E}{N_0}}\right) = Q\left(\sqrt{\frac{2E}{N_0}}\right) = 0.006$$

Example 10.5

Design a matched filter detector for the two baseband signals shown in Fig. 10.27. The additive noise has a power of 0.1 watt/Hz. Find the bit error rate.

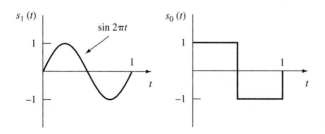

Figure 10.27 Baseband signals for Example 10.5.

Solution: The matched filter detector is shown in Fig. 10.28. Since the signal energies are not equal, the output is compared to a nonzero threshold, namely,

$$y_0 = \frac{E_1 - E_0}{2} = \frac{0.5 - 1}{2} = -0.25$$

The average energy is

$$E = \frac{E_1 + E_0}{2} = 0.75$$

The correlation coefficient is

$$\rho = \frac{2 \int_0^{1/2} \sin 2\pi t \, dt}{0.75} = 0.849$$

Note that this is a very high correlation coefficient because the two signals are not very dissimilar. It represents a very poor design.

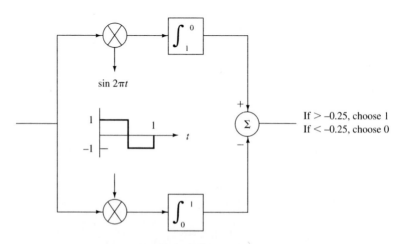

Figure 10.28 Detector for Example 10.5.

The probability of error is then

$$P_e = \frac{1}{2} \text{erfc}\left(\sqrt{0.57}\right) = Q(\sqrt{1.14}) = 0.14$$

PROBLEMS

10.1.1 A first-order phase-locked loop has the following parameter values:

$$A = B = 1; k_0 = 5 \text{ Hz}/V; f_0 = 10 \text{ kHz}$$

Plot the output frequency of the loop as a function of time if a sinusoidal input of frequency 10,010 Hz is applied at time zero.

10.1.2 Repeat Problem 10.1.1 with the input changed from a sinusoid to a square wave.

10.1.3 The clock in a certain system has jitter. Its frequency is given by

$$f(t) = 10^4 + 100 \cos 2\pi \times 300 \, t$$

Design a phased-lock loop that will track the frequency within 10 Hz.

10.1.4 The input to a phase-locked loop is a pulse train of very narrow pulses at a frequency of 10 kHz. The pulse frequency varies slightly by a maximum amount of 5 Hz. This variation occurs at a maximum frequency of 100 Hz. That is, the time it takes for the frequency to deviate by the maximum amount of 5 Hz away from 10 kHz is no shorter than 100 msec.

Design a first-order phase-locked loop to track the input frequency accurately.

10.1.5 Suppose that in a bit synchronization scheme, a known message of 50 bits in length is sent in order to provide for symbol synchronization. The probability of correct reception is as shown in Fig. P10.1.5, where the abscissa is the amount of timing mismatch. That is, when the receiver clock is perfectly synchronized, there is a 0.99 probability of correct reception. With a mismatch of one-half of the bit period, the probability of correct detection drops to 0.5.

The system is considered synchronized if there are 45 or more agreements between the transmitted and detected waveforms.

Find the probability that the system locks to within 1% of correct synchronization.

10.1.6 Assume that the early-late gate of Fig. 10.2 is in perfect lock with a baseband signal at the input. Sketch the waveform at the various points within the block diagram.

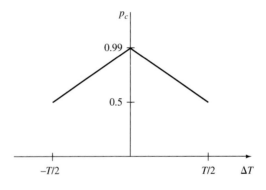

Figure P10.1.5

10.1.7　The DTTL is shown in Fig. 10.3. Repeat the waveform sketches of Fig. 10.4, assuming that the system is one-fourth of a bit period away from the locked position.

10.1.8　The system of Fig. 10.5 is used to recover clock information with a biphase-L coded input signal. Sketch the waveform at each point in this system if the input is a biphase-L signal representing the binary train 00101110.

10.1.9　Find the probability of false word synchronization using a 15-bit pseudonoise (PN) code as a preamble.

10.1.10　A frame synchronization scheme is to be designed with a search mode, a check mode, and a lock mode. Each frame contains 250 bits of information and a preamble of length N bits. The frame must be transmitted within 1 msec. The bit error rate is given by

$$\text{BER} = 0.5 - 0.5e^{-f/2,500}$$

where f is the number of bits per second being transmitted.
(a) If $N = 10$ and frame synchronization must be acquired within 10 frames (with 99% probability), find the threshold T_s to be used in the search mode.
(b) Once the system is in lock, we wish it to remain in this condition with probability 99.9%. Find this threshold value, T_L.
(c) Repeat parts (a) and (b) if N is raised to 100 bits.

10.1.11　Calculate the correlation for the Barker code of length 7.

10.1.12　Calculate the correlation for the Neuman-Hofman code of length 24.

10.1.13　Repeat the design example of Section 10.1.5 for an environment that is no longer noisy. Assume that the bit error rate is 0.1% of the value given by Fig. 10.8.

10.2.1　Find the peak and mean square intersymbol interference for a signal resulting from sending ideal impulse samples through a channel that has a sinusoidal amplitude characteristic as shown in Fig. P10.2.1.

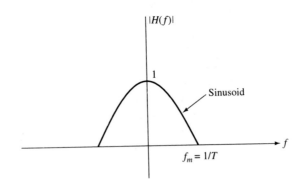

Figure P10.2.1

10.3.1　Binary information is to be transmitted at a rate of 1 kbps on a telephone channel. Assume that the channel passes frequencies between 300 Hz and 3 kHz. White noise of power $N_0 = 10^{-6}$ watt/Hz adds to the signal during transmission. The channel can support voltages between -10 V and $+10$ V.
　　　Design the transmitter and receiver, and evaluate the error probabilities.

10.4.1   A baseband signal is sent using NRZ-L with voltage levels of -2 and $+2$ V for 0's and 1's, respectively. Noise with a power of $N_0 = 10^{-3}$ watt/Hz adds to the signal during transmission. You may assume that the amplitude of the signal at the receiver is the same as that transmitted. Evaluate the performance of a single-sample detector.

10.4.2   **(a)** Design a matched filter detector to decide which of the two signals shown in Fig. P10.4.2 is being received in additive white Gaussian noise.

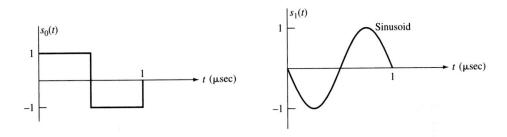

Figure P10.4.2

(b) Find the error probability of the detector if the additive noise has a power of $N_0 = 10^{-6}$ watt/Hz.

10.4.3   You are given the two baseband waveforms shown in Fig. P10.4.3. A matched filter detector is used to decide between the two possible transmitted signals. The additive white Gaussian noise has a power spectral density of $N_0/2 = 0.1$. Find the bit error rate.

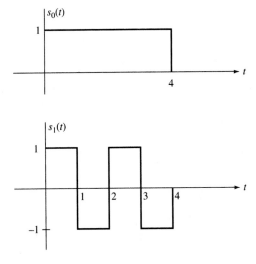

Figure P10.4.3

10.4.4 Given the two signals shown in Fig. P10.4.4, design a matched filter detector, and evaluate its performance as a function of Δ.

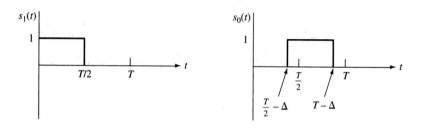

Figure P10.4.4

10.4.5 You are given the two signals shown in Fig. P10.4.5, which are transmitted in the presence of additive Gaussian white noise.

(a) Design a binary matched filter detector to decide which of the two signals is being sent. Your design must include the value of the threshold against which the output is compared.

(b) Find the maximum noise power per Hertz such that the bit error rate does not exceed 10^{-5}.

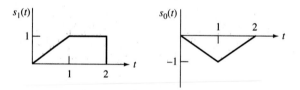

Figure P10.4.5

10.4.6 A biphase baseband communication system uses the two signals shown in Fig. P10.4.6 in order to transmit 1's and 0's. The additive white noise has a power of $N_0 = 10^{-4}$ watt/Hz.

(a) Design a detector to decide between 1's and 0's.

(b) Find the maximum bit rate that can be used in order to achieve a bit error rate below 10^{-4}.

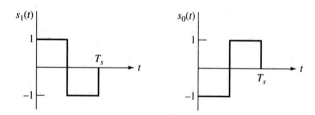

Figure P10.4.6

10.4.7 In Section 10.4.2, we derived the bit error rate of the binary matched filter detector for the case of equal signal energies. This yielded the important result,

$$P_e = \frac{1}{2} \text{erfc} \sqrt{\frac{E(1 - \rho)}{2N_0}}$$

In this equation, E is the energy of each signal. Now prove that the same equation applies to the case of unequal signal energies. In this case, E becomes the average signal energy.

10.4.8 You wish to design a communication system in order to link two stations separated by 5,000 km. The system transmits 1 Mbps for 10 hours each day. Each bit error costs US$1, and each repeater station costs US$3 million (including maintenance costs over its life). A repeater station lasts five years. The bit error rate per link between two repeaters is given by

$$\text{BER} = 10^{-6} e^{-5000/X}$$

where X is the spacing in km between repeaters. Develop formulas, and discuss how you would find the optimum number of repeaters.

11

Digital Modulation

11.0 PREVIEW

What We Will Cover and Why You Should Care

This chapter extends the work of the previous one. In Chapter 10, we developed techniques for transmitting digital information using baseband waveforms. Just as in the case of analog communication, channels often cannot support low-frequency transmissions. For this reason, we now consider modulation techniques parallel to those presented in analog communication.

After reading this material, you will know how to design systems that utilize amplitude shift keying, frequency shift keying, and phase shift keying. You will also understand the hybrid systems that combine amplitude and phase modulation. You will be able to analyze a system's performance and to design systems with desirable performance characteristics. Finally, you will understand modem standards and the techniques used for contemporary data communication.

Necessary Background

Deriving the equations describing the performance of detectors requires a knowledge of Gaussian probability densities. Also, we assume that you have studied the binary matched filter detector and understand how to calculate its performance. The modulated systems of this chapter will have far more meaning if you already understand analog AM, FM, and PM.

11.1 AMPLITUDE SHIFT KEYING

In Chapter 6, we introduced analog AM. Let us now consider the special case where $s(t)$ is an NRZ-L baseband digital signal. The AM waveform is then of the form

$$s_m(t) = [A \pm V]\cos2\pi f_c t \tag{11.1}$$

If we factor out A from the right side of Eq. (11.1), we get

$$s_m(t) = A\left[1 \pm \frac{V}{A}\right]\cos 2\pi f_c t$$

$$= A[1 \pm m]\cos 2\pi f_c t$$

(11.2)

where we have defined the ratio of V to A as the *index of modulation, m*. Therefore, within each bit period, we send a sinusoidal burst with an amplitude equal to either $A[1 + m]$ or $A[1 - m]$, depending on whether we are transmitting a binary 1 or 0 in that interval. A typical waveform might appear as in Fig. 11.1.

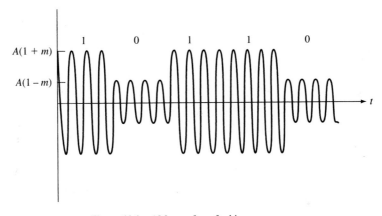

Figure 11.1   AM waveform for binary sequence.

If m were permitted to exceed unity, the amplitude $A[1 - m]$ would go negative. By convention, amplitude is always considered nonnegative, so this reversal would be taken to be a phase shift of 180°. As such, it would no longer be considered amplitude modulation, so we shall consider only cases where m is less than or equal to 1.

In communicating a binary baseband digital signal using amplitude modulation, we see that the amplitude shifts between two levels. Even though this is simply a special case of AM, it is given a separate name: *amplitude shift keying* (ASK). The word "keying" is a carryover from telegraph days, when a key was used to send dots and dashes.

Suppose the index of modulation is 100 percent. The amplitude is then shifting between $2A$ and zero, and the waveform corresponding to Fig. 11.1 would appear as in Fig. 11.2. Note that we transmit a sinusoidal burst of amplitude $2A$ to send a binary 1, and turn off the generator to send a binary 0. This type of communication is given the descriptive name, *on-off keying* (OOK). When we look at system performance later in the chapter, we will find that $m = 1$ is the best choice of modulation index for digital communication. In fact, there are few reasons for choosing any other value of m, so when people refer to ASK, you can automatically assume that they mean OOK unless stated otherwise.

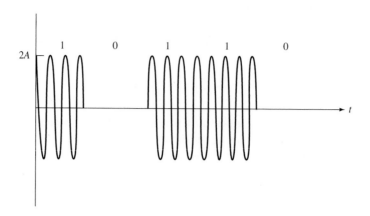

Figure 11.2   ASK with $m = 1$.

11.1.1 ASK Spectrum

We can find the power spectral density of the binary ASK (BASK) signal using a technique similar to that used for baseband signals in Section 8.3. We assume random data and on-off keying. From Eq. (11.2), we see that the carrier amplitude is A, so the carrier power, P_c, is $A^2/2$. The total average power of the ASK signal is found by noting that the transmitted power is $2A^2$ during times that a binary 1 is being sent and zero when a binary 0 is sent. If we assume that 1's and 0's are equally likely, the average transmitted power is A^2. With half of this going to the carrier power, the remainder, $A^2/2$, is the power in the sidebands (information), P_i. This quantity equals the area under the continuous portion of the power spectral density curve (for positive and negative values of f). The power spectral density is sketched in Fig. 11.3. Note the impulse at the carrier frequency, which indicates that this is a form of transmitted carrier AM.

11.1.2 Modulators

There are two approaches to generating the ASK waveform. One starts with the baseband signal and uses this to amplitude modulate a sinusoidal carrier. Since the baseband signal consists of distinct waveform segments, the AM wave consists of distinct modulated segments.

The second approach is to generate the AM wave directly, without first forming the baseband signal. In the binary case, the generator would only have to be capable of formulating one of two distinct AM wave segments. For OOK, we need simply switch an oscillator on and off, as shown in Fig. 11.4.

11.1.3 ASK Demodulation

We now turn our attention to the coherent detector for ASK. If we use the synchronous demodulator for ASK (OOK), the output would be a piecewise constant function. Suppose, for example, that the input to the demodulator is a sinusoidal burst of amplitude A when a

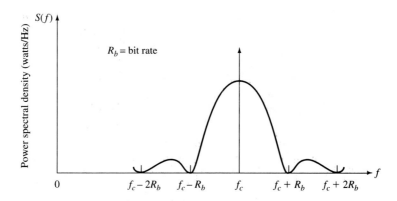

Figure 11.3   OOK power spectral density.

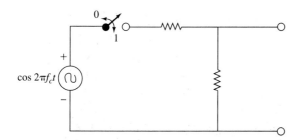

Figure 11.4   Modulator of OOK BASK.

1 is being sent and zero when a 0 is being sent. Then the output of the synchronous demodulator would be a unipolar baseband signal ranging in amplitude between 0 and $A/2$ volts.

Matched filter detector

In our discussion of baseband digital communication in the Section 10.3, we concluded that the matched filter (or correlator) detector was the best receiver for binary digital communication. We can now apply that receiver to the ASK signal. The receiver consists of two legs matched to the two possible received signals. The difference is taken between the two filter outputs and is compared to a threshold. However, for OOK, one of the signals is zero, so one of the legs of the matched filter detector can be eliminated. The resulting detector is shown in Fig. 11.5.

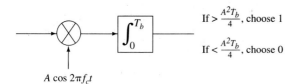

If $> \frac{A^2 T_b}{4}$, choose 1

If $< \frac{A^2 T_b}{4}$, choose 0

Figure 11.5   Matched filter detector for OOK.

Our only design task is to decide what threshold the output must be compared to. The threshold is [see Eq. (10.20) of Section 10.3.2]

$$y_0 = \frac{E_1 - E_0}{2} \tag{11.3}$$

where E_1 and E_0 are the energies of the signals used to send a 1 and a 0, respectively. When a binary 1 is transmitted, a sinusoidal burst of amplitude A and duration T_b (the bit period) is sent. Thus, the energy[1] in this case is given by the power multiplied by time, or

$$E_1 = \frac{A^2 T_b}{2} = \frac{A^2}{2R_b} \tag{11.4}$$

where R_b is the bit rate in bits/sec. Since the signal used to send a binary 0 is zero, the energy in that case is also zero (i.e., $E_0 = 0$). The threshold is then given by

$$y_0 = \frac{A^2 T_b}{4} = \frac{A^2}{4R_b} \tag{11.5}$$

If the detector output exceeds this value, we decide that a 1 is being sent. If the output is less than this value, we decide that a 0 is being sent.

Incoherent Detector

Recovery of the carrier for the matched filter detector is not difficult. However, if the data include long strings of zeros, the receiver experiences relatively long periods when all that is received is random noise. In such cases, the carrier may have to be reacquired when the transmitter again turns on.

Incoherent detection, as used in analog communication, does not require that the carrier be reconstructed. The simplest form of incoherent detector is the envelope detector, shown in Fig. 11.6. The output of this detector is the baseband signal. Once the baseband signal is recovered, we can use the techniques of Section 10.3 to make a decision. This can be done with the matched filter detector, the single-sample detector, or the integrate-and-dump detector.

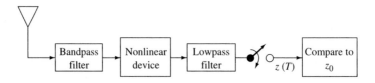

Figure 11.6   Envelope detector for OOK BASK.

---

[1]We remind you that even though we sometimes refer to the transmitted signal, the matched filter detector is pulling the *received* signal out of background noise. We therefore design the detector for the expected *received* signal, which will usually differ from that transmitted in both amplitude and waveshape.

The incoherent detector is much simpler to construct than is the coherent detector. You are cautioned, however, not to make any hasty design decisions until examining the performance of each. As you might expect, the incoherent detector has a higher bit error rate than does the coherent detector.

11.1.4 Detector Performance

We now examine the performance of digital detectors as applied to ASK reception.

Coherent Detection

The matched filter detector is shown in Fig. 11.7. Note that this figure does *not* assume OOK at this time. We borrow the results of Section 10.4.2 to analyze the performance of the detector. We need simply find the average energy per bit and the correlation [see Eq. (10.30)].

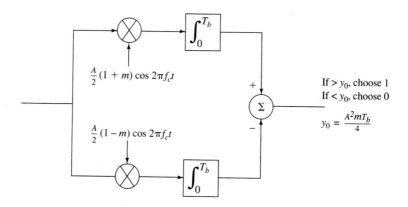

Figure 11.7   Matched filter detector for ASK.

If T_b is the bit transmission period (the reciprocal of the bit rate—i.e., $1/R_b$), the average energy per bit for OOK is

$$E = \frac{E_1 + E_0}{2} = \frac{(A^2 T_b/2) + 0}{2} = \frac{A^2 T_b}{4} \tag{11.6}$$

The correlation coefficient is

$$\rho = \frac{\int_0^{T_b} s_1(t)s_0(t)dt}{E} = 0 \tag{11.7}$$

This quantity is zero, since $s_0(t) = 0$. The bit error rate is found from Eq. (10.30):

$$P_e = \frac{1}{2} \text{erfc}\left(\sqrt{\frac{E(1 - \rho)}{2N_0}}\right) = Q\left(\sqrt{\frac{E(1 - \rho)}{N_0}}\right)$$

$$= \frac{1}{2} \text{erfc}\left(\sqrt{\frac{A^2 T_b}{8N_0}}\right) = Q\left(\sqrt{\frac{A^2 T_b}{4N_0}}\right) \tag{11.8}$$

We shall examine this expression graphically after we evaluate the performance of the incoherent detector.

Example 11.1

Show that an index of modulation of 100 percent (OOK) is the best choice for ASK.

Solution: We can approach this problem from the definition of ASK. That is, we transmit sinusoidal bursts of amplitude $A[1 - m]$ or $A[1 + m]$, depending on whether a binary 0 or 1 is being sent. However, this is not a realistic approach, since the average transmitted power and the maximum signal amplitude both depend on the value of m. A more meaningful approach is to assume that the system limits either the amplitude or the average power. The amplitude limitation would apply in cases where the electronics are being driven to the limits of linear operation. The average power limitation would apply in remote portable applications (e.g., a satellite that must draw all of its energy from solar cells) or if a regulatory body were to limit the average power.

Let us assume that the amplitude is limited and that the sinusoidal burst used to transmit a binary 1 has amplitude A_{max}. Then the amplitude of the burst to transmit a binary 0 is kA_{max}, where k is a constant between zero and unity. Our task is to prove that the best value of k is zero.

The performance of the coherent detector depends on two quantities: the average energy and the correlation coefficient. The energies of the two sinusoidal bursts are

$$E_0 = \frac{(kA_{max})^2 T_b}{2}$$

$$E_1 = \frac{(A_{max})^2 T_b}{2}$$

The average energy is then

$$E = \frac{(k^2 + 1)(A_{max})^2 T_b}{4}$$

The correlation coefficient is

$$\rho = \frac{\displaystyle\int_0^{T_b} s_1(t)s_0(t)dt}{E}$$

$$= \frac{k(A_{max})^2 \displaystyle\int_0^{T_b} \cos^2 2\pi f_c t\, dt}{(k^2 + 1)(A_{max})^2 T_b/4} = 2\frac{k}{k^2 + 1}$$

The probability of error is

$$P_e = \frac{1}{2}\,\text{erfc}\left(\sqrt{\frac{E(1-\rho)}{2N_0}}\right)$$

This quantity decreases as $E(1-\rho)$ increases. To minimize the probability of bit error (for a given noise power per Hz), we want to maximize

$$E(1-\rho) = \frac{(k^2+1)(A_{max})^2 T_b}{2}\,\frac{k^2+1-2k}{k^2+1}$$

We assume that A_{max} and T_b are fixed. Then $E(1-\rho)$ is proportional to

$$k^2 - 2k + 1 = (k-1)^2$$

We wish to maximize this for k between zero and unity. Clearly, the best choice for k is zero, which leads to OOK.

Incoherent Detector

The analysis of the incoherent digital detector is complicated by the fact that the output of the envelope detector is not Gaussian distributed. This is so because a nonlinear operation is being performed. If a binary 0 is being transmitted, the input to the envelope detector is bandpass noise, and the output is

$$z(t) = \sqrt{x^2(t) + y^2(t)} \tag{11.9}$$

where $x(t)$ and $y(t)$ are the low-frequency Gaussian processes in the narrowband noise expansion. The envelope, $z(t)$, then follows a *Rayleigh probability density*.

In the case of a 1 being transmitted, the analysis is more complex, and the output follows a *Ricean density*. Therefore, in contrast to the case represented by Fig. 10.8 (i.e., two bell-shaped conditional probability curves), the two conditional densities have different shapes, and the symmetry we have previously used to simplify the analysis no longer exists. However, using realistic numbers, we operate far out on the tails of the density functions, and the two densities exhibit similar characteristics. Performing the analysis yields the simple result that the bit error rate is approximately

$$P_e = \frac{1}{2}\exp\left(\frac{-E}{2N_0}\right) = \frac{1}{2}\exp\left(\frac{-A^2 T_b}{8N_0}\right) \tag{11.10}$$

(You will be asked to derive this in the problems at the end of the chapter.)

Figure 11.8 shows the bit error rate for coherent detection [Eq. (11.8)] and for incoherent detection [Eq. (11.10)] as a function of the signal to noise ratio in dB. As expected, the incoherent detector does not perform as well as the coherent detector. In system design, you would have to weigh the performance penalty against the advantages gained from the ease of constructing the incoherent detector.

Example 11.2

Binary information is transmitted at 10 kbps using OOK. The carrier frequency is 10 MHz, and the received carrier amplitude is 10^{-2} volt. The additive noise power is 5×10^{-10} watt/Hz.

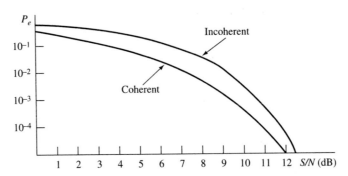

Figure 11.8 Performance of BASK detectors.

(a) Design a coherent detector and find the bit error rate.
(b) Design an incoherent detector and find the bit error rate.
Solution: **(a)** The coherent detector is shown in Fig. 11.9(a). It is simply a matched filter detector that has been simplified because one of the signals is zero. As we have done throughout this chapter, we illustrate the detector operating on the first bit interval, from 0 to 0.1 msec. After that, the detector would shift its attention to the second interval, from 0.1 to 0.2 msec.

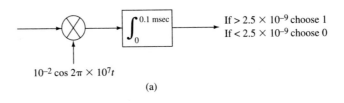

(a)

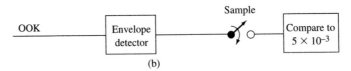

(b)

Figure 11.9 Detectors for Example 11.2.

To find the probability of bit error, either we need to find E and ρ, or we could plug numbers into Eq. (11.8), since we are sure that the derivation of that equation applies in this case. The bit error rate is then

$$P_e = \frac{1}{2}\,\text{erfc}\left(\sqrt{\frac{10^{-4} \times 10^{-4}}{4 \times 10^{-9}}}\right) = \frac{1}{2}\,\text{erfc}(1.58) = 0.013$$

(b) The incoherent (envelope) detector is shown in Fig. 11.9(b). Its performance is, from Eq. (11.10),

$$P_e = \frac{1}{2}\exp\left(-2.5\right) = 0.041$$

Two things should be noted about this result. First, the incoherent detector does not perform as well as the coherent detector, as would be expected. Second, its error rate is extremely high. For the coherent detector, we would expect an average of about 13 errors out of every 1,000 bits transmitted. In fact, the error rate for the incoherent detector is so high that the approximations leading to Eq. (11.10) are of questionable validity. The derivation of that equation requires that we operate far out on the tails of the probability density (far enough so that the Ricean density resembles a Rayleigh density). Therefore, although the error result for the coherent detector is accurate, the result for the incoherent detector should be considered merely a first approximation.

11.2 FREQUENCY SHIFT KEYING

In analog communication, FM is often used in place of AM because of the resulting improvement in performance in the presence of additive noise. We will find additional reasons to use FM for digital communication. For example, we will present some extremely simple implementations of modulators and demodulators.

We start with a system that generates the NRZ baseband signal. Let us assume that this signal is composed of piecewise-constant segments. That is, a binary 1 is transmitted with a square pulse of voltage V_1, and a binary 0 is transmitted with a pulse of V_0 volts, as shown in Fig. 11.10.

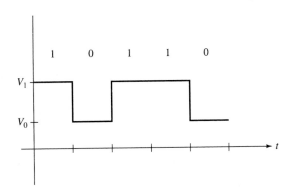

Figure 11.10 Baseband digital signal.

Since we are using *frequency modulation*, it is the *frequency* of the carrier that varies in accordance with the information baseband signal. This contrasts with *amplitude modulation*, wherein the *amplitude* of the carrier is controlled by the baseband signal. Because we assume that the baseband signal takes on only one of two values, it follows that the frequency of the modulated waveform also takes on one of two values, and the modulation process can be thought of as a *keying* operation. The resulting FM waveform is known as *frequency shift keying* (FSK).

If we denote the baseband signal as $s_b(t)$, the resulting FM waveform is

$$\lambda_{fm}(t) = A \cos\left[2\pi f_c t + 2\pi k_f \int s_b(t)dt\right] \tag{11.11}$$

Let us now assume that the baseband signal is bipolar, so that $s_b(t)$ is either $+V$ or $-V$. The instantaneous frequency is then given by the derivative of the phase, or

$$f_i(t) = f_c \pm k_f V \tag{11.12}$$

The maximum frequency deviation, Δf, is $k_f V$. It will prove helpful to define a data signal $d_i(t)$ that is either $+1$ or -1, depending upon whether a 1 or a 0 is being sent. (Think of this signal as the normalized bipolar signal.) With the new signal, the frequency of the FSK waveform is given by

$$f_i(t) = f_c + \Delta f d_i(t) \tag{11.13}$$

Figure 11.11 shows a representative binary FSK waveform.

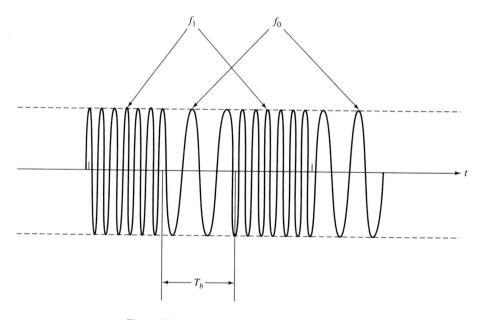

Figure 11.11   Representative binary FSK waveform.

11.2.1 FSK Spectrum

The FSK waveform of Fig. 11.11 can be considered the superposition of two ASK signals. One of these is the ASK signal resulting from modulating a carrier of frequency $f_c + \Delta f$ using OOK from the baseband signal. The other results from amplitude shift keying a carrier of frequency $f_c - \Delta f$ with the *complement* of the baseband signal. Figure 11.12 shows the system that would accomplish this.

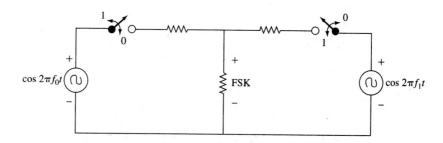

Figure 11.12   FSK as the superposition of two ASK waveforms.

Now that we have decomposed the FSK waveform into two ASK waves, we can borrow the ASK results from Section 11.1 to find the spectrum of the FSK signal. Each of the two components has a frequency spectrum that is of the general shape of $(\sin^2 f)/f^2$ shifted to the carrier frequency. Thus, the power spectrum of the FSK signal is as shown in Fig. 11.13. Note that we are assuming that 0's and 1's are equally probable (i.e., each occurs half of the time, on the average). The power of each carrier is $A^2/8$. The power in the sidebands around each carrier is also $A^2/8$. The total transmitted power is $A^2/2$.

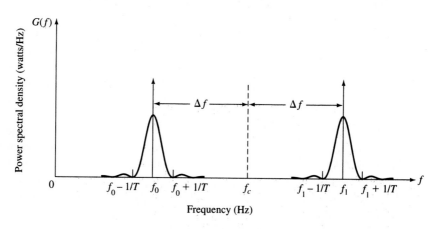

Figure 11.13   FSK power spectral density.

Figure 11.14 shows a modified version of Fig. 11.13 in which we assume that the spacing between the two carrier frequencies is equal to the bit transmission rate. This is known as *orthogonal tone spacing*. Its importance will become clear when we view the performance of FSK systems.

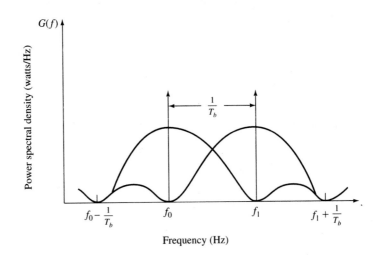

Figure 11.14   FSK power spectral density for orthogonal tone spacing.

The *bandwidth* of the FSK waveform is approximately equal to $2\Delta f + 2R_b$, where we define bandwidth out to the first zero of each lobe of the power spectral density. This is often called the *nominal* bandwidth. In the case of orthogonal tone spacing, the nominal bandwidth is $3R_b$. This contrasts with $2R_b$ for the ASK waveform.

11.2.2 M-ary FSK

Until now, we have been discussing *binary* communication. Let us suppose that in a particular binary communication system (e.g., baseband, ASK, or FSK) the bandwidth of the signal is too large for the channel. One way to reduce the bandwidth is to lower the bit rate. However, this may be unacceptable if information must be transmitted at a specified rate.

A second option is to leave the realm of binary communication altogether, by combining bits into groupings. For example, if we combine bits into pairs, each pair can take on one of four possible values. These values can then be transmitted using FSK with four different frequencies. This is known as *4-ary FSK*. The generator can be visualized as a digital-to-analog converter operating on pairs of bits. The converter is followed by a VCO, as shown in Fig. 11.15. The output of the converter takes on one of four values: 0, 1, 2, or 3. This drives the VCO to one of four frequencies.

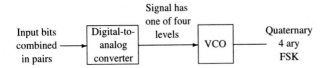

Figure 11.15   4-ary FSK modulator.

The rate at which the frequencies change is one-half of the original bit rate. The power spectral density of an FSK system with four frequencies is the sum of the four waveforms shown in Fig. 11.16. We illustrate this for orthogonal tone spacing, where, in this case, the frequencies are separated by $R_b/2$. Note that the bandwidth is $5R_b/2$, compared to the $3R_b$ required for binary FSK.

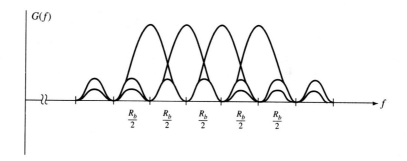

Figure 11.16   Power spectral density of 4-ary FSK.

11.2.3 Modulators

Digital FM can be considered a special case of analog FM in which the information signal takes on one of two discrete values. For this reason, we can consider a digital modulator to be no different from an analog modulator, except that the input is a digital baseband signal. However, a baseband signal that jumps instantaneously between two values is not frequency limited. Hence, the practical implementations of analog FM modulation would not work properly for digital FM. Figure 11.17 illustrates a simple modulator consisting of two oscillators and a switch (key).

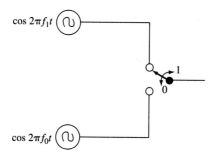

Figure 11.17   FSK modulator.

We have been presenting FSK as if the frequency abruptly changes between the two possible values. That is, taking the baseband modulator approach, we have been assuming that the baseband is composed of perfect square pulses. In practice, the pulses are often shaped (conditioned) prior to modulating the carrier. The frequency transitions are therefore smooth rather than instantaneous. This is necessary to be able to implement FSK using real devices.

11.2.4 Demodulators

FSK demodulation can be either coherent (keeping track of phase) or incoherent.

Matched Filter Detector

We begin with the coherent matched filter detector of Fig. 11.18. As we have done throughout the text, we implement the detector using correlators instead of filters. Note, however, that the matched filter for a sinusoidal burst is a filter whose impulse response is a sinusoidal burst. This is an approximation to a bandpass filter tuned to the frequency of the sinusoid.

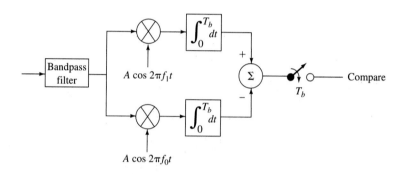

Figure 11.18   Matched filter detector for FSK.

The coherent detector requires the two separate carriers to be reconstructed at the receiver. Recall that FSK can be viewed as a superposition of two ASK waveforms. Since the individual ASK modulations are transmitted carrier, two bandpass filters or two phase lock loops can be used to reconstruct the carriers. However, as in the case of ASK, the carrier is not always present. For example, during periods when a 1 is being transmitted, the carrier at $f_c - \Delta f$ is not being transmitted. Likewise, when 0's are transmitted, the carrier at $f_c + \Delta f$ is shut off. During such periods, the phase lock loop can drift, or if bandpass filters are used, a transient is experienced. These degrading effects get worse as the number of consecutive 1's or 0's increases. The phenomenon is known as *static data*.

Incoherent Detector

The problem of static data leads us to consider the incoherent detector. One simple implementation of this detector is shown in Fig. 11.19. The detector contains two bandpass filters, one tuned to each of the two frequencies used to communicate 0's and 1's. The output of the filter is envelope detected and then baseband detected using an *integrate-and-dump* operation. In intuitive terms, this detector is simply evaluating which of two possible sinu-

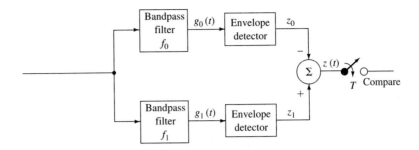

Figure 11.19   Incoherent detector for FSK.

soids is stronger at the receiver. Note that if we take the difference of the outputs of the two envelope detectors, the result is bipolar baseband.

There are other forms of incoherent detector. One example includes a discriminator followed by an envelope detector. The discriminator output has an amplitude proportional to the input frequency. The output of the envelope detector is therefore a baseband signal consisting of two different positive levels. This signal can be detected using any of the baseband techniques.

11.2.5 Detector Performance

Matched Filter Detector

We begin by analyzing the performance of the coherent matched filter, or correlation, detector. The bit error rate is given by

$$P_e = \frac{1}{2}\,\text{erfc}\left(\sqrt{\frac{E(1-\rho)}{2N_0}}\right) \tag{11.14}$$

where E is the average energy of the two signals and ρ is the correlation coefficient.

We need simply find E and ρ. The signal is a sinusoid of amplitude A, regardless of whether a 1 or a 0 is being transmitted. The average energy per bit is then $A^2 T_b/2$. The correlation coefficient is given by

$$\rho = \frac{A^2 \displaystyle\int_0^{T_b} \cos\left[2\pi(f_c - \Delta f)\right]\cos\left[2\pi(f_c + \Delta f)\right]dt}{A^2 T_b/2} \tag{11.15}$$

$$= \frac{1}{T_b}\left(\int_0^{T_b} \cos 4\pi f_c t\,dt + \int_0^{T_b} \cos 4\pi \Delta f t\,dt\right)$$

Since $f_c \gg \Delta f$, we normally assume that the first integral of Eq. (11.15) is negligible (i.e., when a sinusoid is integrated, we divide by the frequency). We then have

$$\rho \approx \frac{1}{T_b}\int_0^{T_b} \cos 4\pi \Delta ft\, dt = \frac{\sin 4\pi \Delta fT_b}{4\pi \Delta fT_b} \tag{11.16}$$

It is desirable to make ρ as small as possible. To minimize ρ, we evaluate the integral and then take its derivative with respect to Δf in order to find the best choice of frequency spacing. The result is

$$4\pi \Delta fT_b = \tan 4\pi \Delta fT_b \tag{11.17}$$

and

$$2\Delta f = \frac{0.715}{T_b} = 0.715\, R_b \tag{11.18}$$

The two frequencies are at $f_c \pm \Delta f$, so the separation is $0.715R_b$. We substitute this frequency separation back into Eq. (11.16) to get $\rho = -0.22$. Other values of frequency separation lead to larger values of ρ. The separation of Eq. (11.18) would therefore seem to be a good design choice. However, we show in Problem 11.2.6 that the resulting error probability is sensitive to phase with this selection of frequency deviation. That is, we assumed in the derivation that the two sinusoids were matched in phase. If they are not perfectly matched, ρ increases and can increase beyond zero to become positive. It is very difficult to match the phases of two oscillators. Additionally, if a VCO is used to produce FSK, the starting phase for each bit interval will vary.

On the other hand, if we choose orthogonal tone spacing—that is, if $\Delta f = R_b/2$—Eq. (11.16) reduces to $\rho = 0$. Note that if Δf is $R_b/2$, the spacing between the two frequencies is R_b. We show in Problem 11.2.7 that this result is true regardless of the phase relationship between the two signals. The performance of the detector is insensitive to phase differences between $s_0(t)$ and $s_1(t)$. Orthogonal tone spacing therefore represents the more conservative approach to design, and it is almost always used. Note that the bandwidth is smaller for orthogonal tone spacing than for the spacing of Eq. (11.18).

For orthogonal tone spacing, the performance is then given by

$$P_e = \frac{1}{2}\,\text{erfc}\left(\sqrt{\frac{E(1-\rho)}{2N_0}}\right) = \frac{1}{2}\,\text{erfc}\left(\sqrt{\frac{A^2T_b}{4N_0}}\right) \tag{11.19}$$

We plot this result in Fig. 11.20. The abscissa of the plot is $A^2T_b/2N_0$, the signal to noise ratio, expressed in dB.

Incoherent Detector

We now turn our attention to the performance of the incoherent FSK detector. The nonlinearities in the incoherent detector make the analysis more complex than that of the coherent detector. The reason is that the various random quantities are no longer Gaussian distributed.

The probability analysis of the incoherent FSK detector exhibits symmetry, unlike that of the incoherent ASK detector. Due to this symmetry, the probability of erroneously detecting a transmitted 1 as a received 0 (a miss) is the same as the probability of detecting a transmitted 0 as a received 1 (a false alarm). Thus, the probability of bit error is equal to either one of these probabilities, and we need only calculate one of them. As we mentioned in the previous section, when we sample the envelope detector outputs, the resulting random variable distributions are Rayleigh and Ricean, depending upon whether or not a signal is present. The bit error rate is

$$P_e = \frac{1}{2}\exp\left(-\frac{A^2 T_b}{2N_0}\right) \tag{11.20}$$

The assumptions necessary to derive this result are that the signal to noise ratio is high enough so that we are operating far out on the tails of the Ricean density and that the bandpass filters of the incoherent detector block the incorrect signal totally. That is, the filter tuned to $f_c + \Delta f$ totally blocks a signal at a frequency of $f_c - \Delta f$.

We now wish to compare the performance of the incoherent detector with that of the coherent detector. That is, we compare Eq. (11.19) with Eq. (11.20). Figure 11.21 shows the two error probabilities as a function of the signal to noise ratio. The curve for the coherent detector is repeated from Fig. 11.20.

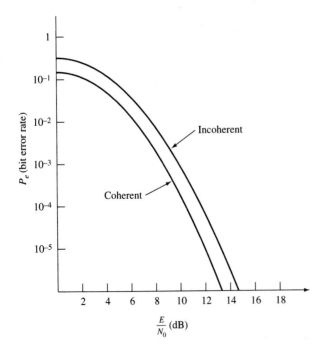

Figure 11.20   Performance of binary FSK with orthogonal tone spacing.

Example 11.3

Find the probability of error of the incoherent detector when it is used for FSK, where the bit period is 2 sec and the signals have an amplitude of 0.4 volt. The two frequencies are 1 kHz and 2 kHz. The additive noise has power of 10^{-2} watt/Hz. Compare the performance of the incoherent detector with that of the coherent detector.

Solution: Let us first determine whether the results of Fig. 11.21 apply to this problem. If they do, we can simply read the answer from the graph. In the derivation of those results, the only assumptions (in the incoherent case) were that the signal frequencies were sufficiently

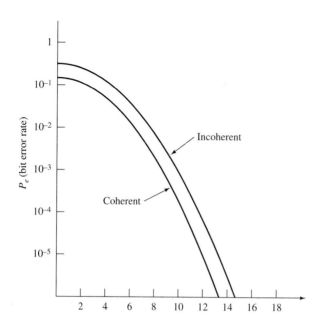

Figure 11.21   Performance comparisons for FSK.

separated that the bandpass filters rejected the alternative frequency and that the signal to noise ratio is high enough to yield a small bit error rate. The bandwidth of the filters would be set to 1 Hz (i.e., $2R_b$), and the frequencies are separated by 1 kHz, so the first assumption is clearly valid. The second assumption will be tested after we derive a bit error rate.

The signal to noise ratio, E/N_0, is $A^2T_b/2N_0$, or 16 (12 dB). We then read from Fig. 11.21 [or Eq. (11.20)] that the probability of error for the incoherent detector is 1.7×10^{-4}.

We now view the curve for the coherent case. The assumption we made in order to derive that curve was orthogonal tone spacing. The power spectral density of each signal passes through zero at spacings of $\frac{1}{2}$ Hz away from the carrier. Since the spacing between the two frequencies is 1 kHz, and this is a multiple of $\frac{1}{2}$, the assumption of orthogonal tone spacing is valid. In actuality, the spacing is so large that, even if it were not a multiple of R_b, the correlation between signals would still be essentially zero. Thus, the probability of error is read from Fig. 11.21 [or Eq. (11.19)] as 3.8×10^{-5}. This is 4.5 times better than that of the incoherent detector.

M-ary FSK Performance

We first analyze the performance of the coherent M-ary FSK, or MFSK, detector. The correlator form of this detector is shown in Fig. 11.22. The system consists of M filters, each matched to one of the M sinusoidal bursts being used. Note that since the matched filter outputs are compared to each other, we could eliminate the amplitude factor A from each of the multipliers without affecting the resulting performance.

The period of integration is T_S, the *symbol period*. This is the duration of a frequency burst, and it consists of more than one bit period. For example, if three bits are combined to form eight different symbols in 8-ary FSK, the symbol period is three times the bit period.

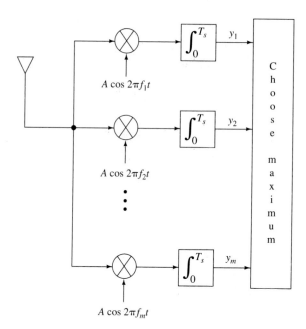

Figure 11.22   Matched filter MFSK detector.

Suppose now that the frequency k is being transmitted. The received waveform is then of the form

$$r(t) = A\cos 2\pi f_k t + n(t) \tag{11.21}$$

The additive noise is assumed to be white and Gaussian. With this as the input to the receiver, the various matched filter outputs are given by

$$y_i = \int_0^{T_s} [A\cos 2\pi f_k t + n(t)]A\cos 2\pi f_i t \, dt \tag{11.22}$$

If we assume that the sinusoidal bursts are orthogonal to each other (i.e., if $f_i - f_k = n/2T_s$), then

$$y_i = \begin{cases} \dfrac{A^2 T_S}{2} + \displaystyle\int_0^{T_s} An(t)\cos 2\pi f_i t \, dt & i = k \\[4mm] \displaystyle\int_0^{T_s} An(t)\cos 2\pi f_i t \, dt & i \neq k \end{cases} \tag{11.23}$$

The y_i are Gaussian random variables whose mean values are given by

$$E\{y\}_i = \begin{cases} \dfrac{A^2 T_S}{2} & i = k \\[4mm] 0 & i \neq k \end{cases} \tag{11.24}$$

The variance is found in the same way we analyzed the binary matched filter detector:

$$\sigma^2 = E\left\{ \int_0^{T_s} \int_0^{T_s} A^2 n(t) n(\tau) \cos 2\pi f_i t \cos 2\pi f_i \tau \, dt \, d\tau \right\}$$

$$= \frac{A^2 N_0}{2} \int_0^{T_s} \int_0^{T_s} \delta(t - \tau) \cos 2\pi f_i t \cos 2\pi f_i \tau \, dt d\tau \qquad (11.25)$$

$$= \frac{A^2 N_0}{2} \int_0^{T_s} \cos^2 2\pi f_i t dt = \frac{A^2 N_0 T_s}{4}$$

The probability density functions are sketched in Fig. 11.23.

The probability that any particular matched filter output y_i (where $i \neq k$) is greater than y_k is

$$\Pr\{y_i > y_k\} = \frac{1}{\sqrt{2\pi}\sigma} \int_{A^2 T_s/4}^{\infty} e^{-y^2/2\sigma^2} \, dx$$

$$= \frac{1}{2} \operatorname{erfc} \sqrt{\frac{A^2 T_s}{4 N_0}} = \frac{1}{2} \operatorname{erfc} \sqrt{\frac{E_S}{2 N_0}} \qquad (11.26)$$

where E_S, the symbol energy, is $\log_2 M$ multiplied by the energy per bit. Thus, for example, in 8-ary FSK, bits are combined in groups of three. The energy per bit is then one-third of the symbol energy.

To find the overall probability of symbol error, we need to look at joint probabilities. For example, assume 16-FSK with frequency number two being transmitted. The probability of error is then

$$\Pr\{y_i > y_2 \text{ OR } y_3 > y_2 \text{ OR } y_4 > y_2\} \qquad (11.27)$$

If the three events in brackets were independent, the overall probability would be the sum of the separate probabilities. If they are not independent, the sum of the separate probabilities provides an upper bound (known as the *union bound*) on the overall probability. We can therefore bound the symbol error probability by

$$P_E \leq \frac{1}{2}(M - 1) \operatorname{erfc} \sqrt{\frac{E_S}{2 N_0}} \qquad (11.28)$$

Notice that we use an uppercase E as the subscript of P to distinguish the symbol error rate from the *bit* error rate. Figure 11.24 shows this bound on the probability of error as a function of the bit signal to noise ratio. The figure is therefore a plot of

$$P_E = \frac{1}{2}(M - 1) \operatorname{erfc} \sqrt{\left(\frac{E_b}{N_0}\right) \frac{\log_2 M}{2}} \qquad (11.29)$$

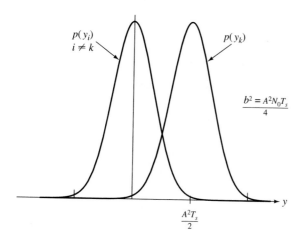

Figure 11.23 Probability density functions of matched filter outputs.

We have been addressing *symbol errors*. We would like to translate their probability into that of *bit errors* so that we can compare MFSK with binary FSK. One symbol error translates to one *or more* bit errors. For example, consider an 8-ary FSK system. Suppose we code the bit sequence 110 into the frequency f_6. Now suppose there is a symbol error and we erroneously detect frequency f_1. We would convert this to 001, thus making three bit errors. On the other hand, if f_6 were erroneously detected as f_7, we would only make one bit error.

Of the seven ways we can make an error, three represent single bit errors, three represent double bit errors, and one represents a triple bit error. If all symbol errors are assumed to be equally likely, the average number of bit errors per symbol error would be

$$\frac{\text{bit errors}}{\text{symbol errors}} = \frac{3}{7} \times 1 + \frac{3}{7} \times 2 + \frac{1}{7} \times 3 = \frac{12}{7} \tag{11.30}$$

Since this is the number of bit errors out of a three-bit combination, the bit error rate is 4/7 times the symbol error rate. In general, the bit error rate is related to the symbol error rate by

$$\frac{\text{bit error rate}}{\text{symbol error rate}} = \frac{P_e}{P_E} = \frac{M/2}{M-1} \tag{11.31}$$

A similar approach is used to find the error rate for the incoherent detector. That detector substitutes bandpass filters and envelope detectors for the matched filters of Fig. 11.22. The symbol error rate is bounded by

$$P_E \le \frac{M-1}{2} \exp\left(-\frac{E_S}{2N_0}\right) \tag{11.32}$$

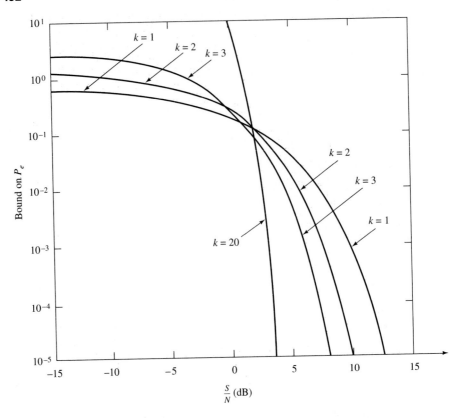

Figure 11.24   Symbol error probability bound.

11.3 PHASE SHIFT KEYING

In *analog communication*, frequency modulation and phase modulation are very similar. The frequency of a waveform is the derivative with respect to time of the instantaneous phase, so varying frequency also varies phase.

In *digital communication*, the distinction between frequency modulation and phase modulation is more significant. This is because digital information signals are drawn from a discrete set of waveforms.

In phase modulation, just as in amplitude and frequency modulation, we start with a sinusoidal carrier of the form

$$s_c(t) = A\cos\theta(t) \tag{11.33}$$

In frequency modulation, the derivative of $\theta(t)$, which is proportional to the instantaneous frequency, follows the baseband or information signal. In phase modulation, it is the phase itself that follows the baseband signal. Thus,

$$\theta(t) = 2\pi[f_c t + k_p s(t)] \tag{11.34}$$

where f_c is the carrier frequency, k_p is the proportionality factor relating the phase shift to the signal voltage, and $s(t)$ is the baseband information signal.

In *binary frequency shift keying* (BFSK), we switch back and forth between sinusoids of two different frequencies, depending upon whether a 1 or a 0 is being transmitted. In *binary phase shift keying* (BPSK), the frequency of the carrier stays constant while the phase shift takes on one of two constant values.

The two signals used to transmit a 0 and a 1 are expressed as

$$s_0(t) = A\cos(2\pi f_c t + \theta_0)$$
$$s_1(t) = A\cos(2\pi f_c t + \theta_1)$$
(11.35)

where θ_0 and θ_1 are constant phase shifts.

Given any values of θ_0 and θ_1, we can add a constant phase shift θ to the carrier and rewrite Eq. (11.35) as

$$\theta = \frac{\theta_0 + \theta_1}{2}$$

$$\Delta\theta = \frac{\theta_1 - \theta_0}{2}$$
(11.36)

$$s_0 = A\cos(2\pi f_c t + \theta - \Delta\theta)$$

$$s_1 = A\cos(2\pi f_c t + \theta + \Delta\theta)$$

We can let $\theta = 0$ and express this signal as

$$s_i(t) = A\cos[2\pi f_c t + \Delta\theta d_i(t)]$$
(11.37)

where $d_i(t)$ is the data sequence consisting of $+1$ or -1 (i.e., normalized NRZ bipolar) and $\Delta\theta$ is the phase deviation, also known as the *modulation index*. We use simple trigonometric expansions to express Eq. (11.37) as

$$s_i(t) = A\cos(2\pi f_c t)\cos[\Delta\theta d_i(t)]$$
$$- A\sin(2\pi f_c t)\sin[\Delta\theta d_i(t)]$$
(11.38)

We use the even and odd properties of the cosine and sine to reduce this further to

$$s_i(t) = A\cos(\Delta\theta)\cos(2\pi f_c t) - A d_i(t)\sin(\Delta\theta)\sin(2\pi f_c t)$$
(11.39)

The first term in Eq. (11.39) is the *residual carrier* (it doesn't depend on the data being sent), and it has a power of

$$P_c = \frac{A^2\cos^2(\Delta\theta)}{2}$$
(11.40)

The second term represents the *modulated information signal*, or sidebands, and the power of this term is

$$P_d = \frac{A^2\sin^2(\Delta\theta)}{2}$$
(11.41)

The total transmitted power is the sum of these two terms, or $A^2/2$. This result is as expected, since we transmit a pure sinusoid of amplitude A regardless of the specific data being sent. The energy contained in the signal used to transmit each bit is

$$E_b = \frac{A^2 T_b}{2} \tag{11.42}$$

For this reason, the amplitude is sometimes written in terms of the bit energy as

$$A = \sqrt{\frac{2E_b}{T_b}} \tag{11.43}$$

If the modulation index, $\Delta\theta$, is set equal to $\pi/2$, then Eq. (11.39) reduces to

$$s_i(t) = Ad_i(t)\sin(2\pi f_c t) \tag{11.44}$$

This represents the *suppressed carrier* case, and the two signals are the negative of each other, that is,

$$s_1(t) = -s_0(t) \tag{11.45}$$

We call such signals *antipodal*. This is the best choice of phase angle, since it achieves a minimum bit error rate. (We prove this later.) In this case, the two signals are 180° out of phase with each other. They can be represented as shown in Fig. 11.25.

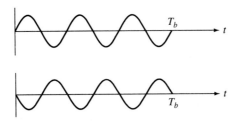

Figure 11.25 Binary PSK (BPSK) waveforms.

As we study complex PSK systems, and in particular, as we generalize from binary to M-ary, an alternative representation will often prove useful. This representation is known as *signal space*. A signal space diagram is a vector representation that illustrates the complex projection of the transmitted signal in the direction of two orthogonal normal signals (i.e., generalized unit vectors). Returning to the general case, we find that the two signals are

$$s_0(t) = A\cos(2\pi f_c t + \theta_0)$$
$$s_1(t) = A\cos(2\pi f_c t + \theta_1) \tag{11.46}$$

These can be expressed in complex notation as

$$s_0(t) = \mathrm{RE}\{Ae^{j2\pi f_c t}e^{j\theta_0}\}$$
$$s_1(t) = \mathrm{RE}\{Ae^{j2\pi f_c t}e^{j\theta_1}\} \tag{11.47}$$

We now shift the frequency down to baseband. This is sometimes known as *despinning*, since you can think of the complex vector [the portion in braces in Eq. (11.47)] as composed of variations around a periodic vector that is spinning around the origin with frequency f_c. The signal space diagram is a plot of the complex part of Eq. (11.47) with the periodic component suppressed.

We illustrate the signal space representation in Fig. 11.26(a). Figure 11.26(b) shows the signal space representation for the suppressed carrier case. The distance of each point from the origin is the square root of the signal energy per bit. We draw the points this way to normalize the diagram. The energy per bit is $A^2 T_b/2$, where T_b is the bit period—the length of each sinusoidal burst used to send a single bit. The distance between the two points in the signal space diagram is $2\sqrt{E}$, or $A\sqrt{2T_b}$. This proves to be an important parameter in measuring the performance of the system. As may be intuitively expected, the greater the distance between points in the signal space representation, the smaller is the probability of bit error. The distance represents the degree of dissimilarity between the two signals.

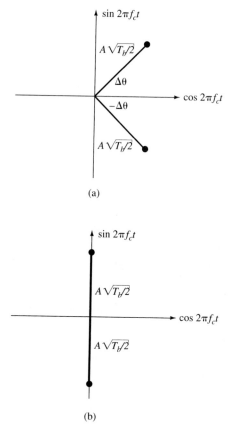

(a)

(b)

Figure 11.26   Signal space representation.

11.3.1 BPSK Spectrum

The BPSK signal can be considered a superposition of two ASK waveforms, and the bandwidth of the resulting waveform can be found by examining the component parts. In the suppressed carrier case, PSK is equivalent to taking the difference between two OOK signals. The first is the OOK signal resulting from ASK using the data signal. The second results from OOK using the complement of the data signal. The frequency spectrum is therefore found in the same manner as illustrated in Section 11.1.1. The power spectral density of the BPSK waveform, assuming random data, is then as shown in Fig. 11.27. Note that the first zero of the spectrum occurs at a distance R_b, the bit rate, away from the carrier frequency. Therefore, the nominal bandwidth of PSK is $2R_b$.

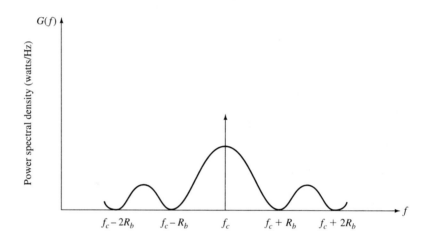

Figure 11.27   BPSK power spectral density.

11.3.2 Quaternary PSK

The nominal bandwidth of a BPSK signal is twice the bit rate. Suppose this bandwidth is too large to transmit through the channel. One approach to dealing with the problem is to reduce the bit rate R_b. But this is often not acceptable. Instead, we can effect a reduction in bandwidth without reducing the bit rate if we leave the realm of binary communication. For example, if bits are combined into pairs, each grouping can take on one of four possible values. We can transmit these values using *quaternary phase shift keying* (QPSK). Since the phase of the transmitted carrier changes once every $2T_b$, the nominal bandwidth is *half that of BPSK*.

QPSK can be viewed as the superposition of two BPSK signals, one modulating a cosine carrier and one modulating a sine carrier. The four transmitted signals are separated by 90° from each other. Figure 11.28(a) shows one technique for generating QPSK. The demultiplexer is a type of decommutator that splits the incoming data waveform into an odd and an even part. That is, the first bit takes the top path, the second takes the bottom, the third takes the top, and so on. The demultiplexer contains a delay so that the pairs of bits are simultaneously fed to the two modulators. Figure 11.28(b) shows the operation of

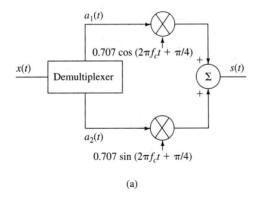

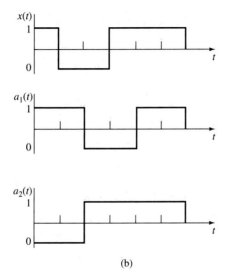

(b)

Figure 11.28   QPSK Generation.

the demultiplexer for a data signal $d(t)$ representing the binary sequence 100111. The output signal, $s_i(t)$, has one of four possible phases. Trigonometric identities should convince you that it takes on the form in the rightmost column of the following table, given the values of d_o and d_e in the other two columns:

d_o	d_e	$s_i(t)$
+1	+1	$+\cos 2\pi f_c t$
+1	-1	$-\sin 2\pi f_c t$
-1	-1	$-\cos 2\pi f_c t$
-1	+1	$+\sin 2\pi f_c t$

The signal space diagram of QPSK is shown in Fig. 11.29. The phase changes after every two input bits. At each transition, the phase can change by either $\pm 90°$ or $\pm 180°$. Transitions of $\pm 180°$ occur if both bits change from one $2T_b$ interval to the next.

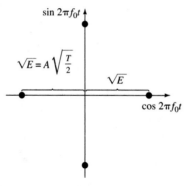

Figure 11.29 Signal space diagram for QPSK.

Abrupt phase changes of as much as 180° are difficult for a demodulator to track. A variation of QPSK is therefore often used. In this variation, the even bits are *not* delayed, allowing the phase to change after *every bit interval* instead of after every second bit interval. Figure 11.30 shows the waveforms corresponding to those of Fig. 11.28(b) where the delay has been eliminated. This approach is known as *offset QPSK* (OQPSK), or *staggered QPSK*.

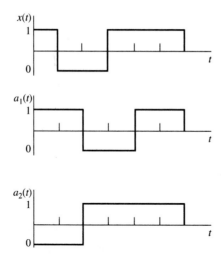

Figure 11.30 Waveforms for OQPSK.

The generator for OQPSK is shown in Fig. 11.31. Typical waveforms are illustrated in Fig. 11.32 for the data stream $d_i(t) = 1, 1, -1, -1, -1, 1, 1, 1$. At each transition, only one of the component signals can change. Thus, the phase can change only by $\pm 90°$ at a transition and not by $\pm 180°$. This suppression of large phase jumps leads to better performance under certain practical conditions. In particular, the modification can significantly improve performance if the channel is bandlimited or nonlinear (e.g., when a hard limiter is present). The reason is that the envelope of the waveform does not go to zero in the OQPSK case, as it does in the QPSK case.

Although the phase in OQPSK changes at a rate of once every T_b, the bandwidth of the resulting signal is still one-half that of binary PSK. OQPSK is the superposition of two half-rate PSK signals. The fact that one of the two signals is offset in time does not affect the power spectral density.

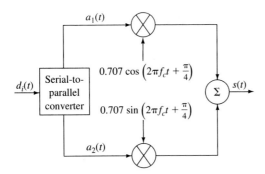

Figure 11.31   OQPSK generator.

11.3.3 Minimum Shift Keying

In either QPSK or OQPSK, abrupt phase changes occur at bit interval transitions. In the case of QPSK, these changes are 90° or 180° in magnitude, and they occur at one-half the bit rate. In OQPSK, the magnitude of the changes is 90°, and they occur at a rate equal to the bit rate. These abrupt changes result in side lobes of the power spectral density. Such signals are distorted when they pass through a bandlimited channel, causing intersymbol interference.

OQPSK exhibits improvement over QPSK because the phase transitions are reduced in magnitude. It would seem intuitively reasonable that further reduction in these abrupt phase changes would improve performance even more. *Minimum shift keying* (MSK) is a method of eliminating the abrupt phase changes occurring in QPSK or OQPSK. The instantaneous phase of the MSK signal is continuous. The method involves a trade-off between the width of the main lobe of the power spectrum and the power content of the side lobe. In fact, the power spectrum of MSK has a main lobe that is 1.5 times as wide as the main lobe in QPSK. The MSK side lobes are much smaller than the QPSK side lobes.

MSK applies sinusoidal weighting to the OQPSK baseband signals prior to modulation. The even and odd waveforms of Fig. 11.32 are therefore modified to become sinusoidal pulses. This is illustrated in Fig. 11.33 for the binary data signal

$$d_i(t) = 1, 1, -1, -1, -1, 1, 1, 1.$$

We then use the system of Fig. 11.31 to obtain the MSK waveform

$$s_i(t) = d_e(t)\cos\left(\frac{\pi t}{2T_b}\right)\cos 2\pi f_c t$$
$$+ d_o(t)\sin\left(\frac{\pi t}{2T_b}\right)\sin 2\pi f_c t \tag{11.48}$$

Figure 11.34 shows the waveform represented by Eq. (11.48). Also shown in the figure is a sketch of the instantaneous phase of the waveform. The instantaneous phase is found by combining the two terms of Eq. (11.48) into a single sinusoid as follows:

$$s_i(t) = d_e(t)\cos\left(2\pi f_c t - d_e(t)d_o(t)\frac{\pi t}{2T_b}\right) \tag{11.49}$$

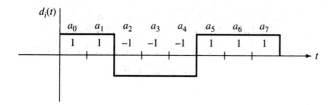

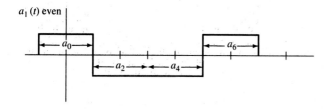

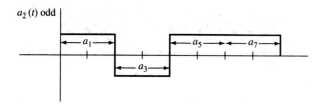

Figure 11.32 Typical waveforms for OQPSK.

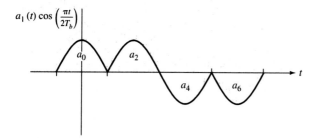

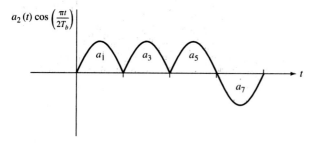

Figure 11.33 MSK generation.

The instantaneous phase is therefore linear (a ramp), as shown in Fig. 11.34(b).

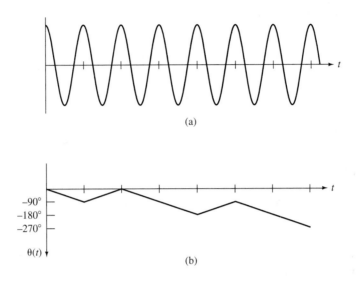

(a)

(b)

Figure 11.34   MSK waveform.

11.3.4 Differential PSK

Demodulation of suppressed carrier PSK is difficult because we must reproduce the exact carrier frequency and phase. An alternative is to compare the received waveform with a delayed version of itself. This is known as *differential PSK* (DPSK). In DPSK, we use the coherence of the carrier in the modulator to observe changes from one bit interval to the next. We therefore are checking for *changes* in the phase of the received signal. This technique does not allow us to observe the absolute phase of the arriving signal. To use the technique, the information must be superimposed upon the carrier in a *differential* manner.

   If the original signal is coded using one of the differential forms (e.g., NRZ-M or NRZ-S), then the digital information is contained in changes between adjacent bit intervals. If we can detect changes in phase between these intervals, we can decode such differential information.

11.3.5 Modulators

Analog phase modulators can be used for PSK modulation. However, practical analog modulators cannot respond instantaneously to discontinuities in the input waveform. The output phase therefore does not abruptly change from one value to another.

   Conceptually, we can envision a keying system as in Fig. 11.35. Here, we show two sinusoidal generators with the two specified phase shifts. The switch is positioned depending on the data signal to be transmitted.

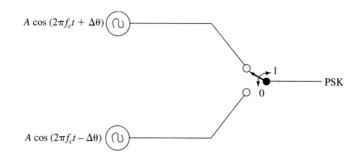

Figure 11.35   Ideal PSK modulator.

In the case of antipodal signals, we can use a suppressed carrier amplitude modulator with an input of $+1$ or -1, as shown in Fig. 11.36. The operation can be accomplished with a gating circuit and a phase splitter. The phase splitter provides two sinusoidal signals of opposite polarity.

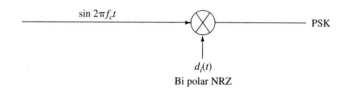

Figure 11.36   AM modulator used for antipodal PSK.

11.3.6 Demodulators

We examine only the coherent detector for PSK; *there is no incoherent form of detector.* This is true because incoherent detectors retain information only on amplitude—they throw away that on phase. If, for example, a PSK waveform forms the input to an envelope detector, the output is a constant that is *independent of the data being transmitted.*

The general form of the coherent detector is the matched filter detector shown in Fig. 11.37. The challenge in building this detector is to reconstruct the correctly phased carriers at the receiver. If the transmitted PSK signal is of the form of transmitted carrier (i.e., $\Delta\theta \neq \pi/2$), the carrier can be recovered using a bandpass filter with a very narrow pass band or a phase-lock loop. This is shown in Fig. 11.38.

The input to the detector is

$$s_i(t) = A\cos\beta \sin(2\pi f_c t) + Ad_i(t)\sin\beta \cos(2\pi f_c t) \qquad (11.50)$$

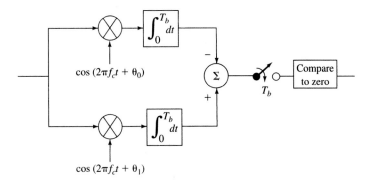

Figure 11.37   Matched filter detector for BPSK.

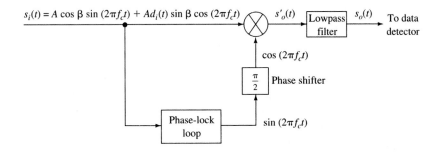

Figure 11.38   Carrier recovery in PSK.

Equation (11.50) is of the form of Eq. (11.39). The output of the lowpass filter is then given by

$$s_o(t) = \frac{A}{2} d_i(t)\sin \beta \qquad (11.51)$$

Note that this is an approximation, since $d_i(t)$ cannot pass undistorted through the lowpass filter because it contains discontinuities that lead to high-frequency components. The output of the filter will be a smoothed version of the data signal.

In the suppressed carrier case, $\beta = \pi/2$, and the output of Eq. (11.51) goes to zero. In this case, we must resort to more complex forms of carrier recovery. One such class of systems is the nonlinear loop; we concentrate on the squaring loop. The signal can be written as

$$s_i(t) = Ad_i(t)\cos (2\pi f_c t) \qquad (11.52)$$

Suppose we square this signal. The result is

$$s_i^2(t) = A^2\cos^2(2\pi f_c t)$$

$$= \frac{A^2}{4} + \frac{A^2}{4} \cos (4\pi f_c t) \qquad (11.53)$$

A bandpass filter or phase-lock loop can be used to extract the pure carrier term at a frequency of $2f_c$ and the exact phase of the transmitted signal [when $d_i(t) = +1$]. Frequency divider circuitry can then be employed to generate a signal at the original carrier frequency. The process is illustrated in Fig. 11.39.

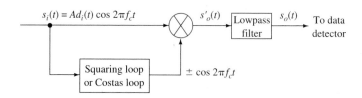

Figure 11.39   Squaring loop for carrier recovery.

The squaring loop introduces a phase ambiguity of π radians, or 180°. This is because the double-frequency recovered carrier has a phase ambiguity of 2π radians. That is, you can add 2π to the phase without changing the waveform. When the frequency is divided by two, the phase is also divided by two, creating a phase ambiguity of π radians. We therefore have not completely recovered the carrier; rather, we have just narrowed it down to two possibilities that differ in sign. This is unacceptable, since an incorrect guess would result in the detector output yielding the exact complement of the binary data signal. Fortunately, the ambiguity can be resolved by sending a known training sequence called a *preamble*. If the first few bits of a message are known, they can be used to resolve the phase ambiguity in the receiver.

A form of squaring loop is the Costas loop, shown in Fig. 11.40. Note that the VCO operates at the carrier frequency, f_c, rather than at twice that frequency, as in the circuit of

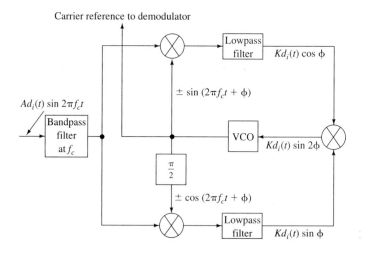

Figure 11.40   Costas loop.

Fig. 11.39. The loop is analogous to the squarer followed by a phase lock loop. The output of the upper lowpass filter is

$$\frac{1}{2} A d_i(t) \sin (- \phi) \tag{11.54}$$

This output is proportional to the sine of the phase difference, ϕ. If the VCO frequency is not matched to the input frequency, ϕ includes a linearly increasing term.

The output of the lower lowpass filter is

$$\frac{1}{2} A d_i(t) \cos (- \phi) \tag{11.55}$$

This output is proportional to the cosine of the phase difference. When the two output terms are multiplied together, the result is the error term,

$$\begin{aligned} e(t) &= \frac{1}{4} A^2 d_i^2(t) \sin (- \phi) \cos (- \phi) \\ &= \frac{1}{4} A^2 \sin (- 2\phi) \end{aligned} \tag{11.56}$$

The loop attempts to drive this error to zero, resulting in a phase difference ϕ of zero. The carrier can then be recovered from the output of the VCO.

The Costas loop can be thought of as comprising two phase lock loops fed by a single VCO. The inputs to the multiplier in the Costas loop are the sine and cosine of the phase difference between the input carrier and the VCO output. Since the loop is basically a squaring loop, it suffers from the π phase ambiguity discussed earlier. The loop also experiences the practical problems associated with all loops, including long acquisition time and the possibility of a false lock. For this reason, we often use DPSKs, as described in Section 11.3.4.

The coherent detector for QPSK is a receiver consisting of four matched filters, or correlators, as shown in Fig. 11.41. Since

$$\cos\left(2\pi f_c t + \frac{\pi}{2}\right) = - \sin (2\pi f_c t) = - \cos\left(2\pi f_c t + \frac{3\pi}{2}\right) \tag{11.57}$$

the detector could have been simplified to contain only two correlators. In that case, a two-step comparison would be required, first, to find the output with the largest magnitude, and second, to test the sign of that output.

While we do not derive performance equations for QPSK in this text, we do present curves of the results in Chapter 12, when we discuss design trade-offs.

DPSK

Figure 11.42 shows a differential PSK demodulator. If the signals being compared by the multiplier are identical, the input to the integrator is the square of the signal segment. Thus, provided that no change occurs between the two adjacent bit intervals, the output of the integrator is

$$\int_0^{T_b} A^2 \cos^2(2\pi f_o t + \theta_i) dt = \frac{A^2 T_b}{2} \tag{11.58}$$

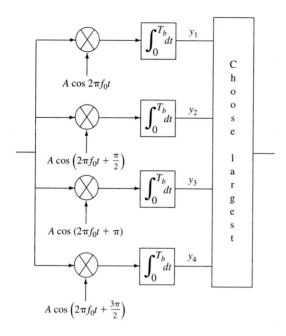

Figure 11.41 Coherent detector for QPSK.

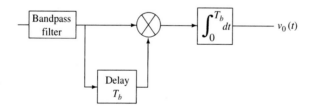

Figure 11.42 Differential PSK demodulator.

If a change does occur between the two adjacent bit intervals, the output of the integrator is

$$A^2 \int_0^{T_b} \cos (2\pi f_0 t + \theta_0) \cos (2\pi f_0 t + \theta_1) dt \approx \frac{A^2 T_b}{2} \cos (\theta_0 - \theta_1) \qquad (11.59)$$

If the phase difference between the two signals is 180° (suppressed carrier), the integrator output becomes $\pm A^2 T/2$. Note that the integrator could be replaced by a lowpass filter with no change in the results.

Thus, the output of the integrator is a bipolar baseband signal, with the positive value obtaining if no change occurs between one interval and the next and the negative value obtaining if a change does occur. If the original data were coded using NRZ-S prior to phase modulation, the output of the demodulator would be the NRZ-L baseband signal.

If the NRZ-M format were used, the output would be the inverse of the original baseband signal. The resulting baseband signal then forms the input to a baseband detector, and decisions are made to recover the original data signal.

11.3.7 Performance

Matched Filter Detector

The performance of the matched filter (coherent) detector is completely specified by the three parameters ρ, N_0, and E. ρ is the correlation coefficient, and E is the average energy of the signals used to transmit a 1 and a 0. The bit error rate of the matched filter detector is

$$P_e = \frac{1}{2}\, \mathrm{erfc}\left(\sqrt{\frac{E}{N_0}}\right)$$

$$= Q\left(\sqrt{\frac{2E}{N_0}}\right) \tag{11.60}$$

In phase shift keying, the two signals are

$$s_0(t) = A\cos(2\pi f_c t + \theta_0) \tag{11.61}$$
$$s_1(t) = A\cos(2\pi f_c t + \theta_1)$$

The average energy is then $A^2 T_b/2$, and the correlation coefficient is

$$\rho = \frac{1}{A^2 T_b/2} \int_0^{T_b} A\cos(2\pi f_c t + \theta_0) A\cos(2\pi f_c t + \theta_1)\, dt$$

$$\approx \cos(\theta_0 - \theta_1) \tag{11.62}$$

The double-frequency term has been omitted from Eq. (11.62) because we assume that T_b is large compared to the period of the sinusoidal signals.

    The complementary error function decreases as its argument increases. Therefore, the probability of error is monotonically increasing in ρ, and we can minimize it by making $\rho = -1$. This value obtains if the phase difference between the two signals is 180°, which occurs in the suppressed carrier case. Such a result is not surprising, since the signals are then as different as possible, one being the negative of each other. The resulting error probability is

$$P_e = \frac{1}{2}\, \mathrm{erfc}\left(\sqrt{\frac{A^2 T_b}{2N_0}}\right)$$

$$= Q\left(\sqrt{\frac{A^2 T_b}{N_0}}\right) \tag{11.63}$$

This curve is plotted in Fig. 11.43.

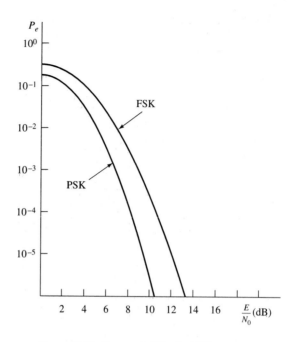

Figure 11.43 Error probability for PSK.

Differentially Coherent Detector Performance

Since the differential detector of Fig. 11.42 multiplies the signal by a delayed version of it-self, it is not surprising to find that we are dealing with nonlinear operations upon the noise. Accordingly, the results follow those of the other nonlinear detectors, such as the in-coherent ASK and FSK detectors.

The simplified detector for DPSK is repeated in Fig. 11.44. Let us assume that the signals in the two consecutive bit intervals are identical. Then the input to the filter (or in-tegrator) is

$$g(t) = [A\cos 2\pi f_c t + n_{nb}(t)][A\cos 2\pi f_c t_d + n_{nb}(t_d)] \tag{11.64}$$

where n_{nb} is the narrowband noise and $t_d = t - T_b$ the time delayed by one bit period. We can expand $g(t)$ into four terms by using the quadrature expansion of the narrowband noise:

$$\begin{aligned}
g(t) = &[A + x(t)]\cos 2\pi f_c t [A + x(t_d)]\cos 2\pi f_c t_d \\
&- [A + x(t)]\cos 2\pi f_c t\, y(t)\sin 2\pi f_c t_d \\
&- y(t)\sin 2\pi f_c t [A + x(t_d)]\cos 2\pi f_c t_d \\
&+ y(t)y(t_d)\sin 2\pi f_c t\, \sin 2\pi f_c t_d
\end{aligned} \tag{11.65}$$

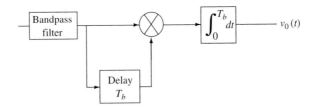

Figure 11.44   DPSK receiver.

Now let us assume that the bit period is a multiple of the period of the carrier. This is necessary for the delayed version to align with the nondelayed version. With this assumption,

$$\cos(2\pi f_c t) = \cos(2\pi f_c t_d)$$
$$\sin(2\pi f_c t) = \sin(2\pi f_c t_d)$$

(11.66)

We make these substitutions in Eq. (11.65) and expand the sinusoidal products using trigonometric identities. We then identify the low-frequency terms that go through the filter (or integrator). The filter output is then

$$s_o(t) = [A + x(t)][A + x(t_d)] + y(t)y(t_d)$$

(11.67)

Since we assumed that the signals in two adjacent intervals were identical, an error is made if the detector output is negative. Thus,

$$P_e = \text{PR}\{s_o(T_b) < 0\}$$

(11.68)

We now need to find the probability density of the filter output. The two noise components are sampled at two different times. To simplify the approach, we define four different variables, using the average and differences of the noise samples.

$$w_1 = A + \frac{x(t) + x(t_d)}{2}$$

$$w_2 = \frac{x(t) - x(t_d)}{2}$$

$$w_3 = \frac{y(t) + y(t_d)}{2}$$

$$w_4 = \frac{y(t) - y(t_d)}{2}$$

(11.69)

Each of the four variables is Gaussian distributed. The first has a mean of A, and the other three have a mean of zero. Using these definitions, we can rewrite the filter output as

$$s_0(t) = (w_1^2 + w_2^2) - (w_3^2 + w_4^2)$$

(11.70)

The probability of error is the probability that this output is less than zero. This is the same as

$$P_e = \text{PR}\left\{ \sqrt{w_1{}^2 + w_2{}^2} < \sqrt{w_3{}^2 + w_4{}^2} \right\} \tag{11.71}$$

We did not need the square roots in Eq. (11.71). We included them in order to make the following observation: The first square root in this equation is a random variable that is Ricean distributed, while the second square root is a random variable that is Raleigh distributed. At this point, then, the approach is identical to that used for incoherent detection of ASK or FSK. The result is

$$P_e = \frac{1}{2} \exp\left(\frac{-E}{N_0}\right) \tag{11.72}$$

The curve of this equation is plotted in Fig. 11.45. We have also plotted the equation for coherent suppressed carrier BPSK on the same set of axes.

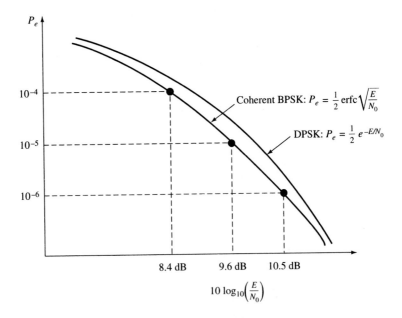

Figure 11.45 Performance of differentially coherent PSK.

Note that the probability of error is higher for the DPSK case than it is for the coherent BPSK case. This is expected, since additional information about the received signal is required for the coherent case. In general, the more you know about a signal, the better the job you can do detecting it.

11.3.8 M-ary PSK

QPSK (see Section 11.3.2) is an example of M-ary PSK, or MPSK, where multiple bits are combined to form symbols. In QPSK, bits are combined in pairs, and one of four possible sinusoidal bursts is transmitted to send each symbol. In 8-PSK, bits are combined in groups of three, and in 16-PSK, they are combined in groups of four. The phases are equally distributed around a circle. The phase separations are 360°/M, where M is the number of possible symbols. A symbol error (also known as a word error) occurs if the detected phase is closer to an incorrect symbol than to a correct symbol. Figure 11.46 presents a representative example. If the received phase is outside of the shaded area, a word error is made.

Calculation of the probability of the phase falling outside the shaded area is beyond the scope of this text. (See, e.g., Lindsey and Simon in the reference section.) The equation is

$$P_E = \frac{M-1}{M} - \frac{1}{2}\,\text{erf}\left(\sqrt{\frac{A^2 T_S}{2N_0}}\,\sin\frac{\pi}{M}\right)$$
$$- \frac{1}{\sqrt{\pi}}\int_0^{\sqrt{\frac{A^2 T_S}{2N_0}}\,\sin\frac{\pi}{M}} e^{-y^2}\text{erf}\left(y\cot\frac{\pi}{M}\right) \tag{11.73}$$

where T_S is the symbol period. As a check on this result, suppose we let $M = 2$. This represents the binary case. The equation then reduces to

$$P_E = \frac{1}{2}\,\text{erfc}\left(\sqrt{\frac{A^2 T_S}{2N_0}}\right) \tag{11.74}$$

which is the correct result for binary PSK.

For relatively large signal to noise ratios (i.e., small word error probabilities) and $M > 2$, Eq. (11.73) can be approximated by

$$P_E \approx \text{erfc}\left(\sqrt{\frac{A^2 T_S}{2N_0}}\,\sin\left(\frac{\pi}{M}\right)\right) \tag{11.75}$$

This equation is for the word error probability. Since M-ary communication involves combining information bits into symbols or words, we are usually more interested in the bit error probability. For example, suppose we are communicating via QPSK. We begin by combining pairs of bits into one of four possible symbols. If one of these symbols is involved in a word error during transmission, this can translate back to either one or two bit errors. While we can get an average bit error rate by examining the symbol assignments, a simpler result is possible with a few assumptions.

First, we assume that if a word error is made in MPSK, the most likely event is that one symbol has been mistaken for an adjacent symbol. Looking at Fig. 11.46, we see that it is more likely to mistake the correct symbol for a symbol separated by 360°/M than it is to mistake it for a symbol separated by more than this minimum angle. If the original encoding is done using a Gray code, adjacent symbols differ in only one bit position. For example, in 8-PSK, going sequentially around the circle, the symbols would represent 000,

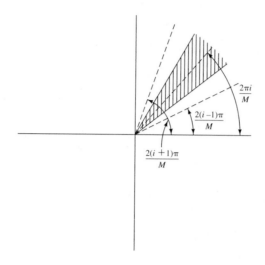

Figure 11.46   Word errors.

001, 011, 111, 101, 100, 110, and 011. With this assignment, a circular movement by 45°
translates to one bit error. Under these assumptions, the bit error rate is the same as the
word error rate. In adapting Eq. (11.73) to become the bit error rate, we need simply recognize that the symbol time is related to the bit period by

$$T_S = T_b \log_2 M \tag{11.76}$$

The bit error rate is then

$$P_E \approx \text{erfc}\left(\sqrt{\frac{A^2 T_b \log_2 M}{2N_0}}\right) \tag{11.77}$$

11.4 HYBRID SYSTEMS

Examination of the results for MFSK and MPSK show that as we move from binary to
M-ary systems, we suffer a degradation in performance. In fact, we can relate the bandwidth to the signal to noise ratio for the various digital signaling techniques described earlier. In one type of practical design, the bit error rate is specified, and we must optimize the
design for that rate. For example, a particular communication system may specify a bit error rate of 10^{-7}, and the engineer is asked to design the system for operation at the maximum transmission rate within a given bandwidth and at a specified signal-to-noise ratio.

A parameter that sums up this scenario is the bit transmission rate per unit of system
bandwidth as a function of signal to noise ratio. To calculate the transmission rate per Hz
of bandwidth, we need to specify a bit error rate. As one example, Fig. 11.47 shows the bit
rate per unit of bandwidth as a function of the signal to noise ratio for a bit error rate of
10^{-4}. The points on the graph were derived from results of the previous sections. (We ask
you to verify other points on the graph in a problem at the end of the chapter.) Note that

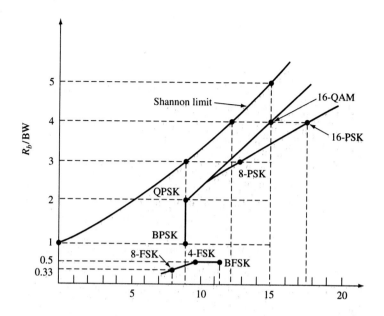

Figure 11.47   Bit rate per unit of bandwidth.

the given bit error rate is typical of a voice transmission system. Lower error rates are often required in data transmission systems.

The curves in the figure represent various trade-offs the engineer must deal with. In general, for a given signal to noise ratio, the higher we operate on the graph, the better is the performance of the system. It is useful to obtain an upper limit for the performance. This comes from Shannon's theorem. We start with the channel capacity expression,

$$C = \text{BW} \log_2\left(1 + \frac{S}{N}\right) \tag{11.78}$$

We can substitute the energy per bit divided by the bit period for the signal power. The noise power is N_0 times the bandwidth. (We assume white noise.) Equation (11.78) then reduces to

$$C = \text{BW} \log_2\left(1 + \frac{E_b R_b}{N_0 \, \text{BW}}\right) \tag{11.79}$$

Substituting R_b for C, we find that

$$\frac{R_b}{\text{BW}} = \log_2\left(1 + \frac{E}{N_0} \frac{R_b}{\text{BW}}\right) \tag{11.80}$$

Equation (11.80) thus relates R_b/BW to E/N_0. This equation is plotted in Fig. 11.47.

The challenge of hybrid systems is to design techniques that come closest to the Shannon limit in performance.

Quadrature Amplitude Modulation

Looking back to the QPSK signal space diagram of Fig. 11.29, we see that the points are equally spaced on the circumference of a circle. If we combined bits into groups of three, there would be eight possible values and eight points on the circumference in the signal space of 8-ary PSK, or 8-PSK. Performance improves when these points are separated as widely as possible. *Quadrature amplitude modulation* (QAM) is one approach toward a different point distribution within the signal space plane.

We present *16-QAM* as an example of this form of modulation. That is, we consider combining bits in groups of four, yielding 16 possible values. These values change every $4T_b$, so we can expect a bandwidth that is one-fourth that of BPSK. If we use 16-ary PSK, the points in the signal space would be equally spaced on the circumference of a circle, and the angular spacing between adjacent points would be 22.5°.

In the case of 16-QAM, both the amplitude and the phase vary, so the points no longer lie on the circumference of a single circle. The signal space diagram consists of 16 points in a uniform square array, as shown in Fig. 11.48. The individual signals are of the form

$$s_i(t) = A_i\cos(2\pi f_c t + \theta_i) \tag{11.81}$$

where the index i takes on values from 0 to 15. We can rewrite this equation as

$$s_i(t) = \mathrm{RE}\{A_i e^{j\theta_i}e^{j2\pi f_c t}\} \tag{11.82}$$

The signal space diagram, or *constellation*, is a plot of $A_i e^{j\theta_i}$ in Eq. (11.88) in the complex plane.

This system combines four input bits to produce one signal burst. Both the phase and the amplitude of the sinusoidal burst are modulated. A block diagram of the modulator is shown in Fig. 11.49. The odd-numbered bits in the input data stream are combined in pairs to form one of four levels that modulate the sine term. The even-numbered bits are similarly combined to modulate the cosine term. The modulated sine and cosine terms are combined. The verification that this system creates the signal space diagram of Fig. 11.48 is left to the problems at the end of the chapter.

Figure 11.50 shows a demodulator for 16-QAM. This system should be compared to that of the modulator of Fig. 11.49 to verify that the original signal is recovered at the output.

While the 16-QAM system may appear complex compared to simple binary systems, it possesses the potential for providing better performance for a given bandwidth.

11.5 MODEMS

Modulator-demodulators, or *MODEMS*, are used to transform the baseband signal into a modulated signal that is capable of transmission through the channel. The data input to the modem is converted into audio frequency signals, which are coupled into the phone lines. Modems are classified as either *asynchronous* or *synchronous*. Asynchronous modems do not have a clock, and their data rates need not be constant. They are the least complex type of modem. In fact, most modems operating below 1,800 bps are asynchronous. Alterna-

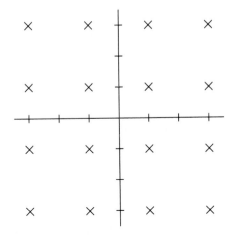

Figure 11.48   16-QAM signal space
diagram.

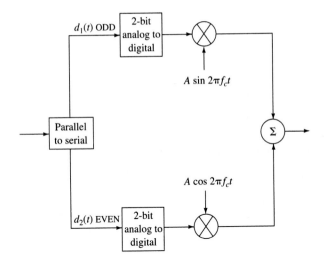

Figure 11.49   16-QAM modulator.

tively, synchronous modems send data at a fixed periodic rate. They can operate at higher data rates than can asynchronous modems.

Standards have been developed in an attempt to make systems compatible. "Type 103" modems (originated as part of the Bell System 103 and 113 series) include asynchronous modems that operate up to 300 baud (bits per second for binary communication). These are *full-duplex modems* (i.e., they permit simultaneous two-way communication). The two directions of transmission are known as *originate* and *answer*. The frequencies are assigned as follows:

	SPACE	MARK
ORIGINATE	1,070 Hz	1,270 Hz
ANSWER	2,025 Hz	2,225 Hz

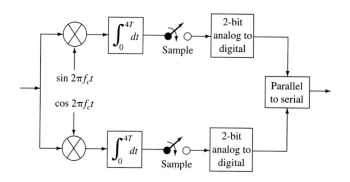

Figure 11.50 16-QAM demodulator.

A different, but related, standard is the Comité Consultatif Internationale de Telegraphique et Telephonique *(CCITT) recommendation V.21*, which provides for the following frequency assignments:

	SPACE	MARK
ORIGINATE	1,180 Hz	980 Hz
ANSWER	1,850 Hz	1,650 Hz

Note that, besides consisting of different frequencies, this standard uses the lower frequency in each pair for the MARK, in contrast to the 103 series, where the reverse is true.

Figure 11.51 shows a block diagram of a low-speed asynchronous modem. The FSK modulator can be a VCO with signal conditioning at the input. The bandpass filter restricts out-of-band signals prior to coupling the signal to the phone line.

The receive portion starts with a *bandpass filter* to decrease noise outside of the band of signal frequencies. An *amplitude limiter* is used to decrease noise effects, since

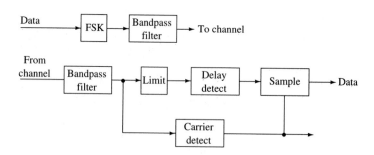

Figure 11.51 Low-speed asynchronous modem

the information in the signal is contained in the frequency of the incoming waveform. The *delay detector* compares the signal to a delayed version of itself. In this manner, it approximates the derivative of the signal, which is proportional to the frequency. It therefore acts as a discriminator yielding an output voltage that is proportional to frequency. The *carrier detect* portion of the system is required because the modem is asynchronous; we would not want the system deciding between 1's and 0's when nothing is being sent.

To achieve higher speeds, modems can be made synchronous instead of asynchronous. Additional improvements in speed are possible by going from full duplex to half duplex (two-way communication, but only one way at a time) and by changing from binary to *M*-ary transmission.

For moderate transmission speeds in the range of 2–5 kbps, MPSK modems are popular. Modems operating at 2,400 bits per second typically use QPSK. The frequency is set at 1,800 Hz, and the band of frequencies occupied by the baseband signal extends from 600 Hz to 3 kHz. That is, the band extends 1,200 Hz above and below the carrier, since bits are combined in pairs.

Modems operating at 4,800 bits per second use 8-PSK. Modems operating above that bit rate typically use hybrid signaling techniques to achieve acceptable error rates. QAM is the most common technique.

As rates move higher and higher (as this text goes to press, modem speeds are approaching 100 kbps), higher order signaling techniques are needed. We have seen that increasing the number of bits that are combined into symbols in an *M*-ary communication system increases the probability of errors.

Trellis-coded modulation techniques are being applied to higher order QAM systems to reduce the probability of errors. The concept of a trellis was examined in Chapter 9 in the context of convolutional codes. We found that by introducing memory into a system, changes occur in a structured way (i.e., transitions are controlled in a manner consistent with the state transition diagram). The same structure can be placed on a QAM system. Without coding, adjacent symbols can take on any values. For example, in an uncoded 16-QAM system, transmitted symbols can jump all over the signal space diagram (see Fig. 11.48). However, in trellis-coded QAM (*TCQAM*), movement from one symbol to another constrained by the coding scheme. If a received symbol sequence could not have been transmitted, error correction is possible.

Table 11.1 summarizes selected modem standards currently in use.

TABLE 11.1 SELECTED MODEM STANDARDS

Standard	Bits/Second	Modulation	Bits/Symbol
Bell 103	300	FSK	1
V.21	300	FSK	1
Bell 212A	1,200	PSK	2
V.22bis	2,400	QAM	4
V.26	2,400	PSK	2
V.27	4,800	PSK	3
V.29	9,600	QAM	4
V.33	14,400	TCQAM	6

PROBLEMS

11.1.1 Binary information is transmitted at 100 kbps using OOK. The carrier frequency is 20 MHz, and the received carrier amplitude is 10^{-3} V. The additive noise power is $N_0 = 10^{-12}$ watt/Hz.

(a) Design a coherent detector and find the bit error rate.

(b) Design an incoherent detector and find the bit error rate.

11.1.2 You are given a binary system with

$$s_1(t) = \cos 2\pi f_c t$$
$$s_0(t) = 0$$
$$T_b = \frac{5}{f_c}$$

The additive noise has a power spectral density in the range between $N_0 = 10^{-2}$ and 10^{-1} watt/Hz.

(a) Plot the probability of error as a function of N_0 for coherent detection.

(b) Repeat for an incoherent detector, and compare your answer to that of part (a).

11.1.3 A baseband signal is as shown in Fig. P11.1.3. The signal amplitude modulates a carrier of 1 MHz. Noise of power $N_0 = 10^{-3}$ watt/Hz is added, and coherent detection is used.

(a) Find the probability of error.

(b) Now assume that the detector treats each received binary signal as a composite of three bits and processes the signal bit by bit. Majority logic is then used to decide which hypothesis is true. That is, if at least two received bits match a particular signal, the corresponding hypothesis is chosen. Find the probability of error, and compare it to your answer to part (a).

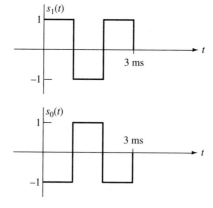

Figure P11.1.3

11.1.4 Find the maximum bit transmission rate for an OOK system in which the peak amplitude of the sinusoidal bursts is 1 V, the additive white Gaussian noise has a power of $N_0 = 10^{-7}$ watt/Hz, and the maximum bit error rate is 10^{-5}.

11.1.5 A binary matched filter detector is used in OOK where the peak amplitude of the sinusoidal burst is 0.5 V and the additive white Gaussian noise has a power of $N_0 = 10^{-6}$ watt/Hz. Information is transmitted at 10 kbps.

   (a) Find the bit error rate.

   (b) A "soft decision" rule is now proposed. To decrease the probability of error, a "dead space" is defined in which no decision is made. That is, if the output is greater than some threshold $y_0 + \Delta$, we decide that a 1 was transmitted. If it is less than $y_0 - \Delta$, a decision is made in favor of a 0. However, if the output is between these levels, no decision is made. (In some systems, a retransmit request is sent.) We wish to use this technique to reduce the bit error rate of part (a) by a factor of 2. Design the detector.

11.1.6 A coherent matched filter detector is used to detect an OOK signal. The carrier frequency is 1 MHz, the additive noise has a power spectral density of 10^{-5} watt/Hz, and the amplitude of the signal is 1 volt. The bit rate is 10 kbps.

   In designing the matched filter detector, there is a phase mismatch of $K°$. Find the maximum value of K such that the coherent detector performs better than an incoherent detector.

11.1.7 Repeat Problem 11.1.6 for a bit transmission rate of 1 kbps.

11.1.8 A random variable x is Ricean distributed with a mean value of zero. Find the probability that $x = x_0$, and compare it to the complementary error function for large x_0.

11.1.9 Prove Eq. (11.16) for the probability of error of an incoherent detector. [*Hint:* Use the result of Problem 11.1.8.]

11.1.10 ASK (OOK) is used to send 100,000 bits per second of binary information. The amplitude of the received signal is 0.2 V. The carrier frequency is 1 MHz. The noise power per Hertz, N_0, ranges from 10^{-8} to 10^{-7} watt/Hz. You are told that the bit error rate must be less than 10^{-3}.

   (a) Design the optimum detector.

   (b) Check whether your design is capable of meeting the error rate specification for the entire range of noise powers. If it is, indicate which part of the range meets the specifications.

11.2.1 A unit-amplitude bipolar baseband signal is shaped by a lowpass filter that passes frequencies up to $1/T_s$, where the sampling period T_s is 1 ms. The shaped signal then frequency modules a carrier of frequency 1 MHz. Find the bandwidth of the resulting signal.

11.2.2 An FSK signal consists of bursts of frequency of 800 and 900 kHz, with the higher frequency used to transmit a binary 1. The bit rate is 2 kbps. Find the bandwidth of the FSK signal.

11.2.3 Find the bandwidth of an FM signal resulting from binary modulation where the pulses are shaped according to a raised cosine. Assume that an alternating train of 0's and 1's is transmitted and that bipolar transmission is used.

11.2.4 What is the nominal bandwidth of 8-ary FSK if orthogonal tone spacing is used and the bit rate is $R_b = 10^4$ bps? Compare your answer to the nominal bandwidth using BPSK.

11.2.5 Find the bit error rate for an FSK system with

$$s_1(t) = \cos{(1{,}100\,\pi t + 30°)}$$
$$s_0(t) = \cos{(1{,}000\,\pi t + 30°)}$$
$$\frac{N_0}{2} = 0.1 \quad \text{and} \quad T_b = 10 \text{ sec}$$

(a) Using a coherent detector.

(b) Using an incoherent detector.

11.2.6 FSK is used to transmit binary information. The frequency deviation is set at the optimum value, $\Delta f = 0.715R_b$, so the correlation coefficient is -0.22. Now suppose that the two signals are mismatched in phase by $\Delta\theta$. Plot the correlation coefficient as a function of the phase mismatch, and find the maximum value of the correlation.

11.2.7 FSK is used to transmit binary information. The frequency deviation is set for orthogonal tone spacing, that is, $\Delta f = R_b/2$, so the correlation coefficient is zero. Now suppose that the two signals are mismatched in phase by $\Delta\theta$. Show that the correlation coefficient is zero regardless of the phase mismatch.

11.2.8 Derive an expression for the maximum allowable phase mismatch between two frequency bursts such that a tone spacing of $0.715/T_b$ is preferable to orthogonal tone spacing. Assume that a coherent detector is used.

11.2.9 Find the probability of error for the incoherent FSK detector of Fig. 11.19, where $A = 5$, $f_1 = 1,200$ Hz, $f_2 = 2,200$ Hz, and the sampling period is 0.1 msec. Noise of $N_0/2 = 10^{-6}$ watt/Hz is added during transmission.

11.2.10 Find the probability of error of the incoherent detector when used for FSK where the bit period is 1 msec. The two signals have an amplitude of 2 V, and the frequencies are 1.2 kHz and 2.2 kHz. The additive noise has a power of $N_0 = 10^{-5}$ watt/Hz. Compare the performance of the incoherent detector with that of the coherent detector.

11.2.11 Verify the three FSK points shown in Fig. 11.47. Assume coherent detection.

11.3.1 Find the power of a suppressed carrier BPSK signal that is bandlimited to a frequency between $f_c - \Delta f < f < f_c + \Delta f$. The cutoff frequencies are related to the bit rate by:

(a) $\Delta f = R_b/2$

(b) $\Delta f = R_b$

(c) $\Delta f = 2R_b$

11.3.2 A bipolar baseband signal is shaped by passing it through a lowpass filter with cutoff frequency $1/T_b$ (i.e., R_b). The filtered baseband signal phase modulates a carrier with maximum phase deviation $\Delta\theta = 180°$. The sampling period is 1 ms. Find the bandwidth of the resulting phase-modulated signal.

11.3.3 Verify the entries in the table showing $s(t)$ for various $a_1(t)$ and $a_2(t)$ in Fig. 11.28.

11.3.4 Find the bandwidth of the OQPSK signal resulting from the signals of Fig. 11.30. Your answer should be given in terms of T_s, the sampling period.

11.3.5 Plot the instantaneous phase of an MSK waveform resulting from:

(a) A train of static data (all 1's or all 0's).

(b) An alternating train of 1's and 0's.

11.3.6 **(a)** Design a coherent detector for a binary transmission system that transmits the following two signals.

$$s_0(t) = A\cos 2\pi f_c t$$
$$s_i(t) = A\cos (2\pi f_c t + 90°)$$

(b) What factors affect the choice of integration time?

(c) Find the detector output under each hypothesis.

11.3.7 **(a)** Design a coherent detector for a binary transmission system that transmits the following two signals.

$$s_0(t) = A\cos 2\pi f_c t$$
$$s_1(t) = A\cos(2\pi f_c t + 180°)$$

Use only one local oscillator and correlator.

(b) Now let the phase of the local oscillator vary from its correct setting by $\Delta\theta$. Find the bit
error rate as a function of $\Delta\theta$.

(c) What values of $\Delta\theta$ would cause decoding errors in the absence of noise?

(d) Show that the detector can be used to detect DPSK for any fixed value of $\Delta\theta$.

11.3.8 Demonstrate the performance of the DPSK detector of Fig. 11.42 for the data input

$$1\ 0\ 1\ 1\ 0\ 1\ 1\ 1\ 0\ 0$$

That is, discuss the form of the transmitted DPSK signal, and find the detector output.

11.3.9 In the DPSK demodulator of Fig. 11.42, the delay path contains an error that causes the delay to vary between $0.9T$ and $1.1T$. Explore the effects of this "timing jitter" upon the demodulator and its performance.

11.3.10 A binary PSK system transmits the following two signals:

$$s_0(t) = 0.01 \cos 2\pi \times 1{,}000\ t$$
$$s_1(t) = 0.01 \cos (2\pi \times 1{,}000\ t + \theta_1)$$

$T = 10\text{ms}$ and $N_0 = 2 \times 10^{-7}$ watt/Hz.

(a) Plot the probability of error of a coherent detector as a function of θ_1.

(b) Repeat part (a), assuming that the local oscillators are mismatched in phase by $45°$.

(c) Assume that $\theta_1 = 180°$. Plot the probability of false alarm, P_{FA}, as a function of $\Delta\theta$, the
phase mismatch.

11.3.11 A discriminator detector is used to detect the following two signals:

$$s_0(t) = 0.05 \cos (2\pi f_c t + 90°)$$
$$s_1(t) = 0.05 \cos (2\pi f_c t - 90°)$$

$T = 0.5$ sec and $N_0 = 10^{-4}$ watt/Hz.

(a) Find the probability of error.

(b) Find the probability of error for a matched filter detector, and compare it to your answer to part (a).

11.3.12 Design a coherent DPSK system to transmit data at a rate of 10 kbps in additive white
Gaussian noise (AWGN) with $N_0 = 10^{-6}$ watt/Hz. The system must achieve a bit error rate
less than 10^{-5}. Find the bandwidth of the system, and specify the minimum amplitude of
the signal at the receiver.

11.3.13 You are given the QPSK system with the following transmitted signals:

$$s_1(t) = 10 \cos (1{,}000\pi t + 30°)$$
$$s_2(t) = 10 \cos (1{,}000\pi t + 120°)$$
$$s_3(t) = 10 \cos (1{,}000\pi t + 210°)$$
$$s_4(t) = 10 \cos (1{,}000\pi t + 300°)$$

The symbol period is 0.1 sec, and the additive noise has power $N_o = 0.01$ watt/Hz. Assume that all four signals are equally likely.

Design a matched filter detector, and find the probability of error and the probability of correct transmission.

11.3.14 Design a coherent detector for a binary transmission system that transmits the following two signals:

$$s_0(t) = A\cos 2\pi f_c t$$
$$s_1(t) = A\cos (2\pi f_c t + 90°)$$

The ratio of signal power to N_0 is 400 Hz. What is the maximum bit transmission rate if the bit error rate is to be less than 10^{-3}?

11.3.15 Design a PSK system to transmit data at a rate of 10 kbps in AWGN with $N_o = 10^{-6}$ watt/Hz. The channel has a bandwidth of 2 kHz. Evaluate the performance of the system.

11.3.16 Verify the BPSK point shown in Fig. 11.47.

11.3.17 PSK is used to send 100,000 bits per second of binary information. The additive noise has a power per hertz of $N_0 = 5 \times 10^{-6}$ watt/Hz. The carrier frequency is 1 MHz. The design specifications call for a maximum bit error rate of 10^{-4}.

(a) Design the optimum detector.

(b) What is the minimum amplitude of the received signal that is needed to meet the error rate specification?

(c) If the transmission is changed from PSK to FSK (orthogonal tone spacing), what changes must you make to your answer of part (b) to meet the specifications?

11.4.1 Derive the function of time for each of the 16 possible signals resulting from the 16-QAM modulation of Fig. 11.49. Plot these as points in a signal space, and compare your answer with Fig. 11.48.

11.4.2 Derive an expression for the signals at the inputs to the two analog-to-digital converters in the 16-QAM demodulator of Fig. 11.50.

11.4.3 You are given a PSK system in which the two signals are

$$s_1(t) = 10 \cos 2 \pi \times 10^6 t$$
$$s_0(t) = 10 \sin 2 \pi \times 10^6 t$$

Bits are transmitted at a rate of 1 million per second. The additive noise has a power $N_0 = 10^{-4}$ watt/Hz.

(a) Design a matched filter detector for this transmission system, and simplify the system as much as possible.

(b) Find the probability of bit error for the detector of part (a).

(c) Repeat parts (a) and (b) for the following signals:

$$s_1(t) = 10 \cos 2 \pi \times 10^6 t$$
$$s_0(t) = 20 \sin 2 \pi \times 10^6 t$$

(Note that the amplitudes do not match each other in part (c).)

11.5.1 A communication channel passes frequencies between 300 Hz and 3 kHz. Voltage amplitudes of up to 0.5 V can be permitted on the channel. Assume that simplex (one-way) transmission is used and that the noise has power $N_o = 10^{-5}$ watt/Hz.

    (a) Design a binary coherent communication system and evaluate its performance.

    (b) If the channel is used to transmit a voice signal of maximum frequency $f_m = 3$ kHz, how many bits of PCM quantization are permitted?

    (c) Repeat part (a) for incoherent detection.

12

Design Considerations

Another name for this chapter might be "The Meaning of It All," but that may be too philosophical. Nonetheless, the chapter attempts to tie the major pieces of the text together.

Some readers may be hoping for a recipe for system design—a type of step-by-step procedure. We shall resist this approach for a couple of reasons. First, it downgrades the teaching of the subject from *educating* to *training*. Second, there are no easy answers; the reason that communication is such an exciting field is that it is challenging. The *trade-off* considerations are manifold, and the inputs to the design process are changing with time. In the past, hardware implementation considerations often provided the most important input trade-off. Now, with advances in electronics—in particular, Very Large Scale Integration (VLSI)—implementation is becoming less of a consideration in system design. Many systems that were previously only of theoretical interest in that they provided an upper bound for system performance (i.e., they were viewed as being too difficult to implement)[1] are now in use. Accordingly, we can only hope to give the tools for communication system design and relegate detailed decisions to the ingenuity of the engineer. Many designs are *open ended*, and there may be more than one correct solution to a particular problem.

The first 11 chapters of the text contain a great deal of information regarding the concepts, implementation, and performance of analog and digital communication systems, as well as the motivation for building them. However, most of the topics are meant to stand alone. On the other hand, the design engineer must assimilate a vast array of information and make trade-off decisions. Fortunately, there are no simple answers to the challenges of meeting system specifications: As mentioned, in the real world, problems have more than one correct answer, and the best approach to a particular design depends upon numerous factors. Were this not the case, communication would represent a very uninteresting specialization within electrical engineering.

Design in any engineering discipline can be broken into the following five major steps:

1. *Define the problem.* This involves carefully stating the purpose of the design. The source of design information might be the customer, market survey specialists, a request for proposal (RFP), orders from your supervisor, or an assignment from your professor.

---

[1]Indeed, in describing such systems, some used the insulting statement, "This is of *academic* interest only." That is no longer the case.

2. *Subdivide the problem.* All but the simplest designs lend themselves to being broken into smaller parts. These parts may include signal selection, transmitter design, receiver design, signal processing, and packaging. In major designs, individual teams are assigned to the various smaller pieces of the overall design problem.

3. *Document your work.* This step pervades all of the other steps. You must communicate your work to others, whether they be members of the team working on the project or the customer who is requesting the design. One of the most serious sins an engineer can commit is to fail to document work. If we have to constantly reinvent the wheel, society would never progress technologically.

4. *Build a prototype.* When an author writes a book, he or she would like to think that the printed page is the *whole truth and nothing but the truth.* One of the most sobering experiences for the new engineer is to construct something according to the theory and find that it does not perform as expected. This does not necessarily mean that the theory is wrong; it may instead be that the model and assumptions used do not match the real-world situation.

   The prototype is sometimes in the form of a computer simulation, but you must make sure that the simulation is duplicating the real world. (Of course, if the final design is implemented by computer, the simulation may well *be* the real world.)

5. *Finalize the design.* This includes *fine-tuning* the design based on experience with the prototype. It also includes the important step of final documentation.

The first two sections of this chapter summarize the critical results of the earlier parts of the text. We relate one type of communication system to another to aid in design decisions.

The final section of the chapter presents five contemporary applications of communication. We include these "case studies" to illustrate the manner in which design decisions have been made.

12.1 ANALOG DESIGN TRADE-OFFS

In this section, we summarize the important results of baseband, AM, FM, PM, and pulse modulation analog communication as those results apply to bandwidth, performance, and system complexity.

12.1.1 Bandwidth

Single-sideband AM has a bandwidth equal to the highest frequency of the information. This is the minimum bandwidth of all of the systems discussed in this text.

   Vestigial sideband AM has a bandwidth greater than the maximum information frequency, but less than twice this quantity. Typically, the bandwidth is on the order or $1.25f_m$.

   Double-sideband AM (transmitted and suppressed carrier) and *narrowband FM and PM* all have bandwidths equal to twice the highest frequency of the information signal.

   The bandwidth of *wideband FM* is proportional to the modulation index. It is typically much higher than that of double-sideband AM.

The bandwidth of *pulse-modulated waveforms* depends on the width of the pulses. For example, in PAM, the maximum pulse width is limited by the sampling period. As the pulse width approaches that period, the bandwidth of the signal (nominal out to the first zeros of the transform) approaches $2f_m$. If time division multiplexing is being used, the pulse width is a small fraction of the sampling period, and the bandwidth is a multiple of $2f_m$.

12.1.2 Performance

For a *single-sideband AM* system using coherent detection, the signal to noise ratio at the output of the receiver is the same as that at the input to the receiver (within the band of the signal).

Double-sideband AM with coherent detection achieves an output signal to noise ratio that is twice that of the input to the receiver. The doubling is realized because of the redundancy between the upper and lower sideband.

The output signal to noise ratio for *transmitted carrier AM* and incoherent detection approaches twice that of the receiver input as the carrier to noise ratio gets larger. However, in such cases, the efficiency decreases.

Wideband FM possesses an improvement in signal to noise ratio that increases as the square of the modulation index. Thus, as the modulation becomes more and more wideband, the noise performance improves.

12.1.3 System Complexity

In Section 7.10, we compared the various analog signals with respect to their complexity. Those observations are repeated here.

Narrowband FM can be generated with a system consisting of a multiplier, an integrator, and a phase shifter. It is demodulated with a discriminator followed by an envelope detector or by a phase lock loop. These are relatively simple implementations. The bandwidth of narrowband FM is $2f_m$, where f_m is the maximum frequency of the information signal. Although the modulation process makes this look very much like a variation of AM, it is definitely different. The modulated waveform has a constant amplitude, which permits us to add a limiter to the receiver, thereby cutting down on additive noise. This is particularly effective if the additive noise is in the form of spikes. The discrimination process in the receiver changes the noise from white to nonwhite, accentuating high frequencies while attenuating low frequencies. This yields a noise advantage over AM if critical information resides at low signal frequencies or if the additive noise contains a nonwhite component at a low frequency.

Narrowband PM is very similar to narrowband FM. The integrator in the modulator is traded for an integrator in the demodulator. The bandwidth is $2f_m$. The amplitude of the PM wave is constant, thus affording the noise advantage realized by adding a limiter to the receiver. The final integration process in the receiver attenuates the high frequencies. This could be advantageous if critical information resides at high signal frequencies or if the additive noise contains a nonwhite component at a high frequency.

Wideband FM is generated either indirectly from narrowband FM (through frequency multiplication) or by a VCO. It is demodulated in the same way as narrowband

FM is. The bandwidth is approximately $2\beta f_m$, which can be considerably greater than that of AM or narrowband angle modulation. The major advantage of wideband FM is its noise reduction ability: The output signal to noise ratio is approximately proportional to β^2.

Wideband PM bears the same relationship to wideband FM as does narrowband PM to narrowband FM. However, unlike wideband FM, in wideband PM the modulation index cannot be increased without limit. The maximum phase deviation is limited to 180°. Beyond that, there is a phase ambiguity, and the original signal cannot be uniquely recovered.

Double-sideband suppressed carrier AM has a bandwidth of $2f_m$ and an efficiency of 100 percent (i.e., no power is wasted in sending a pure carrier). Modulation is performed by a multiplier, while demodulation requires coherent circuits with the accompanying difficulty of reconstructing the carrier at the receiver.

Double-sideband transmitted carrier AM has a bandwidth of $2f_m$ and an efficiency of less than 50 percent because power is being wasted in sending a pure carrier. It enjoys the easiest implementation of a demodulator (envelope detector) of all schemes of modulation. It cannot support a signal with a nonzero dc level, since that information would be lost in demodulation.

Single-sideband AM has the smallest bandwidth of all of the systems we have studied: f_m. It is 100-percent efficient, since no power is wasted in sending a pure carrier. The complexity of implementation of modulators and demodulators is high due to the filtering required in the transmitter and the carrier recovery and coherent detection required in the receiver.

Vestigial sideband suppressed carrier AM has a bandwidth larger than f_m, but less than $2f_m$. The modulator is easier to construct than that of single sideband, but the demodulator requires carrier recovery as well as a carefully controlled filter shape to properly combine the sidebands.

Vestigial sideband transmitted carrier AM has a bandwidth larger than f_m, but less than $2f_m$. The modulator is easier to construct than that of single sideband, and if the carrier is large enough, an envelope detector can be used, thus making demodulation extremely simple.

12.2 DIGITAL DESIGN TRADE-OFFS

12.2.1 Performance Comparisons

We characterize performance by the *bit error rate*, which has been plotted as a function of signal to noise ratio several times in Chapter 11.

Figure 12.1 combines the various results to allow comparison of ASK with PSK. The curve with the lowest bit error rate represents the performance of coherent BPSK with a phase shift of 180°. Recall that this case is characterized by a correlation coefficient equal to -1, which represents the best possible performance. The performance is given by

$$P_e = \frac{1}{2} \operatorname{erfc} \sqrt{\frac{E}{N_0}} \tag{12.1}$$

The next best performance shown in the figure occurs for coherent binary ASK with the same *average* power as coherent BPSK. This refers to the case of OOK; the formula is

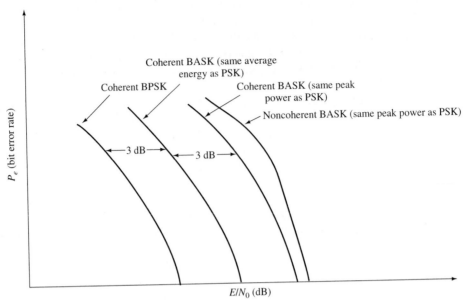

Figure 12.1 Bit error rate comparisons.

$$P_e = \frac{1}{2} \operatorname{erfc} \sqrt{\frac{E}{2N_0}} \qquad (12.2)$$

The signal would have to have twice the power of the signals used for PSK in order to have the same *average* power. This assumption is realistic for energy-limited systems. For example, in a satellite, the energy is normally derived from solar cells. While we have flexibility in choosing peak power, the total energy in each 24-hour period is limited. Since the correlation coefficient for this case is 0 (as opposed to −1 for the previous case), the signal to noise ratio must be twice as large to realize the same bit error rate as obtained with coherent PSK. That is the reason the first two curves are separated by 3 dB.

If the signal power in the binary ASK (OOK) case is the same as that in the PSK case, the *average* power is one-half as large. Therefore, if the *peak* power of the ASK case matches the power of the PSK case, an additional 3-dB degradation occurs, as shown in the third curve of Fig. 12.1.

The preceding may seem to be an unfair comparison between the two systems, but it often represents the true practical situation, where peak power may be limited by the transmitter electronics or by FCC requirements.

If we transmit using OOK with the same peak power as for PSK, but now use an incoherent detector, the rightmost curve of Fig. 12.1 results. This is a plot of

$$P_e = \frac{1}{2} \exp\left(\frac{-E}{2N_0}\right) \qquad (12.3)$$

Although Eq. (12.3) represents yet another degradation in performance, the trade-off between performance and the difficulty of reconstructing the carrier at the receiver may make such a degradation a fair compromise. Incoherent detectors, such as the peak or envelope detector, are extremely simple to build.

Figure 12.2 illustrates another set of performance curves that allows a comparison between PSK and FSK. Once again, we show the coherent PSK performance curve as the best achievable performance for the binary case. If we use FSK with the optimum frequency separation, we can achieve a correlation coefficient of −0.22. Recall that this requires perfect phase coherence, and if such coherence is achievable, one would probably opt for phase modulation. Nonetheless, the second curve shows the performance of this FSK system. The curve is a plot of

$$P_e = \frac{1}{2} \text{erfc} \sqrt{\frac{E(1.22)}{2N_0}} \tag{12.4}$$

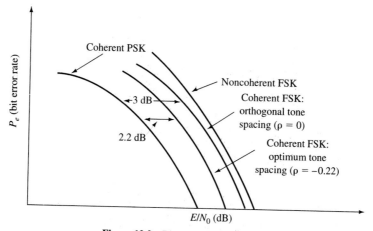

Figure 12.2   Bit error rate comparisons.

The third curve represents FSK with orthogonal tone spacing and is described by

$$P_e = \frac{1}{2} \text{erfc} \sqrt{\frac{E}{2N_0}} \tag{12.5}$$

This form of FSK does not require phase coherence and achieves a correlation coefficient of zero. That is the reason the curve is 3 dB to the right of the optimum PSK curve. To obtain the same performance with the FSK system, the signal to noise ratio must be twice that of the PSK system.

The final curve in Fig. 12.2 represents the performance of an FSK system in which incoherent detection is used, as described by the equation

$$P_e = \frac{1}{2} \exp\left(\frac{-E}{2N_0}\right) \tag{12.6}$$

Figure 12.3 compares MFSK with MPSK. We have not derived any detailed equations for these curves in the text. (In one case, we derived an upper bound on error performance.) However, you should note that MFSK performs better than MPSK as the number

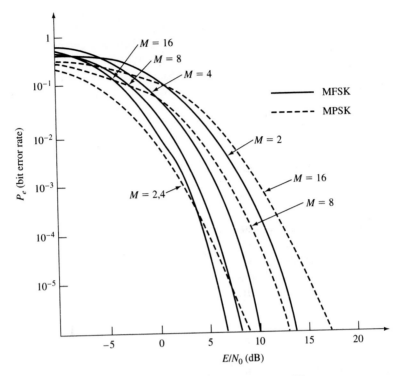

Figure 12.3 Performance of MFSK and MPSK.

of bits that are combined increases. This occurs because, as *M* increases in the *M*-ary system, the phases in PSK get closer and closer together, while the frequencies in FSK are still orthogonally spaced. Before reaching conclusions regarding the better performance of MFSK compared to MPSK, it is important to remember the bandwidth penalty that is paid with MFSK. This is examined in the next section.

12.2.2 Bandwidth Comparisons

We studied bandwidth in Chapters 9, 10, and 11. When we considered the power spectral density of pulsed waveforms, we defined the nominal bandwidth as the distance to the first zero of the density. Thus, we included the main lobe. Many analysts define bandwidth out to half of this amount and call this the *minimum bandwidth*. It can be shown that 91 percent of the total power is contained in the nominal bandwidth, while 78 percent is contained in the minimum bandwidth.

Figure 12.4 shows the ratio of the minimum bandwidth to the bit rate for various forms of modulation. Note that binary PSK has a minimum bandwidth that is equal to the bit rate, so the ratio is plotted as unity. In the case of PSK and ASK, as the number of bits combined into a single symbol *increases*, the bandwidth decreases proportionally. Thus, for 4-ary phase modulation communication (4-PSK), the bandwidth is half as much as for binary (BPSK), provided that the bit rate remains constant. This is true because bits are combined into pairs and the sinusoidal burst doubles in length, cutting the bandwidth in

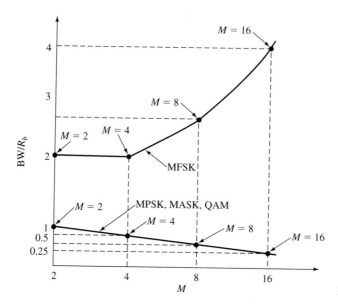

Figure 12.4   Bandwidth vs. bit rate.

half. The bandwidth is reduced by an additional factor of two for every doubling in the number of bits combined into a single symbol.

In the case of FSK, we assume orthogonal tone spacing. For binary FSK, the bandwidth is twice what it is for binary PSK. The bandwidth in the case of MFSK depends upon two quantities: the total number of frequencies and the width of the frequency spectrum of each sinusoidal burst. As M increases, the number of frequencies increases, but the width of each individual spectrum decreases since the sinusoidal bursts become longer. Between $M = 2$ and $M = 4$, these two effects cancel each other. For M greater than 4, the overall bandwidth increases, since it depends more heavily on the number of frequencies (i.e., the spread beyond the highest and lowest frequencies becomes less significant than the spacing between these two frequencies).

12.2.3 bps/Hz Comparisons

Up to this point, we have been concentrating our comparative evaluations of systems upon bit error rate and bandwidth. In one class of practical design situation, the bit error rate is specified, and we must "optimize" the design for that rate. For example, a particular data communication system may specify a bit error rate of 10^{-7}, and the engineer is asked to design for the maximum transmission rate within a given bandwidth and at a given signal to noise ratio. In situations such as these, a parameter of interest is the bit transmission rate per unit of system bandwidth.

Figure 12.5 illustrates the bit rate per unit of bandwidth as a function of the bit signal to noise ratio. The curves are shown for an assigned bit error rate of 10^{-4}. For smaller rates, the curves would move downward, indicating that a slower communication rate is required to reduce the error rate.

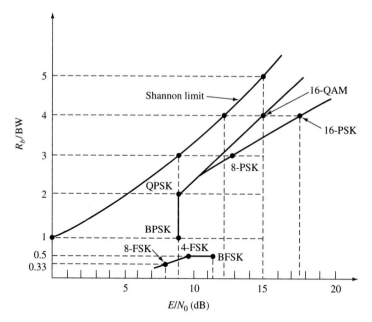

Figure 12.5 bps/Hz vs. signal to noise ratio.

The top curve in the figure represents the theoretical limit provided by Shannon's channel capacity theorem. If we assume that the additive noise is white and Gaussian, the channel capacity is given by

$$C = \mathrm{BW} \log_2 \left(1 + \frac{S}{N} \right) \tag{12.7}$$

where BW is the channel bandwidth and S/N is the signal to noise ratio. The units of the capacity are bits per second. Shannon showed that as long as the bit rate R_b is less than C, an arbitrarily small error rate can be achieved. Doing so may require complex error control coding, but the Shannon limit provides a useful goal against which to measure system performance.

The other points on Fig. 12.5 are obtained from the appropriate curves in Chapter 11. Note that PSK comes closer to the Shannon limit than does FSK and that 16-QAM displays an improvement over PSK. Indeed, of the systems we have presented in the text, 16-QAM comes closest to the Shannon limit for bit signal to noise ratios above about 10 dB. We hasten to point out that the additional complexity of 16-QAM may cancel the benefits associated with this observation.

12.2.4 Digital Communication Design Requirements

In performing a digital system design, one begins with a set of requirements and constraints for the system, perhaps including the following:

1. Required bit transmission rate.
2. Maximum allowable bit error rate.

3. Maximum system bandwidth.
4. Maximum transmitted signal power (and resulting maximum signal to noise ratio).
5. Maximum construction cost (i.e., complexity of the detector).
6. Maximum power utilization of the detector.
7. Maximum acquisition time of the detector.

We expand on each of these requirements and constraints and relate them to the systems examined in the text.

   1.   *Required Bit Transmission Rate*   The required bit transmission rate is a function of how the system is to be used. For example, if a corporation wishes to transmit financial data to its data-processing division, it knows how much data must be sent and how soon the data must reach the receiver.

   If the data contain any redundancies, entropy coding can be used to reduce the number of information bits that must be transmitted. The required bit transmission rate should normally be reduced as much as possible prior to beginning the design of the communication system. Even this reduction bears some trade-off considerations, however: There are a number of costs involved in entropy coding, associated particularly with the hardware. First, while the cost of the special-purpose chips used in entropy encoding is not high, it must be considered in the overall design. Second, the circuitry takes up space on a circuit board, and although the cost of this is small, it is not negligible. Finally, any extra steps that must be performed at the transmitter require power and time. Some remote (e.g., space-based) transmitters have very limited power budgets, and the power used by entropy coders must be considered in their operation. Of course, any reduction in bit rate due to entropy coding results in a reduction of the energy needed to transmit the encoded data.

   The required bit transmission rate for a digitally encoded analog signal (e.g., speech) depends upon the frequency content of the analog waveform, the acceptable levels of the signal-to-quantization-noise ratio, and the properties of the signal. The frequency content determines the minimum sampling rate. The signal-to-quantization-noise ratio determines the number of bits of quantization needed to achieve the required specifications. The properties of the signal determine whether a more sophisticated compression technique (e.g., ADM or DPCM) can be used to reduce the number of bits that must be transmitted.

   We therefore see that the required bit transmission rate is a function of the information that must be transmitted and the amount of work we are willing to do prior to transmission. Although the latter consideration often interacts with other decisions in the design process, we normally consider the transmission rate to be a specified quantity at the beginning of the process.

   2.   *Maximum Bit Error Rate*   The sensitivity of the data normally determines the maximum allowable bit error rate. In a speech or video signal, we know (from experience) the minimum SNR for acceptable reception. For data, the requirements of the system usually provide a maximum bit error rate. For example, military and financial applications would normally require a smaller bit error rate than would entertainment applications. If we find that the maximum rate cannot be achieved within the constraints to be discussed next, we can resort to forward error correction in the form of coding.

3. *Maximum System Bandwidth*   The channel through which the signal is transmitted is an important factor in determining the maximum system bandwidth. This bandwidth is set both by the transmission characteristics of the medium (e.g., the capacitance and inductance of a coaxial cable) and by regulations (e.g., the FCC allocates bandwidth in terrestrial transmissions). When combined with the required bit transmission rate, the maximum system bandwidth constraint helps in the choice of modulation scheme. For example, FSK requires more bandwidth than does PSK. Additional savings in bandwidth can be accomplished by moving toward MPSK, but with a penalty in SNR. If the bit error rate or the transmitted power are not serious considerations, but the bandwidth is, we would probably move toward MPSK transmission or to 16-QAM.

We must also look at the signaling format and its relationships to the system bandwidth. As an example, we saw that biphase baseband formats have frequency spectra that go to zero as the frequency approaches zero. This is in contrast to NRZ systems, where the power spectrum has non-zero value at dc. Some baseband channels do not transmit dc (i.e., they are ac coupled). Similarly, some carrier systems have problems with signals that have significant energy near the carrier frequency. In such cases, biphase formats provide advantages over NRZ formats.

4. *Maximum Transmitted Signal Power*   The maximum transmitted signal power affects the received SNR and, therefore, the bit error rate. With PSK transmission, the bit error rate increases as we move from binary to M-ary systems. We use M-ary systems to achieve efficiency in bandwidth. If the bit error rate provides the overriding constraint, we should stay with the binary systems and attempt to use PSK instead of FSK or ASK. In this manner, we squeeze the minimum bit error rate from a given transmitted signal power. Of course, to do this, we must be able to design a coherent detector, and we must be able to accept the acquisition time, false lock, and drift associated with the control loops in these systems. Such a situation would be an ideal one in which to concentrate upon entropy encoding to reduce the bit transmission rate as much as possible.

Noise reduction can be a means of raising SNR. If we have control over the transmission path, it is sometimes possible to shield the system from additive noise. If fading is causing selective reduction in the SNR, frequency or space diversity techniques might be used. That is, if one *signal path* is subject to strong noise or if the signal fades, we could change the path. If one *frequency range* is subject to such problems, we can vary the frequency range.

5. *Construction Cost*   Consider the following observations:

- Incoherent detectors are inherently simpler than coherent detectors.
- Nonadaptive signal encoders are simpler than adaptive encoders.
- Incoherent ASK detectors are simpler than incoherent FSK detectors.
- Orthogonal tone spacing is easier to achieve in FSK than is optimum tone spacing.
- Binary systems are simpler than M-ary systems.

For all of these statements, we could almost universally substitute the assertion, "Systems with poorer performance are less expensive than systems with better performance." It is always the case that the more work we are willing to perform, the better job we can do in communicating information. Hardware implementation costs are decreasing rapidly, so it is a rare design that will be dictated by such costs. Nonetheless, in designing a small number of systems, it is critical that off-the-shelf components be used. Major integrated electronics manufacturers can supply data sheets describing the systems that have been implemented with LSI and VLSI chips.

6. *Maximum Power Utilization of Detector*   Simply put, the more complex the detector, the more power it needs to operate. Phase lock loops require more power than do passive filters. Quadrature detectors must perform more operations than binary detectors. Some detectors—particularly those used for data acquisition—are in remote locations. Some rely upon solar power for their operation. In such cases, a careful power budget analysis may place constraints upon the system and may require one to live with a higher bit error rate in order to meet the power requirements.

7. *Maximum Acquisition Time of Detector*   Coherent detectors require time to acquire the carrier signal. In the case of transmitted carrier, the phase lock loop in the detector takes time to lock onto the carrier term in the received signal. In the case of suppressed carrier, squaring or Costas loops require even more time to lock onto the carrier. This acquisition time may not pose a serious problem in transmissions that are not time sensitive. That is, we can afford to devote a number of bits at the beginning of the transmitted message to giving the detector time to adjust. In other cases, transmissions may be very short or may require the transfer of data to begin very soon after the transmission begins. In such cases, our options are to do a significant amount of work to provide coherence through external operations (e.g., derive local carriers from an external source locked to the transmitter) or to use incoherent detectors. The latter require a higher SNR to provide the same level of performance as do coherent detectors. They therefore represent a viable alternative if the energy of the signal can be readily increased (through an increase in the signal power or a decrease in the transmission rate) or if the bit error rate is below specifications by a sufficient amount.

12.3 CASE STUDIES

The following five case studies are contemporary applications of communication. They serve to apprise you of the broader aspects of these important applications and to give you an example of how the various design considerations are applied in a practical situation.

12.3.1 Paging Systems

Needs Assessment

As more and more sophisticated communication systems become available, demands increase. In business, people have become more mobile through better communication systems. The down side is that whereas staff members used to be available to answer telephones whenever they rang, more often than not, this is no longer true. Progress in electronics—particularly as it relates to computers—has significantly reduced the amount of time available to react to requests. *Instant access* has become the dream of the decade.

The advantage of this improved capability is that it allows society to react more quickly to emergencies and to changing life-styles. A member of the family is no longer "tied to the home" if he or she must respond to a telephone call.

Telephone answering machines, many with the capability to transmit messages upon request, were the first step in providing flexible communication. But the time delay between recording the message and playing it back was often unacceptably long. This led to a need to alert the individual that a message was waiting. Satisfying that need led to the world of personal mobile communications.

Specifications

The basic requirement is to provide the ability for an individual to send a message to any subscriber anywhere in the world, whether the subscriber is in a New York traffic jam or climbing the Swiss Alps. The more practical current specification in the United States is the ability to transmit up to several dozen alphanumeric symbols to any subscriber situated within a major urban center. (We exclude satellite paging systems from the discussion.) The receiver must be portable, light, inexpensive, and capable of operating at least 48 hours between recharges of the power supply. The service should be available for a *reasonable* price (rental of the receiver and servicing the system well under US$50 per month). An additional specification is to provide the capability of broadcasting a message to a selected group of subscribers. For example, a company may wish to send a message to all of its salespeople without having to cycle through each of them individually.

Design Approach

Satellite transmission is one way to provide wide-area access, but the problems with the receiving antenna have historically been difficult to overcome. Individuals have not wanted to carry around dish antennas or have them on the roofs of their cars. Obstructions (e.g., tunnels, buildings, and bridges) cause additional problems.

The most universal form of broadcasting in use is commercial radio. People in urban areas have ready access to AM and FM radio using inexpensive portable receivers. Commercial radio was a candidate for the kind of need that paging systems were designed to satisfy. VHF radio was another candidate, but it was generally rejected due to the expense of setting up separate transmitters and securing the required licenses.

To provide coverage in most urban areas, a paging service would have to license hundreds of commercial stations devoted to such transmissions. This proves to be prohibitive in terms of cost, even if it would be possible to obtain the necessary licenses.

A vehicle already existed in FM commercial broadcasting. When we studied FM stereo in Section 7.8, we noted that only 106 kHz of the 200-kHz band was used and that the remainder was available for special Subsidiary Carrier Authorization services. This portion of the band can be leased from FM commercial stations at a monthly cost that depends upon the size and population of the listening area. Some FM stations lease these SCA frequencies to music services for professional offices (e.g., doctors and dentists). Others lease them to utilities for power management projects.

Design Details

Although there are several national paging services, the features we describe are common to all of them. The person who wishes to send a message calls a toll-free number that connects with a paging control center. The sender is prompted for two pieces of information: the identification number of the intended receiver and the message. These are entered using *touch tone* signals directly from the telephone.

The paging control center then transmits the information to a satellite through an uplink. One paging system called CUE uses the Westar 4 satellite. The satellite transmits the signal to downlinks at selected FM stations throughout the country. The FM stations then broadcast the information on a subcarrier frequency of 57 kHz. (Recall that the second stereo signal uses a subcarrier of 38 kHz.) The maximum frequency of the information signal is as high as 3 kHz. From Section 7.8, the narrowband FM signal for stereo occupies the frequency band surrounding the carrier and extending 53 kHz on each side of the carrier. With a 3-kHz maximum frequency and 57-kHz subcarrier frequency, the SCA information can range between 54 kHz and 60 kHz from the carrier. This is shown in Fig. 12.6.

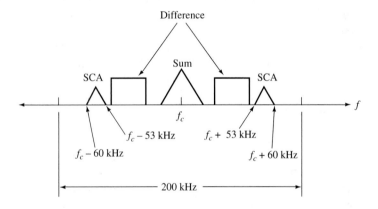

Figure 12.6   Frequency assignment for paging system.

The FM stations first transmit an *attention code* that alerts all receivers that the paging service is sending a message. This is followed by the access code. Each individual receiver has a unique access code. If configured for broadcast capability, it may have more than one such code. For example, a salesperson may have a receiver with one private access code and a second code which is the same as that for all salespeople. The company may then transmit only to that individual or broadcast to all salespeople. The received digits are compared to the access code, and the receiver is *enabled* if they agree. This is followed by the information, which could be a telephone number to call or any prearranged code that transmits information. Some paging receivers have built-in preprogrammed messages. For example, the display on one receiver incorporated into a wristwatch is capable of displaying "call home" when it receives a particular code.

The mobile unit is a receiver that continually scans the FM band looking for an SCA transmission with the correct attention code. The receiver includes a display, usually liquid crystal, to show the 12-digit message. Many receivers contain memory to store several such displays.

12.3.2 Cellular Telephone

Needs Assessment

In 1921, the Detroit police department used mobile radio at a carrier frequency of about 2 MHz in order to communicate between the base station and the police cars. This represents one of the earliest recorded uses of two-way mobile communication.

As the FCC added more and more frequencies to the allocated space for mobile communication (usually in the frequency range of 400–900 MHz), usage proliferated. Ambulance services, taxicab companies, construction companies, and sales forces were among the early competitors for limited frequency allocations. The usage was a restricted one, with base stations communicating with a fixed number of mobile stations.

As society began to spend more and more time in cars (either moving or standing still in traffic), the need for universal communication increased dramatically. Individuals wanted the ability to set up a communication link with anybody in the world, without going through a base station. To meet the need, people began to look at the extensive dial-up telephone system that was already in existence. It was simply necessary to provide mobile stations the ability to tie in to this vast system.

Simple as this sounds, implementation had to wait for a major technological breakthrough, as there simply were nowhere near enough frequencies to go around. Today, multiplexing techniques and sophisticated coding systems can increase the number of simultaneous users, but the capability is still far less than is required.

Throughout history, the emphasis in communication had been to transmit further and further. This meant that a single user tied up a particular frequency band over a large geographic area. The breakthrough that mobile communication was waiting for was an abandonment of the distance goal and an intentional *limiting* of the range of transmission. With such limiting, frequency bands could be allocated to separate users who were not as widely spaced, thereby increasing the total capacity of the system. The required companion development is the ability to build hardware that adaptively selects the particular fixed transmitter with which to communicate. That is, with relatively short communication distances and mobile users, the mobile unit must be in the range of a fixed station. As the unit moves, it must change from one fixed station to another.

The cellular concept represents the solution. It was reported upon in 1979 in the *Bell System Technical Journal* and, in the following decade, experienced a virtual explosion in usage. The concept is to divide a geographic area into *cells* (or smaller units, known as *microcells*), as shown in Fig. 12.7. Each cell contains a fixed station near its center. This station receives messages from mobile transmitters within the cell and also transmits to mobile receivers within the cell. The cell represents the area over which the signals have acceptable power levels. Therefore, it may not be uniformly shaped, and the fixed station may not be in the center; both of these depend upon geographic characteristics.

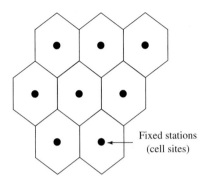

Figure 12.7   Cellular concept.

Specifications

The specifications of a cellular telephone system are deceptively simple. People should be able to use a portable telephone in the same manner that they use a hard-wired telephone. That is, they should be able to transmit and receive voice or modem signals (including facsimile; see Section 12.3.4). They should be able to dial any telephone in the world and also receive calls from any telephone in the world.

The preceding specifications are ideal. In practice, because of the relatively large numbers of fixed stations required, the portable telephone *cannot* be located anywhere in the world. It must be within several miles of a fixed station.

Design Approach

The FCC has allocated frequencies in a manner that is sufficient for a certain number of two-way communication channels. A fraction of these channels are needed to carry control signals between fixed stations and cellular phones. The control aspects are much more diverse than those of hard-wired telephones. Each cell covers a few square miles. A mobile station wishing to transmit must first test which fixed station it is closest to (i.e., the one with the strongest signal) and continually check the strength of the signal as it moves toward another cell. This "hand-off" process must be smooth enough that the user does not realize it is happening.

Design Details

Although frequency bands can be used by multiple cells, the same band cannot be used by adjacent cells (that is, in the current frequency-division multiplexed system; cellular telephony is in the process of converting to a fully digital system that may utilize Code-division multiple access (CDMA), in which frequencies overlap). In other words, cellular technology reduces the transmission range to the point where identical bands can be assigned to transmitters that are relatively close together. Nonetheless, they would not be assigned to adjacent cells. This approach complicates the hand-off process. When a station senses that the signal from a mobile transmitter is getting too weak, it looks for adjacent cells where the reception from that particular mobile transmitter is stronger. When it finds one, it initiates the hand-off process. The process involves changing the frequency assignment, since adjacent cells do not use the same frequencies. Consequently, the mobile unit must be informed of the change, and its oscillators must switch to the new frequency (i.e., change the voltage level into a VCO). All of this happens within a few milliseconds, so the user does not detect it.

Fixed-site cellular transmitters have an output power of 25–35 watts, whereas mobile transmitters have an output of less than 3 watts. (The cellular "revolution" has been the driving force for new standards to assure personal safety relative to exposure to electromagnetic radiation.) The mobile transmitter adjusts the output power to minimize the chance of interfering with other users. If the fixed site receives a signal that is stronger than necessary, it instructs the mobile transmitter to reduce the power. If the signal is too weak (and hand off is not indicated), the fixed site instructs the transmitter to increase the power. Most mobile transmitters are capable of setting the output power to as many as eight different levels.

FM modulation was used in first-generation cellular telephones, leading to a capture effect that serves to minimize interference from nearby cells. That is, just as wideband FM reduces the additive noise, it also reduces other interference that is smaller than the desired signal.

12.3.3 Global-Positioning Satellite

Needs Assessment

Humanity seems to have an insatiable desire to know where and when everything is happening. Watches and timing devices have become increasingly accurate. As the world becomes more crowded, it becomes more important that various locations be precisely known. This, of course, is an oversimplification: The need to know the location of something goes far beyond the avoidance of a collision. Among the many groups that have a need to know the locations of certain things are the following:

- A commercial fishing fleet needs to know the precise location where it found schools of fish the previous day.
- Ocean vessels need to know the proper path to follow for avoiding known dangers.
- Search-and-rescue teams, both on land and sea, need to know the precise locations of various emergencies that arise.
- Aircraft need to know their own location and altitude, and must have the ability to land without visual information.
- Vehicles need navigational help in unfamiliar territory.
- Companies would like to keep track of their fleet of vehicles (hopefully, not of each employee).
- Hikers need to know their location to avoid getting lost.
- Geophysical surveyers need to know precise locations in order to properly survey and map terrain.
- The space shuttle needs precise information on its location preparatory to reentry.
- The Internal Revenue Service needs . . . . (well, use your imagination).

We have omitted military applications from the list, as the use of positioning information for missiles, aircraft, troop movements, etc., should be obvious.

Electronics has been used in navigation for many decades. The long-range navigation Loran-A system helped guide American and British troop transports during World War II. The system operated at 1,850, 1,900 and 1,950 kHz. It was replaced by Loran-C, which provided accuracies on the order of 180 meters at ranges of 800–1000 nautical miles. This system, operated and supported by the U.S. Coast Guard, uses a frequency of 100 kHz.

In addition to Loran, other specialized systems evolved, including the Instrument Landing System (ILS) and the Tactical Air Navigation System (TACAN). In an attempt to establish an acceptable universal system, and to try to reduce the proliferation of specialized systems, the federal Joint Program Office initiated the NAVSTAR global-positioning system (GPS) in 1973. The system is controlled by the Department of Defense, and while the impetus may have been military, channels have been established for commercial and civilian use (perhaps to justify a portion of the investment budget).

Specifications

A global-positioning satellite system must permit an unlimited number of users. It must be accurate to within 30 meters for commercial use and an order of magnitude better for military and special-purpose applications. The system must be usable anywhere on the surface of the earth. Since it may be used for military applications, it must have at least some antijamming capability.

Design Approach

The NAVSTAR global-positioning system would not be possible without several major advances in technology. Among these is the availability of precise space-based atomic clocks, since stable timing is required. Also key to the use of the system are stable (quartz crystal) oscillators, as accurate knowledge of time is needed to establish precise position coordinates on earth.

In order to determine longitude, latitude, and altitude, it would appear that just three independent pieces of information are needed. However, since position on earth is calculated relative to the position of the orbiting satellites, whose locations change continuously, accurate time is required as the fourth variable. It was therefore determined that every location should be within the *footprint* of four satellites at all times. The original system specified 24 satellites—no fewer than four (typically, six to nine) in view from any location. However, in 1978–79, severe budget cuts required that the system operate with only 18 satellites, enough to meet the minimum specifications. These satellites are configured in groups of three in each of six orbital planes. The three satellites are spaced 120° apart.

Design Details

The first decision was on the placement of the satellites. As mentioned previously, 18 satellites are configured in groups of 3. The satellites in each group are spaced 120° apart and orbit such that a satellite passes over the same point once each day (one orbit in 23 hours, 55 minutes, and 58.3 seconds). The planners of the system originally considered a geosynchronous orbit (in which each satellite remains over the same point on earth continuously), but rejected the idea because a number of the satellites would not be visible from the United States. A *global* tracking and control station network would then be needed to be able to communicate with each satellite. Alternatively, with the selected orbits and orbital velocities, each satellite can be reached from U.S. stations for at least some portion of the day. This permits precise tracking of each and allows information on its orbital position to be uploaded on at least a daily basis.

Survivability is a major consideration both for military and commercial applications. The system includes three spare satellites in orbit. In addition, each satellite contains on-board diagnostics with adaptable repair capability, including the capability to correct and reset data.

The satellites transmit a modulated carrier at one of two frequencies in the range of frequencies known as L-band: L_1 at 1575.42 MHz and L_2 at 1227.6 MHz. The precise satellite clock is at 10.23 MHz, so L_1 represents the 154th harmonic and L_2 the 120th harmonic of the clock frequency.

The L_1 carrier is simultaneously modulated with both a precision (P) code at 10.23 Mbps and a coarse acquisition (C/A) code at 1.023 Mbps. The P-code is now known as Precise Positioning Service and provides accuracy of better than 2 meters. A PN sequence is used for the P-code, and it has a period of 267 days (a code length of about 236×10^{12} bits). The C/A code, which is a *Gold* code, repeats every millisecond. It provides accuracy of about 30 meters. The two codes modulate the carrier in phase quadrature.

Each satellite is assigned a unique 7-day segment of the 267-day PN code for precision and its own Gold code for coarse acquisition. Since each satellite has a unique PN code segment, and each PN code has a large distance from every other PN code, the P-code signals can be transmitted from all satellites. A receiver searching for the signal from one satellite would correlate the return with the known PN code. The portion of the return from the desired satellite correlates perfectly with the locally produced PN code, while the contributions from all other satellites are highly uncorrelated. This permits extracting the desired code from the sum. Therefore, although the signals overlap in both time and frequency, we can use the PN correlation process to separate them. This property leads to a form of multiplexing known as *code division multiple access (CDMA)*.

Since the PN code for a particular satellite has a repetition length of seven days (more than 6×10^{12} bits), initial acquisition is quite difficult. It is easier if the receiver has some rough idea of the time and its location. (As a practical matter, acceptable operation requires that the receiver "know" its location within about 3–6 km.) Initial acquisition can be done by first using the C/A code for a rough estimate of position.

Encrypted spread spectrum[2] is used for all ground-to-space telemetry and command-and-control links. It is done to improve antijamming capability.

Because unfriendly groups take advantage of NAVSTAR in tactical situations, the system permits time-varying codes to be used. If a receiver does not "know" the precise code, accurate information on position cannot be obtained.

Let us now examine the specific signal configurations. The P-code is the 7-day segment of a 267-day PN code. The receiver need only correlate this with a locally generated replica of the code to determine the propagation delay between the satellite and the receiver. Of course, without accurate absolute timing, this operation will give only a relative

---

[2]*Spread spectrum* is a technique for reducing the sensitivity of a message to intentional interference. The signal power is spread over a large bandwidth so that an unauthorized receiver can neither demodulate it (i.e., "refocus") nor easily jam the signal.

delay. That is the reason for using four satellites. We triangulate in four dimensions (a misnomer, since the "tri" stands for three, but you get the idea). However, even if we know our precise distance from each of three satellites, how does that tell us where we are? The missing ingredient is yet another piece of multiplexed information from each satellite. The satellite also modulates each signal with navigation data, at 50 bps. This relatively low-speed data signal gives updated information on the location of the satellite, as well as information such as status of the system and atmospheric propagation delays (correction factors).

Clearly, each receiver has a respectable amount of computation to perform before changing the received signal into a readout of longitude, latitude, and altitude. Advances in microcomputers have made this task routine, and hand-held receivers of small size and weight are practical.

The Future

Applications of GPS are limited only by the imagination. One particularly exciting commercial application comes directly from James Bond movies. The dashboard computer/map is quite practical, and as its price falls as sales increase, it will undoubtedly become a common automotive accessory. The concept calls for a display in the automobile that has a local road map with a flashing light indicating the exact location of the vehicle. With the destination preprogrammed, a particularly pleasing voice would say things like "Turn right in 50 meters, just past the 7–11 store," "You just passed the intersection where you wanted to turn. We will reroute you," or, with the added communication link to traffic reports, "There is an accident ahead. We will reroute you to avoid the congestion. Turn left in 100 meters."

"Personal" tracking has many potential applications. For example, tracking of diplomats could prove a deterrent to hostage taking and could be used to guide rescue missions in cases where a hostage is taken.

An even more exciting application in some societies relates to the house arrest of lawbreakers. With overcrowding of jails, we could insist that convicted criminals not leave their houses. If they do, the central computer (affectionately labeled "Big Brother") could track them down and issue an arrest order. As GPS receivers are made smaller and smaller, they can be implanted to avoid tampering. The future is here today, and if you don't like it, blame the communication engineers.

12.3.4 Facsimile

Needs Assessment

Hard-copy two-way communication traditionally was the domain of the postal service. Telephone has provided real-time communication capability, but was traditionally restricted to voice. Computer communication networks (electronic mail) are providing an alternative to postal service for data communication.

As in the case of mobile communication, a need has developed for hard-copy, real-time communication anywhere in the world without necessarily prearranging such communication. The electronic and computer revolutions have led to demands for faster response, and postal services have not been able to respond to those demands. Computer communication requires that information be in the computer (i.e., be generated by it or entered into it). A large number of users must transmit something that is already in hard-copy format, and entering it into a computer would be time consuming.

Companies and individuals therefore need the capability of sending printed material to others possessing a telephone line and the proper receiving equipment. The success of such a system requires that it be widespread; only then could it be cost effective. Therefore, the revolution in facsimile waited until technological advances permitted high-quality transmission at a reasonable price. It also waited for the regulatory changes that allowed such transmission and for the sensible hardware decisions regarding standardization.

Specifications

Facsimile must be capable of transmitting black-and-white copies of standard-size paper with acceptable error control. The most popular systems are *document*-quality facsimile, in which graded densities are not reproduced (i.e., each point on the paper is either black or white). [The class of *photographic*-quality facsimile permits grading of tonal density, but this is not the popular form in widespread use today.] The pages must be transmitted in a reasonable length of time, typically less than 1 minute per page.

Design Approach

The facsimile was invented in 1843, 33 years before the telephone! It was not until the 1920s, however, that practical models were developed. Among the earlier users were news and weather services.

Facsimile transmitters require that the original document be scanned. This is done either in a mechanical drum or by using a flat scanning system. Electronic scanning permits higher speed.

Earlier transmission systems were analog, using AM, VSB and FM transmission. These systems were classified as *Group 1* and *Group 2*, and required as much as 6 minutes to send a page of information.

Group 3 and *Group 4* machines use digital transmission, with Group 3 the standard for dial-up telephones. The advent of digital devices, with the associated sophisticated data compression capability, reduced the time it took to transmit a page to less than a minute (as little as 16 seconds).

Most low-cost receivers use thermal recording on heat-sensitive paper, although plain-paper copying is gaining in usage. Facsimile processing is available on personal computers, and this produces hard copy using whatever technology the computer printer uses.

Design Details

The reading process is most widely done with linear charge-coupled devices (CCDs), which can be thought of as multiple photodiodes on a silicon chip. The standard method of scanning is to use a fluorescent lamp to illuminate the original document, focus the image in the lens to a reduced size, and input this reduced image directly into the face of the

CCD image sensor. The response time of the CCD limits the speed of scanning. (Currently, speeds are about 5 ms/line.)

The writing process is commonly thermal. It uses a line of heating resistors spaced very closely together to reproduce each line as the paper scans the document vertically.

The common data compression technique for Group 3 facsimile is one-dimensional and two-dimensional run-length coding. (See Section 9.1.) Standards exist for both, along with error correction redundancy coding. The actual run-length code used is known as the *modified Huffman* code. The code makes a distinction between runs of up to 63 pixels and runs longer than 63 pixels (known as *make-up*). There is a separate *terminating code* for each run length between 0 and 64. Make-up codes are assigned to run lengths that are multiples of 64 (i.e., 128, 192, 256, and so on up to 1,728). Thus, to transmit a run length of 975, we would send the make-up code for 960 and the terminating code for a run length of 15. The code assignment is shown in Table 12.1.

TABLE 12.1   MODIFIED HUFFMAN FACSIMILE CODE

White Run Length	Code	Black Run Length	Code
0	00110101	0	0000110111
1	000111	1	010
2	0111	2	11
3	1000	3	10
4	1011	4	011
5	1100	5	0011
6	1110	6	0010
7	1111	7	00011
8	10011	8	
.	.	.	.
.	.	.	.
62	00110011	62	000001100110
63	00110100	63	000001100111
64	11011	64	0000001111
128	10010	128	000011001000
EOL	000000000001		

There are several things to note about the table. The first is the variable length of the code, which we can relate to our study of Huffman codes in Section 9.1. Runs of 2 and of 3 black pixels are assigned the shortest code, indicating that a statistical study has shown these to be the most likely runs. The next most likely are runs of 1 and 4 black pixels and, after that, runs of 64 or 128 white pixels, representing the white areas on a page of text.

Note further that the end of each line is indicated by a 12-bit transmission labeled "EOL." Each line begins with white in order to set the color levels properly. Since the actual line may begin with black, the code provides for a zero-length run.

Group 3 standards call for 1,728 pixels per scan line, yielding a resolution of 3.85 lines/mm. Optional modes provide for higher resolution.

We have not commented upon the actual digital transmission technique, since it uses standard modem technology. (See Section 11.5 for details.)

12.3.5 Videotext

Needs Assessment

The average consumer's contact with the outside world through a communication medium is by voice (telephone), one-way voice (radio), one-way voice and video (television), two-way hard copy (postal service or facsimile), and one-way hard copy (newspapers and magazines). The need for timely textual information upon request has not traditionally been met by any of these media. The consumer desires the ability to receive graphic and textual information on a real-time demand basis. Videotext is the response to this need.

Specifications

The ultimate desire is to be able to use a television terminal for complete interactive two-way graphic and textual communication. The more immediate requirement is for a system compatible with broadcast television. The system must provide up to several hundred pages of information upon request by the consumer. The nominal resolution is to provide 24 rows of text (and graphic symbols) with 40 characters per row. The system must have no negative effect upon regular television broadcasting, and must not require a separate channel.

Design Approach

The United Kingdom led the way in videotext. Systems in Britain (as well as in France) came into existence in the mid-1970s. The highest consumer usage has been in the United Kingdom, followed by West Germany, the Netherlands, Sweden, and Austria. Although television standards are not the same throughout the world (e.g., the number of lines varies from 525 to 625), the approach toward videotext has been fairly uniform.

We begin with some historical terminology. *Videotex* (without the *t*) has been used to describe computerized information systems that use telephone lines as their channel. Application of the term has been broadening, and in fact, in most cases, it has merged with the more general term, *videotext*. *Teletext* has traditionally been a term describing systems that use normal television signals to distribute data.

Throughout the history of broadcast television, engineers have faced the challenge of how to multiplex additional information within the 6-MHz bandwidth allocated to each channel. The first challenge was color, then came stereo, and now videotext has arrived.

The solution for videotext was to note that 21 lines of each frame are devoted to vertical retrace. During this process, a vertical blanking pulse turns off the picture. The only necessary activities are to keep the horizontal oscillator in synchronization, to synchronize the color demodulator, and to send some periodic test and reference signals.

U.S. television uses lines 1–9 of the retrace for the vertical synchronization pulse. Lines 17–19 are used for test and reference signals. Line 21 is used for *closed caption* services for the hearing impaired. The timing signals and retrace circuitry on older TV receivers require that lines 10–13 of the retrace not be used for sending data, since this would cause interference patterns. (Those lines occur around the halfway point of the retrace time.) For all of these reasons, TV channels that choose to establish videotext generally look at lines 14 to 21 to send digital information.

As we shall see, there is enough time to send approximately 2,000 bits of information during the eight lines that can be devoted to videotext within each frame. If the transmitted bits control a character generator, and if a full page consists of 24 × 40 = 960 characters, it takes approximately 0.25 second to send an entire page. The specifications call for several hundred pages to be available upon request. Thus, an implementation of demand broadcast is not practical at this time, since it would require a return communication channel and the ability to broadcast different pages to different subscribers.

Therefore, the demand aspect must be met by *sequencing* through the several hundred pages and having the receiver enable the reception only when the desired page is being received. Two hundred pages would take about 50 seconds to cycle through sequentially, which may be an unacceptably long period to wait after making a request (on a push-button pad).

There are two major approaches toward shortening the waiting period. The first, and simplest, is to carefully sequence the pages according to statistical demand projections. The most popular pages could be sent more often. Thus, the average viewer would wait a shorter period of time, while someone with a relatively unpopular request might have to wait longer. Another solution is to build memory into receivers. The electronics can memorize pages as they arrive, thus cutting the access time to near zero. The required memory size can be decreased by selective memorization. For example, if a viewer requested a summary page of sports activities, the memory would then store only related pages as they arrive.

Design Details

The most popular systems transmit information in a burst mode (i.e., only during the time that the correct blanking lines are being broadcast) at 5.727272 Mbps. NRZ coding is used, so in television terms, black and white correspond to 0 and 1, respectively.

Figure 12.8 shows the composition of a single scan line. Each blanking line begins with a line synchronization pulse followed by the color burst. This is in turn followed by

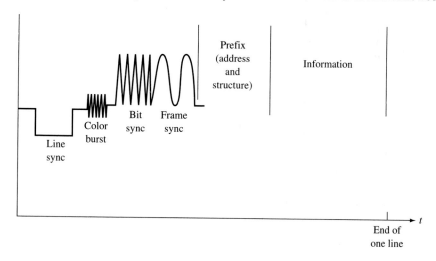

Figure 12.8   Line format.

16 alternating bits used for bit synchronization. The next signal is an 8-bit framing wave-form, also known as *byte synchronization*.

Now that both the bits and frames are in sync, we can start sending information. The next 5 bytes (40 bits) provide information on the *address* and *structure* of the packet. These in turn identify the information to follow. Since this identification is very important (among other things, it sets the orientation of the information within the picture), it con-tains error control in the form of a Hamming code. In fact, in most cases, there are only 20 information bits within the 40, the remainder being redundancy bits. The balance of the line is devoted to information and contains approximately 28 bytes (224 bits) of informa-tion. Each 8-bit byte contains one parity check bit for single-error detection.

The receiver uses a character generator to transform the bytes into characters for dis-play on the screen.

Appendix I

References

GENERAL COMMUNICATION TEXTBOOKS

The following books contain a general approach toward analog and/or digital communication. They may be used to gain additional perspective in various areas. While their coverage of topics is similar and with a great deal of overlap, variations exist in particular subjects. Some of the texts contain added depth in selected areas. If you plan to specialize in communication, you should attempt to build a varied library of texts and reference materials.

A. B. CARLSON, *Communication Systems*, 3rd ed., McGraw-Hill, New York, 1986.

LEON W. COUCH II, *Modern Communication Systems*, Prentice Hall, Englewood Cliffs, NJ, 1995.

R. M. GAGLIARDI, *Introduction to Communications Engineering*, 2d ed., Wiley, New York, 1988.

JERRY D. GIBSON, *Principles of Digital and Analog Communications*, 2d ed., Macmillan, New York, 1993.

S. HAYKIN, *Digital Communications*, Wiley, New York, 1988.

SIMON HAYKIN, *Communication Systems*, 3rd ed., Wiley, New York, 1994.

W. C. LINDSEY AND M. K. SIMON, *Telecommunication Systems Engineering*, Prentice-Hall, Englewood Cliffs, NJ, 1973.

P. Z. PEEBLES, JR., *Digital Communication Systems*, Prentice-Hall, Englewood Cliffs, NJ, 1988.

JOHN G. PROAKIS AND MASOUD SALEHI, *Communication Systems Engineering*, Prentice Hall, Englewood Cliffs, NJ, 1994.

J. G. PROAKIS, *Digital Communications*, 3rd ed., McGraw-Hill, New York, 1995.

M. S. RODEN, *Digital Communication Systems Design*, Prentice-Hall, Englewood Cliffs, NJ, 1988.

M. SCHWARTZ, *Information Transmission, Modulation and Noise*, 4th ed., McGraw Hill, New York, 1990.

B. SKLAR, *Digital Communications*, Prentice-Hall, Englewood Cliffs, NJ, 1988.

H. STARK, F. B. TUTEUR, AND J. B. ANDERSON, *Modern Electrical Communications*, 2d ed., Prentice-Hall, Englewood Cliffs, NJ, 1988.

H. TAUB AND D. SCHILLING, *Principles of Communication Systems*, 2d ed., McGraw-Hill, New York, 1986.

A. M. VITERBI, *Principles of Coherent Communications*, McGraw-Hill, New York, 1966.

R. E. ZIEMER AND R. L. PETERSON, *Introduction to Digital Communications*, Macmillan, New York, 1992.

R. E. ZIEMER AND W. H. TRANTER, *Principles of Communications*, 4th ed., Houghton Mifflin, Boston, 1995.

ARTICLES

Much of the most recent material has not yet been published in textbooks. Of particular importance are IEEE publications, notably those of the IEEE Communications Society. The *IEEE Communications Magazine* is an excellent source of highly readable tutorial articles. From time to time, there are special issues of *IEEE Spectrum Magazine* and *IEEE Proceedings* that focus on particular areas of communication.

You should become familiar with various abstracts, including the *Engineering Index* and *Applied Science and Technology*. Many of the abstracts are available on-line through computer networks. Most university libraries have such network services.

Appendix II

Fourier Transform Pairs

GENERAL PROPERTIES

$$S(f) = \int_{-\infty}^{\infty} s(t)e^{-j2\pi ft}dt$$

$$s(t) = \int_{-\infty}^{\infty} S(f)e^{j2\pi ft}df$$

Function	*Fourier Transform*
$s(t - t_0)$	$e^{-j2\pi ft_0}S(f)$
$e^{j2\pi f_0 t}s(t)$	$S(f - f_0)$
$s(t)\cos 2\pi f_0 t$	$\frac{1}{2}[S(f - f_0) + S(f + f_0)]$
$\dfrac{ds}{dt}$	$j2\pi f S(f)$
$\displaystyle\int_{-\infty}^{t} s(\tau)d\tau$	$\dfrac{S(f)}{j2\pi f}$
$r(t)*s(t)$	$R(f)S(f)$
$r(t)s(t)$	$R(f)*S(f)$
$s(at)$	$\dfrac{1}{a}S\left(\dfrac{f}{a}\right)$
$\dfrac{1}{a}s\left(\dfrac{t}{a}\right)$	$S(af)$

SPECIFIC TRANSFORM PAIRS

Function	*Fourier Transform*
$e^{-at}U(t)$	$\dfrac{1}{a + j2\pi f} \quad a > 0$
$te^{-at}U(t)$	$\dfrac{1}{(a + j2\pi f)^2} \quad a > 0$
e^{-at^2}	$\sqrt{\dfrac{\pi}{a}} \exp\left(-\dfrac{\pi^2 f^2}{a}\right) \quad a > 0$
$\dfrac{\sin at}{\pi t}$	$\begin{cases} 1, & \|f\| < a/2\pi \\ 0, & \text{otherwise} \end{cases}$
$\begin{cases} \frac{1}{2}, & \|t\| < T \\ 0, & \text{otherwise} \end{cases}$	$\dfrac{\sin 2\pi fT}{2\pi f}$
$\begin{cases} 1 - \dfrac{\|t\|}{T}, & \|t\| < T \\ 0, & \text{otherwise} \end{cases}$	$\dfrac{\sin^2 \pi fT}{T\pi^2 f^2}$
$e^{-a\|t\|}$	$\dfrac{2a}{a^2 + 4\pi^2 f^2}$
$\text{sgn}(t)$	$\dfrac{1}{j\pi f}$
$\delta(t)$	1
1	$\delta(f)$
$e^{j2\pi f_0 t}$	$\delta(f - f_0)$
$\cos 2\pi f_0 t$	$\frac{1}{2}[\delta(f - f_0) + \delta(f + f_0)]$
$\sin 2\pi f_0 t$	$\dfrac{j}{2}[\delta(f + f_0) - \delta(f - f_0)]$
$U(t)$	$\frac{1}{2}\delta(f) + \dfrac{1}{j2\pi f}$

Appendix III

Error Function

$$\mathrm{erf}(x) = \frac{2}{\sqrt{\pi}} \int_0^x e^{-u^2} du$$

x	erf (x)	erfc (x)	x	erf (x)	erfc (x)
0	0	1	0.05	0.056	0.944
0.10	0.112	0.888	0.15	0.168	0.832
0.20	0.223	0.777	0.25	0.276	0.724
0.30	0.329	0.671	0.35	0.379	0.621
0.40	0.428	0.572	0.45	0.475	0.525
0.50	0.521	0.479	0.55	0.563	0.437
0.60	0.604	0.396	0.65	0.642	0.358
0.70	0.678	0.322	0.75	0.711	0.289
0.80	0.742	0.258	0.85	0.771	0.229
0.90	0.797	0.203	0.95	0.821	0.179
1.00	0.843	0.157	1.05	0.862	0.138
1.10	0.880	0.120	1.15	0.896	0.104
1.20	0.910	0.0901	1.25	0.923	0.0768
1.30	0.934	0.0659	1.35	0.944	0.0564
1.40	0.952	0.0481	1.45	0.960	0.0400
1.50	0.966	0.0338	1.55	0.972	0.0284
1.60	0.976	0.0238	1.65	0.980	0.0199
1.70	0.984	0.0156	1.75	0.987	0.0128
1.80	0.989	0.0105	1.85	0.991	8.53×10^{-3}
1.90	0.993	6.91×10^{-3}	1.95	0.994	5.57×10^{-3}
2.00	0.995	4.59×10^{-3}	2.05	0.996	3.68×10^{-3}
2.10	0.997	2.98×10^{-3}	2.15	0.998	2.33×10^{-3}
2.20	0.998	1.84×10^{-3}	2.25	0.999	1.44×10^{-3}
2.30	0.999	1.13×10^{-3}	2.35	0.999	8.80×10^{-4}
2.40	0.999	6.82×10^{-4}	2.45	0.999	5.26×10^{-4}
2.50	1.000	4.03×10^{-4}	2.55	1.000	3.08×10^{-4}
2.60	1.000	2.34×10^{-4}	2.65	1.000	1.77×10^{-4}
2.70	1.000	1.33×10^{-4}	2.80	1.000	7.46×10^{-5}
2.90	1.000	4.09×10^{-5}	3.00	1.000	2.20×10^{-5}
3.10	1.000	1.16×10^{-5}	3.20	1.000	6.00×10^{-6}
3.30	1.000	3.06×10^{-6}	3.40	1.000	1.52×10^{-6}
3.50	1.000	7.43×10^{-7}	3.60	1.000	3.56×10^{-7}
3.70	1.000	1.67×10^{-7}	3.80	1.000	7.70×10^{-8}

continued

x	erf (x)	erfc (x)	x	erf (x)	erfc (x)
3.90	1.000	3.48×10^{-8}	4.00	1.000	1.54×10^{-8}
4.10	1.000	6.70×10^{-9}	4.20	1.000	2.86×10^{-9}
4.30	1.000	1.19×10^{-9}	4.40	1.000	4.89×10^{-10}
4.50	1.000	1.97×10^{-10}	4.60	1.000	7.75×10^{-11}
4.70	1.000	3.00×10^{-11}	4.80	1.000	1.14×10^{-11}
4.90	1.000	4.22×10^{-12}	5.00	1.000	1.54×10^{-12}

Appendix IV

Q Function

x	$Q(x)$	x	$Q(x)$
0	0.500	0.05	0.480
0.10	0.460	0.15	0.440
0.20	0.421	0.25	0.401
0.30	0.382	0.35	0.363
0.40	0.345	0.45	0.326
0.50	0.309	0.55	0.291
0.60	0.274	0.65	0.258
0.70	0.242	0.75	0.227
0.80	0.212	0.85	0.198
0.90	0.184	0.95	0.171
1.00	0.159	1.05	0.147
1.10	0.136	1.15	0.125
1.20	0.115	1.25	0.106
1.30	0.097	1.35	0.089
1.40	0.081	1.45	0.074
1.50	0.067	1.55	0.061
1.60	0.055	1.65	0.049
1.70	0.045	1.75	0.04
1.80	0.036	1.85	0.032
1.90	0.029	1.95	0.026
2.00	0.023	2.05	0.020
2.10	0.018	2.15	0.016
2.20	0.014	2.25	0.012
2.30	0.011	2.35	0.009
2.40	0.008	2.45	0.007
2.50	0.006	2.55	0.005
2.60	0.005	2.65	0.004
2.70	0.003	2.75	0.003
2.80	0.003	2.85	0.002
2.90	0.002	2.95	0.002
3.00	0.001	3.05	0.001
3.10	9×10^{-4}	3.15	8×10^{-4}
3.20	7×10^{-4}	3.25	6×10^{-4}
3.30	6×10^{-4}	3.35	5×10^{-4}
3.40	4×10^{-4}	3.45	3×10^{-4}
3.50	3×10^{-4}	3.55	3×10^{-4}

Index

IF YOU ARE
DOING COMMS
AS YOUR PROJECT

U ARE A FUDP!